T0282305

LONDON MATHEMATICAL SOCIETY LECTURE NOTE SERIES

Managing Editor: Professor J.W.S. Cassels, Department of Pure Mathematics and Mathematical Statistics, University of Cambridge, 16 Mill Lane, Cambridge CB2 1SB, England

The books in the series listed below are available from booksellers, or, in case of difficulty, from Cambridge University Press.

London Mathematical Society Lecture Note Series. 205

Ergodic Theory and its Connections with Harmonic Analysis

Proceedings of the 1993 Alexandria Conference

Edited by

Karl E. Petersen
University of North Carolina

and

Ibrahim A. Salama
North Carolina Central University

CAMBRIDGE
UNIVERSITY PRESS

CAMBRIDGE UNIVERSITY PRESS
Cambridge, New York, Melbourne, Madrid, Cape Town, Singapore, São Paulo

Cambridge University Press
The Edinburgh Building, Cambridge CB2 8RU, UK

Published in the United States of America by Cambridge University Press, New York

www.cambridge.org
Information on this title: www.cambridge.org/9780521459990

First published 1995

A catalogue record for this publication is available from the British Library

ISBN 978-0-521-45999-0 paperback

Transferred to digital printing 2007

Cambridge University Press has no responsibility for the persistence or
accuracy of email addresses referred to in this publication.

CONTENTS

PREFACE

Ergodic theory is a crossroads for many branches of mathematics and science, from which it has drawn problems, ideas, and methods, and in which it has found applications.

In recent years the interaction of ergodic theory with several kinds of harmonic analysis has been especially evident: in the Fourier method of proving almost everywhere convergence theorems introduced by Bourgain and developed by Rosenblatt, Wierdl, and others; in the application of the real-variable harmonic analysis of the Stein school to ergodic theory by, for example, Bellow, Boivin, Deniel, Derriennic, Jones, and Rosenblatt; in rigidity theory, where Katok, Margulis, Mostow, Ratner, Spatzier, Zimmer, and others blend ergodic theory with dynamics, noncommutative harmonic analysis, and geometry; in the study of singular measures, spectral properties, and subgroups by Aaronson, Host, Méla, Nadkarni, and Parreau, which has furthered the analysis of nonsingular transformations and has led to progress on old problems like the higher-order mixing conjecture; in the structure theory of single transformations, where concepts like rank, joinings, and approximation are helping workers such as del Junco, King, Lemanczyk, Rudolph, and Thouvenot to classify systems, to understand better the important family of Gaussian processes, and to explore the connections among spectral and dynamical properties; in the combinatorics of adic transformations, introduced by Vershik and pursued by Herman, Kerov, Livshitz, Putnam, and Skau, which makes visible the connections between some invariants of dynamical systems and certain C^*-algebras, especially those associated with group representations; and in the applications of ergodic theory to combinatorial number theory and Diophantine approximation discovered by Furstenberg and developed also by Bergelson, Hindman, Katznelson, and Weiss, which rely in an essential way on techniques from harmonic analysis and spectral theory.

Egypt is a crossroads of many cultures and peoples, convenient to the countries of the Middle East, Africa, and Europe, and roughly equidistant from Asia and America. Our conference, focussed on the interaction of ergodic theory with harmonic analysis, found a natural site in Alexandria, the home of Euclid, Diophantus, Hypatia, Ptolemy, ... (some of whom we could try to claim as early ergodic theorists—Ptolemy, in his celestial mechanics, perhaps even as someone who applied harmonic analysis to dynamics). While no attempt was made to cover the area completely, some important sectors were described in survey talks, and the research talks reported interesting new developments.

Part I of these Proceedings contains three survey articles expanding presentations given at the conference. Each gives an up-to-date description of an area of lively interaction of ergodic theory and harmonic analysis, laying out the background of the field and the sources of its problems, the most important and most interesting recent results, and the current lines of development and outstanding open questions. These papers—by Rosenblatt and Wierdl on Fourier methods in almost everywhere convergence, by Spatzier on rigidity, and by Thouvenot on joinings—should provide convenient starting points for researchers beginning work in these areas.

Part II is a collection of refereed research papers presenting new results on questions related to the theme of the conference. Some of these papers were presented at the meeting, others were contributed later. Two of them (Lesigne, Rudolph) concern the problems in noncommutative harmonic analysis that emerge in trying to understand the nonlinear ergodic averages arising from Furstenberg's diagonal approach to Szemerédi's Theorem, and three others (Forrest, Hendrick, and McCutcheon) treat further developments in this dynamical multiple-recurrence theory. The remainder deal with problems in almost everywhere convergence and a variety of other topics in dynamics.

It is a pleasure to thank the many people and institutions who contributed to the success of the conference and to the preparation of these Proceedings. The National Science Foundation and Institute of Statistics of Cairo University provided financial support. Special thanks are due to Professor Ahmed E. Sarhan and Professor Mahmoud Riad for their hospitality and handling of the official arrangements in Egypt. We also thank Professor Mounir Morsy for welcoming us to his Department of Mathematics at Ain Shams University. Further, we are indebted to the anonymous referees, who improved the papers presented here beyond their original versions, to the many excellent typists who produced the TEX files, and to Lauren Cowles and David Tranah, who smoothly handled the publishing.

Karl Petersen and Ibrahim Salama
Chapel Hill, N.C.
July 11, 1994

TEX and $\mathcal{AMS}$-TEX are trademarks of the American Mathematical Society. This book was prepared in $\mathcal{AMS}$-TEX by the author.

PART I

SURVEY ARTICLES

POINTWISE ERGODIC THEOREMS
VIA HARMONIC ANALYSIS

JOSEPH M. ROSENBLATT AND MÁTÉ WIERDL

CONTENTS

INTRODUCTION

Historical remarks

It has been eighty-five years since Bohl [1909], Sierpiński [1910] and Weyl [1910] proved the now famous equidistribution theorem: if α is an irrational number then the sequence $\alpha, 2\alpha, 3\alpha \ldots$ is uniformly distributed mod 1. This means that for each subinterval I of the unit interval $[0,1)$ we have

$$\lim_{N \to \infty} \frac{\#\{n \mid n \leq N, \langle n\alpha \rangle \in I\}}{N} = |I|, \tag{1}$$

where $\langle x \rangle$ denotes the fractional part of x, that is $\langle x \rangle = x - [x]$, and $|I|$ is the length of the interval I. In fact, Weyl went on to prove, in [Weyl, 1916], that the sequence $\alpha, 2^2\alpha, 3^2\alpha \ldots$ is uniformly distributed mod 1. A bit less than twenty years later Vinogradov [cf. Ellison & Ellison, 1985] proved, as a byproduct of his solution of the 'odd' Goldbach conjecture[1], that the sequence $(p_n\alpha)$, where p_n denotes the n-th prime number, is uniformly distributed mod 1. On the other hand, it is easy to see that for some irrational α the sequence $(2^n\alpha)$ is not uniformly distributed mod 1.

Now the question arises what happens if we replace the interval I in (1) by an arbitrary Lebesgue measurable subset of $[0,1)$. What kind of extensions do the results of Weyl and Vinogradov have in this direction? We cannot expect a word-for-word generalization of their results since the Lebesgue measure of any fixed sequence is 0, and I may even be disjoint from the sequence! In the beginning of the 30's Birkhoff [1931] and Khintchin [1933] proved the appropriate generalization of (1): for any fixed Lebesgue measurable $I \subset [0,1)$, for almost every x we have

$$\lim_{N \to \infty} \frac{\#\{n \mid n \leq N, \langle x + n\alpha \rangle \in I\}}{N} = |I|, \tag{2}$$

where now $|I|$ denotes the Lebesgue measure of I. (It is not at all clear at first sight, but the result in (2) *does* imply the one in (1).) This is an instance of the individual (or pointwise) ergodic theorem. But then it took more than fifty years to obtain similar generalizations of the other result of Weyl and the result of Vinogradov! Bourgain [1988, 1988a, 1989] developed a very powerful method with which he proved in 1987: for any fixed Lebesgue measurable $I \subset [0,1)$, for almost every x we have

$$\lim_{N \to \infty} \frac{\#\{n \mid n \leq N, \langle x + n^2\alpha \rangle \in I\}}{N} = |I|, \tag{3}$$

and

$$\lim_{N \to \infty} \frac{\#\{n \mid n \leq N, \langle x + p_n\alpha \rangle \in I\}}{N} = |I| \tag{4}$$

---

[1]That every large enough odd number is a sum of three prime numbers.

(recall that p_n is the n-th prime). Bourgain's method is a wonderful blend of (analytic) number theory, Fourier analysis and ergodic theory: he uses estimates on the Fourier transforms (or exponential sums)

$$\frac{1}{N} \sum_{n=1}^{N} e^{2\pi i n^2 t}$$

and

$$\frac{1}{N} \sum_{n=1}^{N} e^{2\pi i p_n t}$$

respectively, as they were obtained even by Weyl and Vinogradov.

In this paper we will examine the following general question: for what sequences of integers (a_n) do we have that for any fixed Lebesgue measurable $I \subset [0,1)$, for almost every x

$$\lim_{N \to \infty} \frac{\#\{n \mid n \leq N, \langle x + a_n \alpha \rangle \in I\}}{N} = |I| ? \tag{5}$$

While our primary goal is to introduce the reader to Bourgain's method that led to the results in (3) and (4), we will also discuss other methods and results of related interest, in particular the "early" results of Krengel, Bellow and others. In fact, the problem formulated in (5) is only the starting point of our investigations. To give an idea of some of the questions we shall examine, let us give the analytic reformulation of the results above.

Let f denote the characteristic function of the interval $I \subset [0,1)$. Then (1) can be rewritten as

$$\lim_{N \to \infty} \frac{1}{N} \sum_{n=1}^{N} f(\langle n\alpha \rangle) = \int_0^1 f(x)dx, \tag{1'}$$

and an approximation argument shows that in the above we can take f to be any Riemann integrable function defined on $[0,1)$.

Similarly, if f denotes the indicator function of the Lebesgue measurable set $I \subset [0,1)$, then (2) can be written as

$$\lim_{N \to \infty} \frac{1}{N} \sum_{n=1}^{N} f(\langle x + n\alpha \rangle) = \int_0^1 f(x)dx. \tag{2'}$$

Now Khintchin showed that in the above we can take f to be any Lebesgue integrable funcion. How about a similar generalization of (3) and (4)? That is, writing

$$\lim_{N \to \infty} \frac{1}{N} \sum_{n=1}^{N} f(\langle x + n^2\alpha \rangle) = \int_0^1 f(x)dx, \tag{3'}$$

and

$$\lim_{N\to\infty} \frac{1}{N} \sum_{n=1}^{N} f(\langle x + p_n \alpha \rangle) = \int_0^1 f(x)dx, \tag{4'}$$

can we take f to be any Lebesgue integrable function? The answer is not known! All we know is that (3') and (4') hold for every L^p for $p > 1$. So we arrive at the following question: let (a_n) be a sequence of integers (or real numbers). For what values of p, $1 \le p \le \infty$, do we have that for each $f \in L^p$

$$\lim_{N\to\infty} \frac{1}{N} \sum_{n=1}^{N} f(\langle x + a_n \alpha \rangle) = \int_0^1 f(x)dx \tag{6}$$

for almost every x? If for some p for every $f \in L^p$ we have (6) a.e., then we call the sequence (a_n) a *universally good averaging sequence* (for L^p), because there is no restriction on the irrational α.

We have already mentioned that for some irrational α the sequence $(2^n \alpha)$ is not uniformly distributed mod 1. But we also have a result of Weyl which says that for almost every α the sequence $(2^n \alpha)$ *is* uniformly distributed mod 1. So at least there is something good to be said here. The picture changes dramatically in the ergodic-theoretical setting. It is a result of Bellow that for *every* irrational α there is a characteristic function f for which

$$\frac{1}{N} \sum_{n=1}^{N} f(\langle x + 2^n \alpha \rangle)$$

diverges for almost every x. We can say that in a sense the sequence (2^n) is a *universally bad averaging sequence*.

The methods — mostly developed by Bourgain — to solve the above "subsequence" problems helped to settle other almost everywhere convergence problems. We will mention a number of these problems, but we will develop the method in the special context of these subsequence problems, and usually we refer to other type of applications in the notes after the chapters.

Prerequisites

We do not assume that the reader is familiar with any deeper theories, but this does not mean that she/he has an easy task. The difficulty is that we use tools and results from five branches of mathematics.

Thorough knowledge of the Lebesgue integral is certainly assumed, as well as the elements of functional analysis, such as Baire's category theorem and the uniform boundedness principle. The books [Wheeden, Zygmund, 1977] or [Royden, 1988] contain all the material that is needed from measure theory, and Chapters 7 and 9 of [Royden, 1988] have all the functional analysis we need.

We will use basic facts about harmonic analysis on the classical groups $\mathbb{T}$, $\mathbb{Z}$ and $\mathbb{R}$. We will also use facts about the Hardy-Littlewood maximal function (both on the integers and on the real line), although we will give a proof on $\mathbb{Z}$, since we will use it to prove the ergodic maximal inequality. At certain points — not crucial — we will use the M. Riesz interpolation theorem. Herglotz's theorem on the representation of positive definite sequences is used. All the material that is needed from harmonic analysis can be found in the first two chapters of [Helson, 1991]. Straightforward treatments of the Hardy-Littlewood maximal function can be found in section 4.6 of [Helson, 1991] or in Chapter 9 of [Wheeden, Zygmund, 1977].

From number theory we use the basic properties of congruences. Some of the examples and exercises will refer to deep results of analytic number theory, such as the prime number theorem for arithmetic progressions, but skipping these will not seriously affect the reader's understanding of the other parts of the material. All the material needed from elementary number theory can be found in the first eleven chapters of [Weil, 1979].

The material we assume to be known from probability theory is, in addition to elementary concepts, the moment estimate of Marcinkiewicz and Zygmund. This inequality is a generalization of Khintchin's inequality for the Rademacher functions. The facts we use from probability theory can be found in the first chapter of [Durrett, 1991]. The moment estimate of Marcinkiewicz and Zygmund is in [Garsia, 1970].

Although we will give proofs of both the mean and pointwise ergodic theorems, it is desirable that the reader have some idea of the significance of these results. The reader should know what an aperiodic and an ergodic transformation is, and should know about Rokhlin's tower construction. Strictly speaking, all the material we need from ergodic theory is the first 7 sections (the seventh is "Consequences of ergodicity") and the section "Uniform topology" from [Halmos, 1956]

Summing up: the material is accesible for a third–year (US) graduate student, but she/he may want to skip some of the examples at first reading.

How to use these notes

Our presentation is fairly concise. Although we will try to give careful explanations of the underlying ideas of each proof, we will leave the routine computations to the reader.

There are exercises throughout the text, not just at the ends of the sections. Some of the exercises are difficult, but we provide hints for most of them. The exercises form an integral part of the text, and they often contain interesting developments of the preceding results.

We hope that these notes will motivate the reader to think about some of the unsolved problems of this field (many of which are mentioned in the sequel).

Finally, it is important to note the following. Even with the assumed prerequisites we could not give a full account of Bourgain's method; we give his fundamental inequality without proof, and we refer the interested reader either to Bourgain's original paper [1989] or to Thouvenot's [1990]

Acknowledgements

We want to thank all the mathematicians whose work is discussed in the text; we have freely quoted from the work of Professors M. Akcoglu, A. Bellow, M. Boshernitzan, J. Bourgain, Y. Huang, A. del Junco, R. L. Jones, K. Reinhold-Larsson and H. White. We have benefitted from discussions with Professor M. Lacey.

We also want to thank Professors K. Petersen and I. Salama for inviting us to prepare this article as a more detailed exposition of the survey lecture given at the conference "*Ergodic Theory and Its Connections with Harmonic Analysis*" held in Alexandria, Egypt, in May of 1993.

List of Symbols

We give the page number where the symbol first occurs. We explain the meaning of the symbol or notation only if a brief explanation is possible.

List of Exercises

CHAPTER I

GOOD AND BAD SEQUENCES IN PERIODIC SYSTEMS

The ergodic-theoretical reformulation of the results in (2')-(4') rests on the observation that for a fixed real number α the transformation τ of the unit interval $[0,1)$ defined by

$$\tau(x) = \langle x + \alpha \rangle$$

is Lebesgue measurable, and preserves the (mod 1) Lebesgue measure. To fix the more general terminology to be used in the sequel we introduce

Definition 1.1. Let (X, Σ, m) be a probability measure space, and let τ be a map of X into itself. We say τ is Σ-*measurable* if and only if for each $A \in \Sigma$ we have $\tau^{-1}A \in \Sigma$. We say τ *preserves* m if and only if for each $A \in \Sigma$ we have $m(\tau^{-1}A) = m(A)$. A *measure preserving transformation* of X is a Σ-measurable map of X into itself which preserves m. Finally, the quadruple (X, Σ, m, τ), where τ is a measure-preserving tranformation of X, is called a *dynamical system*.

Let τ be a measure-preserving transformation of the probability space (X, Σ, m), and let $A = (a_n)$ be a sequence of positive integers. For $f : X \to \mathbb{C}$ consider

$$\frac{1}{N} \sum_{n=1}^{N} f(\tau^{a_n} x). \tag{1.1}$$

The purpose of these notes is to examine the almost everywhere convergence of the averages in (1.1) for various sequences A when f is a characteristic function or even when it belongs to some L^p-class. In general, the a.e. limit of the averages in (1.1) is not equal to $\int_X f \, dm$ even if τ is ergodic. We will see simple examples below. Nevertheless, for irrational rotations the a.e. limit will often be the mean of f.

It turns out that instead of the averages in (1.1) it is often more instructive (and easier) to deal with

$$M_t(A, f)(x) = M_t f(x) = \frac{1}{A(t)} \sum_{\substack{1 \le a \le t \\ a \in A}} f(\tau^a x) \tag{1.2,}$$

where $A(t)$ is the counting function of A: $A(t) = \#\{a \mid a \in A, 1 \le a \le t\}$. Note that $A(t)$ is defined for every — large enough — positive real number t. If the sequence (a_n) is strictly increasing then the a.e. behavior of the averages in (1.1) and (1.2) are equivalent. In fact,

in this article, unless we say otherwise, a sequence (a_n) will mean an infinite, strictly increasing sequence of positive integers.

In this chapter we examine the special case when our measure-preserving system is periodic. This means that X consists of q symbols, say $X = \{0, 1, \ldots, q - 1\}$, the measure of each $x \in X$ is $1/q$, and $\tau(x) = x + 1$ mod q. A periodic system is particularly simple, and this chapter is of illustrative nature, but some of the results will be used later. For example, in periodic systems we will even be able to prove rate of convergence results which will have important consequences for Fourier transform estimates.

In this chapter X_q denotes the set $\{0, 1, \ldots, q-1\}$; m_q denotes the measure on X_q specified by $m_q(x) = 1/q$ for $x \in X_q$; and τ_q denotes the shift transformation $\tau(x) = x + 1$ mod q on X_q.

Let (X, Σ, m, τ) be a periodic system on q symbols, so $X = X_q$. Every function on X is a linear combination of characteristic functions of $h \in X$. It follows that the examination of the convergence of the $M_t f$'s reduces to the examination of the convergence of the averages $M_t \delta_h$, where δ_h is the Dirac mass at the point h, i.e. the characteristic function of $\{h\}$. It is clear that it is enough to examine the convergence of the sequences $M_t \delta_h(0)$ for each $h \in X$. In other words, we just need to examine the asymptotic behavior of $A(t, q, h)/A(t)$, where

$$A(t, q, h) = \#\{a \mid a \in A, 1 \le a \le t, a \equiv h \mod q\},$$

that is, $A(t, q, h)$ is the number of a's that are less than or equal to t and are in the arithmetic progression $qs + h$, $s = 0, 1, \ldots$.

1. Ergodic sequences for periodic systems

In this section we give examples for sequences with the nicest property: they are uniformly distributed in residue classes for every modulus q. Since we do not want to mix this concept with the concept of uniform distribution mod 1 discussed in the introduction we use the term "ergodic sequence":

Definition 1.2. We say that the sequence of positive integers $A = (a_n)$ is *ergodic* mod q if and only if for each h we have

$$\lim_{t \to \infty} \frac{A(t, q, h)}{A(t)} = \frac{1}{q}. \tag{1.3}$$

The sequence $A = (a_n)$ is *ergodic for periodic systems* if and only if it is ergodic mod q for every q.

Note that if A is ergodic for periodic systems, then in every periodic system we have

$$\lim_{t \to \infty} M_t f(x) = \int_X f \, dm \tag{1.4}$$

for every x.

Example 1.3. The sequence of positive integers. We take $A = \mathbb{Z}_+$, the sequence of positive integers. We clearly have the estimate, since $A(t) = [t]$,

$$|A(t, q, h) - \frac{A(t)}{q}| \leq 1,$$

which implies (1.3), hence the ergodicity of $\mathbb{Z}_+$ in periodic systems.

Example 1.4. Randomly generated sequences of positive density. Here we consider random sequences. Let (Ω, β, P) be a probability space, and let $0 < \sigma < 1$. Let $Y_1, Y_2, \ldots$ be an i.i.d. sequence of 0-1 valued random variables with $P(Y_n = 1) = \sigma$ and $P(Y_n = 0) = 1 - \sigma$. Then for each $\omega \in \Omega$ we can define a sequence of positive integers A^ω by letting $n \in A^\omega$ if and only if $Y_n(\omega) = 1$.

We want to show that for almost every ω the sequence A^ω is ergodic for periodic systems.

It is enough to show that for each fixed q, with probability 1 the sequence A^ω is ergodic mod q. In anticipating the techniques used later, we give a proof which is not the simplest one; we are going to show that with probability 1 we have

$$\lim_{t \to \infty} M_t(A^\omega, f)(x) = \int_X f \, dm \tag{1.5}$$

for every x. By the strong law of large numbers, since the expectation of Y_n is σ, with probability 1 we have

$$\lim_{t \to \infty} \frac{A^\omega(t)}{\sigma \cdot t} = 1.$$

(The observant reader will notice that this is also incuded in (1.14) to be proved below.) It follows that for almost every ω (1.5) is equivalent with

$$\lim_{t \to \infty} Q_t^\omega f(x) = \int_X f \, dm, \tag{1.6}$$

where

$$Q_t^\omega f(x) = \frac{1}{\sigma t} \sum_{\substack{1 \leq a \leq t \\ a \in A^\omega}} f(\tau^a x).$$

Let us rewrite $Q_t^\omega f$ as

$$Q_t^\omega f(x) = \frac{1}{\sigma t} \sum_{n \leq t} Y_n(\omega) f(\tau^n x). \tag{1.7}$$

Then we are given the idea to compare $Q_t^\omega f$ with its "expectation"

$$V_t f(x) = \frac{1}{\sigma t} \sum_{n \leq t} \sigma f(\tau^n x),$$

which is just the usual ergodic average examined in the previous section. (Here we see the advantage of using the averages of (1.2) instead of (1.3).) Since X has only finitely many elements, it is enough to prove that with probability 1

$$\lim_{t \to \infty} \|Q_t^\omega f(x) - V_t f(x)\|_{L^2(X)} = 0. \tag{1.8}$$

In fact, we shall not prove (1.8), but a weaker version of it. Namely, we shall prove it only when t runs through a subsequence of the positive integers. Our first lemma tells us what subsequence we can take.

Lemma 1.5. *Let (f_n) be a sequence of nonnegative numbers. For $\rho > 1$ denote $I_\rho = \{t \mid t = \rho^k$ for some positive integer $k\}$. Suppose that for each $\rho > 1$ the averages*

$$\frac{1}{t} \sum_{n \leq t} f_n$$

converge to some finite limit as t runs through the sequence I_ρ.
Then for each $\rho > 1$ this limit is the same, say L, and we have

$$\lim_{t \to \infty} \frac{1}{t} \sum_{n \leq t} f_n = L.$$

Proof. Let us set $F_t = \frac{1}{t} \sum_{n \leq t} f_n$. Let $\rho > 1$. For an arbitrary (but large enough) t let us choose the positive ineger m so that $\rho^m \leq t < \rho^{m+1}$. Since the sequence (f_n) is nonnegative, we can estimate

$$F_t = \frac{1}{t} \sum_{n \leq t} f_n \leq \frac{1}{\rho^m} \sum_{n \leq \rho^{m+1}} f_n = \rho \cdot F_{\rho^{m+1}}. \tag{1.9}$$

Similarly, we get

$$\frac{1}{\rho} \cdot F_{\rho^m} \leq F_t. \tag{1.10}$$

Let us denote

$$L_\rho = \lim_{\substack{s \to \infty \\ s \in I_\rho}} F_s.$$

Let us choose a sequence (ρ_k) so that $\rho_k > 1$, $I_{\rho_k} \subset I_{\rho_{k+1}}$ and $\rho_k \to 1$. Clearly, $L_{\rho_k} = L_{\rho_l}$ for each k, l, so set $L = L_{\rho_k}$. Using the estimates in (1.9) and (1.10) with a $\rho \in (\rho_k)$ we obtain

$$\frac{1}{\rho} \cdot L \leq \liminf_{t \to \infty} F_t \leq \limsup_{t \to \infty} F_t \leq \rho \cdot L.$$

Since $\rho \in (\rho_k)$ can be taken as close to 1 as we wish, our proof is complete. $\square$

Exercise 1.6. Show, using the lemma above, that (1.8) follows from
the following: For each $\rho > 1$ there is $\Omega_\rho \subset \Omega$ with $P(\Omega_\rho) = 1$, so that for
each $\omega \in \Omega_\rho$

$$\lim_{\substack{t \to \infty \\ t \in I_\rho}} \|Q_t^\omega f(x) - V_t f(x)\|_{L^2(X)} = 0 \,, \tag{1.11}$$

where I_ρ is defined in the lemma. (Hint: use the lemma for the negative
and positive parts of f separately)

The proof of (1.11) gives us the first opportunity to use Fourier-analysis.

For $f : X_q \to \mathbb{C}$ we define its Fourier transform $\hat f$ as follows. The dual
group of X_q is the set $\hat X_q = \{0, 1/q, \ldots, (q-1)/q\}$ with addition mod 1.
We define $\hat f : \hat X_q \to \mathbb{C}$ by

$$\hat f(b/q) = \int_{X_q} f(x) e(-xb/q) dm = \frac{1}{q} \sum_{x=0}^{q-1} f(x) e(-xb/q) \,,$$

where we set $e(y) = e^{2\pi i y}$. Often we will use $\mathcal{F}f$ to denote $\hat f$.

The inverse Fourier transform of $f : \hat X_q \to \mathbb{C}$ is defined by the formula

$$\mathcal{F}^{-1} f(x) = \check f(x) = \int_{\hat X_q} f(b/q) e(xb/q) d\hat m = \sum_{b=0}^{q-1} f(b/q) e(xb/q) \,.$$

Note that the $\hat m$-measure of each point in $\hat X_q$ is 1. Of course this is so we
can write Parseval's relation in the following form

$$\int_{X_q} |f|^2 dm = \int_{\hat X_q} |\hat f|^2 d\hat m \,.$$

Before we rewrite (1.11) using Fourier transforms we observe that

$$\mathcal{F}(Q_t^\omega f)(b/q) = \mathcal{F}Q_t^\omega(-b/q) \cdot \mathcal{F}f(b/q) \tag{1.12}$$

and

$$\mathcal{F}(V_t f)(b/q) = \mathcal{F}(V_t)(-b/q) \cdot \mathcal{F}f(b/q) \,, \tag{1.13}$$

where

$$\mathcal{F}(Q_t^\omega)(-b/q) = \frac{1}{\sigma t} \sum_{n \le t} Y_n(\omega) e(nb/q)$$

and

$$\mathcal{F}(V_t)(-b/q) = \frac{1}{\sigma t} \sum_{n \le t} \sigma \cdot e(nb/q) \,.$$

The idea behind the formulas in (1.12) and (1.13) is, of course, that $Q_t^\omega f$
and $V_t f$ can be regarded as convolutions, and the Fourier transform of a

convolution is the product of the Fourier transforms. Now using Parseval's formula and (1.12), (1.13) we can estimate as

$$\|Q_t^\omega f(x) - V_t f(x)\|_{L^2(X)} = \|\mathcal{F}\left(Q_t^\omega - V_t\right)(-b/q)\mathcal{F}f(b/q)\|_{L^2(\widehat{X})} \le$$

$$\le \sup_{b/q \in \widehat{X}} \left| \frac{1}{\sigma t} \sum_{n \le t} (Y_n(\omega) - \sigma) e(nb/q) \right| \cdot \|\widehat{f}\|_{L^2(\widehat{X})}.$$

Since there are only finitely many b/q's, we see we just need to prove that for each $b/q \in \widehat{X}$, for a.e. ω we have

$$\lim_{\substack{t \to \infty \\ t \in I_\rho}} \left| \frac{1}{\sigma t} \sum_{n \le t} (Y_n(\omega) - \sigma) e(nb/q) \right| = 0. \tag{1.14}$$

Let us introduce the random variable $Z_n = (Y_n - \sigma)e(nb/q)$. We will get (1.14) if we prove

$$\int_\Omega \left(\sum_{t \in I_\rho} \left| \frac{1}{\sigma t} \sum_{n \le t} Z_n(\omega) \right|^2 \right) dP(\omega) =$$

$$= \sum_{t \in I_\rho} \int_\Omega \left| \frac{1}{\sigma t} \sum_{n \le t} Z_n(\omega) \right|^2 dP(\omega) < \infty. \tag{1.15}$$

We observe, using independence, that the Z_n's are orthogonal: $\int_\Omega Z_n \cdot \overline{Z}_j dP = 0$ if $j \ne n$. It follows that

$$\int_\Omega \left| \frac{1}{\sigma t} \sum_{n \le t} Z_n(\omega) \right|^2 dP(\omega) = \frac{1}{\sigma^2 t^2} \int_\Omega \sum_{n \le t} |Z_n(\omega)|^2 dP(\omega) \le \frac{2}{\sigma^2 t},$$

which implies (1.15).

Let us summarize the main ideas in the above proof. First of all we observe that it is enough to prove the convergence of the averages $M_t(A, f) = M_t f$ when t runs through a sparse sequence. Then we compare the $M_t f$'s to another sequence of averages $V_t f$ for which we previously established convergence. This comparison is done by showing that the L^2-norm of the difference $M_t f - V_t f$ is small. This is achieved — and this is the main step of the proof — by estimating the L^∞-norm of the difference of the Fourier transforms $\mathcal{F}M_t - \mathcal{F}V_t$. In case of a periodic system, since the space X_q has only finitely many points, it was a simple matter to estimate this L^∞-norm; namely, we just needed an estimate at each point $b/q \in \widehat{X}_q$.

Exercises for Example 1.4

Ex.1. Let $A = (a_n)$ be a strictly increasing sequence of positive integers.
a) Show that A is ergodic mod q if and only if for each $b \not\equiv 0 \mod q$

$$\lim_{t \to \infty} \frac{1}{A(t)} \sum_{\substack{a \leq t \\ a \in A}} e(ab/q) = 0.$$

b) Using part a), give another proof that the random sequence A^ω discussed in Example 1.4 is ergodic mod q for every q. (This proof is the same, in principle, as the one given in the text, but the viewpoint is different; it is along the lines of Weyl's proof of his equidistribution theorem.)

Ex.2. Give yet another proof that the random sequence A^ω of Example 1.4 is ergodic mod q by considering the product space $\Omega \times X_q$ with the product measure $P \times m$ etc. (Hint: use the techniques used to establish (1.14).)

Ex.3. a) Show that for every $1 \leq p < \infty$

$$\lim_{t \to \infty} \|Q_t^\omega f(x) - V_t f(x)\|_{L^p(X_q)} = 0.$$

(We use the notation of the text.)
b) Does the conclusion of part a) hold if $p = \infty$?

Ex.4. Let (Ω, β, P) be a probability space, and let $0 \leq \sigma \leq 1$. Let $Y_1, Y_2 \ldots$ be an i.i.d. sequence of $((-1)\text{-}1)$-valued random variables wih $P(Y_n = 1) = \sigma$ and $P(Y_n = -1) = 1 - \sigma$. Denote $a_n(\omega) = \sum_{k \leq n} Y_k(\omega)$. Show that with probability 1, for $f : X_q \to \mathbb{C}$ we have

$$\lim_{t \to \infty} \frac{1}{t} \sum_{n \leq t} f(\tau^{a_n(\omega)} x) = \int_{X_q} f \, dm_q$$

for every $x \in X_q$. (Note that the random sequence $(a_n(\omega))$ is not an increasing sequence of integers anymore even if it has a "positive drift", that is when $\sigma > 1/2$.)

Ex.5. Let $Y_1, Y_2, \ldots$ be a sequence of mean 0, orthogonal and uniformly bounded random variables on the probability space (Ω, β, P). Then $1/t \sum_{n \leq t} Y_n(\omega) \to 0$ almost surely.

Example 1.7. Randomly generated sequences of 0 density. The random sequence A^ω of Example 1.4 had positive density. In this section we examine random sequences of 0 density. In fact, a random sequence of 0 density has the additional property that even its Banach-density is 0.

Definition 1.8. Let $A = (a_n)$ be a strictly increasing sequence of positive integers.

The *upper density*, the *lower density*, and *density* of A are defined respectively as

$$\overline{d}(A) = \limsup_{t\to\infty} \frac{A(t)}{t},$$

$$\underline{d}(A) = \liminf_{t\to\infty} \frac{A(t)}{t},$$

$$d(A) = \lim_{t\to\infty} \frac{A(t)}{t}.$$

The *upper Banach density*, the *lower Banach density*, and *Banach density* of A are defined respectively as

$$\overline{Bd}(A) = \limsup_{|I|\to\infty} \frac{\#\{a \mid a \in A \cap I\}}{|I|},$$

$$\underline{Bd}(A) = \liminf_{|I|\to\infty} \frac{\#\{a \mid a \in A \cap I\}}{|I|},$$

$$Bd(A) = \lim_{|I|\to\infty} \frac{\#\{a \mid a \in A \cap I\}}{|I|},$$

where I denotes (finite) subintervals of $\mathbb{R}_+$.

The random sequence A^ω will have 0 Banach density if we put the integer n into A^ω with probability σ_n, where $\sigma_n \to 0$ as $n \to \infty$.

Let (σ_n) be a sequence of positive numbers satisfying the following properties:

(i) The sequence (σ_n) is decreasing;
(ii) $\lim_{t\to\infty} \sum_{n\le t} \sigma_n = \infty$.

Property (i) will guarantee the ergodicity of the random sequence A^ω; property (ii) guarantees that A^ω has infinitely many elements.

Let (Y_n) be a sequence of independent, (0-1)-valued random variables on the probability space (Ω, β, P) so that $P(Y_n = 1) = \sigma_n$ and $P(Y_n = 0) = 1 - \sigma_n$. In other words, the expectation $E(Y_n)$ of Y_n is σ_n.

We shall show that with probability 1 the random sequence A^ω, defined by $n \in A^\omega$ if and only if $Y_n(\omega) = 1$, is ergodic for periodic systems.

So we will show that with probability 1, if $f : X_q \to \mathbb{C}$ then (1.5) holds. This time we will compare $M_t f$ with

$$V_t f(x) = \frac{1}{\sum_{n\le t} \sigma_n} \sum_{n\le t} \sigma_n f(\tau^n x).$$

The fact that $\lim_{t\to\infty} V_t f(x) = \int_X f\, dm$ follows from the following lemma.

Lemma 1.9. *Suppose the sequence (σ_n) of positive numbers satisfies properties (i) and (ii) above. Suppose that for some sequence (f_n) of complex numbers we have*

$$\lim_{t\to\infty} \frac{1}{t}\sum_{n\leq t} f_n = \alpha.$$

Then we have

$$\lim_{t\to\infty} \frac{1}{\sum_{n\leq t}\sigma_n}\sum_{n\leq t}\sigma_n f_n = \alpha.$$

Proof. It is an exercise in summation by parts.

The key step in Example 1.4 was to establish (1.14), and this time we will prove a similar result, but we need to take a sparser sequence than I_ρ. Indeed, for $\rho > 1$ we take I_ρ to consist of the numbers $t_k, k = 1, 2, \ldots$ defined by $t_k = \min\{t \mid \rho^k \leq \sum_{n\leq t}\sigma_n < \rho^{k+1}\}$. Similarly to Example 1.4 (using an appropriate generalization of Lemma 1.5), we need to prove

$$\lim_{\substack{t\to\infty\\t\in I_\rho}} \left|\frac{1}{\sum_{n\leq t}\sigma_n}\sum_{n\leq t}(Y_n(\omega)-\sigma_n)e(nb/q)\right| = 0. \tag{1.16}$$

We will get (1.16) if we prove

$$\sum_{t\in I_\rho}\int_\Omega \left|\frac{1}{\sum_{n\leq t}\sigma_n}\sum_{n\leq t}(Y_n(\omega)-\sigma_n)e(nb/q)\right|^2 dP(\omega) < \infty. \tag{1.17}$$

By the orthogonality of the random variables $(Y_n-\sigma_n)e(nb/q)$, we have

$$\int_\Omega \left|\frac{1}{\sum_{n\leq t}\sigma_n}\sum_{n\leq t}(Y_n(\omega)-\sigma_n)e(nb/q)\right|^2 dP(\omega) =$$

$$= \frac{1}{(\sum_{n\leq t}\sigma_n)^2}\int_\Omega\sum_{n\leq t}(Y_n-\sigma_n)^2 dP(\omega). \tag{1.18}$$

By the definition of I_ρ, the inequality in (1.17) will follow from (1.18) and the estimate

$$\int_\Omega\sum_{n\leq t}(Y_n-\sigma_n)^2 dP(\omega) \leq \sum_{n\leq t}\sigma_n. \tag{1.19}$$

But this follows since

$$\sum_{n\leq t}\int_\Omega(Y_n-\sigma_n)^2 dP(\omega) = \sum_{n\leq t}\left(\int_\Omega Y_n^2 dP(\omega)-\sigma_n^2\right) \leq$$

$$\leq \sum_{n\leq t}\int_\Omega Y_n dP(\omega) = \sum_{n\leq t}\sigma_n.$$

Exercise 1.10. Suppose that the sequence (σ_n) satisfies property (ii) but not necessarily property (i).

a) Give a necessary and sufficient condition on (σ_n) so that with probability 1 the averages $M_t(A^\omega, f)(x)$ converge for each $f : X_q \to \mathbb{C}$ and $x \in X_q$.

b) Give an example for (σ_n) such that with probability 1 the sequence A^ω is *not* ergodic, but the averages $M_t(A^\omega, f)(x)$ converge for each $f : X_q \to \mathbb{C}$ and $x \in X_q$.

2. Sequences that are good in residue classes

In this section we examine sequences that are not ergodic for periodic systems but $\lim_{t\to\infty} M_t(A, f)(x)$ exists.

Definition 1.11. We say that the sequence of positive integers $A = (a_n)$ is *good* mod q if and only if for each h the limit

$$\Lambda(h) = \Lambda(q, h) = \lim_{t\to\infty} \frac{A(t, q, h)}{A(t)} \tag{1.20}$$

exists. The sequence $A = (a_n)$ is *good for periodic systems* if and only if it is good   mod q for every q.

Note that if A is good for periodic systems, then in every periodic system the limit

$$\lim_{t\to\infty} M_t f(x) \tag{1.21}$$

exists for every $x \in X_q$.

If the sequence A is good mod q, then for each $b/q \in \widehat{X}_q$ the limit

$$\widehat{\Lambda}(b/q) = \lim_{t\to\infty} \frac{1}{A(t)} \sum_{\substack{a \leq t \\ a \in A}} e(ab/q) \tag{1.22}$$

exists. We have seen in Exercise 1 for Example 1.4 that A is ergodic mod q if and only if $\widehat{\Lambda}(b/q) = 0$ for each $b \not\equiv 0$ mod q. Therefore, for a nonergodic sequence we will find a b for which $\widehat{\Lambda}(b/q) \neq 0$. Now, we have the trivial estimate $|\widehat{\Lambda}(b/q)| \leq 1$ for every b. In the examples below we shall see good sequences that are not ergodic mod q; nevertheless,

$$\sup_{\substack{b/q \in \widehat{X}_q \\ \gcd(b,q)=1}} |\widehat{\Lambda}(b/q)| \to 0$$

as $q \to \infty$. So at least in the asymptotic behavior of the Fourier transform we can detect some kind of uniform distribution in residue classes.

Example 1.12. The sequence of squares; $|\widehat{\Lambda}(b/q)| \le C/\sqrt{q}$. Here we examine the sequence of squares $A = \{n^2 \mid n \in \mathbb{Z}_+\}$. This sequence is certainly not ergodic; for example, we have $A(t,3,2) = 0$. To estimate $A(t,q,h)$, let $\theta(h) = \theta(q,h) = \#\{j \mid 0 \le j < q, j^2 \equiv h \mod q\}$. Clearly, for any fixed j with $0 \le j < q$ we have

$$\left| \#\{n \mid 1 \le n \le A(t), n \equiv j \mod q\} - \frac{A(t)}{q} \right| \le 1,$$

hence

$$\left| A(t,q,h) - \theta(h)\frac{A(t)}{q} \right| \le \theta(h). \tag{1.22}$$

It follows that A is good for periodic systems, and $\Lambda(q,h) = \theta(h)/q$.

Exercise 1.13. Let q be an odd prime number, and suppose $\gcd(q,h) = 1$. Then $\theta(q,h)$ is either 0 or 2. (Hint: the fact that $j^2 \equiv h \mod q$ has at most 2 solutions follows if we notice that the residue classes mod q form a field.)

Let us see how the lack of ergodicity is reflected in the asymptotic behavior of the Fourier transform

$$\mathcal{F}M_t(-b/q) = \frac{1}{\sqrt{t}} \sum_{n^2 \le t} e(n^2 b/q).$$

Proposition 1.14. *Suppose $\gcd(b,q) = 1$. Then we have the following equalities*

$$|\widehat{\Lambda}(b/q)| = \lim_{t \to \infty} \left| \frac{1}{\sqrt{t}} \sum_{n^2 \le t} e(n^2 b/q) \right| = \begin{cases} 1/\sqrt{q} & \text{if } q \equiv \pm 1 \mod 4 \\ \sqrt{2/q} & \text{if } q \equiv 0 \mod 4 \\ 0 & \text{if } q \equiv 2 \mod 4. \end{cases}$$

This proposition shows two things. First, we now see in terms of the Fourier transform that the sequence of squares is not ergodic for any $q > 1$. Second, we have some kind of uniformity of the distribution of the squares in residue classes; we have

$$\sup_{\substack{b/q \in \widehat{X}_q \\ \gcd(b,q)=1}} |\widehat{\Lambda}(b/q)| \to 0.$$

Proof of Proposition 1.14. First of all, we have the following explicit formulas for $\widehat{\Lambda}(b/q)$:

$$\widehat{\Lambda}(b/q) = \frac{1}{q} \sum_{n<q} e(n^2 b/q) = \frac{1}{q} \sum_{h<q} \theta(h)e(hb/q).$$

Because of periodicity by q, for any integer m we have

$$\widehat{\Lambda}(b/q) = \frac{1}{q} \sum_{n<q} e((n+m)^2 b/q).$$

Using this, we can write

$$|\widehat{\Lambda}(b/q)|^2 = \overline{\widehat{\Lambda}(b/q)} \cdot \widehat{\Lambda}(b/q) = \frac{1}{q} \sum_{m<q} e(-m^2 b/q) \frac{1}{q} \sum_{n<q} e(n^2 b/q) =$$

$$= \frac{1}{q} \sum_{m<q} e(-m^2 b/q) \frac{1}{q} \sum_{n<q} e\left((n+m)^2 b/q\right) =$$

$$= \frac{1}{q} \sum_{m<q} \frac{1}{q} \sum_{n<q} e\left((n^2 + 2nm)b/q\right) =$$

$$= \frac{1}{q} \sum_{n<q} e(n^2 b/q) \frac{1}{q} \sum_{m<q} e(2nmb/q).$$

Now, $\frac{1}{q} \sum_{m<q} e(2nmb/q)$ is nonzero — and then it is equal to 1 — if and only if q divides $2n$, that is when either $n = 0$ or (in case q is even) $n = q/2$. It follows that for odd q we have

$$|\widehat{\Lambda}(b/q)|^2 = \frac{1}{q},$$

and for even q we have

$$|\widehat{\Lambda}(b/q)|^2 = \frac{1}{q}(1 + e(qb/4)) = \begin{cases} \frac{2}{q}, & \text{if 4 divides } q \\ 08 & \text{if } q \equiv 2 \mod 4. \end{cases}$$

□

Exercises for Example 1.12

Ex.1. Let $A = (a_n)$ be a strictly increasing sequence of positive integers. Show that A is good mod q if and only if for each $b/q \in \widehat{X}_q$ the limit

$$\lim_{t \to \infty} \frac{1}{A(t)} \sum_{\substack{a \leq t \\ a \in A}} e(ab/q)$$

exists.

Ex.2. Let k be a positive integer, and let $A = (n^k)$ be the sequence of k-th powers.

a) Show that the sequence A is good for periodic systems by showing that $\Lambda(q,h) = \theta_k(q,h)/q$, where $\theta_k(q,h)$ is the number of solutions of the congruence $x^k \equiv h \mod q$.

b) Show that

$$\widehat{\Lambda}(b/q) = \frac{1}{q} \sum_{n<q} e(n^k b/q) = \frac{1}{q} \sum_{h<q} \theta_k(h) e(hb/q).$$

Ex.3. Let k be a positive integer, and let $A = (n^k)$ be the sequence of k-th powers. Let q be a prime number and let b be relatively prime to q.

a) Show that for any fixed n the congruence $m^k \equiv n \mod q$ has at most k solutions.

b) Show that

$$\sum_{n<q} |\widehat{\Lambda}(n/q)|^2 \le k^2 .$$

(Hint: Use Parseval's identity for the function $f : X_q \to \mathbb{C}$ defined by $f(x) = \theta_k(x, q)$.)

c) Show that

$$|\widehat{\Lambda}(b/q)|^2 = \frac{1}{q} \sum_{n<q} |\widehat{\Lambda}(n^k b/q)|^2.$$

(Hint: If $\gcd(n, q) = 1$, then $\widehat{\Lambda}(b/q) = \widehat{\Lambda}(n^k b/q)$.)

d) Show that

$$|\widehat{\Lambda}(b/q)| \le \frac{k^{3/2}}{\sqrt{q}} .$$

Ex.4. Let k, A, q, b be as in the previous exercise. Let $d = \gcd(k, q-1)$. Show that

$$|\widehat{\Lambda}(b/q)| \le \frac{d}{\sqrt{q}} .$$

(Hint: Be more careful with the estimations in the previous exercise.)

We remark that the best constant in the above inequality is $d - 1$.

Ex.5. Let (σ_n) be a sequence of positive numbers satisfying the properties:

(i) the sequence (σ_n) is decreasing;
(ii) $\lim_{t \to \infty} \sum_{n \le t} \sigma_n = \infty$.

Let (Y_n) be a sequence of independent, (0-1)-valued random variables on the probability space (Ω, β, P) so that $P(Y_n = 1) = \sigma_n$ and $P(Y_n = 0) = 1 - \sigma_n$.

Let k be a fixed positive number. Show that with probability 1 the random sequence A^ω, defined by $n^k \in A^\omega$ if and only if $Y_n(\omega) = 1$, is good for periodic systems.

Ex.6. Let (σ_n) be a sequence of positive numbers satisfying the properties

(i) $\lim_{n \to \infty} \sigma_n = 0$;
(ii) $\lim_{t \to \infty} \sum_{n \le t} \sigma_n = \infty$.

Let (Y_n) be a sequence of independent, (0-1)-valued random variables on the probability space (Ω, β, P) so that $P(Y_n = 1) = \sigma_n$ and $P(Y_n = 0) = 1 - \sigma_n$.

Show that with probability 1 the random sequence A^ω, defined by $n \in A^\omega$ if and only if $Y_n(\omega) = 1$, is of 0 Banach density.

Example 1.15. The sequence of primes. Let A be the sequence of prime numbers. To describe the distribution of the primes in residue classes we introduce the function

$$li(t) = \sum_{2 \le n \le t} \frac{1}{\log n}.$$

Note that $\frac{li(t)}{t/\log t} \to 1$. The prime number theorem tells us that $\frac{A(t)}{li(t)} \to 1$. If q and h are not relatively prime, then the arithmetic progression $qn + h$, $n = 1, 2, \ldots$, contains at most 1 prime number. On the other hand, the primes are uniformly distributed among the reduced residue classes. This is the content of Siegel-Walfis's theorem [Ellison & Ellison, 1985] on the distribution of primes in arithmetic progressions. Before we state the theorem, we recall the definition of Euler's "ϕ-function": $\phi(q) = \#\{h \mid 0 \le h < q, \gcd(h, q) = 1\}$.

Lemma 1.16 (The prime number theorem for arithmetic progressions). Let $\gcd(q, h) = 1$. Then we have

$$\lim_{t \to \infty} \frac{A(t, q, h)}{li(t)/\phi(q)} = 1.$$

An immediate consequence of this lemma is that the sequence of primes is good for periodic systems and $\Lambda(q, h) = 1/\phi(q)$ if $\gcd(q, h) = 1$ and $\Lambda(q, h) = 0$ otherwise.

Exercises for Example 1.15

Ex.1. Suppose $\gcd(b, q) = 1$. Then we have

$$\widehat{\Lambda}(b/q) = \frac{\mu(q)}{\phi(q)},$$

where μ denotes the Möbius function:

$$\mu(q) = \begin{cases} 1, & \text{if } q = 1 \\ (-1)^k, & \text{if } q \text{ is squarefree, and } k \text{ is the number of} \\ & \quad \text{prime-divisors of } q \\ 0, & \text{otherwise}. \end{cases}$$

(Hint: use the inversion formula of Möbius.)

Ex.2.
a) Show that there is a positive constant c so that for every q

$$\phi(q) > c \cdot \frac{q}{\log q}.$$

b) Show that there is a positive constant c so that for every q

$$\phi(q) > c \cdot \frac{q}{\log \log q}.$$

Ex.3. Let (σ_n) be a sequence of positive numbers satisfying the properties

(i) the sequence (σ_n) is decreasing;
(ii) $\lim_{t\to\infty} \sum_{n\le t} \sigma_n = \infty$.

Let (Y_n) be a sequence of independent, (0-1)-valued random variables on the probability space (Ω, β, P) so that $P(Y_n = 1) = \sigma_n$ and $P(Y_n = 0) = 1 - \sigma_n$.

Let p_n denote the n-th prime number. Show that with probability 1 the random sequence A^ω, defined by $p_n \in A^\omega$ if and only if $Y_n(\omega) = 1$, is good for periodic systems. (Compare this with Exercise 5 for Example 1.12. Is there a common generalization?!)

3. Sequences that are bad for periodic systems

In this section we briefly examine sequences $A = (a_n)$ that are not good for periodic systems. We may distinguish several degrees of bad behavior. It is possible that A is not good mod q for only some values of q. But maybe A is not good mod q for *every* q. If we know that $M_t(A, f)(x)$ does not converge, it may mean only that we do not have convergence only for some $x \in X_q$, but maybe there is no convergence for any $x \in X_q$. Since $0 \le A(t, q, h)/A(t) \le 1$, the worst possible scenario is when

$$\limsup_{t\to\infty} \frac{A(t, q, h)}{A(t)} = 1 \tag{1.23}$$

and

$$\liminf_{t\to\infty} \frac{A(t, q, h)}{A(t)} = 0 \tag{1.24}$$

for every $h \in X_q$.

Definition 1.17. We say that the sequence of positive integers $A = (a_n)$ is *bad* mod q if and only if for each $h \in X_q$ the limit

$$\lim_{t\to\infty} \frac{A(t, q, h)}{A(t)}$$

does not exist. The sequence $A = (a_n)$ is *bad for periodic systems* if and only if it is bad mod q for every q.

We say that the sequence of positive integers $A = (a_n)$ is *sweeping out* mod q if and only if for each $h \in X_q$ the equalities in (1.23) and (1.24) hold. The sequence $A = (a_n)$ is *sweeping out for periodic systems* if and only if it is bad mod q for every q.

Note that A is sweeping out mod q if and only if for every $x \in X_q$ we have

$$\limsup_{t\to\infty} M_t(A, \delta_0)(x) = 1, \qquad \liminf_{t\to\infty} M_t(A, \delta_0)(x) = 0. \tag{1.25}$$

Example 1.18. A sweeping out sequence.

Here we will show that there is a strictly increasing sequence of positive integers which is sweeping out for periodic systems.

The idea of the construction is to make sure that for each q and h there is a $t = t_{(q,h)}$ so that most of the elements $a_1, a_2, \ldots, a_t$ of A are congruent to $h \mod q$. It follows, of course, that for some t' most of the elements $a_1, a_2, \ldots, a_{t'}$ of A are *not* congruent to $h \mod q$. It is more convenient in this proof to consider the averages

$$M_t f(x) = \frac{1}{t} \sum_{n \leq t} f(\tau^{a_n} x),$$

and we assume that t is an integer. For each q let us define $K_q = 1 + 2 + \cdots + q$. Note that each residue class mod q has a representative k satisfying $K_{q-1} < k \leq K_q$. Let us consider a sequence (t_k) of positive integers which satisfy the following condition:

$$\frac{t_k}{t_{k-1}} > q + 1; \quad \text{for } K_{q-1} < k \leq K_q. \tag{1.26}$$

Let us define the strictly increasing sequence $A = (a_n)$ to satisfy, for every k,

$$a_n \equiv k \mod q \quad \text{for} \quad t_{k-1} < n \leq t_k. \tag{1.27}$$

Let us show (1.25); so fix $q > 1$ and let $x \in X_q$. Then there is h with $0 < h \leq q$ such that $x \equiv -h \mod q$. For each positive integer s define the indices $k(s)$ by $k(s) = K_{s \cdot q - 1} + h$. We claim that

$$\lim_{s \to \infty} M_{t_{k(s)}} \delta_0(x) = 1. \tag{1.28}$$

To see this, first we note that by (1.27) we have for every s

$$a_n \equiv h \mod q \quad \text{for} \quad t_{k(s)-1} < n \leq t_{k(s)},$$

and hence

$$x + a_n \equiv 0 \mod q \quad \text{for} \quad t_{k(s)-1} < n \leq t_{k(s)}. \tag{1.29}$$

We also have, by the property in (1.26),

$$\frac{t_{k(s)}}{t_{k(s)-1}} > s \cdot q + 1. \tag{1.30}$$

Now we can estimate, by (1.29) and (1.30),

$$M_{t_{k(s)}} \delta_0(x) \geq \frac{1}{t_{k(s)}} \sum_{t_{k(s)-1} < n \leq t_{k(s)}} \delta_0(x + a_n) = \frac{t_{k(s)} - t_{k(s)-1}}{t_{k(s)}} > 1 - \frac{1}{s \cdot q},$$

which implies (1.28). On the other hand, we claim that

$$\lim_{s\to\infty} M_{t_{k(s)+1}}\delta_0(x) = 0. \tag{1.31}$$

This, together with (1.28), proves (1.25). To see (1.31), note that for every $s > 1$

$$a_n \equiv h+1 \mod q \quad \text{for} \quad t_{k(s)} < n \le t_{k(s)+1},$$

and hence, if $q > 1$,

$$x + a_n \not\equiv 0 \mod q \quad \text{for} \quad t_{k(s)} < n \le t_{k(s)+1}. \tag{1.32}$$

This time we can estimate, by (1.32),

$$M_{t_{k(s)+1}}\delta_0(x) \le \frac{1}{t_{k(s)+1}} \sum_{1 \le n \le t_{k(s)}} \delta_0(x + a_n) \le \frac{t_{k(s)}}{t_{k(s)+1}} < \frac{1}{s \cdot q},$$

finishing the proof of (1.31).

Exercises for Example 1.18

Ex.1. Construct a sequence that is bad mod q for some q but which is good mod q for infinitely many q's.

Ex.2. Is it true that if a sequence is bad mod q for some q then it is bad mod q for infinitely many q?

Ex.3. Is it true that if a sequence is sweeping out mod q for some q then it is sweeping out mod q for infinitely many q?

Ex.4. Construct a sequence of positive density that is sweeping out for some q.

Ex.5. Is there a sequence of positive lower density that is bad for periodic systems?

Ex.6. Let (b_n) be a strictly increasing sequence of positive integers with the property that $b_{n+1} - b_n \to \infty$ as $n \to \infty$. Construct a sequence (a_n) that is bad for periodic systems and satisfies $b_n \le a_n \le b_{n+1}$ for every n.

Ex.7. a) Show that there is a lacunary sequence of integers which is good for periodic systems. (A sequence (a_n) is *lacunary* if and only if there is a constant $\rho > 1$ such that $a_{n+1}/a_n > \rho$ for every large enough n.)

b) Is there a lacunary sequence of integers which is ergodic for periodic systems? (Hint: there is ...)

CHAPTER II

GOOD SEQUENCES FOR MEAN L^2 CONVERGENCE

In this chapter we will examine the L^2-norm convergence of the averages $M_t(A, f)(x)$ for various sequences $A = (a_n)$. We recall that for a given sequence of positive integers $A = (a_n)$ we have defined

$$M_t(A, f)(x) = M_t f(x) = \frac{1}{A(t)} \sum_{\substack{1 \leq a \leq t \\ a \in A}} f(\tau^a x), \qquad (2.1)$$

where τ is a measure-preserving transformation of the probability space (X, Σ, m), and $f : X \to \mathbb{C}$.

In the previous chapter we have seen that Fourier - transform estimates played an important role in establishing convergence in periodic systems, but all the results could have been proved without any reference to Fourier analysis. The role of Fourier analysis will be more significant in this chapter. Indeed, the only known proof of the mean convergence of the averages $M_t(A, f)$ for a random sequence or for the sequence of primes is via Fourier analysis. As we will see, the L^2-convergence of the averages $M_t(A, f)$ is *equivalent* with the convergence of the Fourier transforms

$$\widehat{M_t}(\beta) = \frac{1}{A(t)} \sum_{\substack{1 \leq a \leq t \\ a \in A}} e(a\beta)$$

for every real β.

1. Ergodic sequences

Definition 2.1. Let τ be an ergodic measure-preserving transformation of the probability space (X, Σ, m). We say that the sequence of positive integers $A = (a_n)$ is *ergodic* for the system (X, Σ, m, τ) if and only if for each $f : X \to \mathbb{C}$ we have

$$\lim_{t \to \infty} M_t(A, f) = \int_X f \, dm \qquad (2.2)$$

in L^2-norm. The sequence $A = (a_n)$ is *ergodic* if and only if it is ergodic for every ergodic system (X, Σ, m, T).

Example 2.2. The mean ergodic theorem.

In this section we shall prove that the sequence of positive integers $A = \mathbb{Z}_+$ is ergodic.

We shall give two different proofs. The first is due to von Neumann [1932], and uses Fourier analysis; the second proof is due to F. Riesz [Riesz, Sz.-Nagy, 1990], and it uses geometric (Hilbert space) techniques. Although the first method of proof is more important for our purposes, the main idea of the second proof will also be used later.

Both proofs depend on the observation that the operator T_τ defined by $T_\tau f(x) = f(\tau x)$ is a contraction of the Hilbert space $L^2(X, \Sigma, m)$. Therefore the following should be proved:

Proposition 2.3. *Let U be any contraction of the Hilbert space H. Then for each $f \in H$ we have*

$$\lim_{t \to \infty} M_t f = Pf \,,$$

where $M_t f = (1/t) \sum_{n \leq t} U^n f$ and P is the projection onto the space of U-invariant points of H.

This result implies the ergodicity of the sequence of the natural numbers, for if τ is ergodic then the τ - invariant functions are the constants, and hence, as can be seen easily, $Pf = \int_X f \, dm$.

Both in the Fourier analysis and in the geometric proof we just prove the existence of the limit. For the identification of the limit — that it is Pf — we refer the reader to Exercise 5 below.

Fourier analysis proof. We want to prove that the directed set $\{M_t f \mid t \in \mathbb{R}_+\}$ is "Cauchy", that is

$$\lim_{t,s \to \infty} \|M_t f - M_s f\|_H = 0 \,. \tag{2.3}$$

Let (f, g) denote the inner product on H. Let us define the sequence of operators $(T_n \mid n = 0, \pm 1, \pm 2, \ldots)$ by $T_n = U^n$ for $n \geq 0$ and $T_n = (U^*)^{-n}$ for $n < 0$. Then for a fixed $f \in H$ the sequence of real numbers $(x_n \mid n = 0, \pm 1, \pm 2, \ldots)$ defined by $x_n = (T_n f, f)$ is a positive - definite sequence (cf. Section 9 of the Appendix in [Riesz, Sz.-Nagy, 1990]), hence, by Herglotz's theorem, there is a (nonnegative) Borel measure $\mu = \mu_f$ on the interval $[0, 1)$ so that for every $n \in \mathbb{Z}$

$$(T_n f, f) = \int e\,(n\beta) \, d\mu(\beta) \,. \tag{2.4}$$

Note that we have, in particular, $\int d\mu = \|f\|^2$. A consequence of (2.4) is the following fundamental inequality of von Neumann:

Exercise 2.4. Let $p(x)$ be a polynomial with complex coefficients. Then we have, for every $f \in H$,

$$\|p(\tau)f\|^2 \leq \int |p(e(\beta))|^2 d\mu(\beta) \leq$$
$$\leq \sup_{\beta} |p(e(\beta))|^2 \cdot \|f\|^2 . \tag{2.5}$$

(Hint: Use (2.4) and induction on the degree of $p(x)$.)

By this exercise, we have

$$\|M_t f - M_s f\|^2 \leq \int |\widehat{M}_t(\beta) - \widehat{M}_s(\beta)|^2 d\mu(\beta) ,$$

where

$$\widehat{M}_t(\beta) = \frac{1}{t} \sum_{1 \leq n \leq t} e(n\beta) .$$

Observing that $e(n\beta)$, $n = 1, 2, \ldots$, is a geometric progression, we get that $\widehat{M}_t(\beta) \to 0$ for $\beta \in [0,1) \setminus \{0\}$, hence $\widehat{M}_t(\beta) - \widehat{M}_s(\beta) \to 0$ for every β. But then, by the bounded convergence theorem, we get (2.3). □

Geometric Proof. Let I denote the space of U-invariant points of H, and let C be the set of coboundaries, that is, $C = \{f \mid f = g - Ug$ for some $g \in H\}$. It is clear that for $f \in I$ we have $M_t f \to f$, and if $f \in C$ then $M_t f \to 0$. By the uniform boundedness principle, we just need to prove that the linear span $I + C = \{f + g \mid f \in I, g \in C\}$ of the invariant points and coboundaries is dense in H. We shall prove this by showing that the orthocomplement $C^{\perp}$ of C contains only U-invariant points.

Suppose that h is orthogonal to all points f of the form $f = g - Ug$. We want to show that h is U-invariant. By assumption we have $(g - Ug, h) = 0$ for every $g \in H$. In particular, we have $(h - Uh, h) = 0$. But then, by the Pythagorean theorem, and using that U is a contraction, we get

$$\|h - Uh\|_H^2 = \|Uh\|_H^2 - \|h\|_H^2 \leq 0 ,$$

which implies that $h = Uh$. □

Exercises for Example 2.2

Ex.1. Let $A = (a_n)$ a "block" sequence. This means that A is a union of intervals of consecutive integers: $A = \cup I_k \cap \mathbb{Z}_+$, where (I_k) is a sequence of intervals in $\mathbb{R}_+$.

Show, using both the Fourier - analytic and the geometric method, that A is ergodic provided $|I_k| \to \infty$ as $k \to \infty$.

Ex.2. Let $F(t)$ be a real function satisfying the properties

(i) $F(t) \le t$;

(ii) $\lim_{t \to \infty} F(t) = \infty$.

a) Show that there is a strictly increasing sequence A of positive integers which is ergodic and satisfies

$$\lim_{t \to \infty} \frac{A(t)}{F(t)} = 0.$$

(Hint: use the result of the previous exercise.)

b) Is there is a strictly increasing sequence A of positive integers which is ergodic and satisfies

$$\lim_{t \to \infty} \frac{A(t)}{F(t)} = 1?$$

Ex.3. Let $A = (a_n)$ be a strictly increasing sequence of positive integers. Show that A is ergodic if and only if for each $\beta \in (0, 1)$

$$\lim_{t \to \infty} \frac{1}{A(t)} \sum_{\substack{a \le t \\ a \in A}} e(a\beta) = 0.$$

Ex.4. a) Let $1 \le p < \infty$. Prove that $M_t(\mathbb{Z}_+, f) \to \overline{f}$ in L^p norm, where $\overline{f}$ is τ-invariant. (The $L^p \to L^p$ operator T is defined by $Tf(x) = f(\tau x)$.)

b) Can we take $p = \infty$ in part a)?

Ex.5. Finish the proof of Proposition 2.3 by identifying $\lim_{t \to \infty} M_t f$ as Pf. With the notation of the geometric proof, this means that you have to show that, in fact, $C^{\perp} = I$. (Hint: In the geometric proof we showed that $C^{\perp} \subseteq I$. To show that $I \subseteq C^{\perp}$ write $f \in I$ as $f = g + h$ with $g \in C^{\perp}$ and $h \in \overline{C}$. Then, since $h \in I \cap \overline{C}$, we have $h = \lim_{t \to \infty} M_t h = 0$.)

Example 2.5: Randomly generated sequences of positive density. Here we examine the random sequence considered in Example 1.4. So let (Ω, β, P) be a probability space, and let $0 < \sigma < 1$. Let $Y_1, Y_2, \ldots$ be an i.i.d. sequence of 0-1 valued random variables with $P(Y_n = 1) = \sigma$ and $P(Y_n = 0) = 1 - \sigma$. For $\omega \in \Omega$ we define the sequence of positive integers A^{ω} by letting $n \in A^{\omega}$ if and only if $Y_n(\omega) = 1$.

We want to show that for almost every ω the sequence A^{ω} is ergodic.

As in Example 1.4, we want to show that with probability 1 we have, for every $f \in L^2$,

$$\lim_{t \to \infty} \|Q_t^{\omega} f(x) - V_t f(x)\|_{L^2(X)} = 0,$$

where

$$Q_t^{\omega} f(x) = \frac{1}{\sigma t} \sum_{n \le t} Y_n(\omega) f(\tau^n x) \tag{2.6}$$

and

$$V_t f(x) = \frac{1}{\sigma t} \sum_{n \leq t} \sigma f(\tau^n x) \, .$$

Again, we just need to show that for each fixed $\rho > 1$, with probability 1 we have, for every $f \in L^2$,

$$\lim_{\substack{t \to \infty \\ t \in I_\rho}} \|Q_t^\omega f(x) - V_t f(x)\|_{L^2(X)} = 0 \, , \tag{2.7}$$

where $I_\rho = \{t \mid t = \rho^k \text{ for some positive integer } k\}$. By the second inequality of Exercise 2.4 we need to prove

$$\lim_{\substack{t \to \infty \\ t \in I_\rho}} \sup_{\beta \in [0,1)} |\widehat{Q}_t^\omega(\beta) - \widehat{V}_t(\beta)| = 0 \, , \tag{2.8}$$

where

$$\widehat{Q}_t^\omega(\beta) = \frac{1}{\sigma t} \sum_{n \leq t} Y_n(\omega) e(n\beta), \qquad \widehat{V}_t(\beta) = \frac{1}{\sigma t} \sum_{n \leq t} \sigma e(n\beta) \, .$$

In Example 1.4 we computed second moments to establish what we wanted (cf. (1.15)). Here we need to take 4-th moments; we will prove that

$$\int_\Omega \left(\sum_{t \in I_\rho} \sup_{\beta \in [0,1)} |\widehat{Q}_t^\omega(\beta) - \widehat{V}_t(\beta)|^4 \right) dP(\omega)$$

$$= \sum_{t \in I_\rho} \int_\Omega \sup_{\beta \in [0,1)} |\widehat{Q}_t^\omega(\beta) - \widehat{V}_t(\beta)|^4 dP(\omega) < \infty \, . \tag{2.9}$$

The reason for this complication is that in proving (2.9) we cannot get by with estimates for a single fixed β from the dual group as we did in Example 1.4, because this time the "dual group" — the interval $[0,1)$ — has infinitely many elements. But it turns out we do not really have to get estimations for every $\beta \in [0,1)$. We will use the fact that $\widehat{Q}_t^\omega(\beta) - \widehat{V}_t(\beta)$ is a trigonometric polynomial of degree $[t]$ and hence cannot change too rapidly. As a consequence of Bernstein's theorem on the derivative of a trigonometric polynomial, it is enough to obtain estimates on $|\widehat{Q}_t^\omega(\beta) - \widehat{V}_t(\beta)|$ for roughly t well chosen $\beta's$ to get a uniform estimate for every $\beta \in [0,1)$.

We shall prove the following: there is a constant C (< 100) so that for every β and t

$$\int_\Omega |\widehat{Q}_t^\omega(\beta) - \widehat{V}_t(\beta)|^4 dP(\omega) \leq \frac{C}{t^2} \, . \tag{2.10}$$

Let us see how one can finish the proof, assuming (2.11). Let us denote $S_t = \{0, 1/10t, 2/10t, \ldots, (10t-1)/10t\}$. By (2.11), we have the estimate

$$\int_\Omega \sup_{\beta \in S_t} |\widehat{Q}_t^\omega(\beta) - \widehat{V}_t(\beta)|^4 dP(\omega) \leq \frac{C}{t} \, . \tag{2.11}$$

Now we just observe that for each ω

$$\sup_{\beta\in[0,1)} |\widehat{Q}_t^\omega(\beta) - \widehat{F}V_t(\beta)| \le 3\cdot\sup_{\beta\in S}|\widehat{Q}_t^\omega(\beta) - \widehat{F}V_t(\beta)|. \qquad (2.12)$$

This is a consequence of Bernstein's inequality. Bernstein's inequality says that if f is a trigonometric polynomial with period 1 and of degree at most t, then

$$\sup_\beta |f'(\beta)| \le 4\pi t\cdot\sup_\beta|f(\beta)|.$$

To see (2.12), set $f(\beta) = \widehat{Q}_t^\omega(\beta) - \widehat{F}V_t(\beta)$. Let $\beta_0\in[0,1)$ be a point where the supremum of $|f|$ is taken, and let $\beta\in S_t$ be closest to β_0, so we have $|\beta_0 - \beta|\le 1/20t$. It follows, by Bernstein's inequality, that

$$|f(\beta_0)| \le |f(\beta)| + |f(\beta_0) - f(\beta)| \le |f(\beta)| + 4\pi t\cdot|f(\beta_0)|\cdot|\beta_0-\beta| \le$$
$$\le |f(\beta)| + \frac{4\pi}{20}|f(\beta_0)|,$$

which implies (2.12)

Let us prove (2.11). Let us introduce the random variables $Z_n(\omega) = (Y_n - \sigma)e(n\beta)$. We need to prove

$$\int_\Omega \left|\sum_{n\le t} Z_n\right|^4 dP(\omega) \le Ct^2. \qquad (2.13)$$

We write

$$\left|\sum_{n\le t} Z_n\right|^4 = \sum_I + \sum_{II}.$$

In $\sum_I$ above, we collected those products $Z_i\overline{Z}_jZ_k\overline{Z}_l$ for which either $i = j = k = l$, or $i = j$ and $k = l$, or $i = k$ and $j = l$. It is clear that there are Ct^2 terms in $\sum_I$, each of modulus at most 16; hence we have the estimate

$$\left|\int_\Omega \sum_I\right| \le Ct^2. \qquad (2.14)$$

Now the products $Z_i\overline{Z}_jZ_k\overline{Z}_l$ in $\sum_{II}$ all have the property that at least one of the indices i, j, k, l is not equal to any of the other three. But then, since the Z_n's are independent and have 0 expectation, we get

$$\int_\Omega \sum_I = 0. \qquad (2.15)$$

We see that (2.14) and (2.15) imply (2.13).

Exercises for Example 2.5

Ex.1. a) Show that for every $1 \leq p < \infty$, $Q_t^{\omega} f - V_t f \to 0$ in L^p-norm. (We use the notation of Example 2.5.)
   b) Does the conclusion of part a) hold if $p = \infty$?

Ex.2. Let $0 < \delta < 1/2$. Let (Y_n) be a sequence of independent, (0-1)-valued random variables on the probability space (Ω, β, P) so that $P(Y_n = 1) = 1/n^{\delta}$ and $P(Y_n = 0) = 1 - 1/n^{\delta}$. Show that with probability 1 the random sequence A^{ω}, defined by $n \in A^{\omega}$ if and only if $Y_n(\omega) = 1$, is ergodic.

Ex.3. Let (Ω, β, P) be a probability space, and let $0 \leq \sigma \leq 1$. Let $Y_1, Y_2 \ldots$ be an i.i.d. sequence of $((-1)\text{-}1)$-valued random variables wih $P(Y_n = 1) = \sigma$ and $P(Y_n = -1) = 1 - \sigma$. Denote $a_n(\omega) = \sum_{k \leq n} Y_k(\omega)$. Let T be an ergodic measure-preserving transformation of the probability space (X, Σ, m).
   Show that with probability 1, for every $f \in L^2$ we have

$$\lim_{t \to \infty} \frac{1}{t} \sum_{n \leq t} T_{\tau}^{a_n(\omega)} f = \int_X f \, dm$$

in L^2-norm.
   This is a difficult exercise, but the main idea is again to estimate the fourth moment of the Fourier transform $(1/t) \sum_{n \leq t} e(a_n(\omega)\beta)$ to show that it goes to 0 for noninteger β. More precisely, one should show that for any compact subinterval K of the interval (0,1) we have, with probability 1,

$$\lim_{t \to \infty} \sup_{\beta \in K} \left| \frac{1}{t} \sum_{n \leq t} e(a_n(\omega)\beta) \right| = 0 \,.$$

If you need more help, consult [Blum, Cogburn, 1975].

Example 2.6. Randomly generated sequences of 0 density. In this section, similarly to Example 1.7, we examine random sequences of 0 density.
   Let (σ_n) be a sequence of positive numbers satisfying the following properties:
   (i) the sequence (σ_n) is decreasing;
   (ii) $\lim_{t \to \infty} \frac{\sum_{n \leq t} \sigma_n}{\log t} = \infty$.
We see that property (ii) above is stronger than property (ii) of Example 1.7. If we assume, say,

$$\lim_{t \to \infty} \frac{\sum_{n \leq t} \sigma_n}{\log t} = 1 \,,$$

then — as M. Lacey pointed out to us — with positive probability the random sequence is not going to be ergodic.

Let (Y_n) be a sequence of independent, (0-1)-valued random variables on the probability space (Ω, β, P) so that $P(Y_n = 1) = \sigma_n$ and $P(Y_n = 0) = 1 - \sigma_n$.

We shall show that with probability 1 the random sequence A^ω, defined by $n \in A^\omega$ if and only if $Y_n(\omega) = 1$, is ergodic.

This result is implicitly conatined in [Boshernitzan, 1983], but the proof we give follows [Bourgain, 1988].

By an argument similar to the ones in Example 1.7 and Example 2.5, we conclude that we need to prove

$$\lim_{\substack{t \to \infty \\ t \in I_\rho}} \sup_{\beta \in S_t} \left| \frac{1}{\sum_{n \leq t} \sigma_n} \sum_{n \leq t} (Y_n(\omega) - \sigma_n) e(n\beta) \right| = 0, \qquad (2.16)$$

where $S_t = \{0, 1/10t, 2/10t, \ldots, (10t - 1)/10t\}$, and for $\rho > 1$ we take I_ρ to consist of the numbers $t_k, k = 1, 2, \ldots$, defined by $t_k = \min\{t \mid \rho^k \leq \sum_{n \leq t} \sigma_n < \rho^{k+1}\}$. Introducing $Z_n(\beta, \omega) = (Y_n(\omega) - \sigma_n) e(n\beta)$, we rewrite (2.16) as

$$\lim_{\substack{t \to \infty \\ t \in I_\rho}} \sup_{\beta \in S_t} \left| \frac{1}{\sum_{n \leq t} \sigma_n} \sum_{n \leq t} Z_n(\beta, \omega) \right| = 0. \qquad (2.17)$$

If the reader solved Exercise 2 above, she/he might have realized that if $\sigma_n = 1/\sqrt{n}$ then the method of taking 4-th moments does not work anymore. In general, if $\sigma_n = 1/n^\delta$, then one would have to estimate higher and higher moments as $\delta \to 1$. But even this is not enough if, say, $\sigma_n = \log n / n$. To handle this case as well, we will compute a moment of $\frac{1}{\sum_{n \leq t} \sigma_n} \sum_{n \leq t} Z_n(\beta, \omega)$ that increases with t. Let us assume that the moment we want to compute is the p_t-th moment. It would be very convenient if we had

$$\left\| \sup_{\beta \in S_t} \left| \frac{1}{\sum_{n \leq t} \sigma_n} \sum_{n \leq t} Z_n(\beta, \omega) \right| \right\|_{L^{p_t}(\Omega)} \leq$$

$$\leq C \cdot \sup_{\beta \in S_t} \left\| \frac{1}{\sum_{n \leq t} \sigma_n} \sum_{n \leq t} Z_n(\beta, \omega) \right\|_{L^{p_t}(\Omega)} \qquad (2.18)$$

with an absolute constant C. Comparing (2.18) with the obvious estimate

$$\left\| \sup_{\beta \in S_t} \left| \frac{1}{\sum_{n \leq t} \sigma_n} \sum_{n \leq t} Z_n(\beta, \omega) \right| \right\|_{L^{p_t}(\Omega)} \leq$$

$$\leq (10t)^{1/p_t} \cdot \sup_{\beta \in S_t} \left\| \frac{1}{\sum_{n \leq t} \sigma_n} \sum_{n \leq t} Z_n(\beta, \omega) \right\|_{L^{p_t}(\Omega)}, \qquad (2.18')$$

we see we should choose $p_t = \log t$. Now, suppose we would prove the following for a sequence of random variables $X_1, X_2, \ldots$:

$$\int_\Omega \left(\sum_{t \in I_\rho} |X_t(\omega)|^{p_t} \right) dP(\omega) < \infty.$$

Since $p_t \to \infty$, we can just conclude that for almost every ω the sequence $(X_t(\omega))_{t \in I_\rho}$ is a bounded sequence. Hence, it is not enough to prove

$$\int_\Omega \left(\sum_{t \in I_\rho} \sup_{\beta \in S_t} \left| \frac{1}{\sum_{n \le t} \sigma_n} \sum_{n \le t} Z_n(\beta, \omega) \right|^{p_t} \right) dP(\omega) < \infty.$$

But we will prove the following inequality: there is a constant C so that for each β and t,

$$\left\| \frac{1}{\sum_{n \le t} \sigma_n} \sum_{n \le t} Z_n(\beta, \omega) \right\|_{L^{p_t}(\Omega)} \le C \sqrt{\frac{\log t}{\sum_{n \le t} \sigma_n}}. \tag{2.19}$$

This implies, by (2.18),

$$\left\| \sup_{\beta \in S_t} \left| \frac{1}{\sum_{n \le t} \sigma_n} \sum_{n \le t} Z_n(\beta, \omega) \right| \right\|_{L^{p_t}(\Omega)} \le C \sqrt{\frac{\log t}{\sum_{n \le t} \sigma_n}}. \tag{2.20}$$

Since $\frac{\log t}{\sum_{n \le t} \sigma_n} \to 0$ by assumption (ii), the following lemma would finish the proof.

Lemma 2.7. *Let $X_1, X_2, \ldots$ be a sequence of bounded, complex - valued random variables on our probability space (Ω, β, P).*

Then for any fixed lacunary sequence of integers I, with probability 1 we have

$$\limsup_{\substack{t \to \infty \\ t \in I}} \frac{X_t(\omega)}{\|X_t\|_{L^{p_t}(\Omega)}} \le 1.$$

(Recall that $p_t = \log t$.)

Indeed, by this lemma and the estimate in (2.20), we get that for almost every $\omega \in \Omega$ there is a constant C_ω such that for each $t \in I_\rho$

$$\sup_{\beta \in S_t} \left| \frac{1}{\sum_{n \le t} \sigma_n} \sum_{n \le t} Z_n(\beta, \omega) \right| \le C_\omega \cdot \sqrt{\frac{\log t}{\sum_{n \le t} \sigma_n}}.$$

Proof of Lemma 2.7. The proof follows immediately from the remark that for any fixed $c > 1$ we have

$$\int_\Omega \sum_{t \in I} \left(\frac{|X_t(\omega)|}{c \cdot \|X_t\|_{L^{p_t}(\Omega)}} \right)^{p_t} dP(\omega) < \infty .$$

$\square$

Let us turn to the proof of (2.19). For the estimation of moments we will use the following fundamental result of Marczinkiewic and Zygmund [Garsia, 1970].

Lemma 2.8. *Let* $X_1, X_2, \ldots$ *be a sequence of independent, bounded, mean 0, complex valued random variables on our probability space* (Ω, β, P). *Then for every* $1 < p < \infty$ *and* t *we have*

$$\left\| \sum_{n \leq t} X_n \right\|_{L^p(\Omega)} \leq 2\sqrt{p} \cdot \left\| \sqrt{\sum_{n \leq t} |X_n|^2} \right\|_{L^p(\Omega)} .$$

The inequality in (2.19) follows if we show that for $2 \leq p \leq \sum_{n \leq t} \sigma_n$ we have

$$\left\| \sum_{n \leq t} Z_n \right\|_{L^p(\Omega)} \leq 11 \sqrt{p \cdot \sum_{n \leq t} \sigma_n} . \qquad (2.21)$$

(We are using the simplified notation $Z_n(\omega) = Z_n(\beta, \omega)$, since our estimates will be independent of β.) First we estimate, using Lemma 2.8,

$$\left\| \sum_{n \leq t} Z_n \right\|_{L^p(\Omega)} \leq 2\sqrt{p} \cdot \left\| \sqrt{\sum_{n \leq t} |Z_n|^2} \right\|_{L^p(\Omega)} = 2\sqrt{p} \cdot \left\| \sqrt{\sum_{n \leq t} (Y_n - \sigma_n)^2} \right\|_{L^p(\Omega)}$$

$$\leq 2\sqrt{p} \cdot \left(\left\| \sqrt{\sum_{n \leq t} Y_n^2} \right\|_{L^p(\Omega)} + \sqrt{\sum_{n \leq t} \sigma_n^2} \right) .$$

Since $\sigma_n^2 \leq \sigma_n$ and

$$\left\| \sqrt{\sum_{n \leq t} Y_n^2} \right\|_{L^p(\Omega)} = \sqrt{\left\| \sum_{n \leq t} Y_n^2 \right\|_{L^{p/2}(\Omega)}} \leq \sqrt{\left\| \sum_{n \leq t} Y_n \right\|_{L^p(\Omega)}}$$

(recall that Y_n is 0-1-valued), we are done if we prove

$$\left\| \sum_{n \leq t} Y_n \right\|_{L^p(\Omega)} \leq 9 \sum_{n \leq t} \sigma_n . \qquad (2.22)$$

To see this, we again use Lemma 2.8, with $Y_n - \sigma_n$ replacing Z_n in the above argument:

$$\sum_{n \leq t} \|Y_n\|_{L^p(\Omega)} \leq \sum_{n \leq t} \|Y_n - \sigma_n\|_{L^p(\Omega)} + \sum_{n \leq t} \sigma_n \leq$$

$$\leq 2\sqrt{p \cdot \sum_{n \leq t} \|Y_n\|_{L^p(\Omega)}} + 3\sum_{n \leq t} \sigma_n \, ,$$

which, solving for $\sum_{n \leq t} \|Y_n\|_{L^p(\Omega)}$ · gives (2.22). In the above we repeatedly used that $p \leq \sum_{n \leq t} \sigma_n$.

Exercises for Example 2.6

Ex.1. Prove the following weaker version of Bernstein's theorem: if f is a 1-periodic trigonometric polynomial of degree at most t, then

$$\sup_\beta |f'(\beta)| \leq 4\pi t^2 \sup_\beta |f(\beta)| \, .$$

Note that in the method introduced in Example 2.5 even this weak version of Bernstein's inequality is sufficient.

Ex.2. Let $X_1, X_2, \ldots$ be a sequence of bounded, complex - valued random variables on the probability space (Ω, β, P). Let $p_t = \log t$. Prove that with probability 1

$$\limsup_{t \to \infty} \frac{X_t(\omega)}{\|X_t\|_{L^{p_t}(\Omega)}} \leq e \, .$$

Ex.3. Let (σ_n) be a sequence of positive numbers satisfying the properties

(i) the sequence (σ_n) is decreasing;

(ii) $\lim_{t \to \infty} \sum_{n \leq t} \sigma_n = \infty$.

Let (Y_n) be a sequence of independent, (0-1)-valued random variables on the probability space (Ω, β, P) so that $P(Y_n = 1) = \sigma_n$ and $P(Y_n = 0) = 1 - \sigma_n$. Define the random sequence A^ω by $n \in A^\omega$ if and only if $Y_n(\omega) = 1$.

a) Show that for any fixed irrational α, with probability 1 the sequence $(a \cdot \alpha)_{a \in A^\omega}$ is ergodic for the transformation $Tx = \langle x + \alpha \rangle$ of the unit interval (equipped with the mod 1 Lebesgue measure). (Hint: Use the method of Example 1.7.)

b) Show that for any fixed irrational α with probability 1 the sequence $(a \cdot \alpha)_{a \in A^\omega}$ is uniformly distributed mod 1.

Ex.4. a) Show that if the sequence of positive integers A is ergodic, then for each irrational number α the sequence $(a \cdot \alpha)_{a \in A}$ is uniformly distributed mod 1.

b) Is the converse of part a) true?

Ex.5. Prove that in fact the optimal choice for p_t is $p_t = 2 \log t$. (Hint: Compare the inequalities in (2.18′) and (2.21).)

2. Sequences that are good in the mean

In this section we examine sequences A that are not ergodic, but for which $\lim_{t\to\infty} M_t(A, f)$ exists in L^2.

Definition 2.9. Let τ be a measure-preserving transformation of the probability space (X, Σ, m). We say that the sequence of positive integers $A = (a_n)$ is *good in the mean* for the system (X, Σ, m, τ) if and only if for each $f : X \to \mathbb{C}$ the limit

$$\lim_{t\to\infty} M_t(A, f) = \overline{f} \tag{2.23}$$

exists in L^2-norm.

The sequence $A = (a_n)$ is *good in the mean* if and only if it is good in the mean for every measure-preserving system (X, Σ, m, τ).

It is best to start with the following exercise:

Exercise 2.10. a) Let $A = (a_n)$ be a strictly increasing sequence of positive integers.

Show that A is good in the mean if and only if for each β the limit

$$\lim_{t\to\infty} \frac{1}{A(t)} \sum_{\substack{a \leq t \\ a \in A}} e(a\beta) \overset{\text{def}}{=} \widehat{\Lambda}(\beta)$$

exists.

b) Show that the sequence of even integers $2\mathbb{Z}_+$ is good in the mean but is not ergodic.

Example 2.11. The sequence of squares. Here we examine the sequence of squares $A = \{n^2 \mid n \in \mathbb{Z}_+\}$. As we already mentioned in Example 1.12, this sequence is not ergodic; for example, we have $A(t, 3, 2) = 0$. In Example 1.12, we saw that if $\beta = b/q \in X_q$ then

$$\widehat{\Lambda}(b/q) = \frac{1}{q} \sum_{n<q} e(n^2 b/q).$$

Our main objective here is to prove that $\widehat{\Lambda}(\beta) = 0$ for irrational β, therefore establishing that the sequence of squares is good in the mean.

For our purposes it is best to use Dirichlet's characterization of an irrational number in terms of its rational approximations.

Lemma 2.12. *Let β be a real number. Then for each real number Y there exists a rational number b/q with $\gcd(b, q) = 1$, $1 \leq q \leq Y$ and*

$$|\beta - b/q| \leq \frac{1}{qY}.$$

Note that if $b(Y)/q(Y)$ is a rational number satisfying the conclusion of the lemma, then for an irrational β we have $q(Y) \to \infty$ as $Y \to \infty$.

Let $Q(t) = t^{2/3}$. For a given β, by Dirichlet's theorem, there are $b = b_t, q = q_t$ such that $\gcd(b, q) = 1$, $q \le Q(t)$ and $|\beta - b/q| \le 1/qQ(t)$. We will show that then we have the estimate

$$\left| \frac{1}{\sqrt{t}} \sum_{n^2 \le t} e(n^2\beta) \right| \le C \cdot \left(\frac{1}{\sqrt{q}} + \frac{\sqrt{\log t}}{t^{1/6}} \right). \tag{2.24}$$

Note that for irrational β this implies $\widehat{\Lambda}(\beta) = 0$, for $q_t \to \infty$ as $t \to \infty$. The estimate in (2.24) will be achieved by estimating the Fourier transform $\left| \frac{1}{\sqrt{t}} \sum_{n \le t} e(n^2 b/q) \right|$ at b/q. As we shall see, we will need two kinds of estimates on the Fourier transform at b/q. The first estimate is used when q is small compared to t, the second one is used in the remaining case. More precisely, letting $P(t) = t^{1/3} = t/Q(t)$, the first estimate is effective when $1 \le q \le P(t)$, the second estimate is effective in the remaining case, that is, when $P(t) \le q \le Q(t)$. This distinction between rational approximations with small denominators and with large ones is a fundamental feature of the so-called circle method to be discussed in more detail later .

Let us denote

$$U_t(\beta) = \sum_{n^2 \le t} e(n^2\beta)$$

and

$$V_t(\beta) = \sum_{n \le t} \frac{1}{2\sqrt{n}} e(n\beta).$$

It would be tempting to try to estimate $U_t - V_t$ similarly to the case of random sequences, because naively we would think of the squares that are close to the random sequence we get by putting the integer n into the random sequence with probability $1/(2\sqrt{n})$. But we have seen that the sequence of squares is not even close to a random sequence: it is not uniformly distributed in residue classes. This is reflected in the asymptotic behaviour of the Fourier transform at rational points: we have, for example, if q is odd, $\gcd(b, q) = 1$, that $|\widehat{\Lambda}(b/q)| = 1/\sqrt{q}$.

First estimate: $q \le P(t)$. Here we are going to show that if $|\beta - b/q| \le 1/qQ(t)$ then

$$\left| U_t(\beta) - \widehat{\Lambda}(b/q) \cdot V_t(\beta - b/q) \right| \le C \cdot (q + P(t)). \tag{2.25}$$

Now, if $q \le P(t) = t^{1/3}$, then this implies

$$\left| \frac{1}{\sqrt{t}} \sum_{n^2 \le t} e(n^2\beta) \right| \le C \cdot \left(\frac{1}{\sqrt{q}} + \frac{1}{t^{1/6}} \right), \tag{2.26}$$

which is even better than what we promised in (2.24). To get (2.26) from (2.25), divide both sides of (2.25) by $\sqrt{t}$ and use that, by Proposition 1.14, $|\widehat{\Lambda}(b/q)| \leq C/\sqrt{q}$.

To prove (2.25), first note that for every positive y we have

$$\left| U_y(b/q) - \widehat{\Lambda}(b/q) \cdot V_y(0) \right| \leq C \cdot q. \tag{2.27}$$

Indeed, we can write, noting that $V_y(0) = \sqrt{y} + O(1)$,

$$U_y(b/q) = \sum_{s < [\sqrt{y}/q]} \sum_{r < q} e((qs+r)^2 b/q) + O(q) = \frac{\sqrt{y}}{q} \cdot \sum_{r < q} e(r^2 b/q) + O(q) =$$

$$= V_y(0) \cdot \widehat{\Lambda}(b/q) + O(q),$$

where we used that $(qs+r)^2 \equiv r^2 \pmod{q}$. The estimate in (2.25) is a consequence of the following lemma on summation by parts.

Lemma 2.13. *Suppose $D_1, \ldots, D_t$ are complex numbers satisfying*

$$|D_y| \leq H, \qquad y = 1, \ldots, t.$$

Then for each real number δ we have

$$\left| \sum_{j \leq t} (D_j - D_{j-1}) e(j\delta) \right| \leq 7H \cdot (t|\delta| + 1),$$

where $D_0 = 0$.

To obtain (2.25), first write

$$U_t(\beta) - \widehat{\Lambda}(b/q) \cdot V_t(\beta - b/q) = \sum_{n^2 \leq t} e(n^2 \beta) - \widehat{\Lambda}(b/q) \sum_{j \leq t} \frac{1}{2\sqrt{j}} e(j(\beta - b/q)) =$$

$$= \sum_{n^2 \leq t} e(n^2 b/q) e(n^2(\beta - b/q)) - \sum_{j \leq t} \widehat{\Lambda}(b/q) \frac{1}{2\sqrt{j}} e(j(\beta - b/q)) =$$

$$= \sum_{j \leq t} (D_j - D_{j-1}) e(j(\beta - b/q)),$$

where we set

$$D_y = U_y(b/q) - \widehat{\Lambda}(b/q) \cdot V_y(0)$$

for positive y and $D_0 = 0$. Now take $\delta = \beta - b/q$, $H = Cq$ in the lemma, note that $|\delta| \leq 1/qQ(t)$, and recall that $t/Q(t) = P(t)$.

Proof of Lemma 2.13. We write, using summation by parts,

$$\sum_{j \leq t} (D_j - D_{j-1}) e(j\delta) = \sum_{j < t} D_j(\alpha) \left(e(j\delta) - e((j+1)\delta) \right) + D_{t-1}(\alpha) e(t\delta).$$

Hence, also recalling the estimate $|1 - e(\delta)| \leq 2\pi|\delta| \leq 7|\delta|$,

$$|\sum_{j\leq t}(D_j - D_{j-1})e(j\delta)| \leq \sum_{j<t}|D_j| \cdot |1 - e(\delta)| + |D_t| \leq$$
$$\leq t \cdot H \cdot 7|\delta| + H = 7H \cdot (t|\delta| + 1),$$

and we are done. $\square$

Second estimate: $P(t) \leq q \leq Q(t)$. Here we are going to show that if $|\beta - b/q| \leq 1/qQ(t)$ and $q \geq P(t)$ then

$$|U_t(\beta)| \leq C \cdot \left(\frac{\sqrt{t}}{\sqrt{q}} + \sqrt{q \log t}\right). \tag{2.28}$$

For $q \leq Q(t) = t^{2/3}$ we obtain, dividing both sides of (2.28) by $\sqrt{t}$,

$$\left|\frac{1}{\sqrt{t}}\sum_{n^2\leq t}e(n^2\beta)\right| \leq C \cdot \left(\frac{1}{\sqrt{q}} + \frac{\sqrt{\log t}}{t^{1/6}}\right),$$

which, together with (2.26), completes the proof of the inequality in (2.24). Our main estimate is

$$|U_y(b/q)| \leq C \cdot \left(\frac{\sqrt{y}}{\sqrt{q}} + \sqrt{q \log y}\right). \tag{2.29}$$

which is valid for every y and b, q with $\gcd(b, q) = 1$.

Exercise 2.14. Show that (2.29) implies (2.28).(Hint: Use Lemma 2.13 with $D_y = U_y(b/q)$, $\delta = \beta - b/q$ and $H = C \cdot (\sqrt{t}/\sqrt{q} + \sqrt{q \log t})$. Note that $|\delta| \leq 1/qQ(t) \leq 1/t$ since $q \geq P(t)$.)

The proof of (2.29) begins as that of (2.27), but we need to treat the remainder more carefully. We write, with $x = [\sqrt{y}]$,

$$U_y(b/q) = \sum_{s<[x/q]}\sum_{r<q}e((qs + r)^2b/q) + \sum_{n\leq x-[x/q]\cdot q}e(([x/q] \cdot q + n)^2b/q) =$$
$$= \frac{x}{q} \cdot \sum_{r<q}e(r^2b/q) + O(1) + \sum_{n\leq x-[x/q]\cdot q}e(n^2b/q) =$$
$$= x \cdot \hat{\Lambda}(b/q) + O(1) + \sum_{n\leq x-[x/q]\cdot q}e(n^2b/q).$$

Since by Proposition 1.14 $|\hat{\Lambda}(b/q)| \leq C/\sqrt{q}$, we see all we need to prove is that for $z \leq q$ we have

$$\left|\sum_{n\leq z}e(n^2b/q)\right| \leq C \cdot \sqrt{q \log z}, \tag{2.30}$$

where C is an absolute constant.

The idea is the same as in the proof of Proposition 1.14, but we cannot use periodicity so extensively. So the property of the sequence of squares the proof depends on is that the difference sequence $((n+j)^2 - n^2)$ is an arithmetic progression for each fixed j.

First we write

$$\left| \sum_{n=0}^{z} e(n^2 b/q) \right|^2 = \sum_{n=0}^{z} \sum_{m=0}^{z} e\left((-n^2 + m^2)b/q \right) =$$

$$= \sum_{n=0}^{z} \sum_{j=-n}^{z-n} e\left((-n^2 + (n+j)^2)b/q \right) = \sum_{|j| \leq z} \sum_{n \in I_j} e\left((j^2 + 2nj)b/q \right) \leq$$

$$\leq \sum_{|j| \leq z} \left| \sum_{n \in I_j} e(2njb/q) \right|,$$

where I_j is an interval of length at most z; hence we have the trivial estimate

$$\left| \sum_{n \in I_j} e(2njb/q) \right| \leq z + 1 \leq 2z.$$

But sometimes we can do better than this. For fixed j, the numbers $e(2njb/q)$, $n \in I_j$, form a geometric progression, and summing this geometric progression gives the estimate

$$\left| \sum_{n \in I_j} e(2njb/q) \right| \leq \frac{2}{|1 - e(2jb/q)|} \leq \frac{2}{\parallel 2jb/q \parallel},$$

where $\parallel \alpha \parallel$ denotes the distance of α from the nearest integer. It follows that we have the estimate

$$\left| \sum_{n=0}^{z} e(n^2 b/q) \right|^2 \leq 2 \sum_{|j| \leq z} \min\left(z, \frac{1}{\parallel 2jb/q \parallel} \right). \qquad (2.31)$$

Since $z \leq q$ and $\gcd(b, q) = 1$, as j runs through the integers between $-z$ and z the numbers $2jb$ run through each mod q residue class at most four times. As a consequence, the numbers $\parallel 2jb/q \parallel$ run through at most eight times the numbers $1/q, 2/q, \ldots, [q/2]/q$. It follows that

$$\sum_{|j| \leq z} \min\left(z, \frac{1}{\parallel 2jb/q \parallel} \right) \leq 8 \cdot \sum_{r \leq q} \min\left(z, \frac{q}{r} \right). \qquad (2.32)$$

Since $\min\left(z, \frac{q}{r} \right) = z$ only if $r \leq q/z$, we can estimate

$$\sum_{r \leq q} \min\left(z, \frac{q}{r} \right) \leq \sum_{r \leq q/z} z + \sum_{q/z \leq r \leq q} \frac{q}{r} \leq C \cdot q \log z.$$

This, together with (2.32) and (2.31), imply (2.30), completing the proof.

Exercises for Example 2.11

Ex.1. Show that the sequence of squares is ergodic for irrational rotations, that is, when the transformation τ is the map of the unit interval defined by $\tau(x) = \langle x + \alpha \rangle$ for some irrational α, and the measure m is the mod 1 Lebesgue measure.

Ex.2. Let (σ_n) be a sequence of positive numbers satisfying the properties

(i) the sequence (σ_n) is decreasing;

(ii) $\lim_{t \to \infty} \frac{\sum_{n \le t} \sigma_n}{\log n} = \infty$.

Let (Y_n) be a sequence of independent, (0-1)-valued random variables on the probability space (Ω, β, P) so that $P(Y_n = 1) = \sigma_n$ and $P(Y_n = 0) = 1 - \sigma_n$.

Show that with probability 1 the random sequence A^ω, defined by $n^2 \in A^\omega$ if and only if $Y_n(\omega) = 1$, is good in the mean.

Ex.3. a) Let A be a sequence good in the mean. Suppose that for some q and h the limit $\lim_{t \to \infty} A(t, q, h)/A(t)$ exists and is positive. Show that the sequence $(a \mid a \in A, a \equiv h \mod q)$ is also good in the mean. (Hint: consider the product space $X \times X_q$ with the product transformation $U = \tau \times \tau_q$, where τ_q is the shift on X_q.)

b) Let h be a quadratic residue mod q. Show that the sequence $(n^2 \mid n^2 \equiv h \mod q)$ is good in the mean.

Example 2.15. the sequence of primes. Let A be the sequence of prime numbers. According to Example 1.15, for $b/q \in X_q$ we have

$$\widehat{\Lambda}(b/q) = \frac{1}{\phi(q)} \sum_{\substack{n < q \\ \gcd(n,q)=1}} e(nb/q).$$

Here we will show that $\widehat{\Lambda}(\beta) = 0$ for irrational β, and this shows that the sequence of primes is good in the mean.

Our argument is similar to the one for the squares in Example 2.11. The main idea is again to derive estimates at rational points of the interval $[0,1)$, and doing this we need to make a distinction between rational approximations with small (compared to t) denominators and large ones.

Since $A(t)/li(t) \to 1$, we need to prove that for irrational β

$$\lim_{t \to \infty} \frac{1}{li(t)} \sum_{\substack{a \le t \\ a \in A}} e(a\beta) = 0.$$

Let us denote

$$U_t(\beta) = \sum_{\substack{a \le t \\ a \in A}} e(a\beta)$$

and

$$V_t(\beta) = \sum_{n \leq t} \frac{1}{\log n} e(n\beta).$$

Let $B > 1$, and define $P(t) = (\log t)^B$ and $Q(t) = t/P(t)$. For our purposes any $B \geq 10$ will do. For a given β, by Dirichlet's theorem there are b, q such that $\gcd(b, q) = 1$, $q \leq Q(t)$ and $|\beta - b/q| \leq 1/qQ(t)$. Our first estimate of $U_t(\beta)$ is used when q is small, namely when $q \leq P(t)$.

Lemma 2.16. *There are positive constants C and c so that whenever $1 \leq b \leq q \leq P(t)$, $\gcd(b, q) = 1$, one has*

$$\left| U_t(b/q) - \widehat{\Lambda}(b/q) li(t) \right| \leq C \cdot \frac{li(t)}{e^{c\sqrt{\log t}}}. \tag{2.33}$$

Exercise 2.17. Prove that there are positive constants C and c so that whenever $1 \leq b \leq q \leq P(t)$, $\gcd(b, q) = 1$, and $|\beta - b/q| \leq 1/qQ(t)$ one has

$$\left| U_t(\beta) - \widehat{\Lambda}(b/q) V_t(\beta - b/q) \right| \leq C \cdot \frac{li(t)}{e^{c\sqrt{\log t}}}.$$

(Hint: Use Lemma 2.13 with $\delta = \beta - b/q$. Note that $V_t(0) = li(t)$.)

As we will see, the proof of Lemma 2.16 is quite straightforward, but we have to use a quantitative version of the prime number theorem for arithmetic progressions. In fact, this is a general principle which we already used in our argument for the squares: a rate of convergence result about the distribution of a given sequence in residue classes yields an estimate on the Fourier transform of the sequence which is effective at points that are close to rational points with small denominator.

Our second estimate of $U_t(\beta)$ is used for the remaining cases, that is, when $P(t) \leq q \leq Q(t)$.

Lemma 2.18. *There is a constant C so that whenever $1 \leq q \leq t$, $\gcd(b, q) = 1$, then one has*

$$|U_t(b/q)| \leq C \cdot (\log t)^3 \left(\frac{t}{\sqrt{q}} + t^{4/5} + \sqrt{tq} \right).$$

We shall not prove this estimate here. This is a deep result of Vinogradov [cf. Ellison & Ellison, 1985], although nowadays the proof can be written up on three pages.

Exercise 2.19. Show that there is a constant C so that whenever $P(t) \leq q \leq Q(t)$, $\gcd(b, q) = 1$, and $|\beta - b/q| \leq 1/qQ(t)$, then one has

$$|U_t(\beta)| \leq C \cdot \frac{li(t)}{(\log t)^{B/2-4}}.$$

(Hint: Use Lemma 2.13 with $\delta = \beta - b/q$. Note that $|\delta| \leq 1/t$.)

Exercise 2.20. Prove that for irrational β we have $\widehat{\Lambda}(\beta) = 0$. (Hint: By Dirichlet's theorem, there are b, q such that $\gcd(b, q) = 1$, $q \leq Q(t)$ and $|\beta - b/q| \leq 1/qQ(t)$. If $q \leq P(t)$ use Exercise 2.17, and Exercises 1 and 2 to Example 1.15. If $P(t) \leq q \leq Q(t)$, then use Exercise 2.19.)

Proof of Lemma 2.17. The proof follows from the following quantitative version of the prime number theorem for arithmetic progressions, which we state without proof [cf. Ellison & Ellison, 1985].

Lemma 2.21. *There are positive constants C and c so that whenever $1 \leq h \leq q \leq P(t)$, $\gcd(h, q) = 1$, we have, for every t,*

$$\left| A(t, q, h) - \frac{1}{\phi(q)} \cdot li(t) \right| \leq C \cdot \frac{li(t)}{e^{c\sqrt{\log t}}}.$$

Now note that

$$U_t(b/q) = \sum_{\substack{a \leq t \\ a \in A}} e(ab/q) = \sum_{h < q} A(t, q, h) e(hb/q).$$

It follows, since $A(t, q, h)$ is at most 1 if q and h are not relatively prime, that

$$\left| U_t(b/q) - \sum_{\substack{h < q \\ \gcd(h,q)=1}} A(t, q, h) e(bh/q) \right| \leq q \leq P(t),$$

and so we just need to prove

$$\left| \sum_{\substack{h < q \\ \gcd(h,q)=1}} A(t, q, h) e(bh/q) - \widehat{\Lambda}(b/q) \cdot li(t) \right| \leq C \cdot \frac{li(t)}{e^{c\sqrt{\log t}}}.$$

But this is an immediate consequence of Lemma 2.21 if we recall that

$$\widehat{\Lambda}(b/q) = \frac{1}{\phi(q)} \sum_{\substack{n < q \\ \gcd(n,q)=1}} e(nb/q).$$

□

Exercises for Example 2.15

Ex.1. Show that the sequence of primes is ergodic for irrational rotations, that is when the transformation τ is the map of the unit interval defined by $\tau(x) = \langle x + \alpha \rangle$ for some irrational α, and the measure m is the mod 1 Lebesgue measure.

Ex.2. Let (σ_n) be a sequence of positive numbers satisfying the properties

(i) the sequence (σ_n) is decreasing;

(ii) $\lim_{t\to\infty} \frac{\sum_{n\le t} \sigma_n}{\log n} = \infty$.

Let (Y_n) be a sequence of independent, (0-1)-valued random variables on the probability space (Ω, β, P) so that $P(Y_n = 1) = \sigma_n$ and $P(Y_n = 0) = 1 - \sigma_n$.

Show that with probability 1 the random sequence A^ω, defined by $p_n \in A^\omega$ if and only if $Y_n(\omega) = 1$, is good in the mean. (Here p_n denotes the n-th prime number.)

Ex.3. Suppose $\gcd(q, h) = 1$. Show that the sequence $(p \mid p$ is prime, $p \equiv h \mod q)$ is good in the mean.

Ex.4. Suppose that for some q and every h the sequence (a_{nq+h}) is good in the mean. Then the full sequence (a_n) is good in the mean.

Ex.5. Suppose that the sequence (a_n) is good in the mean. Is it true that for each q and h the sequence (a_{nq+h}) is good in the mean? (Hint: It is not true even if we assume that (a_n) is ergodic.)

3. Notes on sequences bad in the mean

In this section, for a given sequence $A = (a_n)$ of integers it is more convenient to examine the averages

$$M_t(A, f)(x) = \frac{1}{t} \sum_{n=1}^{t} f(\tau^{a_n} x).$$

Note that t is an integer. In Section 3 of the first chapter we saw several examples of sequences that were bad for periodic systems, and as a consequence these sequences are not good in the mean. So one may ask if there are sequences that are bad in the mean in the sense that for *every* measure-preserving system (X, Σ, m, τ) we can find $f \in L^2$ so that the averages $M_t(A, f)$ do not converge in L^2-norm. It is a most surprising fact that such a sequence does not exist! With respect to Exercise 2.10, the following theorem of Weyl disproves the existence of a bad sequence.

Theorem 2.22. *Let (a_n) be a strictly increasing sequence of integers. Then for Lebesgue-almost-every β we have*

$$\lim_{t\to\infty} \frac{1}{t} \sum_{n=1}^{t} e(a_n \beta) = 0.$$

Exercise 2.23. Let (a_n) be a strictly increasing sequence of integers. Then (a_n) is ergodic for Lebesgue-almost-every irrational rotation of the unit interval.

Proof of Theorem 2.22. Although our theorem is a special case of Exercise 5 to Example 1.4, we write out the proof here. First of all, the reader should verify that it is enough to prove that for almost every $\beta \in [0, 1)$

$$\lim_{t \to \infty} \frac{1}{t^2} \sum_{n=1}^{t^2} e(a_n \beta) = 0 \,.$$

This follows if we prove

$$\int_0^1 \left(\sum_{t=1}^{\infty} \left| \frac{1}{t^2} \sum_{n=1}^{t^2} e(a_n \beta) \right|^2 \right) d\beta < \infty \,. \tag{2.34}$$

But the functions $f_n(\beta) = e(a_n \beta)$ are orthonormal, so we have for each t

$$\int_0^1 \left| \frac{1}{t} \sum_{n=1}^{t} e(a_n \beta) \right|^2 d\beta = \frac{1}{t} \,,$$

which implies (2.34). □

This theorem is a great contrast to Example 1.18, and we will see that if we consider almost everywhere convergence instead of convergence in L^2-norm, then there are again "universally bad" sequences.

Example 2.24: Lacunary sequences. Here we examine lacunary sequences. A lacunary sequence has the property that it is bad for at least one irrational rotation. (It is interesting to recall here Exercise 7 to Example 1.18.)

Here we just show that if the sequence (a_n) satisfies $a_{n+1}/a_n \geq 8$ for each n, then for some irrational rotation of the unit interval it is not good in the mean.

The case of an arbitrary lacunary sequence is deferred to the exercises (Exercise 10 below). To show the underlying idea more clearly, first we just show the existence of a real rotation for which (a_n) is not good in the mean. The proof of the existence of an *irrational* rotation is given in Exercise 2.26 below.

So suppose the sequence (a_n) is lacunary enough, that is, it satisfies $a_{n+1}/a_n \geq 8$ for each n. We want to show that for some β there is a sequence (t_k) of indices with $t_k \to \infty$ so that

$$\liminf_{k \to \infty} \int_0^1 \left| \frac{1}{t_{2k}} \sum_{n=1}^{t_{2k}} e(x + a_n \beta) - \frac{1}{t_{2k+1}} \sum_{n=1}^{t_{2k+1}} e(x + a_n \beta) \right|^2 dx \geq 2 \,.$$

This is equivalent with

$$\liminf_{k\to\infty} \left| \frac{1}{t_{2k}} \sum_{n=1}^{t_{2k}} e(a_n\beta) - \frac{1}{t_{2k+1}} \sum_{n=1}^{t_{2k+1}} e(a_n\beta) \right| \geq \sqrt{2}. \qquad (2.35)$$

Choose the strictly increasing sequence (t_k) of indices to satisfy

(i) $\frac{t_{k+1}}{t_k} > k$ for each k.

For example, $t_k = k!$ will do. The inequality in (2.35) clearly follows if we prove that for some β for each even k we have $\langle a_n\beta \rangle \leq 1/4$, $t_k \leq n < t_{k+1}$; and for each odd k we have $1/2 \leq \langle a_n\beta \rangle \leq 3/4$, $t_k \leq n < t_{k+1}$. In fact, we have the following general result, which expresses the fundamental property of lacunary sequences.

Lemma 2.25. *Suppose the sequence (a_n) satisfies the lacunarity condition $a_{n+1}/a_n \geq 8$ for each n. For each n let I_n be a closed subinterval of the unit interval $[0,1)$ of length at least $1/4$.*

Then there is a β which satisfies $\langle a_n\beta \rangle \in I_n$ for each n.

Proof. Because of periodicity by 1, we need to show that there exists a β with $a_n\beta \in I_n + \mathbb{Z}$ for every n, that is, we need to show that the set $\bigcap_n (I_n + \mathbb{Z})/a_n$ is nonempty, where for $J \subseteq \mathbb{R}$ and $r \in \mathbb{R}$ we define $J \cdot r = \{jr \mid j \in J\}$.

Let us set $J_1 = I_1$, and $k_1 = 1$. Because of the lacunarity assumption there is an integer k_2 so that $k_2/a_2, (k_2 + 1)/a_2 \in J_1$. It follows that $J_2 = I_2 + k_2/a_2 \subseteq J_1$. Repeating this with J_1 replaced by J_2, we obtain an integer k_3 so that $J_3 = I_3 + k_3/a_3 \subseteq J_2$. Continuing this way, we get a nested sequence (J_n) of closed intervals such that $J_n \subseteq (I_n + \mathbb{Z})/a_n$, so we can take $\beta \in \bigcap_n J_n$. $\square$

Exercise 2.26. a) Let (f_k) be a sequence of continuous functions defined on $[0,1)$. Suppose that for some positive δ there is a dense set of β's so that $\liminf_{k\to\infty} f_k(\beta) \geq \delta$, and suppose that for a dense set of β's $\limsup_{k\to\infty} f_k(\beta) \leq 0$. Then for an uncountable, dense set of β's the limit $\lim_{k\to\infty} f_k(\beta)$ does not exist. (Hint: although this can be done by hand, the proof becomes particularly clear if you use Baire's category theorem.)

b) Suppose the sequence (a_n) satisfies the condition $a_{n+1}/a_n \geq 8$ for each n. For each n, let I_n be a closed subinterval of the unit interval $[0,1)$ of length at least $1/4$.

Show that the set of β's satisfying $\langle a_n\beta \rangle \in I_n$ for each large enough n, say $n \geq n(\beta)$, is dense.

c) Suppose the sequence (a_n) satisfies $a_{n+1}/a_n \geq 8$ for each n. Then for some irrational rotation of the unit interval, (a_n) is not good in the mean.

Exercises for Section 3

In the exercises below (a_n) is always a strictly increasing sequence of integers.

Ex.1. Suppose that for some $\rho > 4$ the sequence (a_n) satisfies $a_{n+1}/a_n \geq \rho$ for each n. Then for some irrational rotation of the unit interval, (a_n) is not good in the mean.

Ex.2. Let β be an irrational number, and let $F(n)$ be an arbitrary function. Construct a sequence (a_n) which is ergodic for the rotation by β and satisfies $a_{n+1} - a_n \geq F(n)$ for every n. (Hint: just make sure that $\langle a_n \beta \rangle$ distributes evenly in intervals with diadic endpoints.)

Ex.3. Let β be an irrational number, and let (b_n) be a strictly increasing sequence of positive integers with the property that $b_{n+1} - b_n \to \infty$ as $n \to \infty$. Construct a sequence (a_n) that is not good for the rotation by β and satisfies $b_n \leq a_n \leq b_{n+1}$ for every n.

Ex.4. Suppose that the sequence (a_n) is good in the mean for some irrational rotation β. Is it true that for each q and h the sequence (a_{nq+h}) is good in the mean for the rotation by β? (Hint: It is not true, and there is even an example for (a_n) that is ergodic for β.)

Ex.5. Let β be irrational, and suppose that the sequence (a_n) satisfies $\lim_{n\to\infty}\langle a_n\beta \rangle = 0$. Prove that (a_n) is not good in the mean. (Hint: for each fixed integer k, we have $\lim_{n\to\infty}\langle a_n k\beta \rangle = 0$. Hence, for a dense set of α's we have $\lim_{n\to\infty} e(a_n\alpha) = 1$. Use now the method of Exercise 2.27.)

Ex.6. Suppose that the sequence (a_n) has the property that for each integer q there is $n(q)$ such that q divides a_n for $n > n(q)$. Show that (a_n) is not good in the mean. Note that this proves that the sequence (2^n) is not good in the mean. (Hint: Use the method of Exercise 2.27.)

Ex.7. Show that the sequence $(2^n 3^m)$ arranged in increasing order is not good in the mean. (Hint: Modify your argument of the previous exercise.)

Ex.8. a) Let I be a fixed subinterval of the unit interval, and let β be irrational. Then the sequence $\{n \mid \langle n\beta \rangle \in I\}$ (arranged in increasing order) is good in the mean. (Hint: Consider a suitable product system.)

b) Let I be a fixed subinterval of the unit interval, and let β be irrational. Then the sequence $\{n^2 \mid \langle n^2\beta \rangle \in I\}$ (arranged in increasing order) is good in the mean.

c) Generalize parts a) and b).

Ex.9. Show that for Lebesgue-almost-every β the sequence $(a_n\beta)$ is uniformly distributed mod 1. (Hint: Use Theorem 2.23 and Weyl's criterion for uniform distribution mod 1.)

Ex.10. This exercise serves to prove that a given lacunary sequence is not good in the mean for some irrational rotation. Let $\rho > 1$, and suppose the sequence of integers (a_n) satisfies $a_{n+1}/a_n \geq \rho$ for every n. Let q be a positive integer large enough to satisfy $\log \rho / \log(2q) > 2/q$.

a) Show that for a dense set of β's we have

$$\liminf_{t\to\infty} \frac{1}{t} \sum_{n\leq t} 1_{(0,1/q)}(a_n\beta) \geq \frac{\log \rho}{\log(2q)} \, .$$

(Hint: Let m be the smallest integer to satisfy $\rho^m \geq 2q$, that is $m \geq \log(2q)/\log \rho$, and let $b_n = a_{mn}$. Show that for a dense set of β's we have $\langle b_n\beta \rangle \in (0, 1/q)$.)

b) Show that for a dense set of β's we have

$$\liminf_{t\to\infty} \frac{1}{t} \sum_{n\leq t} 1_{(0,2/q)}(a_n\beta) = \frac{2}{q} \, .$$

(Hint: Use Exercise 9.)

c) Show that for an uncountable, dense set of β's we have

$$\liminf_{t\to\infty} \frac{1}{t} \sum_{n\leq t} 1_{(0,2/q)}(a_n\beta) \leq \frac{2}{q}$$

and

$$\limsup_{t\to\infty} \frac{1}{t} \sum_{n\leq t} 1_{(0,1/q)}(a_n\beta) \geq \frac{\log \rho}{\log(2q)} \, .$$

d) Show that for an irrational rotation the sequence (a_n) is not good in the mean.(Hint: Prove a suitable modification of part a) of Exercise 2.26.)

Ex.11. Let $\rho > 1$, and suppose the sequence of integers (a_n) satisfies $a_{n+1}/a_n \geq \rho$ for every n. Then for a residual set of β's the sequence (a_n) is not good in the mean for the rotation of the unit interval by β. (A set of real numbers is called *residual* if and only if its complement is of first (Baire) category.)

Notes to Chapter II

1. The geometric proof of the mean ergodic theorem of F. Riesz appeared first in Hopf's book [1937]. In English, see Riesz and Szőkefalvi-Nagy's book [1990]. Our proof is a simplified version of Riesz's original one.

2. The fundamental result of Exercise 2.10 does not appear in the literature. Though it seems to have been first stated explicitly in [Wierdl, 1989], it is no doubt that von Neumann was aware of it — and all other ergodic theorists ever since.

3. It was Salem and Zygmund [1954] who gave the first treatment of random trigonometric polynomials. Their method is applicable only in Example 2.5, where we give a simplified treatment because it is sufficient to our purposes. A colorful display of results about random trigonometric polynomials and other random functions can be found in Kahane's book [1985]. The notes from Marcus and Pisier [1981] use deep methods from probability theory and are less accesible for the nonexpert, but the results are often "best possible" in nature. Further ergodic theoretical applications of random trigonometric polynomials can be found in [Rosenblatt, 1988].

4. Exercise 3 for Example 2.5 is from [Blum, Cogburn, 1975].

5. There is a far-reaching generalization of Example 2.5, the theorem of Wiener and Wintner [1941]. This theorem says the following. *Let (Ω, B, P, τ) be an ergodic dynamical system. Suppose $P(E) > 0$. Then with probability 1 the sequence A^ω defined by $n \in A^\omega$ if and only if $\tau^n \omega \in E$ is good in the mean.*

We get Example 2.5 from this by taking the usual representation of an i.i.d. sequence of random variables in a Bernoulli shift. The original demonstration of the Wiener-Wintner theorem was — to put it mildly — difficult. See Halmos's comment at the end of his book [1956]. A relatively simple proof of the Wiener-Wintner theorem can be found in the survey article [Bellow, Losert, 1985].

Let us state the WW theorem in the following form. *Let (Ω, B, P, τ) be an ergodic dynamical system. Suppose $P(E) > 0$. Let g be the characteristic function of E. Then there is $\Omega_1 \subset \Omega$ with $P(\Omega_1) = 1$ so that whenever $\omega \in \Omega_1$ the averages*

$$\frac{1}{t} \sum_{n \leq t} g(\tau^n \omega) e(n\beta)$$

converge for **every** *real β.*

E. Lesigne [1993] gave a generalization of the WW theorem to subsequences. He proved the following, using a Hilbert space version of Van der Corput's inequality obtained by V. Bergelson. *Let (Ω, B, P, τ) be an ergodic dynamical system. Suppose $P(E) > 0$. Let g be the characteristic function of E. Then there is $\Omega_1 \subset \Omega$ with $P(\Omega_1) = 1$ so that whenever*

$\omega \in \Omega_1$ *the averages*

$$\frac{1}{t} \sum_{n \leq t} g(\tau^n \omega) e(n^2 \beta)$$

converge for **every** *real* β.

While the sequence of squares above can be replaced by the values of a given polynomial at (positive) integer points, it is not known if a similar theorem is true involving the sequence of prime numbers (cf. Exercise 2 for Example 2.11 and Exercise 2 for Example 2.15).

6. (Here we will use concepts from the next chapter.) A problem related to the WW theorem is the boundedness of the rotated Hilbert transform or helical transform. The *ergodic Hilbert transform* is an average of the form

$$H_t f(x) = \sum_{|n| \leq t}' \frac{f(\tau^n x)}{n} ,$$

where the prime signifies that in the above sum we omit the term corresponding to $n = 0$. Of course we now assume that τ is invertible. The ergodic Hilbert transform was introduced in [Cotlar, 1955]. Cotlar gives a difficult treatment, because he thought that in the ergodic theoretical setting he does not have the tools (such as Fourier transform) used to handle the real Hilbert transform[2]. It was Calderón [1968] who showed, via his tranference principle to be discussed in the next chapter, that all the properties of the ergodic Hilbert transform are direct consequences of the corresponding properties of the real-variable Hilbert transform. In particular, the operators H_t satisfy a weak maximal inequality in L^1, and for $f \in L^1$ the averages $H_t f(x)$ converge a.e. as $t \to \infty$ — expressing a complicated cancellation property.

Now, Petersen and Campbell [1989] considered the rotated Hilbert transform (or helical transform)

$$H_t^\beta f(x) = \sum_{|n| \leq t}' e(n\beta) \frac{f(\tau^n x)}{n} .$$

In their paper they prove many interesting results about the helical transform. In particular, they show that the "double" maximal function

$$H^* f(x) = \sup_\beta \sup_t |H_t^\beta f(x)|$$

satisfies a weak inequality in L^2. This is a deep result: as they point out, the fact that H^* satisfies a weak L^2 inequality is a close relative of Carleson's theorem on the a.e. convergence of Fourier series!

---

[2]Nevertheless, Cotlar developed a nice theory the elements of which (such as "Cotlar's lemma") are widely used in harmonic analysis today.

They also ask the question if the Hilbert-transform analogue of the WW theorem is true: whether for a.e. x the averages $H_t^\beta f(x)$ converge for *every* β. The L^2 boundedness of H^* and its equivalence with the Carleson-Hunt Theorem and various other inequalities are proved in [Assani, Petersen, 1992]. The L^p boundedness of H^* for $p > 1$ is proved in [Assani, 1993]; and in [Assani, Petersen, 1992] and [Assani, 1992] it is shown that H^* does not satisfy a weak inequality in L^1, and for some $f \in L^1$ there is a set of x's of positive measure for which $H_t^\beta f(x)$ does not converge for every β.

A complete and correct proof for the full analog of the WW theorem (that is, the question of convergence) for the averages $H_t^\beta f(x)$ if $f \in L^p$, $p > 1$, has not appeared yet in the literature.

Another interesting observation on the helical transform is the following. Suppose that in the definition of H^* we take the supremum in β only from a fixed lacunary sequence. For definiteness, let $I = \{\beta \mid \beta = 2^n$ for some $n \in \mathbb{Z}\}$, and define

$$M f(x) = \sup_{\beta \in I} \sup_t |H_t^\beta f(x)| \, .$$

Then the L^2 boundedness of M is equivalent with the L^2 boundedness of the usual ergodic maximal function, and in fact it even implies convergence, via an oscillation inequality which we will discuss in Chapter IV. As M. Lacey points out, this is an instance of a general principle, namely that boundedness of singular integral operators leads to the boundedness of certain ergodic maximal functions, and in fact to convergence of ergodic averages.

For an overview on the helical transform see [Assani, Petersen, White, 1991]. There are results about helical transforms along algebraic surfaces announced in [Stein, Wainger, 1990], but no proofs appeared yet.

7. Our treatment of the result in Example 2.6 closely follows Bourgain's original one in [Bourgain, 1988]. Boshernitzan [1983] follows an entirely different and more elementary path, but his method does not apply to almost everywhere convergence results — which is our main purpose here.

8. The method of proving the estimate in (2.24) is, in essence, from Weyl's paper [1916].

9. V. Bergelson gave a non-Fourier-analytical, "Hilbert-space"-proof of the result of Example 2.11. See [Furstenberg, 1990].

10. For the shortest proof of Lemma 2.18, see [Vaughan, 1981], but for a nonexpert it is probably more advisable to consult the extremely well-written book [Ellison & Ellison, 1985].

11. Lemma 2.21 is due to Siegel and Walfis. For a proof see [Ellison & Ellison, 1985].

12. Theorem 2.22 is in [Weyl, 1916].

13. The result of Example 2.24 is due to J. Rosenblatt, but in our proof we use ideas from [Bellow, 1983].

14. Let us mention finally an unsolved problem, the problem of mean L^1 convergence of averages related to triple recurrence.

Let τ, σ and ν be commuting measure-preserving transformations on the probability space (X, Σ, m). Then for $f, g, h \in L^3$ do the averages

$$\frac{1}{t} \sum_{n \le t} h(\nu^n x) g(\sigma^n x) f(\tau^n x)$$

converge in L^1-norm?

The answer to the corresponding problem for double recurrence (when we just have τ and σ and f and g above) is in [Conze, Lesigne, 1984].

CHAPTER III

UNIVERSALLY GOOD SEQUENCES FOR L^1

We finally have arrived to the main course of these notes: almost everywhere convergence. The main purpose of this chapter is to give a proof of the pointwise ergodic theorem that can be applied to many other situations. This proof — regrettably — is not the most elementary proof known, but close to it, and it serves our purpose the best, which is to prove almost everywhere convergence along subsequences other than the whole set of positive integers. Our point of view is this: we easily find a class of functions for which the ergodic averages converge almost everywhere, and this class is dense in L^1. Then a maximal inequality will prove a. e. convergence for every function in L^1. We will point out that a maximal inequality for an arbitrary measure-preserving system is equivalent with a maximal inequality on the integers. But this is not the only way to prove almost everywhere convergence results! In the next chapter we will show an alternative method to prove almost everywhere convergence for functions in L^2; namely, we will prove a so-called oscillation inequality. This alternative approach will be very important to us when we examine a. e. convergence along the squares or primes.

Analogously to the first two chapters, we start with the following definition.

Definition 3.1. Let $1 \leq p \leq \infty$. Let τ be a measure-preserving transformation of the probability space (X, Σ, m). We say that a sequence of positive integers $A = (a_n)$ is L^p-*good* for the system (X, Σ, m, τ) if and only if for each $f \in L^p(X)$ the limit

$$\lim_{t \to \infty} M_t(A, f)(x)$$

exists for almost every $x \in X$, where

$$M_t(A, f)(x) = M_t f(x) = \frac{1}{A(t)} \sum_{\substack{1 \leq a \leq t \\ a \in A}} f(\tau^a x) \,.$$

The sequence $A = (a_n)$ is *universally L^p-good*, or simply L^p-*good*, if and only if it is L^p-good for every system (X, Σ, m, T).

First let us say a few words about a general scheme for proving almost everywhere convergence for the averages $M_t f(x)$. Let us set $B = L^p(X, \Sigma, m)$, and let $\|f\|$ denote the L^p norm of the function f. By the uniform boundedness principle, to prove the B-norm convergence of the $M_t f$'s it is enough to prove the convergence for a dense set of functions. The method of proof of a. e. convergence in many cases (but not always, as we shall see later) is similar: first we prove the a. e. convergence for a dense (in B) set of functions, and then a so-called maximal inequality ensures the a. e. convergence for all functions of B. A typical proof runs as follows. Let Wf be the oscillation of the $M_t f$'s:

$$Wf(x) = \limsup_{t,s \to \infty} |M_t f(x) - M_s f(x)| \,.$$

We want to show that Wf is 0 almost surely. One way to show this is to prove that for each fixed $\lambda > 0$

$$m(Wf > \lambda) = 0 \,. \tag{3.1}$$

Now suppose that for g in a dense subset C of B we know that $Wg = 0$ almost everywhere. Let $\eta > 0$, and choose $g \in C$ so that $\|f - g\| < \eta$. Then almost surely we have

$$Wf = W(f - g) \,. \tag{3.2}$$

Suppose we know that the operator W is continuous at 0 in measure. This means that for any $\epsilon, \lambda > 0$ there is $\eta > 0$ such that $m(Wh > \lambda) < \epsilon$ whenever $\|h\| < \eta$. Letting $h = f - g$, we get

$$m(W(f - g) > \lambda) < \epsilon \,,$$

which implies, by (3.2), that

$$m(Wf > \lambda) < \epsilon \,.$$

Since ϵ is arbitrary, (3.1) follows.

Thus the problem is reduced to checking the continuity of W in measure. Let us define the so-called *maximal function* by

$$M^* f(x) = \sup_t |M_t f(x)| \,.$$

Since $Wf \le 2M^* f$, it is enough to show that the maximal operator M^* is continuous in measure. What we have gained is that instead of dealing with the oscillation we need to control the maximal function, which is much easier to handle, mostly because its definition involves only "sup" while the definition of the oscillation involves "limsup". It seems quite a daring step to go from the oscillation to the maximal operator, but we will see in the chapter on universally bad sequences that the a. e. finiteness of Wf for every f implies the continuity in measure of the maximal operator M^*.

In what form does one prove the continuity of the maximal operator in measure? In these notes we will encounter only two forms. Sometimes we will prove the strongest possible property: that M^* is a bounded $B \to B$ operator. This means that for a constant C, for every $f \in B$ we have

$$\|M^*f\| \leq C\|f\|.$$

In this case we say that the operators M_t satisfy a *strong maximal inequality in* L^p. Sometimes we will prove a weaker statement: we will prove — for $p < \infty$ — the existence of a constant C so that for each $f \in B$ and $\lambda > 0$ we have

$$m(M^*f > \lambda) \leq \frac{C}{\lambda^p}\|f\|^p.$$

In this case we say that the operators M_t satisfy a *weak maximal inequality in* L^p.

Example 3.2. The pointwise ergodic theorem. Here we will prove the individual or pointwise ergodic theorem.

In other words, we prove that the sequence of positive integers is universally L^1-*good.*

The plan follows the scheme above: first we prove a. e. convergence for a dense (in L^1) set of functions, and then we show that the maximal operator satisfies a weak inequality in L^1.

The dense set of functions will be the same one we used when proving mean L^2 convergence, namely the linear span of the τ-invariant functions and the coboundaries, that is, functions f that can be written in the form $f = g - T_\tau g$ with some $g \in L^2$. (Note that in case of a finite measure space a set of functions which is dense in L^2 is also dense in L^1.)

It is clear that the averages $M_t f(x)$ converge a.e. for τ-invariant functions. Suppose now that $f = g - T_\tau g$ for some $g \in L^2$. Since we just need a dense set of functions we can even assume that g is bounded. Then, since $M_t f$ is a telescoping sum, we get the a. e. uniform (in x) estimate $|M_t f(x)| = |\frac{1}{t}(g(x) - g(\tau^{[t]+1}x)| \leq 2K/t$, where K is the almost sure bound of g. But this estimate implies that $M_t f(x) \to 0$ for a. e. x.

We are left to prove the weak L^1 maximal inequality, that is, the continuity of the maximal operator in measure.

Lemma 3.3: The ergodic maximal inequality. *For every* $\lambda > 0$ *we have*

$$m(M^*f > \lambda) \leq \frac{2}{\lambda} \cdot \|f\|_{L^1}, \tag{3.3}$$

where now $M^*f(x) = \sup_t 1/t \sum_{n \leq t} f(\tau^n x)$.

Proof. The main idea of our proof is to first establish the corresponding inequality on the integers. The set of integers $\mathbb{Z}$, with the counting measure and the shift transformation $\tau(x) = x + 1$, is a measure-preserving system, hence we could say that a maximal inequality on the integers is a special case of the lemma to be proved here — except that in the enunciation of the lemma we implicitly assumed a probability space. But the reader will check easily that in the proof below we use the finiteness of the measure nowhere. So, interestingly, the general case will follow from the simplest one!

To distinguish the functions defined on the integers from the ones defined on X, the functions on $\mathbb{Z}$ will be denoted by greek letters: φ, ϕ etc. To further emphasize the structure of the operators on $\mathbb{Z}$ we introduce

Definition 3.4. The convolution of the functions $\varphi, \phi : \mathbb{Z} \to \mathbb{C}$ is defined by

$$\phi * \varphi(j) = \sum_{k \in \mathbb{Z}} \overline{\phi}(k)\varphi(j+k) \,.$$

On $\mathbb{Z}$ the ergodic averages $M_t(A, f)$ along some sequence of integers A are convolutions:

$$M_t(A, \varphi)(j) = M_t(A) * \varphi(j) \,,$$

where the function $M_t(A)$ on $\mathbb{Z}$ is defined by

$$M_t(A)(j) = \frac{1}{A(t)} \sum_{\substack{1 \le a \le t \\ a \in A}} \delta_a(j) \,.$$

Lemma 3.5. *For every* $\lambda > 0$ *and for every* $\varphi : \mathbb{Z} \to \mathbb{C}$ *of finite support, we have*

$$\#\{j \mid \sup_t |M_t * \varphi(j)| > \lambda\} \le \frac{2}{\lambda} \cdot \|\varphi\|_{\ell^1} \,. \qquad (3.4)$$

Proof. In this proof, by $[a, b]$ we mean the integers n satisfying $a \le n \le b$.

We can clearly assume that φ is real-valued and nonnegative. Since the inequality we want to prove is homogeneus in λ, we can assume that $\lambda = 1$. Let us denote $E = \{j \mid \sup_t |M_t * \varphi(j)| > 1\}$. First of all we observe that the set E is finite. This follows easily since φ has finite support, and if j is "far away" from the support of φ, then $|M_t * \varphi(j)|$ is small. We want to prove that

$$\#E \le 2 \cdot \|\varphi\|_{\ell^1} \,. \qquad (3.5)$$

Notice that $j \in E$ if and only if for some positive integer t the arithmetic average of the numbers $\varphi(j+1), \varphi(j+2), \dots, \varphi(j+t)$ is greater than 1. In other words,

$$t < \sum_{n=j+1}^{j+t} \varphi(j+n) = \sum_{n \in j+[1,t]} \varphi(n) \,.$$

To be specific, for each j let t_j denote the smallest of the t's satisfying the above. Our plan is to select j's from E, say $j_1, \dots, j_K$, so that the sets $j_1 + [1, t_{j_1}], \dots, j_K + [1, t_{j_K}]$ are pairwise disjoint, and

$$\#E \le 2 \cdot (t_{j_1} + \cdots + t_{j_K}) \,. \qquad (3.6)$$

This would finish the proof, since for each i we have the inequality (cf. (3.5))

$$t_{j_i} < \sum_{n \in j_i + [1, t_{j_i}]} \varphi(n),$$

and, summing this for $i = 1, \ldots, K$, we get

$$t_{j_1} + \cdots + t_{j_K} < \sum_{i=1}^{K} \sum_{n \in j_i + [1, t_{j_i}]} \varphi(n) \leq \sum_{n} \varphi(n) = \|\varphi\|_{\ell^1}.$$

So let us see how we can select $j_1, \ldots, j_K$ with the property that the sets $j_1 + [1, t_{j_1}], \ldots, j_K + [1, t_{j_K}]$ are pairwise disjoint while (3.6) holds. Let us put $E_1 = E$ and $F_1 = \{t_j \mid j \in E_1\}$. It is convenient to introduce the notation $I_j = [1, t_j]$. Since F_1 is finite (for E_1 is finite), there is a $j_1 \in E_1$ such that t_{j_1} is a maximal element of F_1. Suppose that for some $j \in E_1$ the sets $j + I_j$ and $j_1 + I_{j_1}$ are not disjoint. This implies that $j \in j_1 + I_{j_1} - I_j$. But by the maximality of t_j we have $I_{j_1} - I_j \subseteq I_{j_1} - I_{j_1}$, and hence $j \in j_1 + I_{j_1} - I_{j_1}$. Let us set $E_2 = E_1 \setminus (j_1 + I_{j_1} - I_{j_1})$ and $F_2 = \{t_j \mid j \in E_2\}$. Note that E_2 is a set strictly smaller than E_1, since $j_1 \notin E_2$. Let j_2 be an element of E_2 such that t_{j_2} is a maximal element of F_2. By the above, the sets $j_1 + I_{j_1}$ and $j_2 + I_{j_2}$ are disjoint. Continuing this way, we find $j_1, \ldots, j_K$ with the property that the sets $j_1 + I_{j_1}, \ldots, j_K + I_{j_K}$ are pairwise disjoint and

$$E = E_1 \subseteq (j_1 + I_{j_1} - I_{j_1}) \cup \cdots \cup (j_K + I_{j_K} - I_{j_K}).$$

Now, if we observe that

$$\# (j + I_j - I_j) = \# (I_j - I_j) \leq 2 \cdot \# I_j = 2 \cdot t_j,$$

then we see that (3.6) follows. $\square$

How does Lemma 3.5 imply the ergodic maximal inequality (3.3)? The idea is to apply the inequality on the integers to the "incomplete" periodic system formed by a finite initial segment $x, \tau x, \ldots, \tau^K x$ of the orbit of each $x \in X$, and then take the average (integral) of these inequalities over the space X. We may visualize all this by taking $K + 1$ copies of X, say $X_0, \ldots, X_K$, and we imagine that $f(\tau^j x)$ is a function of $x \in X_j$.

We can assume that $f \geq 0$ and $\lambda = 1$. By the monotone convergence theorem, it is enough to prove that for every fixed N

$$m(\max_{t \leq N} |M_t f| > 1) \leq 2 \cdot \|f\|_{L^1}. \tag{3.7}$$

Let $K > N$, and for $x \in X$ define the function $\varphi = \varphi_x : \mathbb{Z} \to \mathbb{C}$ by

$$\varphi(j) = \begin{cases} f(\tau^j x) & \text{for } 0 \leq j \leq K \\ 0 & \text{otherwise}. \end{cases}$$

We have, for $0 \leq j \leq K - N$,

$$\max_{t \leq N} |M_t f(\tau^j x)| = \max_{t \leq N} |\frac{1}{t} \sum_{n \leq t} f(\tau^{j+n} x)| = \max_{t \leq N} |M_t * \varphi(j)| .$$

Thus, letting F and E be the sets $F = \{\max_{t \leq N} |M_t f| > 1\}$ and $E = \{\max_{t \leq N} |M_t * \varphi| > 1\}$, we have

$$\sum_{0 \leq j \leq K-N} 1_F(\tau^j x) = \sum_{0 \leq j \leq K-N} 1_E(j) \leq \#E .$$

This is the point where we use Lemma 3.5: by (3.4), we have $\#E \leq 2\|\varphi\|_{\ell^1}$, hence

$$\sum_{0 \leq j \leq K-N} 1_F(\tau^j x) \leq 2 \cdot \|\varphi\|_{\ell^1} = 2 \cdot \sum_{0 \leq j \leq K} |f(\tau^j x)| .$$

Integrating both ends of the above inequalities with respect to $x \in X$ and using that τ is measure-preserving, we obtain

$$(K - N + 1) \cdot m(F) = (K - N + 1) \int_X 1_F \, dm \leq 2(K + 1) \int_X f \, dm =$$
$$= 2(K + 1) \cdot \|f\|_{L^1} .$$

Divide both sides by K and let K go to infinity to get (3.7). $\square$

Exercise 3.6. The weak transference principle. Let $M_1, M_2, \ldots$ be a sequence of functions in $\ell^1(\mathbb{Z}_+)$, and let $1 \leq p < \infty$. Suppose that there is a constant C so that for every $\varphi : \mathbb{Z} \to \mathbb{C}$ of finite support we have

$$\#\{\sup_t |M_t * \varphi| > 1\} \leq C \cdot \|\varphi\|_{\ell^p}^p .$$

Then in every dynamical system (X, Σ, m, T) the operators M_t defined by

$$M_t f(x) = \sum_n M_t(n) f(\tau^n x)$$

satisfy a weak maximal inequality in L^p with the same constant C on the right-hand side.

Exercises for Example 3.2

Ex.1. Let $I_1, I_2, \ldots$ be a sequence of finite sets of integers with the following properties:

(i) $I_t \subseteq I_{t+1}$ for every t;

(ii) There is a constant C such that $\#(I_t - I_t) \leq C \cdot \#I_t$ for every t.

On a dynamical system (X, Σ, m, τ), consider the averages on the sets I_t:

$$M_t f(x) = \frac{1}{\#I_t} \sum_{n \in I_t} f(\tau^n x) \,.$$

Show that the operators M_t satisfy a weak maximal inequality in L^1 with the constant C of property (ii) on the right-hand side.

Ex.2. This exercise will show that there are universally L^1-good sequences of 0 density. The sequence A will be a block sequence. Let $[v_1, w_1), [v_2, w_2), \ldots$ be a sequence of intervals of positive numbers, with $w_n \leq v_{n+1}$ for every n. Let $A_n = [v_n, w_n) \cap \mathbb{Z}$ and $A = \bigcup_n A_n$.

a) Suppose $\#A_n \to \infty$. Then in every dynamical system (X, Σ, m, τ), for a dense (in L^2 and hence in L^1) set of functions the averages $M_t(A, f)$ converge almost everywhere.

b) Suppose that the sequence $(\#A_n)$ is increasing, and that for every n we have $\#A_n \geq v_{n-1}$. Show that in every dynamical system (X, Σ, m, τ) the averages $M_t(A, f)$ satisfy a weak maximal inequality in L^1. (Hint: Use the result of Ex.1.)

c)Let $F(t)$ be a real function satisfying the properties

(i) $F(t) \leq t$;

(ii) $\lim_{t \to \infty} F(t) = \infty$.

Show that there is a strictly increasing sequence A, universally good for L^1, which satisfies

$$\lim_{t \to \infty} \frac{A(t)}{F(t)} = 0 \,.$$

Ex.3. Show that there is a block sequence A which is ergodic but for which the operators $M_t * \varphi$ do not satisfy a weak maximal inequality in $\ell^1(\mathbb{Z})$. (We shall see later that the failure of a maximal inequality on $\mathbb{Z}$ implies the failure of a maximal inequality in *every* aperiodic dynamical system, which in turn implies the existence of a function f for which $\limsup_{t \to \infty} M_t(A, f)(x) = \infty$ almost everywhere.)

The next exercise gives an alternative proof of Lemma 3.5, and hence the ergodic maximal inequality of Lemma 3.3. The method is a bit simpler than the one in the text and it gives a sharper result, but it is not applicable in more general situations like in Exercise 2 above.

Ex.4. We use the notation of the proof of Lemma 3.5.

a) Suppose that the integer i satisfies $j \leq i < j + k_j$ for some $j \in E$. Show that $i \in E$, and in fact $k_i = k_j - (i - j)$.

b) Show that there are $j_1, \ldots, j_K \in E$ so that the sets $j_1 + [1, t_{j_1}], \ldots, j_K + [1, t_{j_K}]$ are pairwise disjoint, and

$$E = (j_1 + [1, t_{j_1})) \cup \cdots \cup (j_K + [1, t_{j_K})) \, .$$

(Hint: Let j_1 be the smallest element of $E = E_1$. Set $E_2 = E_1 \backslash (j_1 + [1, t_{j_1}))$, and let j_2 be the smallest element of E_2, etc. Then use part a).)

c) Show that for every $\varphi : \mathbb{Z} \to \mathbb{C}$ of finite support we have

$$\#E \leq \sum_{j \in E} |\varphi(j)| \, ,$$

where $E = \{j \mid \sup_t |M_t * \varphi(j)| > 1\}$.

d) Prove the following strengthening of the ergodic maximal inequality (Lemma 3.3): In every dynamical system (X, Σ, m, τ), for $f \in L^1$ and $\lambda > 0$ we have

$$m(F_\lambda) \leq \frac{1}{\lambda} \int_{F_\lambda} |f(x)| dm \, ,$$

where $F_\lambda = \{x \mid \sup_t |M_t f(x)| > \lambda\}$.

Ex.5. Let $1 \leq p < \infty$, and let M_t denote the usual ergodic averages, that is, along the sequence $A = \mathbb{Z}_+$. Show that in every dynamical system the operators M_t satisfy a weak maximal inequality in L^p. (Hint: By Hölder's inequality we have $|M_t f| \leq (M_t |f|^p)^{1/p}$.)

Ex.6. Let $t_1 \leq t_2 \leq \ldots$ be a sequence of positive numbers. Show that in every dynamical system (X, Σ, m, τ), for $f \in L^1$ and $\lambda > 0$ we have

$$\sum_{n=1}^{\infty} m \left(|M_{t_n} f(x)| > n\lambda \right) \leq \frac{2}{\lambda} \cdot \|f\|_{L^1} \, ,$$

where the $M_t f$'s are the usual ergodic averages, that is, along the sequence $A = \mathbb{Z}_+$. (Hint: Prove the corresponding inequality on the integers, and then use a suitable version of the transference principle.) Note that a measurable function f on a finite measure space is integrable if and only if for every positive λ we have $\sum_{n=1}^{\infty} m(|f| > n\lambda) < \infty$.

Ex.7. Show that there is a constant C such that in every dynamical system (X, Σ, m, τ), for $f \in L^1$ and $\lambda > 0$ we have

$$\sum_{t=1}^{\infty} m \left(|M_{t+1} f(x) - M_t f(x)| > \lambda \right) \leq \frac{C}{\lambda} \cdot \|f\|_{L^1} \, ,$$

where the $M_t f$'s are the usual ergodic averages, that is, along the sequence $A = \mathbb{Z}_+$. (Hint: use the previous exercise.)

Ex.8. Prove the following common generalization of the ergodic maximal inequality and Exercise 6. Let $t_1 \leq t_2 \leq \ldots$ be a sequence of positive numbers. Then in every dynamical system (X, Σ, m, τ), for $f \in L^1$ and $\lambda > 0$ we have

$$\sum_{n=1}^{\infty} m \left(\sup_{t_n \leq t \leq t_{n+1}} |M_t f(x)| > n\lambda \right) \leq \frac{2}{\lambda} \cdot \|f\|_{L^1},$$

where the $M_t f$'s are the usual ergodic averages, that is, along the sequence $A = \mathbb{Z}_+$.

Example 3.7. Randomly generated sequences of positive density. Here we examine the random sequence considered in Examples 1.4 and 2.5. Let (Ω, β, P) be a probability space, and let $0 < \sigma \leq 1$. Let $Y_1, Y_2, \ldots$ an i.i.d. sequence of (0-1)-valued random variables with $P(Y_n = 1) = \sigma$ and $P(Y_n = 0) = 1 - \sigma$. For $\omega \in \Omega$, we define the sequence of positive integers A^ω by letting $n \in A^\omega$ if and only if $Y_n(\omega) = 1$.

We want to show that for almost every ω the sequence A^ω is universally good in L^1.

Since with probability 1 the density of the sequence A^ω is $\sigma > 0$, the weak maximal inequality in L^1 for the averages $M_t(A^\omega, f)$ follows from the following exercise.

Exercise 3.8.
a) Suppose that the sequence A of positive integers has positive lower density. Then in every dynamical system (X, Σ, m, τ) the averages $M_t(A, f)$ satisfy a weak maximal inequality in L^1.
b) Is there a universally L^1-good sequence A with positive upper density and 0 lower density? (Hint: Yes, there is — for example, consider a certain block sequence in a periodic system.)

With the maximal inequality in hand, we just have to find a dense set of functions for which $M_t(A^\omega, f)$ converges a. e. This dense set is going to be the whole L^2! The almost-everywhere convergence of $M_t(A^\omega, f)$ is equivalent with the convergence of

$$Q_t^\omega f(x) = \frac{1}{\sigma t} \sum_{n \leq t} Y_n(\omega) f(\tau^n x).$$

According to Lemma 1.5, it is enough to prove the convergence of $Q_t^\omega f(x)$ along the subsequence $I_\rho = \{t \mid t = \rho^k \text{ for some positive integer } k\}$ for each fixed $\rho > 1$. We want to compare $Q_t^\omega f(x)$ — along I_ρ — with its "expectation"

$$V_t f(x) = \frac{1}{\sigma t} \sum_{n \leq t} \sigma f(\tau^n x);$$

we want to show that with probability 1 we have, for every $f \in L^2$,

$$\lim_{\substack{t \to \infty \\ t \in I_\rho}} |Q_t^\omega f(x) - V_t f(x)| = 0$$

almost everywhere. This will follow if we show that with probability 1 we have, for every $f \in L^2$,

$$\int_X \left(\sum_{t \in I_\rho} |Q_t^\omega f(x) - V_t f(x)|^2 \right) dm < \infty.$$

By Exercise 2.4, this inequality follows from

$$\sup_{\beta \in [0,1)} \left(\sum_{t \in I_\rho} |\widehat{Q}_t^\omega(\beta) - \widehat{V}_t(\beta)|^2 \right) < \infty,$$

where

$$\widehat{Q}_t^\omega(\beta) = \frac{1}{\sigma t} \sum_{n \leq t} Y_n(\omega) e(n\beta), \qquad \widehat{V}_t(\beta) = \frac{1}{\sigma t} \sum_{n \leq t} \sigma e(n\beta).$$

Interchanging the "sup" with the summation, we see we need to show

$$\int_\Omega \left(\sum_{t \in I_\rho} \sup_{\beta \in [0,1)} |\widehat{Q}_t^\omega(\beta) - \widehat{V}_t(\beta)|^2 \right) dP(\omega) =$$

$$= \sum_{t \in I_\rho} \int_\Omega \left(\sup_{\beta \in [0,1)} |\widehat{Q}_t^\omega(\beta) - \widehat{V}_t(\beta)|^2 \right) dP(\omega) < \infty. \quad (3.8)$$

But by (2.11) and (2.12), we have the fourth-moment estimate

$$\int_\Omega \left(\sup_{\beta \in [0,1)} |\widehat{Q}_t^\omega(\beta) - \widehat{V}_t(\beta)|^4 \right) dP(\omega) \leq \frac{C}{t},$$

which implies, by Hölder's inequality, the second-moment inequality

$$\int_\Omega \left(\sup_{\beta \in [0,1)} |\widehat{Q}_t^\omega(\beta) - \widehat{V}_t(\beta)|^2 \right) dP(\omega) \leq \frac{C}{\sqrt{t}}.$$

This proves (3.8). □

Exercises for Example 3.7

Ex.1. Let (Ω, β, P) be a probability space, and let $Y_1, Y_2, \ldots$ be an i.i.d. sequence of *bounded* random variables on Ω. Then with probability 1 we have the following: in a dynamical system (X, Σ, m, τ), for $f \in L^1$ the random averages

$$M_t^\omega f(x) = \frac{1}{t} \sum_{n \le t} Y_n(\omega) f(\tau^n x)$$

converge for almost every $x \in X$. (Hint: The weak L^1 maximal inequality is clear. To prove convergence for $f \in L^2$, assume first that Y_n is nonnegative.)

Ex.2. Let $p \ge 1$, and let q be the conjugate exponent: $1/p + 1/q = 1$. Suppose that (Ω, β, P) is a probability space and $Y_1, Y_2, \ldots$ is an i.i.d. sequence of random variables on Ω with finite p-th moment. Then with probability 1 we have the following: in a dynamical system (X, Σ, m, τ), for $f \in L^q$ the random averages $M_t^\omega f(x)$ of the previous exercise converge for almost every $x \in X$. (Hint: For $q < \infty$ the weak L^q maximal inequality follows from Hölder's. Using truncation, reduce the problem to the previous exercise.)

Ex.3. Let β be an irrational number, and let I be a nondegenerate subinterval of the unit interval. Then the sequence A defined by $\{n \mid \langle n\beta \rangle \in I\}$ (arranged in increasing order) is universally L^1-good. (Hint: The weak L^1 maximal inequality follows from Exercise 3.8. To prove a. e. convergence for L^2 functions, first prove a. e. convergence for averages of the form

$$\frac{1}{t} \sum_{n \le t} e(nk\beta) f(\tau^n x),$$

with $k \in \mathbb{Z}$ fixed, by considering an appropriate product system. Then approximate the characteristic function of I — both from above and from below — by trigonometric polynomials.)

Ex.4. Let A be a sequence of positive lower density. Then there is an aperiodic, ergodic system for which A is L^1-good. (Hint: The weak L^1 maximal inequality in *any* system follows from Exercise 3.8. Now appeal to Weyl's result, Theorem 2.22.)

Ex.5. Let $1 \le p < \infty$. Denote by Θ the collection of those integer sequences A which have the property that in each dynamical system the averages $M_t(A, f)$ satisfy the weak maximal inequality in L^p. Define the pseudometric[3] ϱ on Θ by

$$\varrho(A, B) = \limsup_{t \to \infty} \frac{\#\{n \mid n \le t, \, a_n \ne b_n\}}{t},$$

---

[3]The "pseudo" part means that $\varrho(A, B) = 0$ does not imply $A = B$.

where $A = (a_n)$ and $B = (b_n)$.

a) Show that the set of universally L^p-good sequences forms a closed subset of Θ with respect to ϱ.

b) Suppose that for each n the sequence A_n is universally L^∞-good. Show that if $\varrho(A_n, A) \to 0$, and A has positive lower density, then A is universally L^1-good.

Ex.6.

a) Let Q be a finite set of positive integers, and define A to be the sequence we get by arranging the set $\{n \mid \text{no } q \in Q \text{ divides } n\}$ in increasing order. Then A is universally L^1-good.

b) Suppose $Q = \{q_1, q_2, \dots\}$ is a set of positive integers with $\gcd(q_i, q_j) = 1$ if $i \neq j$, and $\sum_n \frac{1}{q_n} < \infty$. Then the sequence $A = \{n \mid \text{no } q \in Q \text{ divides } n\}$ is universally L^1-good. (Hint: Prove first that A has positive density, and then use Exercise 5.)

c) Show that the sequence of squarefree integers is universally L^1-good.

Notes to Chapter III

1. The pointwise ergodic theorem was first proved by Birkhoff [1931], but it was Khintchine [1933] who gave its abstract formulation used today.

2. The ergodic maximal inequality was first proved by Kakutani and Yosida [1939]. Our proof follows Wiener's [1939], but aimed at a more abstract formulation due to Tempelman, so Exercise 1 for Example 3.2 is due to Tempelman.

3. The transference principle of Exercise 3.6 is due to Calderón [1968], who stated and used it in a more general context, but the origin of the transference principle is in [Wiener, 1939].

4. The result of Exercise 2 for Example 3.2 is from [Bellow, Losert, 1984].

5. The idea of Exercise 4 for Example 3.2 is F. Riesz's. This is definitely his way of proving the weak L^1 maximal inequality for the Hardy-Littlewood maximal function [Riesz, 1938], but it seems that it was R. Jones who combined this with the transference principle explicitly.

6. In [Akcoglu, del Junco, 1975] Akcoglu and del Junco examined "moving averages"; they consider averages of the following form: Let (v_n) and (l_n) be two sequences of positive integers, and in a dynamical system (X, Σ, m, T) let

$$M_{(v_t, l_t)} f(x) = \frac{1}{l_t} \sum_{n=v_t}^{v_t + l_t - 1} f(\tau^n x).$$

   Akcoglu and del Junco prove that if $v_n = n^2$ and $l_n = n$, then the averages $M_{(v_t, l_t)} f(x)$ do not necessarily converge a. e., even for bounded functions. (Note, on the other hand, that we have mean convergence.) In [Bellow, Jones, Rosenblatt, 1990], a necessary and sufficient condition, the so-called "cone condition", is given for $M_{(v_t, l_t)} f(x)$ to converge a.e.. They observe that if both sequences (v_n) and (l_n) are increasing, then either $M_{(v_t, l_t)} f(x)$ converges a. e. for every $f \in L^1$ or there is a counterexample even in L^∞. Their work is closely related to the work of Nagel and Stein [1984] on boundary behaviour of harmonic functions. In [Rosenblatt, Wierdl, 1992] a more general condition than the cone condition is established. In particular, one can handle not necessarily increasing sequences (v_n) and (l_n) as well. This general method allows one to compute the "correct" factor q_t with which one should divide $M_{(v_t, l_t)} f(x)$ to have a.e. convergence for every $f \in L^1$. This means that one can determine a sequence (q_n) in terms of the given sequences (v_n) and (l_n) so that the averages $\frac{1}{q_t} M_{(v_t, l_t)} f(x)$ converge a.e. for every $f \in L^1$, but if (c_n) is any sequence with $\lim_{n \to \infty} c_n = \infty$, then for some $f \in L^1$ we have $\limsup_{t \to \infty} \frac{c_t}{q_t} M_{(v_t, l_t)} f(x) = \infty$ almost everywhere.

7. Exercises 6, 7 and 8 for Example 3.2 are from [Rosenblatt, Wierdl, 1992]. In this long article many other large deviation inequalities are established, and an intrinsic connection with moving averages (cf. above) is

shown. At the end of the paper several unsolved problems are noted; here we mention just one. *Let A be the sequence of squares. Let $t_1 \leq t_2 \leq \dots$ be a sequence of positive numbers. Is it true that there is a constant C so that in every dynamical system (X, Σ, m, τ) for $f \in L^1$ and $\lambda > 0$ we have*

$$\sum_{n=1}^{\infty} m\left(|M_{t_n}(A, f)(x)| > n\lambda\right) \leq \frac{C}{\lambda} \cdot \|f\|_{L^1} ?$$

Further developments can be found in [Wittmann].

8. There are some extensions of the result of Example 3.7 in [Huang, 1992].

9. There is a generalization of the result of Example 3.7, the so-called re-turn time theorem. The return time theorem is an "almost everywhere"-version of the Wiener Wintner theorem mentioned in the Notes to Chapter II. The return time theorem says the following. *Let (Ω, B, P, τ) be an ergodic dynamical system. Suppose $P(E) > 0$. Then with probability 1 the sequence A^ω defined by $n \in A^\omega$ if and only if $\tau^n \omega \in E$ is universally good in L^1.*

This theorem was first proved by Bourgain, but he never published his proof, because shortly after learning about Bourgain's discovery Fursten-berg, Katznelson and Ornstein discovered a much more elementary proof. The proof of Furstenberg, Katznelson and Ornstein appeared in an ap-pendix to [Bourgain, 1989]. But Bourgain did not abandon his proof; he put the elements of it into good use in [Bourgain, 1989]. Another interesting proof of the return time theorem is Rudolph's [1994] using the theory of joinings.

The following "almost everywhere" analogue of Lesigne's theorem men-tioned in the Notes to Chapter II is presently just a conjecture. *Let $p > 1$. Let (Ω, B, P, τ) be an ergodic dynamical system. Suppose $P(E) > 0$. Let g be the characteristic function of E. Then there is $\Omega_1 \subset \Omega$ with $P(\Omega_1) = 1$ so that whenever $\omega \in \Omega_1$ the following is true: In any dy-namical system (X, Σ, m, σ), if $f \in L^p$ the averages*

$$\frac{1}{t} \sum_{n \leq t} g(\tau^n \omega) f(\sigma^{n^2} x)$$

converge for m-almost-every $x \in X$.

10. The reader might have been wondering what happened to the almost-everywhere version of Exercise 3 for Example 2.5. Well, the problem is dicussed in the paper [Lacey, Petersen, Rudolph, Wierdl], but the solution is too difficult to be included here as an exercise. The following is proved in [Lacey, Petersen, Rudolph, Wierdl]:

Let (Ω, β, P) be a probability space, and let $1/2 \leq \sigma \leq 1$. Let $Y_1, Y_2 \dots$ be an i.i.d. sequence of $((-1)\text{-}1)$-valued random variables wih $P(Y_n =$

1) $= \sigma$ and $P(Y_n = -1) = 1 - \sigma$. Denote $a_n(\omega) = \sum_{k \leq n} Y_k(\omega)$. With probability 1 we have the following:

a) If $p > 1$, $\sigma > 1/2$ and (X, Σ, m, τ) is a dynamical system, then for every $f \in L^p$ the averages

$$\lim_{t \to \infty} \frac{1}{t} \sum_{n \leq t} f(\tau^{a_n(\omega)} x)$$

converge for a.e. $x \in X$.

b) If $\sigma = 1/2$ and (X, Σ, m, τ) is an aperiodic dynamical system, then for some $f \in L^\infty$ the above averages do not converge for a.e. $x \in X$.

As for part a), it is not known what happens in L^1. In [Lacey, Petersen, Rudolph, Wierdl] a fundamental connection — via a nice tower construction of Rudolph — with the return time theorem is revealed.

11. Let us mention here another unsolved problem, the problem of almost everywhere convergence of averages related to double recurrence. Let τ and σ be commuting measure-preserving transformations on the probability space (X, Σ, m). Then for $f, g \in L^2$ do the averages

$$\frac{1}{t} \sum_{n \leq t} g(\sigma^n x) f(\tau^n x)$$

converge for almost every $x \in X$?

Again, Bourgain [1990a] has partial results.

12. Exercise 5 for Example 3.7 is due to M. Boshernitzan.

13. In this chapter we have mentioned only a small percentage of the results on the almost everywhere convergence of averages along sequences of positive density. The survey article of Bellow and Losert [1985] gives a fuller account up to 1985.

CHAPTER IV

UNIVERSALLY GOOD SEQUENCES FOR L^2

In the previous chapter we saw several examples for universally L^1 good sequences. Most of these sequences were of positive density, but in Exercise 2 for Example 3.2 we saw an example with 0 density. But this latter example was a block sequence, hence the sequence had Banach density 1. In this chapter we will see examples of universally good sequences with 0 Banach density. We will prove, in particular, that some of the random sequences considered in Examples 1.7 and 2.6, or the sequence of squares, are universally good. We will concentrate on proving "L^2 goodness", but in certain simple cases we will prove that the sequence is universally L^p good for every $p > 1$. On the other hand, we will not see any example for a universally L^1 good sequence with 0 Banach-density. This is because no such example is known! In fact, we risk the following conjecture

Conjecture 4.1. *Suppose that the sequence A has 0 Banach density, and let (X, Σ, m, τ) be an aperiodic dynamical system. Then for some $f \in L^1$ the averages $M_t(A, f)$ do not converge almost everywhere.*

One might think that there should be a general result saying that if a sequence is L^p good for every $p > 1$ then maybe it is neccessarily L^1 good. But the distinction between universally L^p good sequences by the value of p is not redundant. Not only are there sequences which are L^p good for every $p > 1$ and not L^1 good, but there are sequences which are L^∞ good and not L^p good for finite p — and anything in between. These results are discussed in Chapter VII.

Let us start with a result that shows the first difference between the behavior of ergodic averages in L^1 and L^p, $p > 1$. Let (X, Σ, m, τ) be a dynamical system. We saw in Exercise 5 for Example 3.2 that the usual ergodic averages $M_t(\mathbb{Z}_+, f)$ satisfy a weak maximal inequality in L^p for $1 \leq p < \infty$. For $p > 1$ we can strengthen this to

Proposition 4.2. The strong ergodic maximal inequality. *Let $p > 1$. Then in any dynamical system (X, Σ, m, τ) the averages $M_t(\mathbb{Z}_+, f)$ satisfy a strong maximal inequality in L^p. In fact, we have*

$$\left\| \sup_t M_t(\mathbb{Z}_+, f) \right\|_{L^p} \leq \frac{p}{p-1} \|f\|_{L^p}.$$

Exercise 4.3. Show that the averages $M_t(\mathbb{Z}_+) * \varphi$ do not satisfy a strong maximal inequality in ℓ^1. Indeed, show that if φ is nonnegative and not identically 0, then

$$\left\| \sup_t M_t(\mathbb{Z}_+) * \varphi \right\|_{\ell^1} = \infty.$$

(We will see later, using Rokhlin's tower construction, that the lack of a maximal inequality on $\mathbb{Z}$ implies that there is no maximal inequality in any aperiodic system either.)

Proof of Proposition 4.2. Although we could prove the proposition directly in any dynamical system, we prove it first on the integers, so we give the opportunity for the reader to familiarize her/himself with the transference principle for strong maximal inequalities.

Lemma 4.4. *For any* $\varphi : \mathbb{Z} \to \mathbb{C}$ *of finite support, we have*

$$\left\| \sup_t |M_t(\mathbb{Z}_+) * \varphi| \right\|_{\ell^p} \le \frac{p}{p-1} \|\varphi\|_{\ell^p} .$$

Proof. We can assume that $\varphi \ge 0$ and $p < \infty$. To ease our notation, we introduce the $\mathbb{Z} \to \mathbb{C}$ function M defined by $M(j) = \sup_t M_t(\mathbb{Z}_+) * \varphi(j)$. Our starting point is the inequality we proved in part c) of Exercise 4 for Example 3.2. An immediate consequence of that inequality is the following: for each $\lambda > 0$ we have

$$\#E_\lambda \le \frac{1}{\lambda} \sum_{j \in E_\lambda} \varphi(j), \tag{4.1}$$

where $E_\lambda = \{j \mid M(j) > \lambda\}$. Now rewrite (4.1) as

$$\sum_j 1_{E_\lambda}(j) \le \frac{1}{\lambda} \sum_j \varphi(j) \cdot 1_{E_\lambda}(j). \tag{4.2}$$

Notice that for each fixed $j \in \mathbb{Z}$ and $r > -1$ we have

$$\int_0^\infty \lambda^r \cdot 1_{E_\lambda}(j) d\lambda = \int_0^{M(j)} \lambda^r d\lambda = \frac{M(j)^{r+1}}{r+1}. \tag{4.3}$$

If we multiply both sides of the inequality in (4.2) by λ^{p-1} and integrate with respect to λ we obtain, using (4.3) with $r = p - 1$ on the left-hand side and with $r = p - 2$ on the right,

$$\frac{1}{p} \sum_j M(j)^p \le \frac{1}{p-1} \sum_j \varphi(j) \cdot M(j)^{p-1} . \tag{4.4}$$

By Hölder's inequality, we can estimate

$$\sum_j \varphi(j) \cdot M(j)^{p-1} \le \left(\sum_j \varphi(j)^p \right)^{1/p} \cdot \left(\sum_j M(j)^p \right)^{1-1/p} .$$

Comparing this with (4.4), we see that the proof of our lemma is complete. □

It is now up to the reader to finish proving the proposition by working out the next exercise.

Exercise 4.5. The strong transference principle. Let $M_1, M_2, \ldots$ be a sequence of functions in $\ell^1(\mathbb{Z}_+)$, and let $p \geq 1$. Suppose that there is a constant C so that for every $\varphi : \mathbb{Z} \to \mathbb{C}$ of finite support we have

$$\left\| \sup_t |M_t * \varphi| \right\|_{\ell^p} \leq C \|\varphi\|_{\ell^p} .$$

Then in every dynamical system (X, Σ, m, τ) the operators M_t defined by

$$M_t f(x) = \sum_n M_t(n) f(\tau^n x)$$

satisfy a strong maximal inequality in L^p with the same constant C on the right-hand side. □

Exercise 4.6. Prove Proposition 4.2 directly by using part d) of Exercise 4 for Example 3.2.

1. Ergodic, universally L^2 good sequences

Example 4.7. Randomly generated sequences of 0 density. In this section we examine random sequences of 0 density. In Example 1.7 we considered random sequences of the following sort.

Let (σ_n) be a sequence of positive numbers satisfying the following properties:

(i) the sequence (σ_n) is decreasing;
(ii) $\lim_{t \to \infty} \sum_{n \leq t} \sigma_n = \infty$.

Let (Y_n) be a sequence of independent, (0-1)-valued random variables on the probability space (Ω, β, P) so that $P(Y_n = 1) = \sigma_n$ and $P(Y_n = 0) = 1 - \sigma_n$. We showed that with probability 1 the random sequence A^ω defined by $n \in A^\omega$ if and only if $Y_n(\omega) = 1$ is ergodic for periodic systems. In Example 2.6 we had to strengthen the assumption in (ii) to

$$\lim_{t \to \infty} \frac{\log t}{\sum_{n \leq t} \sigma_n} = 0$$

in order to conclude that with probability 1 the random sequence A^ω is ergodic. For almost everywhere convergence we further have to strengthen the assumption on the rate of growth of $\sum_{n \leq t} \sigma_n$.

We will be able to prove that with probability 1 the random sequence A^ω is universally L^2 good under the assumption that for each $\rho > 1$

$$\sum_{t \in I_\rho} \frac{\log t}{\sum_{n \leq t} \sigma_n} < \infty, \qquad (ii)'$$

where we take I_ρ to consist of the numbers $t_k, k = 1, 2, \ldots$, defined by $t_k = \min\{t \mid \rho^k \leq \sum_{n \leq t} \sigma_n < \rho^{k+1}\}$.

Exercise 4.8. Suppose that the function $F(t) = \log t / \sum_{n \leq t} \sigma_n$ is decreasing (as $t \to \infty$). Then it is enough to check property $(ii)'$ for one $\rho > 1$. In other words, $\sum_{t \in I_2} \log t / \sum_{n \leq t} \sigma_n < \infty$ implies $(ii)'$ for every $\rho > 1$.

The next exercise indicates the thinnest L^2 good sequences that can be constructed with our random method.

Exercise 4.9. Let $\delta > 1$.

a) Let $\sigma_n = \frac{(\log \log n)^\delta}{n}$. Then property $(ii)'$ is satisfied.

b) Let $\sigma_n = \frac{(\log \log n) \cdot (\log \log \log n)^\delta}{n}$. Then property $(ii)'$ is satisfied. — etc.

Suppose now that properties (i) and $(ii)'$ are satisfied. As in Example 3.7, it is enough to prove that for each fixed $\rho > 1$, with probability 1 the averages

$$Q_t^\omega f(x) = \frac{1}{\sum_{n \leq t} \sigma_n} \sum_{n \leq t} Y_n(\omega) f(\tau^n x)$$

converge a. e. as $t \to \infty$, $t \in I_\rho$. This follows if we prove that with probability 1 we have, for every $f \in L^2$,

$$\int_X \left(\sum_{t \in I_\rho} |Q_t^\omega f(x) - V_t f(x)|^2 \right) dm < \infty,$$

where

$$V_t f(x) = \frac{1}{\sum_{n \leq t} \sigma_n} \sum_{n \leq t} \sigma_n f(\tau^n x).$$

Repeating the steps of Example 3.7, we conclude that we should prove

$$\sum_{t \in I_\rho} \int_\Omega \left(\sup_{\beta \in [0,1)} |\widehat{Q}_t^\omega(\beta) - \widehat{V}_t(\beta)|^2 \right) dP(\omega) < \infty, \tag{4.5}$$

where

$$\widehat{Q}_t^\omega(\beta) = \frac{1}{\sum_{n \leq t} \sigma_n} \sum_{n \leq t} Y_n(\omega) e(n\beta), \qquad \widehat{V}_t(\beta) = \frac{1}{\sum_{n \leq t} \sigma_n} \sum_{n \leq t} \sigma_n e(n\beta).$$

Now it follows from inequality (2.20) of Example 2.6 and from Bernstein's inequality (or from a weaker inequality proved in Exercise 1 for Example 2.6) that

$$\left\| \sup_{\beta \in [0,1)} |\widehat{Q}_t^\omega(\beta) - \widehat{V}_t(\beta)| \right\|_{L^{p_t}(\Omega)} \leq C \sqrt{\frac{\log t}{\sum_{n \leq t} \sigma_n}}, \tag{4.6}$$

where $p_t = \log t$. This implies, since $\|f\|_r \le \|f\|_p$ for $r \le p$ on a probability space,

$$\left\| \sup_{\beta \in [0,1)} |\widehat{Q}_t^\omega(\beta) - \widehat{V}_t(\beta)| \right\|_{L^2(\Omega)} \le C \sqrt{\frac{\log t}{\sum_{n \le t} \sigma_n}} .$$

And now we see that this estimate with property $(ii)'$ proves (4.5). $\square$

Exercises for Example 4.7

In the exercises below we are using the notation of Example 4.7.

Ex.1. Suppose that property (i) and

$$\sum_{t \in I_2} \frac{\log t}{\sum_{n \le t} \sigma_n} < \infty$$

are satisfied. Then with probability 1, in every dynamical system the averages $M_t(A^\omega, f)$ satisfy a strong maximal inequality in L^2.

Ex.2. Let (σ_n) be a sequence of positive numbers ≤ 1 which does not neccessarily satisfy (i) but does satisfy (ii). Then with probability 1 there is a constant C_ω so that in every dynamical system, for each t the norm of the $L^2 \to L^2$ operator $Q_t^\omega - V_t$ is at most

$$C_\omega \cdot \sqrt{\frac{\log t}{\sum_{n \le t} \sigma_n}} .$$

(Hint: Use the estimate in (4.6) and Exercise 2 for Example 2.6.)

Ex.3. Let $1 < p \le 2$. Suppose that (i) holds and for every $\rho > 1$

$$\sum_{t \in I_\rho} \left(\frac{\log t}{\sum_{n \le t} \sigma_n} \right)^{p-1} < \infty$$

hold. Then with probability 1 the sequence A^ω is universally L^p good. (Hint: find a bound for the $L^p \to L^p$ operator $Q_t^\omega - V_t$ by interpolating (using the M. Riesz-Thorin theorem) between the L^2 norm obtained in the previous exercise and the trivial L^1-norm.)

Ex.4.
a) Let $1 < p \le 2$ and $\delta > 1/(p-1)$. Let $\sigma_n = \frac{(\log \log n)^\delta}{n}$. Then with probability 1 the random sequence A^ω is universally L^p good, and in every dynamical system the averages $M_t(A^\omega, f)$ satisfy a strong L^p maximal inequality.

b) Let $\delta > 0$ and let $\sigma_n = \frac{(\log n)^\delta}{n}$. Then for every $p > 1$ with probability 1 the random sequence A^ω is universally L^p good, and in every dynamical system the averages $M_t(A^\omega, f)$ satisfy a strong maximal inequality in L^p.

Ex.5. Suppose that for the sequence (σ_n) of probabilities property (i) is satisfied and $\lim_{t \to \infty} \frac{\log t}{\sum_{n \le t} \sigma_n} = 0$. Let (Y_n) be a sequence of independent, $(0\text{-}1)$-valued random variables on the probability space (Ω, β, P) so that $P(Y_n = 1) = \sigma_n$ and $P(Y_n = 0) = 1 - \sigma_n$. Let the (random) sequence A^ω be defined by $n \in A^\omega$ if and only if $Y_n(\omega) = 1$.

a) Let $1 < p \le 2$. Show that there is a sequence $t_1 < t_2 < \ldots$ going to infinity so that with probability 1, in every ergodic dynamical system (X, Σ, m, τ) for every $f \in L^p$ we have $\lim_{k \to \infty} M_{t_k}(A^\omega, f)(x) = \int_X f \, dm$ for a. e. $x \in X$.

b) Show that there is a sequence $t_1 < t_2 < \ldots$ going to infinity so that with probability 1, for *every* p, $1 < p \le 2$, we have the conclusion of part a).

c) Let $\sigma_n = \log \log \log n$, and let $1 < p \le 2$. What is the slowest sequence $t_1 < t_2 < \ldots$ you can find so that the conclusion of part a) holds?

d) Let $\sigma_n = \log \log \log n$. What is the slowest sequence $t_1 < t_2 < \ldots$ you can find so that the conclusion of part a) holds for every p, $1 < p \le 2$?

2. The oscillation inequality

In this paragraph we introduce a new tool to prove almost everywhere convergence. This tool is called the oscillation inequality. Given a sequence (M_n) of $L^2 \to L^2$ operators, the oscillation inequality proves that for $f \in L^2$ the sequence $(M_n f(x))$ is a Cauchy sequence for almost every x. In this section we shall prove the oscillation inequality only in the context of the usual ergodic averages $M_t(\mathbb{Z}_+ f)$, giving a good opportunity for the reader to become familiar with this technique in a rather simple case.

Recall from Example 3.2 that in Birkhoff's ergodic theorem it was not at all hard to deduce a.e. convergence once we had the maximal inequality. In retrospect, the crucial observation was that the limit — if it existed at all — had to be τ-invariant. This gave us a dense (in L^1) set of functions for which the a.e. convergence obviously occurs; thus, having the maximal inequality already established, we just needed to check a.e. convergence for τ-invariant functions, and for functions f that can be written in the form $f = g - T_\tau g$ with some bounded g. We observed in Examples 1.12 and 1.15, respectively, that the sequences of squares and primes are not ergodic for periodic systems, hence the ergodic averages along these sequences do not, in general, converge to a τ-invariant function. As a consequence, it does not seem easier to prove convergence for a function of the form $f = g - T_\tau g$ than for an arbitrary function $f \in L^2$. The a.e. convergence of the ergodic aver! ages along the sequence of square

As we said earlier, here we will introduce the oscillation inequality only in the context of the usual ergodic averages $M_t(\mathbb{Z}_+, f)$. Unless we say otherwise, for the rest of this paragraph we use the simplified notation $M_t f = M_t(\mathbb{Z}_+, f)$. This paragraph is devoted to the proof of the following result:

Theorem 4.10. *For any $\rho > 1$ there is a constant $C = C(\rho)$ so that for any sequence $t_1 \le t_2 \le \ldots$ of positive numbers in any dynamical system (X, Σ, m, τ) we have the inequality*

$$\sum_n \left\| \sup_{\substack{t_n \le t \le t_{n+1} \\ t \in I_\rho}} \left| M_t f(x) - M_{t_{n+1}} f(x) \right| \right\|_{L^2}^2 \le C \cdot \|f\|_{L^2}^2 , \qquad (4.7)$$

where $I_\rho = \{t \mid t = \rho^k \ for \ some \ positive \ integer \ k\}$.

How does the inequality in (4.7) imply a.e. convergence of $M_t f(x)$? By Lemma 1.5, it is enough to prove that for each fixed $\rho > 1$ the averages $M_t f(x)$ converge a.e. as t goes to infinity through the elements of I_ρ. So to see that the inequality of the theorem implies a.e. convergence of $M_t f(x)$, the reader should work out the following exercise.

Exercise 4.11. Let (f_k) be a sequence of functions in L^2. Suppose that for each sequence $k_1 \le k_2 \le \ldots$ of positive integers we have

$$\sum_{n \le N} \left\| \sup_{k_n \le k \le k_{n+1}} \left| f_k(x) - f_{k_{n+1}}(x) \right| \right\|_{L^2}^2 \le C(N) ,$$

where the function $C(N)$ satisfies

$$\lim_{N \to \infty} \frac{C(N)}{N} = 0 .$$

Then the sequence $(f_k(x))$ converges for almost every x.

Before we turn to the proof of the theorem, we note that the theorem *implies* the strong ergodic maximal inequality in L^2 (although not with the constant of Proposition 4.2 on the right hand-side). But after learning the proof, this fact will not be so exciting, for the proof *uses* the strong L^2 maximal inequality. Let us also mention the following open problem

Problem 4.12. *Can we take the constant C in (4.7) independent of $\rho > 1$?*

Proof of Theorem 10. First of all, we just have to prove the corresponding inequality on $\mathbb{Z}$: for any $f : \mathbb{Z} \to \mathbb{C}$ of finite support, we have

$$\sum_n \left\| \sup_{\substack{t_n \le t \le t_{n+1} \\ t \in I_\rho}} \left| M_t f(x) - M_{t_{n+1}} f(x) \right| \right\|_{\ell^2}^2 \le C \cdot \|f\|_{\ell^2}^2 . \qquad (4.8)$$

Exercise 4.13. Prove — by establishing an apropriate version of the transference principle — that the inequality in (4.8) implies the one in (4.7) for any measure-preserving system.

Not surprisingly, we will achieve the inequality in (4.8) using Fourier transform estimates — in addition to the strong maximal inequality in ℓ^2. Though one could give a proof directly analyzing the Fourier transform of M_t, we first replace the operator M_t by another operator V_t, the Fourier transform of which is particularly suitable to prove the oscillation inequality (4.8). In fact, we will define the operator V_t via its Fourier transform.

Let us consider the Fourier transform $\widehat{M_t}$ of M_t; it is

$$\widehat{M_t}(\beta) = \frac{1}{t} \sum_{n \leq t} e(-n\beta) .$$

Since we can take any interval of length one as the range for β, here we take $\beta \in [-1/2, 1/2)$ because it will prove to be convenient. Noting that $e(n\beta)$, $n \leq t$, is a geometric progression, we get the estimate

$$\left| \frac{1}{t} \sum_{n \leq t} e(n\beta) \right| \leq \frac{2}{t \cdot |1 - e(\beta)|} .$$

Elementary calculus shows that $|1 - e(\beta)| \geq C|\beta|$; and so we have the estimate

$$|\widehat{M_t}(\beta)| \leq \frac{C}{t \cdot |\beta|} , \tag{4.9}$$

which is effective when $|\beta| \geq \frac{1}{t}$. For β close to 0, we have the estimate

$$|1 - \widehat{M_t}(\beta)| \leq C \cdot t \cdot |\beta| . \tag{4.10}$$

This is because

$$|1 - \widehat{M_t}(\beta)| \leq \frac{1}{t} \sum_{n \leq t} |1 - e(n\beta)| ,$$

and $|1 - e(\alpha)| \leq C|\alpha|$ for every α. Roughly speaking, the inequalities in (4.9) and (4.10) show that $\widehat{M_t}(\beta)$ is small if $|\beta| \geq 1/t$, and it is close to 1 if $|\beta| \leq 1/t$. Let now V_t be the $\ell^2 \to \ell^2$ operator the Fourier transform of which is the characteristic function of the interval $(-1/t, 1/t)$, that is

$$\widehat{V_t}(\beta) = 1_{(-1/t, 1/t)}(\beta) .$$

Defining V_t by its Fourier transform means that we consider the operator V_t which for every f on $\mathbb{Z}$ of finite support satisfies

$$V_t f(x) = \int_{-1/2}^{1/2} 1_{(-1/t, 1/t)}(\beta) \widehat{f}(\beta) e(x\beta) d\beta .$$

That V_t extends to an $\ell^2 \to \ell^2$-operator is an exercise for the reader.

Now, it is enough to prove the inequality in (4.8) with M_t replaced by V_t, because we have the estimate

$$\sum_{t \in I_\rho} \|M_t f(x) - V_t f(x)\|_{\ell^2}^2 \le C \cdot \|f\|_{\ell^2}^2 , \tag{4.11}$$

where the constant C depends on ρ only. Postponing the proof of this, let us see how we prove the theorem. The main step is the oscillation inequality for the V_t's: with the assumptions of the theorem, we have

$$\sum_n \left\| \sup_{\substack{t_n \le t \le t_{n+1} \\ t \in I_\rho}} |V_t f(x) - V_{t_{n+1}} f(x)| \right\|_{\ell^2}^2 \le C \cdot \|f\|_{\ell^2}^2 , \tag{4.12}$$

with C depending on ρ only.

Exercise 4.14. Show that the inequalities in (4.12) and (4.11) together imply the one in (4.8).

To prove (4.12), we need another consequence of the estimate in (4.11). Namely, we have the maximal inequality for the V_t's when the supremum is taken over the set I_ρ; so we have with some $C = C(\rho)$,

$$\left\| \sup_{t \in I_\rho} |V_t f(x) f(x)| \right\|_{\ell^2} \le C \cdot \|f\|_{\ell^2} . \tag{4.13}$$

Exercise 4.15. Show that the inequality in (4.12) and the (strong L^2) maximal inequality for the operators M_t imply (4.13).[4]

The property of the V_t's that makes it easy to prove the oscillation inequality for them is that for $t \ge s > 0$ we have

$$V_t V_s = V_s V_t = V_t . \tag{4.14}$$

Indeed, for f of finite support we have

$$V_t(V_s f)(x) = \mathcal{F}^{-1}(1_{(-1/t,1/t)} \cdot 1_{(-1/s,1/s)} \cdot \widehat{f})(x) = \mathcal{F}^{-1}(1_{(-1/t,1/t)} \cdot \widehat{f})(x)$$
$$= V_t f(x) ,$$

---

[4]Although it is not relevant to our purpose here, one may ask if the constant C of (4.13) depends on $\rho > 1$, or — the same thing — whether we have the ℓ^2 maximal inequality for the V_t's with no restriction on the range of t. The answer happens to be that one *can* take C independent of ρ — but this is a deep fact; it is equivalent with the theorem of Carleson on the a.e. convergence of the partial sums of the Fourier series of an L^2 function. So we see that the operators V_t and M_t are very different, since a maximal inequality for the M_t's when t is restricted to the set, say, I_2 *does* imply the maximal inequality for the whole sequence (M_t).

and this computation[5] also shows that V_t and V_s commute. Because of (4.14) we have, for $t_n \le t \le t_{n+1}$, that $V_t - V_{t_{n+1}} = V_t(V_{t_n} - V_{t_{n+1}})$. Hence we can estimate

$$\sup_{\substack{t_n \le t \le t_{n+1} \\ t \in I_\rho}} \left| V_t f(x) - V_{t_{n+1}} f(x) \right| \le \sup_{t \in I_\rho} \left| V_t \left((V_{t_n} - V_{t_{n+1}})f \right)(x) \right| .$$

Using the maximal inequality in (4.13) with $f = (V_{t_n} - V_{t_{n+1}})f$, we get

$$\left\| \sup_{\substack{t_n \le t \le t_{n+1} \\ t \in I_\rho}} \left| V_t f(x) - V_{t_{n+1}} f(x) \right| \right\|_{\ell^2} \le C \cdot \left\| (V_{t_n} - V_{t_{n+1}})f \right\|_{\ell^2} .$$

As a consequence, we have

$$\sum_n \left\| \sup_{\substack{t_n \le t \le t_{n+1} \\ t \in I_\rho}} \left| V_t f(x) - V_{t_{n+1}} f(x) \right| \right\|_{\ell^2}^2 \le C \cdot \sum_n \left\| V_{t_n} f(x) - V_{t_{n+1}} f(x) \right\|_{\ell^2}^2 .$$

Using Parseval's formula, we get

$$\sum_n \left\| V_{t_n} f(x) - V_{t_{n+1}} f(x) \right\|_{\ell^2}^2 = \sum_n \left\| (\widehat{V}_{t_n} - \widehat{V}_{t_{n+1}}) \cdot \widehat{f} \right\|_{L^2(-1/2,1/2)}^2 .$$

We estimate

$$\sum_n \left\| (\widehat{V}_{t_n} - \widehat{V}_{t_{n+1}}) \cdot \widehat{f} \right\|_{L^2(-1/2,1/2)}^2 =$$

$$= \int_{-1/2}^{1/2} \left(\sum_n |\widehat{V}_{t_n}(\beta) - \widehat{V}_{t_{n+1}}(\beta)|^2 \cdot |\widehat{f}(\beta)|^2 \right) d\beta \le$$

$$\le \left(\sup_\beta \sum_n |\widehat{V}_{t_n}(\beta) - \widehat{V}_{t_{n+1}}(\beta)|^2 \right) \cdot \left\| \widehat{f} \right\|_{L^2(-1/2,1/2)}^2 =$$

$$= \left(\sup_\beta \sum_n 1_{(-1/t_n, -1/t_{n+1})}(\beta) + 1_{(1/t_{n+1}, 1/t_n)}(\beta) \right) \cdot \left\| f \right\|_{\ell^2}^2 ,$$

using Parseval's formula again at the end. The proof of (4.12) is finished, since

$$\sum_n 1_{(-1/t_n, -1/t_{n+1})} + 1_{(1/t_{n+1}, 1/t_n)} \le 1_{(-1/t_1, 1/t_1)} ,$$

---

[5] At this point the suspicious reader should consult Exercises 1 and 2 at the end of this paragraph.

and without any harm we can assume that $t_1 \geq 2$.

To prove the theorem we need to prove (4.11), but before we get to it let us extract an estimate from the above computations. Namely, if K_n is a sequence of $\ell^2 \to \ell^2$ operators, then we have the estimate

$$\sum_n \|K_n f(x)\|_{\ell^2}^2 \leq \left(\sup_\beta \sum_n |\widehat{K}_n(\beta)|^2 \right) \cdot \|f\|_{\ell^2}^2 . \qquad (4.15)$$

The reader will notice that in earlier sections we got similar estimates using the spectral theorem, and indeed (4.15) is a special case of those more general inequalities.

By the estimate in (4.15), we get (4.11) if we prove that

$$\left(\sup_\beta \sum_{t \in I_\rho} |\widehat{M}_t(\beta) - 1_{(-1/t,1/t)}(\beta)|^2 \right) < \infty . \qquad (4.16)$$

(Recall that the Fourier transform of V_t is $1_{(-1/t,1/t)}$.) Let us fix $\beta \in (-1/2, 1/2)$. We can assume that $\beta \neq 0$. Let us write

$$\sum_{t \in I_\rho} |\widehat{M}_t(\beta) - 1_{(-1/t,1/t)}(\beta)|^2 = \sum_{\substack{1/t > |\beta| \\ t \in I_\rho}} + \sum_{\substack{1/t \leq |\beta| \\ t \in I_\rho}} .$$

Let K be a positive integer so that $\rho^{-(K+1)} \leq |\beta| < \rho^{-K}$. To estimate the first sum above, we use the estimate of (4.10); we obtain

$$\sum_{\substack{1/t > |\beta| \\ t \in I_\rho}} |\widehat{M}_t(\beta) - 1_{(-1/t,1/t)}(\beta)|^2 \leq$$

$$\sum_{\substack{1/t > |\beta| \\ t \in I_\rho}} |\widehat{M}_t(\beta) - 1|^2 \leq C \cdot \sum_{\substack{1/t > |\beta| \\ t \in I_\rho}} (t \cdot |\beta|)^2 \leq$$

$$\leq \sum_{k \leq K+1} \left(\rho^k \cdot \rho^{-K} \right)^2 \leq C(\rho) .$$

To estimate the second sum, we use the estimate in (4.9); we get

$$\sum_{\substack{1/t \leq |\beta| \\ t \in I_\rho}} |\widehat{M}_t(\beta) - 1_{(-1/t,1/t)}(\beta)|^2 =$$

$$= \sum_{\substack{1/t \leq |\beta| \\ t \in I_\rho}} |\widehat{M}_t(\beta)|^2 \leq C \cdot \sum_{\substack{1/t \leq |\beta| \\ t \in I_\rho}} \left(\frac{1}{t \cdot |\beta|} \right)^2 \leq$$

$$\leq \sum_{k \geq K} \left(\frac{1}{\rho^k \cdot \rho^{-(K+1)}} \right)^2 \leq C(\rho) ,$$

and we are done. □

Exercises for Section 2

In the exercises below we are using the notation of the paragraph.

Ex.1. The computations that led to the formulas in (4.14) used that, for f of finite support and for every $x \in \mathbb{Z}$,

$$V_t(V_s f)(x) = \mathcal{F}^{-1}(1_{(-1/t,1/t)} \cdot 1_{(-1/s,1/s)} \cdot \widehat{f})(x).$$

Prove this.

Ex.2. For each positive t, the operator V_t is a projection of ℓ^2. Identify the subspace of ℓ^2 onto which V_t projects.

Ex.3. Assume the maximal inequality

$$\left\| \sup_{t>0} |V_t f(x)| \right\|_{\ell^2} \le C \cdot \|f\|_{\ell^2} . \tag{4.17}$$

(As we mentioned in the footnote to Excercise 4.15, this maximal inequality is equivalent with the result of Carleson on the a.e. convergence of the Fourier series of an $L^2[0,1)$ function.) Prove that with the constant C of (4.17), for *any* sequence $t_1 \le t_2 \le \ldots$ of real numbers we have the oscillation inequality

$$\sum_n \left\| \sup_{t_n \le t \le t_{n+1}} |V_t f(x) - V_{t_{n+1}} f(x)| \right\|_{\ell^2}^2 \le C \cdot \|f\|_{\ell^2}^2 .$$

Ex.4. Prove that there is a constant C so that for any sequence $t_1 \le t_2 \le \ldots$ of real numbers we have

$$\sum_n \left\| M_{t_n} f(x) - M_{t_{n+1}} f(x) \right\|_{\ell^2}^2 \le C \cdot \|f\|_{\ell^2}^2 .$$

(Hint: Do *not* try to prove this by somehow using the previous exercise. Instead, try this. We can assume that $t_{n+1} - t_n \ge 1$ for every n. By (4.15), we need to show

$$\sup_\beta \sum_n |\widehat{M_{t_n}}(\beta) - \widehat{M_{t_{n+1}}}(\beta)|^2 < \infty .$$

For a fixed $\beta \in [-1/2, 1/2)$, divide the sum above into two sub-sums according to $|(t_n - t_{n+1})\beta| \ge 1$ or $|(t_n - t_{n+1})\beta| < 1$.)

Ex.5. a) Prove that there is a constant C so that for any sequence $t_1 \le t_2 \le \ldots$ of real numbers satisfying $t_{n+1} > (t_n)^{10}$ we have

$$\sum_n \left\| \sup_{t_n \le t \le t_{n+1}} |M_t f(x) - M_{t_{n+1}} f(x)| \right\|_{\ell^2}^2 \le C \cdot \|f\|_{\ell^2}^2 . \tag{4.18}$$

b) Prove that for any $\sigma > 1$ there is $C = C(\sigma)$ so that for any sequence $t_1 \le t_2 \le \ldots$ of real numbers satisfying $t_{n+1} > (t_n)^\sigma$ we have (4.18).

3. The sequence of squares, I: The maximal inequality

In this and the following paragraph we will show that the sequence of squares is universally good for L^2. In other words, we want to show that in any dynamical system (X, Σ, m, τ) the averages

$$\frac{1}{\sqrt{t}} \sum_{n^2 \leq t} f(\tau^{n^2} x) \tag{4.19}$$

converge a.e. for $f \in L^2$.

The proof is divided into two major parts.

Here we will prove the strong maximal inequality for L^2, and in the next paragraph we will show almost everywhere convergence via an oscillation inequality.

The reader will see that it is not necessary to separate these two proofs. In fact, a certain inequality (inequality (4.27)) proved in this section is enough to prove the oscillation inequality, and hence convergence. But we feel that it is easier to understand the method in the context of the maximal inequality.

To avoid certain technical difficulties we consider a weighted version of the averages in (4.19); we set

$$M_t f(x) = \frac{1}{t} \sum_{n^2 \leq t} (2n - 1) f(\tau^{n^2} x).$$

It is not accidental that $(2n - 1)$ is the number of integers between $(n-1)^2$ and n^2.

Exercise 4.16. Show that the a.e. convergence of the averages in (4.19) is equivalent with the a.e. convergence of $M_t f(x)$.

Our task is to prove that for some constant C

$$\left\| \sup_t |M_t f(x)| \right\|_{L^2} \leq C \cdot \|f\|_{L^2}.$$

By an apropriate version of the transference principle and a familiar "subsequence argument", it is enough to prove that there is a constant C such that for every $f : \mathbb{Z} \to \mathbb{C}$ of finite support we have

$$\left\| \sup_{t \in I} |M_t f(x)| \right\|_{\ell^2} \leq C \cdot \|f\|_{\ell^2}, \tag{4.20}$$

where $I = I_2 = \{t \mid t = 2^k \text{ for some } k \in \mathbb{Z}_+.\}$. Similarly to the previous paragraph, we will prove (4.20) with M_t replaced by another $\ell^2 \to \ell^2$ operator, R_t, the Fourier transform of which is close to that of M_t and easier to handle. So for this operator R_t we will prove

$$\left\| \sup_{t \in I} |R_t f(x)| \right\|_{\ell^2} \leq C \cdot \|f\|_{\ell^2}. \tag{4.21}$$

and

$$\sup_{\beta} \sum_{t \in I} |\widehat{M}_t(\beta) - \widehat{R}_t(\beta)|^2 < \infty. \tag{4.22}$$

Exercise 4.17. Show that (4.21) and (4.22) imply (4.20).

In fact, first we will have an intermediate operator $S_t : \ell^2 \to \ell^2$ which satisfies

$$\sup_{\beta} \sum_{t \in I} |\widehat{S}_t(\beta) - \widehat{R}_t(\beta)|^2 < \infty$$

and

$$\sup_{\beta} |\widehat{M}_t(\beta) - \widehat{S}_t(\beta)| \leq C \cdot \frac{\sqrt{\log t}}{t^{1/6}}.$$

By now it is clear that these two inequalities imply the one in (4.22).

The operator S_t is defined by its Fourier transform, as it is suggested by the analysis of $\widehat{M}_t$. To analyze $\widehat{M}_t$, we will use the so-called circle method[6] of Hardy and Littlewood. We already used the elements of the circle method in Example 2.11, where we proved that the sequence of squares is good for mean convergence. To describe $\widehat{M}_t$ according to the circle method, one divides the interval $[0, 1)$ into two sets. The main part of $\widehat{M}_t$ is carried on the first set, which is the union of the so-called *major intervals*. The major intervals consist of those numbers which are close to a rational number with small (compared to t) denominator. On these intervals we will have an explicit description of $\widehat{M}_t$. The second set is the complement of the major intervals; this is the union of the so-called *minor intervals*. On the second set, $\widehat{M}_t$ is small. Let us explain this in more detail. Let us denote, as in Example 2.11, $P(t) = t^{1/3}$ and $Q(t) = 2t/P(t) = 2t^{2/3}$.

First set (the major intervals): As we said, this is the set of β's which are close to rational numbers with small denominator. Precisely, this means that β is in a major interval if and only if either

$$0 \leq \beta < \frac{1}{Q(t)}$$

or there is a rational number b/q with

$$1 \leq b \leq q \leq P(t), \quad \gcd(b, q) = 1, \quad \text{and} \quad |\beta - b/q| < \frac{1}{qQ(t)}.$$

---

[6]It is called "circle" method because Hardy and Littlewood carried out their analysis on the unit circle.

Exercise 4.18. Show that if b/q and c/r are two distinct rational numbers in reduced terms, and $1 \leq q, r \leq P(t)$, then the intervals $(b/q - 1/Q(t), b/q + 1/Q(t))$ and $(c/r - 1/Q(t), c/r + 1/Q(t))$ are disjoint.

Hence this first set is a union of disjoint intervals of length $2/Q(t)$, except the ones at 0 and 1; they are of length $1/Q(t)$. If β is in a major interval, then we shall have the following estimate;

$$\left| \widehat{M_t}(\beta) - \widehat{\Lambda}(b/q) \cdot v_t(\beta - b/q) \right| \leq C \cdot \frac{1}{t^{1/6}}, \qquad (4.23)$$

where

$$\widehat{\Lambda}(b/q) = \frac{1}{q} \sum_{n < q} e(n^2 b/q)$$

and v_t is (up to conjugation) the Fourier transform of the usual ergodic averages,

$$v_t(\alpha) = \frac{1}{t} \sum_{n \leq t} e(n\alpha).$$

So we see that in the neighborhood of b/q the Fourier transform of M_t behaves like the Fourier transform of the usual ergodic averages around 0 tempered by the "multiplier" $\widehat{\Lambda}(b/q)$. It may be worthwhile to point out here that if we had sticked to the unweighted averages of (4.19) instead of M_t, then we would have had an estimate like the one in (4.23), but instead of v_t we would have had $\frac{1}{\sqrt{t}} \sum_{n \leq t} \frac{1}{\sqrt{n}} \cdot e(n\alpha)$, which is not so pleasant to deal with, because of countless application of summation by parts.

Second set (the minor intervals): This set consists of all β which do not belong to the first set. By Dirichlet's theorem, Lemma 2.13 (taking $Y = Q(t)$), if β is not in a major interval then there is a rational approximation b/q to β with

$$1 \leq b \leq q, \quad \gcd(b,q) = 1, \quad P(t) < q \leq Q(t), \quad \text{and} \quad |\beta - b/q| < \frac{1}{q \cdot Q(t)}.$$

If β is in a minor interval, then $\widehat{M_t}(\beta)$ is small; we will have the estimate

$$\left| \widehat{M_t}(\beta) \right| \leq C \cdot \frac{\sqrt{\log t}}{t^{1/6}}. \qquad (4.24)$$

We discusss the proofs of the estimates in (4.23) and (4.24) in Exercise 4.21 below. We shall see there that the argument for (4.23) is similar to that for (2.25), and (2.24) is a straightforward consequence of (2.29).

Let us see now how the proof of the maximal inequality (4.20) goes. Encouraged by the estimates (4.23) and (4.24), we define the operator S_t by its Fourier transform as follows. Let B_t be the collection of those rational numbers that are centers of major intervals; $B_t = \{b/q \mid 1 \le b \le q \le P(t), \ \gcd(b,q) = 1\}$. Define now

$$\widehat{S}_t(\beta) = \sum_{b/q \in B_t} \widehat{\Lambda}(b/q) \cdot v_t(\beta - b/q) \cdot \eta_{t,q}(\beta - b/q),$$

where $\eta_{t,q}$ is the indicator function of the interval $(-1/qQ(t), 1/qQ(t))$. Note that the major interval at 0 is taken care of with the term corresponding to $b/q = 1$. By (4.23) and (4.24), we have the estimate

$$\sup_\beta |\widehat{M}_t(\beta) - \widehat{S}_t(\beta)| \le C \cdot \frac{\sqrt{\log t}}{t^{1/6}}.$$

We now define the operator R_t by simply replacing $v_t \cdot \eta_{q,t}$ by $1_{(-1/t,1/t)}$ in the definition of S_t:

$$\widehat{R}_t(\beta) = \sum_{b/q \in B_t} \widehat{\Lambda}(b/q) \cdot 1_{(-1/t,1/t)}(\beta - b/q). \qquad (4.25)$$

Let us show that the operators S_t and R_t are close to each other, that is,

$$\sup_\beta \sum_{t \in I} |\widehat{S}_t(\beta) - \widehat{R}_t(\beta)|^2 < \infty.$$

In other words, we need to prove that

$$\sup_\beta \sum_{t \in I} \sum_{b/q \in B_t} \left|\widehat{\Lambda}(b/q) \cdot \left(v_t \cdot \eta_{t,q} - 1_{(-1/t,1/t)}\right)(\beta - b/q)\right|^2 < \infty.$$

Let us fix β. Since for each t there is at most one $b/q \in B_t$, say b_t/q_t, for which $\left(v_t \cdot \eta_{t,q} - 1_{(-1/t,1/t)}\right)(\beta - b/q)$ is nonzero, we need to find an estimate for

$$\sum_{t \in I} \left|\widehat{\Lambda}(b_t/q_t) \cdot \left(v_t \cdot \eta_{t,q_t} - 1_{(-1/t,1/t)}\right)(\beta - b_t/q_t)\right|^2$$

which is independent of β. Denote $B = B(\beta) = \{b_t/q_t \mid t \in I\}$, and estimate the above sum from above by

$$\sum_{b/q \in B} \left|\widehat{\Lambda}(b/q)\right|^2 \sum_{t \in I} \left|\left(v_t \cdot \eta_{t,q} - 1_{(-1/t,1/t)}\right)(\beta - b/q)\right|^2.$$

An argument almost identical to the one used to prove (4.16) shows that

$$\sum_{t \in I} \left|\left(v_t \cdot \eta_{t,q} - 1_{(-1/t,1/t)}\right)(\beta - b/q)\right|^2 \le C,$$

with an absolute constant C. It remains to prove that $\sum_{b/q \in B} \left| \widehat{\Lambda}(b/q) \right|^2$ is bounded independently of β. We can bound this sum by

$$\left| \widehat{\Lambda}(1) \right|^2 + \sum_{j=1}^{\infty} \sup_{b/q \in D_j} \left| \widehat{\Lambda}(b/q) \right|^2,$$

where $D_j = \{b/q \mid b/q \in B_{2^j}, \quad q > P(2^{j-1})\}$. We now just remark that by Proposition 1.14 we have

$$\sup_{b/q \in D_j} \left| \widehat{\Lambda}(b/q) \right|^2 \leq C \cdot \frac{1}{P(2^{j-1})} \leq C \cdot \frac{1}{2^{j/3}},$$

which finishes our estimation of (4.26).

As a consequence of the above, we have the inequality in (4.22), and, by Exercise 4.17, we are left to prove (4.21). To do this we make a minor adjustment in the definition of $P(t)$; namely we assume that it is a dyadic number, so $P(t) = 2^p$, where p is the greatest integer not exceeding $\log_2(t^{1/3})$. It is clear that this adjustment does not affect the essence of what we said before. Now, in proving (4.21) the size of $\widehat{\Lambda}(b/q)$ (which is intimately connected with the distribution of the squares in mod q residue classes) plays a crucial role. We want to group together those b/q for which $\widehat{\Lambda}(b/q)$ is of the same size. In Proposition 1.14 we proved that $|\widehat{\Lambda}(b/q)| \leq C/\sqrt{q}$, and so if we group together those b/q for which q is between the same two consecutive dyadic numbers, then the size of $\widehat{\Lambda}(b/q)$ does not change significantly. For each nonegative integer p, let us define

$$E_p = \{b/q \mid 1 \leq b \leq q, \quad \gcd(b,q) = 1, \quad 2^p \leq q < 2^{p+1}\}.$$

With this, we can write

$$\widehat{R}_t(\beta) = \sum_{p \leq p(t)} \sum_{b/q \in E_p} \widehat{\Lambda}(b/q) \cdot 1_{(-1/t,1/t)}(\beta - b/q),$$

where the integer $p(t)$ is defined by $2^{p(t)} = P(t)$, and it is the greatest integer not exceeding $\log_2(t^{1/3})$. For each p and t, let us define the operator $R_{p,t}$ by its Fourier transform as

$$\widehat{R}_{p,t}(\beta) = \sum_{b/q \in E_p} \widehat{\Lambda}(b/q) \cdot 1_{(-1/t,1/t)}(\beta - b/q).$$

Then, of course, we can write

$$R_t = \sum_{p \leq p(t)} R_{p,t}.$$

Noting that the b/q's in E_p will be centers of major arcs when $p \leq p(t)$, that is, when — basically — $t \geq 2^{3p}$, we can estimate

$$\sup_{t \in I} |R_t f(x)| \leq \sum_p \sup_{\substack{t \geq 2^{3p} \\ t \in I}} |R_{p,t} f(x)|,$$

and hence, by the triangle inequality,

$$\left\| \sup_{t \in I} |R_t f(x)| \right\|_{\ell^2} \leq \sum_p \left\| \sup_{\substack{t \geq 2^{3p} \\ t \in I}} |R_{p,t} f(x)| \right\|_{\ell^2}.$$

The proof of the maximal inequality (4.21) will be complete if we show

$$\left\| \sup_{\substack{t \geq 2^{3p} \\ t \in I}} |R_{p,t} f(x)| \right\|_{\ell^2} \leq C \cdot \frac{p^3}{\sqrt{2^p}} \cdot \|f\|_{\ell^2}. \tag{4.26}$$

This estimate is a consequence of the following fundamental result of Bourgain:

Lemma 4.19. *Let* $\beta_1, \ldots, \beta_S$ *be distinct points in the interval* $[0, 1)$, *and let* δ *be the smallest of the numbers* $|\beta_i - \beta_j|$, $i \neq j$. *Define the* $\ell^2 \to \ell^2$ *operators* V_t *by their Fourier transforms as follows:*

$$\widehat{V_t} = \sum_{s=1}^{S} 1_{(-1/t, 1/t)} (\beta - \beta_s).$$

Then we have the following maximal inequality

$$\left\| \sup_{\substack{t \geq \delta^{-1} \\ t \in I}} |V_t f(x)| \right\|_{\ell^2} \leq C \cdot (\log S)^3 \|f\|_{\ell^2}.$$

While this lemma is the single most important result of our notes, we are not going to prove it, because it would require more background than we assume here. For the proof we refer the reader to the expository article [Thouvenot, 1990], or to the original paper [Bourgain, 1989].

Here is how we shall use the lemma above to prove (4.26). With p fixed, let us define the operator U_t by its Fourier transform as

$$\widehat{U_t}(\beta) = \sum_{b/q \in E_p} 1_{(-1/t, 1/t)} (\beta - b/q).$$

So the Fourier transform of U_t is just the Fourier transform of $R_{p,t}$ without the multipliers $\widehat{\Lambda}(b/q)$. We have the following inequality as a consequence of Lemma 4.19: there is a constant C so that for any $g \in \ell^2$

$$\left\| \sup_{\substack{t \geq 2^{2p} \\ t \in I}} |U_t g(x)| \right\|_{\ell^2} \leq C \cdot p^3 \|g\|_{\ell^2}. \tag{4.27}$$

Exercise 4.20. Show that in fact (4.27) is a consequence of Lemma 4.19. (Hint: use the lemma with $\{\beta_1, \ldots, \beta_S\} = E_p$. Then $\delta^{-1} \leq \#E_p \leq 2^{2p}$.)

The idea to prove (4.26) using (4.27) is to move the multipliers $\widehat{\Lambda}(b/q)$ from the operator $R_{p,t}$ into the function f. Let us denote by η_p the indicator function of the interval $(-2^{-2p-1}, 2^{-2p-1})$. Let us define $g \in \ell^2$ by its Fourier transform as

$$\widehat{g}(\beta) = \sum_{b/q \in E_p} \widehat{\Lambda}(b/q) \cdot \eta_p(\beta - b/q) \cdot \widehat{f}(\beta).$$

Note that the supports of the functions $\eta_p(\beta - b/q)$, $b/q \in E_p$, are disjoint. Since $t \geq 2^{3p}$, we have that $1_{(-1/t,1/t)} \cdot \eta_p = 1_{(-1/t,1/t)}$. It follows that for each $t \geq 2^{3p}$,

$$\widehat{R}_{p,t} \cdot \widehat{f} = \widehat{U}_t \cdot \widehat{g}.$$

As a consequence, we have $R_{p,t} f = U_t g$, hence

$$\left\| \sup_{\substack{t \geq 2^{3p} \\ t \in I}} |R_{p,t} f(x)| \right\|_{\ell^2} = \left\| \sup_{\substack{t \geq 2^{3p} \\ t \in I}} |U_t g(x)| \right\|_{\ell^2}.$$

Using (4.27), we get

$$\left\| \sup_{\substack{t \geq 2^{3p} \\ t \in I}} |R_{p,t} f(x)| \right\|_{\ell^2} \leq C \cdot p^3 \, \|g\|_{\ell^2}.$$

The proof of (4.26) is complete if we show

$$\|g\|_{\ell^2} \leq C \cdot \frac{1}{\sqrt{2^p}} \cdot \|f\|_{\ell^2}.$$

This is because, starting with Parseval's formula,

$$\|g\|_{\ell^2}^2 = \|\widehat{g}\|_{L^2}^2 =$$

$$= \int_{[0,1)} \left| \sum_{b/q \in E_p} \widehat{\Lambda}(b/q) \cdot \eta_p(\beta - b/q) \cdot \widehat{f}(\beta) \right|^2 d\beta =$$

(since the supports of $\eta_p(\beta - b/q)$, $b/q \in E_p$, are disjoint)

$$= \int_{[0,1)} \sum_{b/q \in E_p} \left| \widehat{\Lambda}(b/q) \cdot \eta_p(\beta - b/q) \cdot \widehat{f}(\beta) \right|^2 d\beta \leq$$

$$\leq \max_{b/q \in E_p} \left| \widehat{\Lambda}(b/q) \right|^2 \cdot \int_{[0,1)} \sum_{b/q \in E_p} \left| \eta_p(\beta - b/q) \cdot \widehat{f}(\beta) \right|^2 d\beta \leq$$

(again, the supports of $\eta_p(\beta - b/q)$, $b/q \in E_p$, are disjoint)

$$\leq C \cdot \frac{1}{2^p} \cdot \left\| \widehat{f} \right\|_{L^2}^2 = C \cdot \frac{1}{2^p} \cdot \|f\|_{\ell^2}^2,$$

using Parseval's formula again at the end.

Exercise 4.21. (To prove (4.23) and (4.24).)
Throughout this exercise we use the following notation:

$$U_y(\beta) = \sum_{n^2 \le y} (2n-1) \cdot e(n^2\beta),$$

$$V_y(\beta) = \sum_{n \le y} e(n\beta),$$

$$D(y,q,h) = \sum_{\substack{n^2 \le y \\ n \equiv h \bmod q}} (2n-1).$$

a) Prove that there is a constant C so that for every y,q,h we have

$$\left| D(y,q,h) - \frac{y}{q} \right| \le C \cdot \sqrt{y}.$$

b) Prove that there is a constant C so that for every y,b,q with $\gcd(b,q) = 1$ we have

$$\left| U_y(b/q) - \widehat{\Lambda}(b/q) \cdot V_y(0) \right| \le C \cdot q \cdot \sqrt{y}.$$

(Hint: Note that $U_y(b/q) = \sum_{h<q} D(y,q,h)e(h^2 b/q)$.)
c) Prove that if $\gcd(b,q) = 1$ and $|\beta - b/q| \le 1/q \cdot Q(t)$, then

$$\left| U_t(\beta) - \widehat{\Lambda}(b/q) \cdot V_t(\beta - b/q) \right| \le C \cdot \left(t^{5/6} + q \cdot \sqrt{y} \right),$$

with an absolute constant C. This implies the estimate in (4.23), since then we have $q \le P(t) = t^{1/3}$. (Hint: Use part b) and Lemma 2.13 with $D_y = U_y(b/q) - \widehat{\Lambda}(b/q) \cdot V_y(0)$ and $\delta = \beta - b/q$.)
d) Prove that there is a constant C so that for every y,b,q with $\gcd(b,q) = 1$ we have

$$|U_y(b/q)| \le C \left(\frac{y}{\sqrt{q}} + \sqrt{y \cdot q \cdot \log y} \right).$$

(Hint: Use the estimate in (2.29) and summation by parts.)
e) Prove that if $\gcd(b,q) = 1$, $q \ge P(t) = t^{1/3}$ and $|\beta - b/q| \le 1/q \cdot Q(t)$, then

$$|U_t(b/q)| \le C \left(\frac{t}{\sqrt{q}} + \sqrt{t \cdot q \cdot \log t} \right).$$

This implies the estimate in (4.24), because then $q \le Q(t) = 2t^{2/3}$ as well. (Hint: Use part d) and Lemma 2.13 with $D_y = U_y(b/q)$ and $\delta = \beta - b/q$. Note that $|\delta| \le 2/t$.)

Exercises for Section 3

Ex.1. What happens with our proof if we take $P(t)$ to be $t^{1/2}$ or t^{100} or $(\log t)^{100}$?

Ex.2. What changes should be made in our proof of the maximal inequality (4.20) if we replace the averages $M_t f$ of the text by the unweighted averages of (4.19)?

The next three exercises serve to help the reader to prove the L^2 maximal inequality for the ergodic averages along the prime numbers. We shall use the following notations: A denotes the set of prime numbers;

$$M_t f(x) = \frac{1}{t} \sum_{\substack{a \leq t \\ a \in A}} (\log a) \cdot f(\tau^a x)\,;$$

$$D(t,q,h) = \sum_{\substack{a \leq t \\ a \in A \\ a \equiv h \bmod (q)}} \log a\,;$$

$P(t)$ denotes the largest dyadic number not exceeding $(\log t)^{10}$; $Q(t) = 2t/P(t)$. The major and minor intervals are defined with the help of $P(t)$ and $Q(t)$ as in the text.

Ex.3. **a)** Show that there are positive constants C and c so that whenever $1 \leq h \leq q \leq P(t)$, $\gcd(h,q) = 1$, we have, for every t,

$$\left| D(t,q,h) - \frac{1}{\phi(q)} \cdot t \right| \leq C \cdot \frac{t}{e^{c\sqrt{\log t}}}\,.$$

Note that in particular we have

$$\lim_{t \to \infty} \frac{\sum_{\substack{a \leq t \\ a \in A}} \log a}{t} = 1\,. \tag{4.32}$$

(Hint: Use Lemma 2.22 and summation by parts.)

b) Show that the sequence of primes is universally good if and only if in every dynamical system, for every $f \in L^2$ the averages $M_t f(x)$ converge a. e. (Hint: Use summation by parts and (4.32).)

Ex.4. **a)** Show that if β is in a major interval with center b/q, then

$$\left| \widehat{M_t}(\beta) - \frac{\mu(q)}{\phi(q)} v_t(\beta - b/q) \right| \leq C \cdot \frac{1}{e^{c\sqrt{\log t}}}\,,$$

where

$$\widehat{M_t}(\beta) = \frac{1}{t} \sum_{\substack{a \leq t \\ a \in A}} (\log a) \cdot e(a\beta)\,,$$

and v_t is as in the text:

$$v_t(\alpha) = \frac{1}{t} \sum_{n \leq t} e(n\alpha).$$

(Compare with Exercise 2.18.)

    b) Show that if β is in a minor interval, then one has the estimate

$$|\widehat{M_t}(\beta)| \leq C \cdot \frac{1}{\log t}.$$

(Compare with Exercise 2.20.)

    Ex.5. Prove that in every dynamical system we have the maximal inequality

$$\left\| \sup_t |M_t f(x)| \right\|_{L^2} \leq C \cdot \|f\|_{L^2} .$$

(Hint: Do not forget about Exercise 2 for Example 1.15.)

4. The sequence of squares, II: The oscillation inequality

In this section we finish proving that the sequence of squares is universally good for L^2 by proving an oscillation inequality. We continue with the notation

$$M_t f(x) = \frac{1}{t} \sum_{n^2 \leq t} (2n-1) f(\tau^{n^2} x) .$$

By Exercises 4.16 4.11 and by Lemma 1.5, we just need to prove

    Theorem 4.23. *For any $\rho > 1$ there is a constant $C = C_\rho$ such that for any sequence $t_1 \leq t_2 \leq \ldots$ of positive numbers, in any dynamical system (X, Σ, m, τ) we have the inequality*

$$\sum_{n=1}^{\infty} \left\| \sup_{\substack{t_n \leq t \leq t_{n+1} \\ t \in I_\rho}} |M_t f(x) - M_{t_{n+1}} f(x)| \right\|_{L^2}^2 \leq C \cdot \|f\|_{L^2}^2 ,$$

where $I_\rho = \{t \mid t = \rho^k \; for \, some \, positive \, integer \, k\}$.

    Proof. By an apropriate version of the transference principle, we need to prove the oscillation inequality on the integers: for $f : \mathbb{Z} \to \mathbb{C}$ of finite support

$$\sum_{n=1}^{\infty} \left\| \sup_{\substack{t_n \leq t \leq t_{n+1} \\ t \in I_\rho}} |M_t f(x) - M_{t_{n+1}} f(x)| \right\|_{\ell^2}^2 \leq C \cdot \|f\|_{\ell^2}^2 . \tag{4.33}$$

As we mentioned in the previous paragraph, we will not use the maximal inequality (4.20). Rather, we use a slight generalization of the inequality in (4.26) and the method used to prove the oscillation inequality (4.12), which was the main step in proving the oscillation inequality for the usual ergodic averages.

Let R_t be the $\ell^2 \to \ell^2$ operator as defined in (4.25). First the reader should solve

Exercise 4.24. Show that with some constant $C = C_\rho$, we have

$$\sum_{t \in I_\rho} \|R_t f(x) - M_t f(x)\|_{\ell^2}^2 \leq C \cdot \|f\|_{\ell^2}^2 .$$

(Hint: use the techniques used to prove (4.22).)

By this exercise, it is enough to prove the oscillation inequality for the R_t's; we will prove that with some constant $C = C_\rho$, for any sequence $t_1 \leq t_2 \leq \ldots$ of positive numbers we have

$$\sum_{n=1}^{\infty} \left\| \sup_{\substack{t_n \leq t \leq t_{n+1} \\ t \in I_\rho}} \left| R_t f(x) - R_{t_{n+1}} f(x) \right| \right\|_{\ell^2}^2 \leq C \cdot \|f\|_{\ell^2}^2 . \tag{4.34}$$

For notational convenience we make a formal modification in the definition of $R_{p,t}$. Namely, for $p \leq p(t)$ we still define $R_{p,t}$ by its Fourier transform as

$$\widehat{R_{p,t}}(\beta) = \sum_{b/q \in E_p} \widehat{\Lambda}(b/q) \cdot 1_{(-1/t,1/t)}(\beta - b/q) ,$$

where

$$E_p = \{ b/q \mid 1 \leq b \leq q, \quad \gcd(b,q) = 1, \quad 2^p \leq q < 2^{p+1} \} ,$$

but for $p > p(t)$ we define $R_{p,t} = 0$. Then we can write

$$R_t = \sum_{p=1}^{\infty} R_{p,t} .$$

To prove (4.34), first we estimate, using the triangle inequality first for the "sup" then for the norm $\sqrt{\sum_n \| \cdot \|_{\ell^2}^2}$,

$$\sqrt{\sum_{n=1}^{\infty} \left\| \sup_{\substack{t_n \leq t \leq t_{n+1} \\ t \in I_\rho}} \left| R_t f(x) - R_{t_{n+1}} f(x) \right| \right\|_{\ell^2}^2} \leq$$

$$\leq \sum_{p} \sqrt{\sum_{n=1}^{\infty} \left\| \sup_{\substack{t_n \leq t \leq t_{n+1} \\ t \in I_\rho}} \left| R_{p,t} f(x) - R_{p,t_{n+1}} f(x) \right| \right\|_{\ell^2}^2} .$$

We see we just need to prove the estimate

$$\sum_{n=1}^{\infty} \left\| \sup_{\substack{t_n \le t \le t_{n+1} \\ t \in I_p}} |R_{p,t}f(x) - R_{p,t_{n+1}}f(x)| \right\|_{\ell^2}^2 \le C \cdot \frac{p^6}{2^p} \cdot \|f\|_{\ell^2}^2 , \qquad (4.35)$$

with a constant not depending on p (and, of course, f). The same way we derived (4.26) from (4.27), we get (4.35) from the estimate

$$\sum_{n=1}^{\infty} \left\| \sup_{\substack{t_n \le t \le t_{n+1} \\ t \in I_p}} |U_t g(x) - U_{t_{n+1}}g(x)| \right\|_{\ell^2}^2 \le C \cdot p^6 \cdot \|g\|_{\ell^2}^2 , \qquad (4.36)$$

where the operator $U_t = U_{p,t}$ is defined by its Fourier transform as

$$\widehat{U_t}(\beta) = \sum_{b/q \in E_p} 1_{(-1/t,1/t)}(\beta - b/q)$$

if $p \le p(t)$ and $U_t = 0$ otherwise. To prove (4.36), we need an apropriate generalization of the inequality in (4.27). Namely, with some constant $C = C_\rho$, we have, for every p,

$$\left\| \sup_{t \in I_p} |U_{p,t}g(x)| \right\|_{\ell^2} \le C \cdot p^3 \cdot \|g\|_{\ell^2} . \qquad (4.37)$$

The proof of this is asked for in Exercise 1 below. In proving (4.36), because of (4.37) we can assume that $p \le p(t_1)$. Now, the oscillation inequality in (4.36) follows from (4.37) the same way as the maximal inequality in (4.13) implied the oscillation inequality in (4.12). Indeed, the U_t's are projections of ℓ^2, and for $t \ge s \ge t_1$ they satisfy

$$U_t U_s = U_s U_t = U_t.$$

Then, as in the proof of (4.12),

$$\sum_n \left\| \sup_{\substack{t_n \le t \le t_{n+1} \\ t \in I_p}} |U_t g(x) - U_{t_{n+1}}g(x)| \right\|_{\ell^2}^2 \le$$

$$\le C \cdot p^6 \cdot \sum_n \|U_{t_n}g(x) - U_{t_{n+1}}g(x)\|_{\ell^2}^2 .$$

The proof of (4.36) is complete by showing

$$\sum_n \|U_{t_n}g(x) - U_{t_{n+1}}g(x)\|_{\ell^2}^2 \le \|g\|_{\ell^2}^2 . \qquad (4.38)$$

Exercise 4.25. Prove the inequality in (4.38) above.

Exercises for Section 4

Ex.1. a) Let (t_n) be a sequence of numbers with $t_{n+1}/t_n \geq 2$ for each n, and denote $J = \{t_n\}$. With the assumptions and notation of Lemma 4.19, we have

$$\left\| \sup_{\substack{t \geq \delta^{-1} \\ t \in J}} |V_t f(x)| \right\|_{\ell^2} \leq C \cdot (\log S)^3 \, \|f\|_{\ell^2} \,, \tag{4.41}$$

where C depends only on the constant of the maximal inequality of Lemma 4.19. (Hint: for each n, let $p(n)$ be the largest integer p with $2^p \leq t_n$, and let $J_0 = \{2^{p(n)} \mid n = 1, 2, \dots \}$. Then we have

$$\sup_{\beta} \sum_{\substack{n \\ 2^{p(n)} \geq \delta^{-1}}} \sum_{s=1}^{S} \left| 1_{(-1/t_n, 1/t_n)}(\beta - \beta_s) - 1_{(-1/2^{p(n)}, 1/2^{p(n)})}(\beta - \beta_s) \right|^2$$

$$\leq C < \infty,$$

with an absolute constant C.)

b) Let $\rho > 1$. Let (t_n) be a lacunary sequence of numbers with $t_{n+1}/t_n \geq \rho$ for each n, and denote $J = \{t_n\}$. Then again we have the conclusion of part a), inequality (4.41), but with a constant C that may depend on $\rho > 1$ as well.

c) Prove the inequality in (4.37).

Ex.2. Show that if τ is an irrational rotation of the unit interval, that is, if for some irrational α the transformation τ is the map of the unit interval defined by $\tau(x) = \langle x + \alpha \rangle$, and the measure m is the mod 1 Lebesgue measure, then for $f \in L^2$ we have

$$\lim_{t \to \infty} \frac{1}{t} \sum_{n^2 \leq t} f(\tau^{n^2} x) = \int_0^1 f \, dm$$

for almost every x. (Hint: See Exercise 1 for Example 2.11.)

Ex.3. Let (σ_n) be a sequence of positive numbers satisfying the properties;

(i) the sequence (σ_n) is decreasing;

(ii) For each $\rho > 1$ we have $\sum_{t \in I_\rho} \frac{\sum_{n \leq t} \sigma_n}{\log n} < \infty$, where I_ρ is as defined in the text.

Let (Y_n) be a sequence of independent, (0-1)-valued random variables on the probability space (Ω, β, P) so that $P(Y_n = 1) = \sigma_n$ and $P(Y_n = 0) = 1 - \sigma_n$. Show that with probability 1 the random sequence A^ω, defined by $n^2 \in A^\omega$ if and only if $Y_n(\omega) = 1$, is universally L^2 good.

Ex.4. a) Show that if (a_n) is a universally good sequence for L^2 then for any fixed integer a the sequence $a_n + a$ is also L^2 good.

b) Let (Y_n) be a sequence of independent, identically distributed, (0-1)-valued random variables on the probability space (Ω, β, P) so that $P(Y_n = 1) = P(Y_n = 0) = 1/2$. Show that with probability 1 the random sequence $(n^2 + Y_n(\omega))$ is L^2 good. (Hint: Write

$$\frac{1}{t} \sum_{n^2 \leq t} f(\tau^{n^2 + Y_n(\omega)} x) =$$

$$= \frac{1}{t} \sum_{n^2 \leq t} X_n(\omega) \cdot f(\tau^{n^2} x) + \frac{1}{t} \sum_{n^2 \leq t} (1 - X_n)(\omega) \cdot f(\tau^{n^2 + 1} x),$$

where (X_n) is an i.i.d. sequence with $P(X_n = 1) = P(X_n = 0) = 1/2$.)

c) Let (Y_n) be a sequence of independent, identically distributed, integer-valued random variables with finite first moment. Show that with probability 1 the random sequence $(n^2 + Y_n(\omega))$ is L^∞ good.

Ex.5. a) Let A be an L^2 good sequence of integers. Suppose that for some q and h the limit $\lim_{t \to \infty} A(t, q, h)/A(t)$ exists and is positive. Show that the sequence $(a \mid a \in A, a \equiv h \mod q)$ is also good in the mean.

b) Let h be a quadratic residue mod q. Show that the sequence $(n^2 \mid n^2 \equiv h \mod q)$ is L^2 good.

Ex.6. Solve Exercises 2-5 above with the sequence of squares replaced by the sequence of primes.

Ex.7. Let k be a positive integer. Solve Exercises 2-5 above with the sequence of squares replaced by the sequence of k-th powers, (n^k). (Hint: you need estimates on the Fourier transforms of the corresponding averages. Either derive them yourselves using the ideas of Example 2.11, or look them up say in [Vaughan, 1981].)

Ex.8. Suppose that for some q and every h the sequence (a_{nq+h}) is L^2 good. Then the full sequence (a_n) is L^2 good.

Notes to Chapter IV

1. The strong ergodic maximal inequality is due to Wiener [1939].

2. The result of Example 4.7 is due to Bourgain, and our treatment is just a slight variation of his in [Bourgain, 1988].

3. Further refinements on randomly generated sequences can be found in [Huang, 1992].

4. As a result of a conversation with M. Lacey, we can point out that the condition

$$\lim_{t \to \infty} \frac{\log t}{\sum_{n \le t} \sigma_n} = 0$$

   is not enough to conclude that with probability 1 the random sequence is universally good in L^2. A condition similar to $(ii)'$ in Example 4.7 is necessary for a. e. convergence, but the exact condition is not known yet.

5. The oscillation inequality of Theorem 4.10 is from [Bourgain, 1988]. Our approach is from [Wierdl, 1989] although our presentation in terms of projections was suggested to us by M. Lacey.

6. Exercise 4 for section 2 is from [Jones, Ostrogradsky, Rosenblatt].

7. Exercise 5 for section 2 is from H. White's Master thesis at The University of North Carolina; it got published in [Assani, Petersen, White, 1991].

8. The strong L^2 maximal inequality for the averages (4.19) along the squares was first proved in [Bourgain, 1988], but our approach is based on [Bourgain, 1989]. The methods in [Bourgain, 1988] are more elementary, but the details are very complicated, since a much weaker inequality is used in place of the one in Lemma 4.19. The method of [Bourgain, 1989] uses deep results from martingale theory, but the underlying idea shows more clearly with the use of Lemma 4.19. It is interesting to note that in [Bourgain, 1988] additional complications appeared for the sequence of k-th powers when $k > 3$, while with Lemma 4.19 these extra complications disappear.

9. It was Bourgain [1988] who proved that the squares are universally good for L^2. But he proved a weaker version of our oscillation inequality in Theorem 4.23. He just proved an oscillation inequality of the form

$$\sum_{n \le N} \left\| \sup_{\substack{t_n \le t \le t_{n+1} \\ t \in I_\rho}} |M_t f(x) - M_{t_{n+1}} f(x)| \right\|_{L^2}^2 \le o(N) \cdot \|f\|_{L^2}^2 ,$$

   which, by Exercise 4.11, is enough to prove convergence. The stronger version of the oscillation inequality of Theorem 4.23 is due to M. Lacey.

10. That the sequence of primes is L^2 good was first proved by Bourgain. In [Wierdl, 1989] a general theorem is given which implies not only that the sequence of squares or primes are L^2 good, but also that sequences like

the squares of primes, $\{p^2 \mid p \quad \text{prime}\}$, are L^2 good. That the sequence of squares of primes is L^2 good was proved independently in [Nair, 1991].

11. The following sequences are L^p good for every $p > 1$:
   a) the sequence of squares [Bourgain, 1989];
   b) the sequence of primes [Wierdl, 1988];
   c) sequences like $(n^{3/2})$, $([n \log n])$, $([n^2 / \log n])$. In fact, these sequences are ergodic [Wierdl, 1989]. We note that the sequence $(n[\log n])$ is not good even for L^∞, cf. Proposition 6.14. below.

12. The sequence of integers the decimal expansion of which contains 0's and 1's only is L^∞ good [Bourgain, 1990].

13. Exercise 4 for section 4 is due to D. Schneider; it will be part of his Ph. D. thesis at the University of Strasbourg, France.

14. It is proved in [Jones, Olsen, Wierdl, 1992] that the ergodic averages along the squares remain a.e. convergent even if the operator on L^2 induced by the measure-preserving transformation τ is replaced by a positive contraction on L^2. In fact, a more general theorem is proved in [Jones, Olsen, Wierdl, 1992]: whenever the ergodic averages along a sequence satisfy some kind of oscillation inequality — hence these averages converge a. e. — we have a.e. convergence of the same averages with T replaced by an arbitrary positive contraction of L^2. In particular, all the subsequence ergodic theorems mentioned above can be generalized to the setting of positive contractions on L^2.

CHAPTER V

UNIVERSALLY BAD SEQUENCES

1. Banach principle and related matters

The Banach principle is an essential abstract feature of all almost everywhere convergence results for a sequence (T_n) of operators on a Banach space $\mathcal{B}$. This principle is that when $\sup_{n \geq 1} |T_n f(x)| < \infty$ a.e., then the set of functions $f \in \mathcal{B}$ for which $\lim_{n \to \infty} T_n f(x)$ exists a.e. forms a closed subspace in the norm topology of $\mathcal{B}$. As in [Garsia, 1970], very mild hypotheses on (T_n) are needed to guarantee that this principle holds.

Let (X, Σ, m) be a probability space. Assume m is non-atomic; this is not needed in all places, but it guarantees applications of the Rokhlin Lemma are valid. Let $L(X)$ be the linear space of m-equivalence classes of measurable functions on X which are a.e. real-valued. Suppose that $(\mathcal{B}, \varrho)$ is a metric space and $T : \mathcal{B} \to L(X)$. Then T is said to be *continuous in measure* if whenever (b_i) is a sequence in $\mathcal{B}$ and $\lim_{i \to \infty} \varrho(b_i, b) = 0$ for some $b \in \mathcal{B}$, then $\lim_{i \to \infty} m\{|Tb_i - Tb| \geq \varepsilon\} = 0$ for all $\varepsilon > 0$. The theorems in this section all deal with a sequence of maps (T_n), where $T_n : \mathcal{B} \to L(X)$ for some suitable metric space $\mathcal{B}$. For $N \geq 1$ and $b \in B$, we will use the notation $T_N^* b = \sup\{|T_n b| : 1 \leq n \leq N\}$ and $T^* b = \sup\{|T_n b| : n = 1, 2, 3, \dots\}$.

Exercise 5.1. Show that if each T_n is continuous in measure, then each T_N^* is also continuous in measure.

The theorems in this section have similar proofs, with a common prototype of Garsia's proof of Banach's principle in [Garsia, 1970].

Consider first the Banach principle in its simplest form.

Theorem 5.1. *Suppose $\mathcal{B}$ is a Banach space in the norm $\| \cdot \|$ and (T_n), $T_n : \mathcal{B} \to L(X)$, is a sequence of linear maps which are continuous in measure. Assume that $\sup_{n \geq 1} |T_n f(x)| < \infty$ a.e. for all $f \in \mathcal{B}$. Then the set of functions $f \in \mathcal{B}$ such that $\lim_{n \to \infty} T_n f(x)$ exists a.e. forms a norm-closed subspace of $\mathcal{B}$.*

Proof. We first prove that the finiteness of $\sup_{n \geq 1} |T_n f|$ a.e. forces $m\{\sup_{n \geq 1} |T_n f| \geq C\}$ to go to zero uniformly over the unit ball of $\mathcal{B}$ as $C \to \infty$.

Consider the set $E_\ell(k)$ of $b \in B$ such that $m\{\sup_{n \geq 1} |T_n b| > k\} \leq 1/\ell$. It is easy to see that $b \in E_\ell(k)$ if and only if

$$m\{ \sup_{1 \leq n \leq m} |T_n b| > k \} \leq 1/\ell \text{ for all } m \geq 1.$$

It is an easy exercise, using the linearity of (T_n) and the fact that T_m^* is continuous in measure, to see that

$$E_\ell(k, N) = \{b \in \mathcal{B} : m\{P \sup_{1 \leq n \leq N} |T_n b| > k\} \leq 1/\ell\}$$

is closed in the norm topology on $\mathcal{B}$. Hence, $E_\ell(k) = \bigcap_{N=1}^{\infty} E_\ell(k, N)$ is closed too. Now for each ℓ, by the Baire Category Theorem, there exists $k \geq 1$, $\varepsilon > 0$, and $b_0 \in \mathcal{B}$ such that for all $b \in \mathcal{B}$, $\|b - b_0\| \leq \varepsilon$, $b \in E_\ell(k)$. But then if $\|b\| \leq 1$, we can apply the inherent estimate to $\varepsilon b + b_0$, and we see that

$$\{\sup_{n \geq 1} |T_n b| > 2k/\varepsilon\} = m\{\sup_{n \geq 1} |T_n(\varepsilon b)| > 2k\}$$

$$\leq m\{\sup_{n \geq 1} |T_n(\varepsilon b + b_0)| > k\} + m\{\sup_{n \geq 1} |T_n b_0| > k\}$$

$$\leq 2/\ell.$$

With $C(\ell) = 2k/\varepsilon$, this gives a gauge $C(\ell)$ such that $m\{\sup_{n \geq 1} |T_n C| > C(\ell)\} \to 0$ as $\ell \to \infty$, uniformly for $b \in B$, $\|b\| \leq 1$.

Now we argue that the elements b for which $\lim_{n \to \infty} T_n b$ exists a.e. form a norm-closed subspace. Since T_n is linear for each n, such elements form a subspace. Suppose $b_N \to b$ in norm in $\mathcal{B}$ and $\lim_{n \to \infty} T_n b_N$ exists a.e. for all $N \geq 1$. The uniformity principle above says that if ℓ is fixed, then there exists $C(\ell)$ with

$$m\{\sup_{n \geq 1} |T_n(b - b_N)| > C(\ell)(\|b - b_N\|)\} \leq 1/\ell.$$

Hence, with $\ell = r^2$, and $N_r \geq 1$ such that for $N \geq N_r$, $C(\ell)\|b - b_N\| \leq 1/r$, we see if $N \geq N_r$, $M\{\sup_{n \geq 1} |T_n(b - b_N)| > 1/r\} \leq 1/r^2$. That is, $\sup_{n \geq 1} |T_n(b - b_N)| \to 0$ a.e. as $r \to \infty$. Relable b_{N_r} as b_r for convenience.

We have

$$\limsup_{n \to \infty} T_n b \leq \limsup_{n \to \infty} T_n(b - b_r) + \limsup_{n \to \infty} T_n(b_r)$$

$$\leq \sup_{n \geq 1} |T_n(b - b_r)| + \limsup_{n \to \infty} T_n b_r$$

$$= \sup_{n \geq 1} |T_n(b - b_r)| + \liminf_{n \to \infty} T_n b_r$$

$$\leq \sup_{n \geq 1} |T_n(b - b_r)| + \liminf_{n \to \infty} T_n(b_r - b) + \liminf_{n \to \infty} T_n b$$

$$\leq 2 \sup_{n \geq 1} |T_n(b - b_r)| + \liminf_{n \to \infty} T_n b.$$

Now letting $r \to \infty$, we see that a.e.

$$\limsup_{n \to \infty} T_n b \leq \liminf_{n \to \infty} T_n b.$$

But $\sup_{n\geq 1} |T_n b| < \infty$ a.e., so this says that $\lim_{n\to\infty} T_n b$ exists a.e.   □

Remark. This principle is at the heart of all convergence theorems for (T_n). If one wants a.e. convergence to hold on all of $B = L_p(X)$ for some p, then besides the obviously necessary fact that $\sup_{n\geq 1} |T_n f(x)| < \infty$ a.e. for all $f \in L_p(X)$, one only also needs the necessary fact that $\lim_{N\to\infty} T_n f(x)$ exists a.e. for some dense set of functions $f \in L_p(X)$. Sometimes, as in the Birkhoff Individual Ergodic Theorem, Example 3.2, the dense class part of this is easy and it is the finiteness (or maximal inequality) which is difficult. Sometimes, as in Bourgain's result for averages along squares in Chapter IV, both aspects are not trivial. And then, for cases like averaging along sequences with a positive density, the maximal inequality is easy (given Birkhoff's Theorem) and the dense class becomes the issue.

This basic result is certainly enough to explain why classical maximal inequalities are useful in proving a.e. convergence results. Indeed, if for all $f \in B$, $\mu\{\sup_{n\geq 1} |T_n f| > \lambda\} \leq \frac{C(f)}{\varphi(\lambda)}$, where $C(f)$ is a constant depending on f and $\varphi(\lambda) \to \infty$ as $\lambda \to \infty$, then the essential hypothesis of this result is satisfied. A very interesting feature of the connection between classical maximal inequalities and this theorem was discovered by Sawyer [1966] and Stein [1961a]. They show that under general assumptions on (T_n), with additionally the hypothesis that the (T_n) commute with a "mixing" family of operators, a classical maximal inequality is always a consequence of assuming $\sup_{n\geq 1} |T_n f(x)| < \infty$ a.e. for all $f \in B$. The proofs of these results will not be given here; usually we get the best maximal inequality possible when trying just to get some inequality. However, in the arguments about strong sweeping out in 5.5, and in the proof that Bourgain gives for his entropy result [Bourgain, 1988b], the idea of the operators (T_n) commuting with a "mixing" family of operators appears again in an essential fashion.

Here now are some refinements of the Banach principle. In the applications of the following theorem, B is often taken to be the (mean zero) functions in $C[0,1]$ or $L_p[0,1]$ for some p, $1 \leq p \leq \infty$.

Theorem 5.2. *Suppose B is a Banach space in the norm $\|\cdot\|$ and (T_n), $T_n : B \to L(X)$, is a sequence of linear maps which are continuous in measure. Assume that for all $\varepsilon > 0$ and $K > 0$, there exists $f \in B$, $\|f\| \leq 1$, such that $\mu\{T^* f > K\} \geq 1 - \varepsilon$. Then there is a dense G_δ subset $\mathcal{R} \subset B$ such that if $f \in \mathcal{R}$, then $T^* f = \infty$ a.e.*

Proof. For $0 < \gamma < 1$, let

$$\mathcal{E}_\gamma = \{f \in B : \mu\{T^* f = \infty\} < 1 - \gamma\}.$$

We will show that $\mathcal{E}_\gamma$ is always a set of first category. Then the set $\mathcal{E} = \bigcup_{k=1}^{\infty} \mathcal{E}_{1/k}$ is a set of first category, and $B \setminus \mathcal{E}$ contains a dense G_δ subset $\mathcal{R}$ as claimed in the theorem.

To show $\mathcal{E}_\gamma$ is first category, let $\mathcal{E}_\gamma(m)$ be $\{f \in \mathcal{B} : \mu\{T^*f > m\} < 1-\gamma\}$. Since $\{T^*f = \infty\}$ is the intersection of the decreasing sequence of sets ($\{T^*f > m\}$), we have $\mathcal{E}_\gamma = \bigcup_{m=1}^{\infty} \mathcal{E}_\gamma(m)$. Let $\mathcal{C}_\gamma(m)$ be $\{f \in \mathcal{B} : \mu\{T^*f > m\} \le 1-\gamma\}$ and $\mathcal{C}_\gamma(m, N)$ be $\{f \in \mathcal{B} : \mu\{T_N^*f > m\} \le 1-\gamma\}$. Since $\{T^*f > m\}$ is the union of the increasing sequence of sets ($\{PT_N^*f > m\}$), we have $\mathcal{C}_\gamma(m) = \bigcap_{N=1}^{\infty} \mathcal{C}_\gamma(m, N)$ for all $m \ge 1$. Now, because T_N^* is continuous in measure, each of the sets $\mathcal{C}_\gamma(m, N)$ is a closed set in $\mathcal{B}$. Hence, $\mathcal{C}_\gamma(m)$ is a closed set for all $m \ge 1$. The proof is completed by showing that $\mathcal{C}_\gamma(m)$ has empty interior, and so $\mathcal{E}_\gamma(m)$ is nowhere dense for all $m \ge 1$.

Suppose some $\mathcal{C}_\gamma(m)$ did have non-empty interior. Then there would be a $\delta > 0$ and $g_0 \in \mathcal{C}_\gamma(m)$ such that for all $f \in \mathcal{B}$, $\|f\| \le 1$, $\mu\{T^*(g_0 + \delta f) > m\} \le 1 - \gamma$. Choose $\varepsilon > 0$ so that $\gamma - \varepsilon > 0$ and choose a whole number M such that $M(\gamma - \varepsilon) \ge 1$. By the hypothesis of the theorem, there exists $f_0 \in \mathcal{B}$, $\|f_0\| \le 1$, such that $G_0 = \{T^*f_0 > 2mM/\delta\}$ has measure $\mu(G_0) \ge 1 - \varepsilon$. For each $i = 0, \dots, M$, let

$$F_i = \{T^*(g_0 - (i/M)\delta f_0) \le m\}.$$

Because T^* is subadditive, for any $i, j = 0, \dots, M$,

$$T^*((i/M)\delta f_0 - (j/M)\delta f_0) \le T^*(g_0 - (i/M)\delta f_0) + T^*(g_0 - (j/M)\delta f_0).$$

Hence, on $F_i \cap F_j$, $2m \ge (|i-j|/M)\delta T^*f_0$, and on $G_0 \cap F_i \cap F_j$, $2m > 2m|i-j|$. Therefore, the sets $G_0 \cap F_i$, $i = 0, \dots, M$, are pairwise disjoint. By the choice of g_0 and δ, $\mu(F_i) \ge \gamma$ and $\mu(G_0 \cap F_i) \ge \gamma - \varepsilon$ for all $i = 0, \dots, M$. Hence,

$$1 \ge \sum_{i=0}^{M} \mu(G_0 \cap F_i) > (\gamma - \varepsilon)M \ge 1.$$

This contradiction means that $\mathcal{C}_\gamma(m)$ has empty interior for any $0 < \gamma < 1$ and $m \ge 1$. $\square$

In several cases, it will be worthwhile to have a theorem like Theorem 5.2, but one which has a domain $\mathcal{B}$ consisting only of characteristic functions. Let $(Y, \mathcal{G}, \nu)$ be a probability space and define $\varrho(E, F) = \nu(E \triangle F)$ for $E, F \in \mathcal{G}$. If $\mathcal{N}$ is the class of ν-null sets in $\mathcal{G}$, then ϱ induces a metric on $\mathcal{G}/\mathcal{N}$ which will also be denoted by ϱ. For simplicity of notation, the metric space $(\mathcal{G}/\mathcal{N}, \varrho)$ will be denoted by $(\mathcal{G}, \varrho)$ from here on out. We say that $T : \mathcal{G} \to L(X)$ is $linear$ if whenever $E = F_1 \cup F_2$ for disjoint $F_1, F_2 \in \mathcal{G}$, then $TE = TF_1 + TF_2$. In this next theorem, the metric space $\mathcal{B}$ is taken to be $\mathcal{G}$. The proof of this theorem is very much like the previous one and is left as an exercise.

Exercise 5.3. Suppose (X, Σ, μ) is separable, and let (T_n), $T_n : \mathcal{G} \to L(X)$, be a sequence of linear maps which are continuous in measure. Assume that for all $\varepsilon > 0$ and $K > 0$ there exists $A \in \mathcal{G}$, $\nu(A) < \varepsilon$, such that $\mu\{T^*A > K\} \geq 1 - \varepsilon$. Then there is a dense G_δ subset $\mathcal{R} \subset \mathcal{G}$ such that if $A \in \mathcal{R}$, then $T^*A = \infty$ a.e.

There is yet another version of these two theorems which will be useful later. Let $L^+(X)$ be the cone of positive equivalence classes in $L(X)$. A mapping $T : \mathcal{G} \to L^+(X)$ *is monotone if* $E \subset F$ *implies* $TE \leq TF$ for $E, F \in \mathcal{G}$. This proof contains the important complementation principle.

Theorem 5.4. *Let* (T_n), $T_n : \mathcal{G} \to L^+(X)$, *be a sequence of monotone linear maps which are continuous in measure and such that* $T_n Y = 1$ *for all* n. *Assume that for all* $\varepsilon > 0$ *and* $M \geq 1$, *there exists* $A \in \mathcal{G}$, $\nu(A) < \varepsilon$, *such that* $\mu\left\{ \sup_{n \geq M} T_n A > 1 - \varepsilon \right\} \geq 1 - \varepsilon$. *Then there is a dense* G_δ *subset* $\mathcal{R} \subset \mathcal{G}$ *such that if* $A \in \mathcal{R}$, *then* $\limsup_{n \to \infty} T_n A = 1$ *a.e. and* $\liminf_{n \to \infty} T_n A = 0$ *a.e..*

Proof. For $0 < \gamma < 1$ and $M \geq 1$, let $\mathcal{E}_{\gamma, M}$ be

$$\left\{ A \in \mathcal{G} : \mu\left\{ \sup_{n \geq M} T_n A = 1 \right\} < 1 - \gamma \right\}.$$

It is easy to see as an exercise that $\mathcal{E}_{\gamma, M}$ is a set of first category. Then $\mathcal{E} = \bigcup_{M=1}^{\infty} \bigcup_{k=1}^{\infty} \mathcal{E}_{(1/k), M}$ is a set of first category, and $\mathcal{D} = \{A \in \mathcal{G} : Y \backslash A \in \mathcal{E}\}$ is also a set of first category, because $A \to Y \backslash A$ is an isometry of $(\mathcal{G}, \varrho)$. Hence, the complement of $\mathcal{E} \cup \mathcal{D}$ contains a dense G_δ subset $\mathcal{R}$; if $A \in \mathcal{R}$, then $\limsup_{n \to \infty} T_n A = 1$ a.e. and $\limsup_{n \to \infty} T_n(Y \backslash A) = 1$ a.e. Since $T_n Y = 1$ and T_n is linear for all n, $T_n(Y \backslash A) = T_n Y - T_n A = 1 - T_n A$. Hence, for all $A \in \mathcal{R}$, both $\limsup_{n \to \infty} T_n A = 1$ a.e. and $\liminf_{n \to \infty} T_n A = 0$ a.e.. $\square$

Remark. Because of the property that the typical set A has in Theorem 5.4, (T_n) is said to be *strong sweeping out*.

It is interesting that the hypotheses of Theorem 5.4 can be significantly weaker in regard to the property that the sets A must have to get strong sweeping out. To do this, we need some details of terminology from theorems of Sawyer and Stein.

A family of measure-preserving transformations $\{S_\alpha\}$ is said to be *mixing* if for each pair of sets A and B in Σ, and $\rho > 1$, there exists S_α in the family such that $m(A \cap S_\alpha^{-1}(B)) < \rho \cdot m(A) \cdot m(B)$.

The following variant of Sawyer's Theorem will be needed to study the 'strong sweeping out property'.

Theorem 5.5. *Let* (X, Σ, m) *denote a probability space. Assume that* (T_k) *is a sequence of linear operators,* $T_k : L^1 \to L^1$, *with the properties*

(i) $T_k \geq 0$;

(ii) $T_k 1 = 1$;

(iii) the T_k's commute with a family $\{S_\alpha\}$ which is a mixing family of measure-preserving transformations.

For each n define $M_n f = \sup_{k \geq n} |T_k f|$. Assume that for each $\varepsilon > 0$ and $n \in \mathbb{N}$, there exists a sequence of sets (A_p) such that if

$$E_p = \{M_n 1_{A_p} \geq 1 - \varepsilon\} \text{ then } \sup_p \frac{m(E_p)}{m(A_p)} = +\infty.$$

Then the strong sweeping out property holds: given $\varepsilon > 0$, we can find a set B with $m(B) < \varepsilon$ such that

$$\limsup_k T_k 1_B = 1 \ a.e.$$

and

$$\liminf_k T_k 1_B = 0 \ a.e..$$

Proof. By Theorem 5.4, it is enough to show that for each $\varepsilon > 0$ and $n \in \mathbb{N}$, we can find a set B, $m(B) < \varepsilon$, such that

$$M_n 1_B \geq 1 - \varepsilon \ a.e..$$

Now we argue as in Sawyer's proof. We may assume without loss of generality that $m(E_p)/m(A_p) \geq p^2$, otherwise we relabel the sequence. Let h_p be a natural number such that $1 \leq h_p m(E_p) \leq 2$ and so

$$h_p m(A_p) \geq \frac{h_p m(E_p)}{p^2} \leq \frac{2}{p^2}.$$

Take $A_p^1, A_p^2, \ldots, A_p^h$ to be identical copies of A_p, i.e., $A_p^j = A_p$ for $j = 1, 2, \ldots h_p$. Then $E_p^j = E_p$ for $j = 1, 2, \ldots, h_p$. By Sawyer's auxiliary lemma (see [de Guzmán, 1981, p. 20]), there are $S_p^j \in \{S_\alpha\}$, $p = 1, 2, \ldots$, $j = 1, 2, \ldots h_p$, such that almost every x belongs to infinitely many $(S_p^j)^{-1}(E_p^j)$. Choose p_0 so that

$$\sum_{p=p_0}^{\infty} h_p m(A_p) \leq \sum_{p=p_0}^{\infty} \frac{2}{p^2} < \varepsilon.$$

Define

$$F(x) = \sup_{p \geq p_0, 1 \leq j \leq h_p} S_p^j 1_{A_p^j}(x) = 1_B(x).$$

Then

$$\|F\|_1 \leq \sum_{p \geq p_0} \sum_{j=1}^{h_p} \int_X S_p^j 1_{A_p^j} \, dm = \sum_{p \geq p_0} \sum_{j=1}^{h_p} m(A_P^j) = \sum_{p \geq p_0} h_p m(A_p) < \varepsilon,$$

and

$$M_n F(x) \geq M_n(S_p^j 1_{A_p^j})(x) = M_n(1_{A_p^j})(S_p^j x).$$

Now if $x \in (S_p^j)^{-1}(E_p^j)$, then $S_p^j x \in E_p^j$ and hence $M_n(1_{A_p^j})(S_p^j x) \geq 1-\varepsilon$, which implies $M_n 1_B(x) \geq 1 - \varepsilon$. Thus $M_n 1_B(x) \geq 1 - \varepsilon$ for almost every $x \in X$, and the theorem is proved.   □

Remark. The assumptions on the sequence of operators (T_k) used in the above theorem can be weakened and the conclusion can be strengthened. It is enough for these operators to satisfy the assumptions of Theorem 5.4, and the conclusion can be strengthened to show that an entire dense G_δ collection of subsets of Σ will work.

There are many interesting applications of the result in Theorem 5.4 in diophantine approximation and ergodic theory: see [del Junco, Rosenblatt, 1979]. It should be noticed here that the existence of the set A in the hypothesis can easily give the conclusion that for all $\varepsilon > 0$ there exists A such that $m(A) < \varepsilon$ and $\limsup_{n \to \infty} T_n A = 1$ a.e., because $T_n \geq 0$. However, to get the oscillation requires a further iteration of this construction, which can be quite cumbersome. The use of the Baire category theorem here removes the need for such a construction entirely. Here are some examples of the use of these results.

One of the important summability problems in diophantine approximation concerns Cesaro summability. This problem is to determine the functions $f \in L_\infty[0,1]$ such that $\lim_{N \to \infty} (1/N) \sum_{k=1}^{N} f\{nx\}$ exists a.e.. For example, Koksma [1952] proved that if $f \in L_2[0,1]$ and the Fourier coefficients (c_k) of f satisfy $\sum_{k \geq 3} |c_k|^2 (\log\log|k|)^3 < \infty$, then the limit $\lim_{n \to \infty} (1/n) \sum_{k=1}^{n} f\{kx\} = \int_0^1 f(x)dx$ for a.e. x. However, Marstrand [1970] showed that for all $\varepsilon > 0$, there exists an open set $V \subset [0,1]$ with $\mu(V) < \varepsilon$ such that $\limsup_{n \to \infty}(1/n) \sum_{k=1}^{n} 1_V\{kx\} = 1$ for all x. Hence, Theorems 5.2 and 5.4 give the following corollary to Marstrand's example.

Proposition 5.6. *There is a dense G_δ subset $\mathcal{R}$ of the measurable sets in $[0,1]$ modulo null sets such that if $A \in \mathcal{R}$, then $\limsup_{n \to \infty}(1/n) \sum_{k=1}^{n} 1_A\{kx\} = 1$ a.e. and $\liminf_{n \to \infty}(1/n) \sum_{k=1}^{n} 1_A\{kx\} = 0$ a.e.. Also, there is a dense G_δ subset of $L_p[0,1]$ such that if $f \in \mathcal{R}$, then $\sup_{n \geq 1}(1/n)\left|\sum_{k=1}^{n} f\{kx\}\right| = \infty$ a.e..*

We now consider another application of these category theorems, this time in ergodic theory. Let (b_n) be an increasing sequence of positive integers. Then for any ergodic measure-preserving transformation $\tau : [0,1] \to$

$[0,1]$, and for any $f \in L_p[0,1]$, $1 \leqq p < \infty$, the averages

$$T_n f = (1/b_n) \sum_{k=n+1}^{n+b_n} f \circ \tau^k$$

converge in L_p-norm to $\int f \, dm$. In [Bellow, Jones, and Rosenblatt, 1990] and [Rosenblatt, Wierdl 1992], it was shown that if $b_n/n \to 0$, then $T_n f$ will not generally converge a.e. to $\int f \, dm$. We can use Theorem 5.4 to improve this theorem in the same manner as our previous examples.

Corollary 5.7. *Let (b_n) be an increasing sequence of positive integers with $\lim_{n\to\infty} b_n/n = 0$. Let $\tau\colon [0,1] \to [0,1]$ be an invertible measure-preserving ergodic transformation of $[0,1]$. There is a dense G_δ subset $\mathcal{R}$ of Σ modulo null sets such that if $A \in \mathcal{R}$ then*

$$\limsup_{n\to\infty} (1/b_n) \sum_{k=n+1}^{n+b_n} 1_A(\tau^k x) = 1 \text{ a.e.}$$

and

$$\liminf_{n\to\infty} (1/b_n) \sum_{k=n+1}^{n+b_n} 1_A(\tau^k x) = 0 \text{ a.e..}$$

Also, there is a dense G_δ subset $\mathcal{R}$ of $L_p[0,1]$, $1 \leqq p < \infty$, such that if $f \in \mathcal{R}$, then

$$\sup_{n\geq 1} (1/b_n) \left| \sum_{k=n+1}^{n+b_n} f(\tau^k x) \right| = \infty \text{ a.e..}$$

Proof. For $n \geqq 1$ and $\varepsilon > 0$, by Rokhlin's lemma there is a measurable set E such that $(\tau^k E\colon k = 1, \ldots, n + b_n)$ are pairwise disjoint and $\mu\left(\bigcup_{k=1}^{n+b_n} \tau^k E \right) \geqq 1 - \varepsilon$. Let $A = \bigcup_{k=n+1}^{n+b_n} \tau^k E$. Then $\mu(A) \leqq b_n/(n+b_n)$. Fix N, $1 \leqq N \leqq n$. For $x \in \tau^\ell E$ with $\ell = 0, \ldots, n - N$, we have $T_{n-\ell}1_A(x) = (1/b_{n-\ell}) \sum_{k=n-\ell+1}^{n-\ell+1+b_{n-\ell}} 1_A(\tau^k x) = (1/b_{n-\ell})b_{n-\ell}$, because $b_n \geqq b_{n-\ell}$. Hence $\mu\left\{ \sup_{N \leqq k \leqq n} T_k 1_A = 1 \right\} \geqq (n - N)\mu(E) \geqq (1 - \varepsilon)(n - N)/(n + b_n)$. Letting $n \to \infty$ and $\varepsilon \to 0$ shows that for all $\varepsilon > 0$, there exists a measurable set $A \subset [0,1]$ with $\mu(A) \leqq \varepsilon$ such that $\mu\left\{ \sup_{N \leqq k} T_k 1_A \geqq 1 - \varepsilon \right\} \geqq 1 - \varepsilon$. Therefore, the hypotheses of Theorem 5.4 are satisfied and the dense G_δ subset $\mathcal{R}$ of Σ modulo null sets exists. But then if $\varepsilon > 0$ and $K > 0$, there exists $A \in \Sigma$ such that $0 < \mu(A) < (1 - \varepsilon)^p/K^p$ and $\mu\{T_n^* 1_A \geqq 1 - \varepsilon\} \geqq 1 - \varepsilon$.

If $f = 1_A/\mu(A)^{1/p}$, then $\|f\|_p \leq 1$ and $\mu\{T_n^* f \geq K\} \geq \mu\{T_n^* 1_A \geq \mu(A)^{1/p}K\} \geq \mu\{T_n^* 1_A \geq 1-\varepsilon\} \geq 1-\varepsilon$. Hence, the hypotheses of Theorem 5.2 are satisfied too. $\square$

Here is a third example of the use of these category results. In this case we are interested in the behavior of averages $\mu f(x) = \sum \mu(k) f(\tau^k x)$ for probability measures μ on $\mathbb{Z}$ and measure-preserving transformations τ. The case of $\mu = \frac{1}{2}(\delta_0 + \delta_1)$ is the prototype of this. But by using the Central Limit Theorem, it is also possible to show that the bad behavior of the convolution powers μ^n of this prototype is typical, at least if μ has second moment $m_2(\mu) = \sum_{k=-\infty}^{\infty} k^2 \mu(k)$ finite and expectation $E(\mu) = \sum_{k=-\infty}^{\infty} k\mu(k)$ not equal to 0.

Theorem 5.8. *If μ has $m_2(\mu) < \infty$ and $E(\mu) \neq 0$, and if (X, Σ, m, τ) is an aperiodic dynamical system, then there exists a residual set $\mathcal{R}$ of Σ in the symmetry pseudo-metric such that if $E \in \mathcal{R}$, then $\limsup_{n\to\infty} \mu^n 1_E = 1$ a.e. and $\liminf_{n\to\infty} \mu^n 1_E = 0$ a.e..*

Proof. As is typical for proving the strong sweeping out above, by Theorem 5.4 it suffices to show that for all $\varepsilon > 0$ there is $E \in \Sigma, m(E) < \varepsilon$, such that $\limsup_{n\to\infty} \mu^n 1_E \geq 1-\varepsilon$ a.e.. Also, without loss of generality $\text{supp}(\mu)$ is not a singleton. Indeed, if $\mu = \delta_k$, $k \neq 0$, then this sweeping out follows from [Ellis, Friedman 1978].

The idea is to use the Central Limit Theorem to get enough information about the distribution of (μ^n) so that Theorem 5.4 can give this sweeping out result. Let $(X_i : i \geq 1)$ be a sequence of independent identically-distributed random variables on some probability space $(\Omega, \mathcal{M}, p)$ with $E(X_i) = E(\mu)$, $var(\mu) = m_2(\mu) - E(\mu)^2$, and each X_i having a discrete distribution on $\mathbb{Z}$ given by μ, i.e. $p\{X_i = k\} = \mu(k)$ for all $k \in \mathbb{Z}$, $i \geq 1$. Then $S_n = \sum_{i=1}^{n} X_i$ has a discrete distribution given by μ^n, again in the sense that $p\{S_n = k\} = \mu^n(k)$ for all $k \in \mathbb{Z}$, $n \geq 1$. Note that $var(\mu) \neq 0$ because the support of μ is not a single point.

First, for any $\varepsilon > 0$, there exists $b > 0$ such that for $n \geq n_0$,

$$\sum\{\mu^n(k) : |k - nE(\mu)| \leq b\sqrt{var(\mu)n}\} > 1 - \varepsilon. \tag{5.1}$$

To see this, choose $b > 0$ so that $\frac{1}{\sqrt{2\pi}} \int_{-b}^{b} e^{-\frac{u^2}{2}} du > 1 - \frac{\varepsilon}{2}$. Then by the Central Limit Theorem, there exists $n_0 \geq 1$ such that if $n \geq n_0$,

$$|p\{\frac{S_n - nE(\mu)}{\sqrt{var(\mu)n}} \in [-b,b]\} - \frac{1}{\sqrt{2\pi}} \int_{-b}^{b} e^{\frac{-u^2}{2}} du| \leq \frac{\varepsilon}{2}.$$

Thus, $p\{\frac{S_n - nE(\mu)}{\sqrt{var(\mu)n}} \in [-b,b]\} > 1 - \varepsilon$. Since $\sum\{\mu^n(k) : |k - nE(\mu)| \leq b\sqrt{var(\mu)n}\} = p\{S_n - nE(\mu) \in [-b\sqrt{var(\mu)n}, b\sqrt{var(\mu)n}]\}$. This proves (5.1).

But also, there exists a constant $A < \infty$ such that for all $n \geq 1$,

$$(5.2) \qquad \sup_k |\mu^n(k)| \leq \frac{A}{\sqrt{n}} \,.$$

This is proved in [Chung, Erdös, 1951] when μ satisfies special restrictions on its support, namely if $\{u_i : i \geq 1\} = \mathrm{supp}(\mu)$,
a)   the u_i are not all the same sign,
b)   the greatest common divisor of $u_i - u_j, i, j \geq 1$, is equal to 1.
It is not hard to extend this to any μ which is not supported at a point.

The estimates (5.1) and (5.2) say that μ^n distributes in a manner similar to b^n. To be precise, as in Corollary 9 in [Bellow, Jones, and Rosenblatt, 1990], for any $\varepsilon' > 0$, there exists E, $m(E) < \varepsilon'$, such that for a.e. x there are infinitely many n such that,

$$\frac{1}{(2[b\sqrt{var(\mu)n}] + 1)} \sum_{|k| \leq [b\sqrt{var(\mu)n}]} 1_E(\tau^{E(\mu)n+k}x) \geq 1 - \varepsilon' \,.$$

Let $K_n = \{k : |k| \leq [b\sqrt{var(\mu)n}]$ and $1_E(\tau^{E(\mu)n+k}x) = 0\}$. Then by the above, for a.e. x, $\#K_n \leq \varepsilon'(2[b\sqrt{var(\mu)n}]+1)$ for infinitely many n. Thus, for a.e. x there are infinitely many n with

$$\begin{aligned}
\mu^n 1_E(x) &\geq \sum \{\mu^n(k)1_E(\tau^k x) : |k - E(\mu)n| \leq [b\sqrt{var(\mu)n}], \ k \notin K_n\} \\
&\geq \sum \{\mu^n(k) : |k - E(\mu)n| \leq [b\sqrt{var(\mu)n}]\} \\
&\quad - \#K_n \frac{A}{\sqrt{n}}
\end{aligned}$$

by (5.2). Hence, by (5.1) and the estimate for $\#K_n$, for a.e. x there are infinitely many n with

$$\mu^n 1_E(x) \geq 1 - \varepsilon - \frac{A}{\sqrt{n}}\varepsilon'(2[b\sqrt{var(\mu)n}] + 1) \,.$$

Since ε' can be chosen arbitrarily small, this shows that for all $\varepsilon > 0$ there exists $E \in \Sigma$, $m(E) < \varepsilon$, such that $\limsup_{n \to \infty} \mu^n 1_E(x) \geq 1 - \varepsilon$ for a.e. x.□

Remark. The same argument shows that many subsequences (μ^{n_m}) also still satisfy the strong sweeping out property. Indeed, $n_m = m^2$ or $n_m = 2^m$ do this. This is seen by using strong sweeping out as above for the corresponding subsequence (C_{n_m}), where $C_k = \frac{1}{\sqrt{k}} \sum_{j=k+1}^{k+[\sqrt{k}]+1} \delta_j$. See [Bellow, Jones, and Rosenblatt, 1990] for the details needed to modify the argument above.

Another way in which the idea of strong sweeping out has appeared has been in the detailed study of sequences which are bad for pointwise convergence results. The problem is to determine for which subsequences (n_k) of the sequence $(1, 2, 3, \ldots)$ the averages $T_n f = (1/n) \sum_{k=1}^{n} f \circ \tau^{n_k}$ converge a.e. or in L_p norm for a function $f \in L_p[0, 1]$. Krengel [1971] was the first to show how to construct an increasing sequence (n_k) such that these averages generally fail to converge a.e.. The essential idea of the construction is that there is an increasing sequence (n_k) such that for any $\varepsilon > 0$ and $N \geq 1$ there is a measurable A with $\mu(A) < \varepsilon$ such that $\mu \left\{ \sup_{N \leq n} T_n 1_A \geq 1 - \varepsilon \right\} \geq 1 - \varepsilon$.

Then, the hypotheses of Theorem 5.4 are satisfied. Also, with $\mathcal{B} = L_p[0, 1]$, $1 \leq p < \infty$, the hypotheses of Theorem 5.2 are satisfied. Check that the averages along the sequence we constructed in Example 1.18 satisfy the hypotheses of Theorem 5.4, and hence its conclusion.

Since the publication of Krengel's result, there has been much attention to universally bad sequences like the one that he built. Bellow showed that any lacunary sequence is going to be bad to a certain extent. The degree to which this is true was later improved by Rosenblatt. Then finally, in a definitive article on the subject, it has been shown in [Akcoglu, et al, 1994] that any lacunary sequence is strong sweeping out. We do not present this new result here. However, in the next chapter, using the entropy method, it is shown that lacunary sequences are at least sweeping out to some degree.

2. Types of bad sequences

Here is a short discussion of the different types of concepts of bad sequences and averaging methods. We will follow it with some more interesting examples. However, before beginning this discussion, we will need to have the Conze principle available to us.

The Conze principle (see [Conze, 1973]) says essentially that a maximal inequality holds for one ergodic dynamical system if and only if it does for every ergodic dynamical system. There are two ways, seemingly quite different, to prove this principle. One is via transfer to $\mathbb{Z}$, and the other is via weak approximation of transformations. However, at the core of both methods is the Rokhlin Lemma, and this is where the similarity between the methods lies. Let us state the principle in this generality. Let μ be a probability measure on $\mathbb{Z}$. Let $f: X \to \mathbb{C}$ be a measurable function. Let τ be an invertible measure-preserving transformation. Then $\mu^\tau f(x) = \sum_{k=-\infty}^{\infty} \mu(k) f(\tau^k x)$ for all $x \in X$.

Theorem 5.9. *Let (μ_N) be a sequence of probability measures on $\mathbb{Z}$. Suppose that for some ergodic τ there is a universal constant C such that*

$m\left\{\sup_{N\geq 1}|\mu_N^\tau f|>\lambda\right\}\leq\frac{C}{\lambda^p}\|f\|_p^p$ for all $f\in L_p(X,m)$. Then the same holds for any measure-preserving transformation σ in place of τ.

Proof. We use Halmos' Theorem; since τ is ergodic, the conjugates $\rho\tau\rho^{-1}$ of τ by invertible measure-preserving transformations are dense in the weak topology in the set of all invertible measure-preserving transformations. That is, if $A_1,\ldots,A_k$ are measurable sets, there is a sequence (ρ_n) with $\lim_{n\to\infty}m(\rho_n\tau\rho_n^{-1}A_i\Delta\sigma A_i)=0$ for all $i=1,\ldots,k$. It follows easily that $m\left\{\sup_{1\leq N\leq k}|\mu_N^{\rho_n\tau\rho_n^{-1}}f|>\lambda\right\}\to m\left\{\sup_{1\leq N\leq k}|\mu_N^\sigma f|>\lambda\right\}$ for all $\lambda\geq 0$ and $f\in L_p(X)$. But

$$m\left\{\sup_{1\leq N\leq k}|\mu^{\rho_n\tau\rho_n^{-1}}f|>\lambda\right\}=m\left\{\sup_{1\leq N\leq k}|\mu_N^\tau f\circ\rho_n|>\lambda\right\}$$

by the invariance of m under ρ_n and because

$$\mu_N^{\rho_n\tau\rho_n^{-1}}f(x)=\sum\mu_N(k)f((\rho_n\tau\rho_n^{-1})^kx)=\sum\mu_N(k)f(\rho_n\tau^k\rho_n^{-1}x)$$

for $x\in X$.

Hence, the maximal inequality says that

$$m\left\{\sup_{1\leq N\leq k}|\mu_N^{\rho_n\tau\rho_n^{-1}}f|>\lambda\right\}=m\left\{\sup_{1\leq N\leq k}|\mu_N^\tau f\circ\rho_n|>\lambda\right\}$$
$$\leq\frac{C}{\lambda^p}\|f\circ\rho_n\|_p^p=\frac{C}{\lambda^p}\|f\|_p^p.$$

Letting $n\to\infty$, the weak approximation says $m\left\{\sup_{1\leq N\leq k}|\mu_N^\sigma f|>\lambda\right\}\leq$ $\frac{C}{\lambda^p}\|f\|_p^p$. Now let $M\to\infty$, to see $m\left\{\sup_{N\geq 1}|\mu_N^\sigma f|>\lambda\right\}\leq\frac{C}{\lambda^p}\|f\|_p^p$ for all $f\in L_p(X)$. $\square$

Exercise 5.10. The alternative proof that one can give by transfer goes as follows. First, by use of Rokhlin's Lemma and simple functions in place of f, the maximal inequality implies that for all $\varphi\in\ell_p(\mathbb{Z})$,

$$\#\left\{\sup_{N\geq 1}|\mu_N*\varphi|>\lambda\right\}\leq\frac{C}{\lambda}\|\varphi\|_{\ell_p}^p.$$

But then, via the Calderón transfer principle, for any measure-preserving transformation σ,

$$\#\left\{\sup_{N\geq 1}|\mu_N^\sigma f|>\lambda\right\}\leq\frac{C}{\lambda^p}\|f\|_p^p.$$

Moreover, τ being ergodic is not necessary. It suffices to assume that τ is aperiodic, since then the Rokhlin Lemma and Halmos' Theorem hold. However, τ has to have some non-degeneracy since nothing significant could result from the case $\tau = Id$, for example.

There are different variations on the Conze principle. As an exercise the reader should check the following:

Exercise 5.11. 1) Let (μ_N), (ν_N) be probability measures on $\mathbb{Z}$. Suppose there is a constant C such that

$$\left\| \left(\sum_{n=1}^{\infty} |\mu_N^\tau f - \nu_N^\tau f|^2 \right)^{1/2} \right\|_2 \leq C\|f\|_2$$

for all $f \in L_2(X)$. If τ is ergodic, then the same inequality holds with any measure-preserving transformation σ in place of τ.

2) Let (μ_N) be probability measures on $\mathbb{Z}$ and suppose there is a constant C such that for some fixed constants $C_n > 0$

$$\sum_{n=1}^{\infty} m\{|\mu_n^\tau f| > C_n\} \leq C\|f\|_1 .$$

If τ is ergodic, then the same inequality holds with any measure-preserving transformation σ in place of τ.

By using a different argument than the ones in [Bellow, 1983] and [Bellow, Losert, 1985], specifically Bourgain's entropy theorem in [Bourgain, 1988], results that are in some ways stronger than those in [Bellow, 1983] and [Bellow, Losert, 1985] can be obtained. To clarify these results, we need some definitions.

Definition 5.12. A sequence (n_k) is L_p *universally bad* if for all ergodic dynamical systems, there is $f \in L_p(X)$ such that $\lim_{N\to\infty} \frac{1}{N} \sum_{k=1}^{N} f(\tau^{n_k}x)$ fails to exist for all x in a set of positive measure. The sequence is L_p *universally good* if for all ergodic dynamical systems and all $f \in L_p(X)$, $\lim_{N\to\infty} \frac{1}{N} \sum_{k=1}^{N} f(\tau^{n_k}x)$ exists a.e..

It is not hard to see, using the Conze principle [Conze, 1973] and the Banach principle of Sawyer, that for $1 \leq p < \infty$, (n_k) is L_p universally bad if and only if there is no constant $C < \infty$ such that for all ergodic dynamical systems and all $f \in L_p(X)$, $m\{\sup_{N\geq 1} |\frac{1}{N} \sum_{k=1}^{N} f(\tau^{n_k}x)| > \lambda\} \leq \frac{C}{\lambda}\|f\|_p$. This is why the definition here when $p = 1$ is the same as the one in [Bellow, 1983]. Also, because of the examples in [Bellow, 1989], for any $p, 1 < p < \infty$, there exists (n_k) which is L_p universally bad, but L_q universally good for all $p < q \leq \infty$.

Some universally bad sequences are not beyond redemption. For those that are, there is

Definition 5.13. A sequence (n_k) is L_p *persistently universally bad* if for all sequences (N_s), $N_s < N_{s+1}$ for $s \geq 1$, and all ergodic dynamical systems, there is $f \in L_p(X)$ such that $\lim\limits_{s\to\infty} \frac{1}{N_s} \sum\limits_{k=1}^{N_s} f(\tau^{n_k} x)$ fails to exist for all x in a set of positive measure.

For example, it is easy to construct a sequence (n_k) so that for two disjoint sequences (M_s) and (N_s), the discrete measures on $\mathbb{Z}$, $\mu_s = \frac{1}{M_s} \sum\limits_{k=1}^{M_s} \delta_{n_k}$

and $\nu_s = \frac{1}{N_s} \sum\limits_{k=1}^{N_s} \delta_{n_k}$, satisfy both

a) $\sum_{s=1}^{\infty} \| \mu_s - \frac{1}{M_s} \sum_{k=1}^{M_s} \delta_k \|_1 < \infty$

and

b) $\sum_{s=1}^{\infty} \| \nu_s - \frac{1}{N_s} \sum_{k=1}^{N_s} \delta_{2^k} \|_1 < \infty$.

Then (n_k) is L_∞ universally bad (see Chapter VI), but for all dynamical systems and all $f \in L_p(X)$, $1 \leq p \leq \infty$, $\lim\limits_{s\to\infty} \frac{1}{M_s} \sum\limits_{k=1}^{M_s} f(\tau^{n_k} x)$ exists a.e..

So (n_k) is not L_1 persistently universally bad. This sequence (n_k) will have upper density 1 and lower density 0; also, the limit $\lim\limits_{N\to\infty} \frac{1}{N} \sum\limits_{k=1}^{N} f \circ \tau^{n_k}$ will not generally exist in any L_p norm. Another example, but now of density zero, is given by the sequence (η_k) composed of blocks $(2^m+1, \ldots, 2^m+m)$, $m \geq 1$. Bellow and Losert [1985] show that this sequence is L_p universally bad, $1 \leq p < \infty$. But because of the lengthening blocks, if $\mu_N = \frac{1}{N} \sum\limits_{k=1}^{N} \delta_{\eta_k}$, then $\hat{\mu}_N \to 0$ uniformly on each $S_\delta = \{\gamma : |\gamma| = 1, \delta \leq |\gamma - 1|\}$. Thus, the averages $\mu_N f = \frac{1}{N} \sum\limits_{k=1}^{N} f \circ \tau^{\eta_k}$ converge in L_p norm, $1 \leq p < \infty$; but also, by [Bellow, Jones, and Rosenblatt, 1988], there is a subsequence (N_s) such that $\lim\limits_{s\to\infty} \mu_{N_s} f(x)$ exists a.e. for all $f \in L_p(X), 1 < p \leq \infty$. That is, (η_k) is not L_p persistently universally bad for $1 < p \leq \infty$. It is not clear whether this sequence is L_1 persistently universally bad. In Chapter VI, we will show that there certainly are L_∞ persistently universally bad sequences.

Moreover, some L_∞ persistently universally bad sequences can be worse than others. A sequence (μ_N) of probability measures on $\mathbb{Z}$ is called *dissipative* if $\lim\limits_{N\to\infty} \mu_N(k) = 0$ for all $k \in \mathbb{Z}$.

Definition 5.14. A sequence (n_k) is L_p *inherently universally bad* if for all dissipative sequences (μ_N) with supports in $\{n_k\}$, and all ergodic dynamical systems, there is $f \in L_p(X)$ such that $\lim\limits_{N\to\infty} \mu_N f(x)$ fails to exist for a set of x of positive measure.

For example, in the next chapter it is shown that $(n_k) = (2^k)$ is L_∞ inherently universally bad. But not all L_∞ persistently universally bad sequences are L_1 inherently universally bad. To see this, note that by the subsequence theorem in [Rosenblatt, Wierdl, 1992], there is a sequence of blocks $B_s = (m_s + 1, \ldots, m_s + n_s)$ with $m_s, n_s \to \infty$ as $s \to \infty$ and $\lim_{s \to \infty} \frac{n_s}{\log m_s} = 0$ such that for all $f \in L_1(X)$, $\lim_{s \to \infty} \frac{1}{n_s} \sum_{k=m_s+1}^{m_s+n_s} f(\tau^k x)$ exists a.e.. Take the lacunary sequence $(\eta_k) = (2^k)$ and add to it a subsequence of the sequence (B_s) which is chosen rarely enough so that the new sequence (ξ_k) has

$$\sum_{N=1}^{\infty} \frac{\#\{\xi_k \in \bigcup\limits_{s=1}^{\infty} B_s : k \leq N\}}{N} < \infty.$$

Then (ξ_k) is L_∞ persistently universally bad because (η_k) is, but (ξ_k) is not L_1 inherently universally bad because of the presence of the blocks. Indeed, there is a dissipative sequence (μ_N) supported in (ξ_k) such that, for any dynamical system,

$$\lim_{N \to \infty} \mu_N f(x) \text{ exists a.e. for all } f \in L_1(X).$$

These three definitions show that there are levels of bad behavior among universally bad sequences. Another sense in which this occurs for L_∞ universally bad sequences is in the degree of oscillation exhibited by the averages $\frac{1}{N} \sum_{k=1}^{N} f(\tau^{n_k} x)$. It is clear that (n_k) is L_∞ universally bad, even persistently or inherently so, if and only if the corresponding failure of convergence occurs for a characteristic function f. A gauge of the degree of oscillation in the averages is given by the sweeping out concept.

Definition 5.15. Given $\delta > 0$, a sequence of probability measures (μ_N) is δ *sweeping out* if for all ergodic dynamical systems and all $\varepsilon > 0$, there is $E \in \Sigma$ such that $\mu(E) < \varepsilon$ and $\limsup\limits_{N \to \infty} \mu_N 1_E \geq \delta$ a.e.. The sequence is *strongly sweeping out* if it is 1 sweeping out. A sequence (n_k), is δ *sweeping out* (strongly sweeping out) if the sequence (μ_N), where $\mu_N = \frac{1}{N} \sum_{k=1}^{N} \delta_{n_k}$, is δ sweeping out (strongly sweeping out).

It is also convenient to extend Definitions 5.10 and 5.11 to the context above. A sequence (μ_N) is L_p *(persistently) universally bad* if for all ergodic dynamical systems (and all subsequences (N_s)), there exists $f \in L_p(X)$ such that $\lim\limits_{N \to \infty} \mu_N f(x) \left(\lim\limits_{s \to \infty} \mu_{N_s} f(x) \right)$ fails to exist for all x in a set of positive measure.

Examples later will show that (n_k) exists which is L_∞ universally bad, but which is only δ sweeping out for some $0 < \delta \leq \frac{1}{2}$. However, it is not clear whether an L_∞ universally bad sequence must be δ sweeping out for some $\delta > 0$. On the other hand, by Theorem 5.4, if (μ_N) is strongly sweeping out, then for a residual subset $\mathcal{R}$ of Σ in the symmetry pseudo-metric, if $E \in \mathcal{R}$, then $\limsup_{n \to \infty} \mu_N 1_E = 1$ a.e. and $\liminf_{N \to \infty} \mu_n 1_E = 0$ a.e.. This is the worst possible oscillation that can occur, but in any case δ sweeping out always implies the failure of a.e. convergence. We will need the following simple results.

Exercise 5.16. If (μ_N) is δ sweeping out, then for all $\delta' < \delta$ and $\varepsilon > 0$, and all ergodic dynamical systems, there exists $E \in \Sigma$, $m(E) < \varepsilon$, and a set $X' \subset X$ with $m(X \backslash X') \leq \varepsilon$ such that

$$\limsup_{N \to \infty} \mu_N 1_E - \liminf_{N \to \infty} \mu_N 1_E \geq \delta'$$

for all $x \in X'$.

Corollary 5.17. *If a sequence is δ sweeping out, then it is L_∞ universally bad.*

Notes to Chapter V

1. Theorem 5.1 is an extension of the principle discussed at the beginning of Chapter III.
2. Theorems 5.2, 5.3, 5.4 and 5.7 appeared in [del Junco, Rosenblatt, 1979].
3. Theorem 5.5 was stated in [Bellow, Jones, and Rosenblatt, 1990].
4. Theorem 5.8 is in [Bellow, Jones, and Rosenblatt, 1994].
5. The term "strong sweeping out" was introduced by A. Bellow because of the term "sweeping out" used in [Ellis, Friedman, 1978] for what is essentially the hypothesis of Theorem 5.4.
6. The discussion at the end of Chapter V is from [Rosenblatt, 1991].

CHAPTER VI

THE ENTROPY METHOD

1. Bourgain's Entropy Method

The central idea in this section is to apply the main theorem in [Bourgain, 1988b] in order to prove the failure of almost everywhere convergence in several contexts. Stating this theorem for one special case used here, let (μ_N) be a sequence of probability measures on $\mathbb{Z}$. For a given dynamical system, to say that the L_2 *entropy of* (μ_N) *is finite* means that for all $\beta > 0$ there is $M_\beta < \infty$ such that if $f \in L_2(X)$, $\|f\|_2 \leq 1$, and $N_1, \ldots, N_M \geq 1$ are distinct and such that $\|\mu_{N_i} f - \mu_{N_j} f\|_2 \geq \beta$ for all $i \neq j$, then $M \leq M_\beta$.

Theorem 6.1 (Bourgain). . *If* (μ_N) *is a sequence of probability measures on* $\mathbb{Z}$ *and* (X, Σ, m, τ) *is an ergodic dynamical system, then the* L_2 *entropy of* (μ_N) *is finite whenever* $\lim_{N \to \infty} \mu_N f(x)$ *exists a.e. for all* $f \in L_p(X)$ *for some fixed* p, $1 \leq p \leq \infty$.

Remark. A stronger fact is actually proved in [Bourgain, 1988b]. It is remarked there that if $\lim_{N \to \infty} \mu_N f(x)$ exists a.e. for all $f \in L_\infty(X)$, then $\int \sup_{N \geq 1} |\mu_N f| dm \to 0$ as $\int |f| dm \to 0$ among $f \in L_\infty(X)$, $\|f\|_\infty \leq 1$. This is not hard to prove here, because our operators (μ_N) satisfy $\mu_N 1 = 1$ and $\mu_N \geq 0$. But it is also true in the generality of [Bourgain, 1988b]. Then it is shown in [Bourgain, 1988b] that the L_2 entropy of (μ_N) is finite if $\int \sup_{N \geq 1} |\mu_N f| dm \to 0$ as $\int |f| dm \to 0$ among $f \in L_\infty(X)$, $\|f\|_\infty \leq 1$. This version of Bourgain's theorem is very useful for stating sweeping out results.

In order to handle a few details later that give sweeping out results, consider the following definitions for a sequence (μ_N) of probability measures on $\mathbb{Z}$. Given a dynamical system and a sequence (μ_N) of probability measures, let $\varepsilon > 0$ and define $\mathcal{D}_{\tau,\varepsilon}$ to be the supremum of $\int \sup_{N \geq 1} |\mu_N f(x)| dm(x)$ taken over $f \in L_\infty(X)$, $|f| \leq 1$, $\int |f| dm \leq \varepsilon$. Let $\Delta_{\tau,\varepsilon}$ be the supremum of $\int \sup_{N \geq 1} \mu_N 1_E(x) dm(x)$ taken over $E \in \Sigma$, $m(E) \leq \varepsilon$.

Proposition 6.2. *The maximal quantities* $\mathcal{D}_{\tau,\varepsilon}$ *and* $\Delta_{\tau,\varepsilon}$ *satisfy* $\mathcal{D}_{\tau,\varepsilon} \geq \Delta_{\tau,\varepsilon}$ *and* $\Delta_{\tau,\varepsilon/\delta} \geq \mathcal{D}_{\tau,\varepsilon} - \delta$.

Proof. The first inequality is trivial. For the second, notice that if $\delta > 0$ and $|f| \leq 1$, then

$$
\int \sup_{N \geq 1} |\mu_N f(x)| dm(x) \leq \int \sup_{N \geq 1} \mu_N |f| dm
$$

$$
\leq \int \sup_{N \geq 1} \mu_N 1_{\{|f| \geq \delta\}} dm + \int \sup_{N \geq 1} \mu_N (\delta 1_{\{|f| < \delta\}}) dm
$$

$$
\leq \int \sup_{N \geq 1} \mu_N 1_{\{|f| \geq \delta\}} dm + \delta,
$$

and

$$
\int |f| dm \geq \delta m\{|f| \geq \delta\}.
$$

Hence, if $\int |f| dm \leq \varepsilon$ and $\int \sup_{N \geq 1} |\mu_N f| dm \geq \mathcal{D}_{\tau,\varepsilon} - \varepsilon'$ for some $\varepsilon' > 0$, then $m\{|f| \geq \delta\} \leq \varepsilon/\delta$ and $\int \sup_{N \geq 1} |\mu_N 1_{\{|f| \geq \delta\}}| dm \geq \mathcal{D}_{\tau,\varepsilon} - \varepsilon' - \delta$. Thus, $\Delta_{\tau,\varepsilon/\delta} \geq \mathcal{D}_{\tau,\varepsilon} - \delta$.   □

Exercise 6.3. We have $\lim_{\varepsilon \to 0} \mathcal{D}_{\tau,\varepsilon} = 0$ if and only if $\lim_{\varepsilon \to 0} \Delta_{\tau,\varepsilon} = 0$.

Remark. Both $\Delta_{\tau,\varepsilon}$ and $\mathcal{D}_{\tau,\varepsilon}$ are decreasing as ε decreases, and Proposition 10 shows $\lim_{\varepsilon \to 0} \Delta_{\tau,\varepsilon} = \lim_{\varepsilon \to 0} \mathcal{D}_{\tau,\varepsilon}$.

The next lemma is just the Conze principle in a special form; see [Conze, 1973] and our earlier remarks. The proof of this proposition is left as an exercise.

Proposition 6.4. *Given two dynamical systems $(X_1, \Sigma_1, m_1, \sigma)$ and (X, Σ, m, τ), where τ is an aperiodic invertible measure-preserving transformation, $\mathcal{D}_{\sigma,\varepsilon} \leq \mathcal{D}_{\tau,\varepsilon}$ and $\Delta_{\sigma,\varepsilon} \leq \Delta_{\tau,\varepsilon}$ for all $\varepsilon > 0$.*

Remark. This shows that $\Delta_{\tau_1,\varepsilon} = \Delta_{\tau_2,\varepsilon}$ and $\mathcal{D}_{\tau_1,\varepsilon} = \mathcal{D}_{\tau_2,\varepsilon}$ if τ_1 and τ_2 are ergodic.

Proposition 6.5. *If there is a dynamical system $(X_1, \Sigma_1, m_1, \sigma)$ with $\lim_{\varepsilon \to 0} \mathcal{D}_{\sigma,\varepsilon} \neq 0$, then there exists $\delta > 0$ such that (μ_N) is δ sweeping out.*

Proof. From Proposition 6.3, there is a $\delta > 0$ such that for all ergodic dynamical systems (X, Σ, m, τ) and all $\varepsilon > 0, \mathcal{D}_{\tau,\varepsilon} \geq 2\delta$. By the proof of Corollary 6.3, we have $\Delta_{\tau,\varepsilon} \geq \delta$ for all $\varepsilon > 0$. We proceed with the analysis on the dynamical system (X, Σ, m, τ).

For any $\delta' > 0$, $\varepsilon > 0$, there exists $E \in \Sigma$, $m(E) < \varepsilon$, $\int \sup_{N \geq 1} \mu_N 1_E dm \geq \delta - \delta'$. Since $\varepsilon > 0$ is arbitrary, it is easy to see that for any $M \geq 1$, there exists $E_M \in \Sigma$, $m(E_M) \leq \varepsilon/2^M$, with $\int \sup_{N \geq M} \mu_N 1_{E_M} dm \geq \delta - 2\delta'$. But then $E_0 = \bigcup\limits_{M=1}^{\infty} E_M$ has $m(E_0) \le \varepsilon$, and

$$\int \limsup_{N\to\infty} \mu_N 1_{E_0} dm = \lim_{M\to\infty} \int \sup_{N\ge M} \mu_N 1_{E_0} dm$$
$$\ge \delta - 2\delta'.$$

Fix δ_0, $0 < \delta_0 < \delta - 2\delta'$. Then if $L_{\delta_0} = \{\limsup\limits_{N\to\infty} \mu_N 1_{E_0} \ge \delta_0\}$,

$$\delta - 2\delta' \le \int \limsup_{N\to\infty} \mu_N 1_{E_0} dm$$
$$\le \int_{L_{\delta_0}} 1 dm + \int_{X\backslash L_{\delta_0}} \delta_0 dm.$$

Hence, $m(L_{\delta_0}) \ge (\delta - 2\delta' - \delta_0)/(1 - \delta_0)$. If one lets $\delta' = \delta/8$ and $\delta_0 = \delta/4$, then, this shows that for all $\varepsilon > 0$ there exists $E \in \Sigma$, $m(E) \le \varepsilon$, with $m\{\limsup\limits_{N\to\infty} \mu_N 1_E \ge \delta/4\} \ge (\delta/2)/(1 - \delta/4) \ge \delta/2$.

Since $\delta > 0$ is fixed and $\varepsilon > 0$ is arbitrary, as in the sweeping out theorem, Theorem 5.5, we can show that for all $\varepsilon > 0$ there exists $E \in \Sigma$, $m(E) \le \varepsilon$, with $\limsup\limits_{N\to\infty} \mu_N 1_E \ge \delta/4$ a.e.. $\square$

These propositions give this basic variation on Bourgain's theorem.

Theorem 6.6. *Given a sequence (μ_N) of probability measures on $\mathbb{Z}$, if there is an ergodic dynamical system $(X_1, \Sigma_1, m_1, \sigma)$ for which the L_2 entropy of (μ_N) is not finite, then (μ_N) is δ sweeping out for some $\delta > 0$.*

Proof. The remarks after Theorem 6.1 show $\lim\limits_{\varepsilon\to 0} \mathcal{D}_{\sigma,\varepsilon} \ne 0$. Proposition 6.5 finishes the proof. $\square$

Remark 6.7. It is possible to take a different viewpoint to the above transfer from a fact about one ergodic dynamical system to a fact about all ergodic dynamical systems. For a given (μ_N) and dynamical system $(X_1, \Sigma_1, m_1, \sigma)$, let $\mathcal{M}_{\sigma,\beta}$ be the supremum over all M in the definition of the L_2 entropy. As in the proof of Proposition 6.3, for any aperiodic dynamical system (X, Σ, m, τ), if $\varepsilon > 0$ then $\mathcal{M}_{\tau,\beta} \ge \mathcal{M}_{\sigma,\beta+\varepsilon}$. Thus, if $\mathcal{M}_{\sigma,\beta}$ is not finite for some $\beta > 0$, then there is some $\beta > 0$ such that $\mathcal{M}_{\tau,\beta}$ is infinite for all ergodic dynamical systems (X, Σ, m, τ). So Theorem 6.6 could be proved by applying the Conze principle at a different point in the argument.

Before proving the main result, consider this analog of the well-known metric theorem of Weyl, Theorem 2.22, in the theory of uniformly distributed sequences. Here, as in several places in this chapter, the Fourier transform $\hat\mu$ of μ will be very important. This is defined by $\hat\mu(\gamma) = \sum\limits_{k=-\infty}^{\infty} \mu(k)\gamma^{-k}$ for all $\gamma \in \mathbb{T}$.

Proposition 6.8. *Given a sequence of probability measures on* $\mathbb{Z}$, *if* $\lim_{N \to \infty} \max_k \mu_N(k) = 0$, *then for some subsequence* (N_s), $\lim_{s \to \infty} \hat{\mu}_{N_s}(\gamma) = 0$ *for a.e.* γ.

Proof. Here $\hat{\mu}_N(\gamma) = \sum_{k=-\infty}^{\infty} \mu_N(k)\gamma^{-k}$. It follows that $\int_{\mathbb{T}} |\hat{\mu}_N(\gamma)|^2 \, d\gamma = \sum_{k=-\infty}^{\infty} \mu_N(k)^2 \leq \max_k \mu_N(k)$. Choose (N_s) such that $\sum_{s=1}^{\infty} \max_k \mu_{N_s}(k) < \infty$. Then $\int_{\mathbb{T}} \sum_{s=1}^{\infty} |\hat{\mu}_{N_s}(\gamma)|^2 \, d\gamma = \sum_{s=1}^{\infty} \int_{\mathbb{T}} |\hat{\mu}_{N_s}(\gamma)|^2 \, d\gamma \leq \sum_{s=1}^{\infty} \max_k \mu_{N_s}(k) < \infty$. So $\sum_{s=1}^{\infty} |\hat{\mu}_{N_s}(\gamma)|^2$ converges to a finite value for a.e. γ, and for a.e. γ, $\lim_{s \to \infty} \hat{\mu}_{N_s}(\gamma) = 0$. □

Remark 6.9. In the case $\mu_N = \frac{1}{N} \sum_{k=1}^{N} \delta_{n_k}$ for some sequence $(n_k) \in \mathbb{Z}$, Theorem 2.22 is stating the stronger fact that $\lim_{N \to \infty} \hat{\mu}_N(\gamma) = 0$ a.e.. In [Rosenblatt, 1994] an example is given to show a subsequence may be needed in general.

Theorem 6.10. *Suppose* (μ_N) *is a dissipative sequence and there exists* $b > 0$ *such that* $\liminf_{N \to \infty} |\hat{\mu}_N(\gamma)| \geq b$ *for all* $\gamma \in D$, *where* D *is a dense subset of* $\mathbb{T}$. *Then* (μ_N) *is* δ *sweeping out for some* $\delta > 0$.

Proof. There is no loss in assuming $\lim_{N \to \infty} \max_k \mu_N(k) = 0$. Otherwise, because (μ_N) is dissipative, there exists a $c > 0$ and increasing sequences (n_k) and (N_k) such that $\mu_{N_k} \geq c\delta_{n_k}$ for all $k \geq 1$. Then for any ergodic dynamical system, if $\varepsilon > 0$, by the sweeping out theorem of [Ellis, Friedman, 1978], there is $E \in \Sigma$ such that $m(E) < \varepsilon$ and $\limsup_{k \to \infty} 1_{\tau^{n_k} E} = 1$ a.e.. Hence, $\limsup_{k \to \infty} \mu_{N_k} 1_E \geq c$ a.e. and $\limsup_{N \to \infty} \mu_N 1_E \geq c$ a.e..

So we can assume $\lim_{N \to \infty} \max_k \mu_N(k) = 0$, and Proposition 6.8 applies; thus, by passing to a subsequence, it may be assumed also that $\lim_{N \to \infty} \hat{\mu}_N(\gamma) = 0$ a.e.. Theorem 6.6 says that only one ergodic dynamical system in which the L_2 entropy is infinite is needed to conclude the proof. It will be shown that if $\sigma(x) = \theta + x \mod 1$, where θ is irrational, then $[0, 1]$, with Lebesgue measure λ, with the transformation σ, is such a dynamical system.

First, fix $M \geq 1$ and let $(\sigma_{ij} : i = 1, \ldots, 2^M, j = 1, \ldots, M)$ be an arbitrary matrix of 0's and 1's. By induction it is possible to choose $\gamma_1, \ldots, \gamma_{2^M} \in \mathbb{T}$ and $N_1, \ldots, N_M$ distinct such that

a) $|\hat{\mu}_{N_j}(\gamma_i)| < b/4$ if $\sigma_{ij} = 0$

and

b) $|\hat{\mu}_{N_j}(\gamma_i)| > b/2$ if $\sigma_{ij} = 1$.

To see this, let $V \subset \mathbb{T}$ be a dense set such that $\lim\limits_{N \to \infty} \hat{\mu}_N(\gamma) = 0$ for all $\gamma \in V$. Recall $\liminf\limits_{N \to \infty} |\hat{\mu}_N(\gamma| \geq b$ for all $\gamma \in D$, where D is a dense set in $\mathbb{T}$. Choose γ_i^1 in V or D as σ_{i1} is 0 or 1, respectively. There exists $N_0 \geq 1$ such that if $N \geq N_0$ then $|\hat{\mu}_N(\gamma_i^1)| < b/4$ if $\sigma_{i1} = 0$ and $|\hat{\mu}_N(\gamma_i^1)| > b/2$ if $\sigma_{i1} = 1$. Thus, we can choose N_1 so that a), b) are satisfied for all (i,j) with $j = 1$, when γ_i^1 is used in place of γ_i. Assume now that $(\gamma_i^J : i = 1, \ldots, 2^M, J = 1, \ldots, M_0)$ and $N_1, \ldots, N_{M_0}$ exist so that for each $J \leq M_0$, a), b) hold for all (i,j) with $j \leq J$ when γ_i^J replaces γ_i. Choose $\gamma_i^{M_0+1}$ in V or D as σ_{i,M_0+1} is 0 or 1. By continuity of the $\hat{\mu}_{N_i}$, $i = 1, \ldots, M_0$, and the density of V and D, $\gamma_i^{M_0+1}$ can also be chosen so that a), b) still hold for all (i,j), $j \leq M_0$, with $\gamma_i^{M_0+1}$ replacing $\gamma_i^{M_0}$. Then choose N_{M_0+1} so that a), b) hold for all (i,j) with $j = M_0 + 1$. This completes the induction step of the choice of $(\gamma_i^J : i = 1, \ldots, 2^M, J = 1, \ldots, M)$ such that a), b) hold for all (i,j) with $j \leq J$ when γ_i^J replaces γ_i. Let $\gamma_i = \gamma_i^M$ for $i = 1, \ldots, M$. This gives a), b) for all $(i,j), i = 1, \ldots, 2^M, j = 1, \ldots, M$.

Choose any $\gamma_0 \in \mathbb{T}$ of infinite order. Then $\{\gamma_0^\ell : \ell \in \mathbb{Z}\}$ is dense in $\mathbb{T}$. Choose $(\sigma_{ij} : i = 1, \ldots, 2^M, j = 1, \ldots, M)$ to be 0's and 1's in such way that if $\mathcal{L}_j = \{i : \sigma_{ij} = 1\}$, then $\#\mathcal{L}_j = 2^{M-1}$ and the $\mathcal{L}_j$ are independent in that $\#(\mathcal{L}_{j_1} \cap \mathcal{L}_{j_2}) = 2^{M-2}$ for all $j_1 \neq j_2$, $1 \leq j_1, j_2 \leq M$. Let (γ_i) and (N_i) satisfy a), b) above for this choice of (σ_{ij}). Then choose $\ell_1, \ldots, \ell_{2^M}$ so that $\gamma_0^{\ell_i}$ is sufficiently close to γ_i in $\mathbb{T}$ that a), b) still hold with $\gamma_0^{\ell_i}$ replacing γ_i throughout. Let $f(\gamma) = \frac{1}{\sqrt{2^M}} \sum\limits_{i=1}^{2^M} \overline{\gamma}^{\ell_i}$ for all $\gamma \in \mathbb{T}$. Then $\|f\|_2 = 1$ and, with respect to the dynamical system of $\mathbb{T}$ with Lebesgue measure, and $\sigma(\gamma) = \gamma_0 \gamma$,

$$
\begin{aligned}
\mu_N f(\gamma) &= \sum_{k=-\infty}^{\infty} \mu_N(k) f(\gamma_0^k \gamma) \\
&= \frac{1}{\sqrt{2^M}} \sum_{i=1}^{2^M} \sum_{k=-\infty}^{\infty} \mu_N(k) \overline{\gamma}_0^{\ell_i k} \overline{\gamma}^{\ell_i} \\
&= \frac{1}{\sqrt{2^M}} \sum_{i=1}^{2^M} \hat{\mu}_N(\gamma_0^{\ell_i}) \overline{\gamma}^{\ell_i} .
\end{aligned}
$$

Hence, for any N_{j_1}, N_{j_2},

$$
\| \mu_{N_{j_1}} f - \mu_{N_{j_2}} f \|_2^2 = \frac{1}{2^M} \sum_{i=1}^{2^M} |\hat{\mu}_{N_{j_1}}(\gamma_0^{\ell_i}) - \hat{\mu}_{N_{j_2}}(\gamma_0^{\ell_i})|^2 ,
$$

by the orthonormality of the functions $\gamma \mapsto \overline{\gamma}^p$. But then if $N_{j_1} \neq N_{j_2}$,

$$\| \mu_{N_{j_1}} f - \mu_{N_{j_2}} f \|_2^2 \geq \frac{1}{2M} \sum_{\mathcal{L}_{j_1} \cap \mathcal{L}_{j_2}^c} |\hat{\mu}_{N_{j_1}}(\gamma_0^{\ell_i}) - \hat{\mu}_{N_{j_2}}(\gamma_0^{\ell_i})|^2$$

$$+ \frac{1}{2M} \sum_{\mathcal{L}_{j_2} \cap \mathcal{L}_{j_1}^c} |\hat{\mu}_{N_{j_1}}(\gamma_0^{\ell_i}) - \hat{\mu}_{N_{j_2}}(\gamma_0^{\ell_i})|^2$$

$$\geq \frac{1}{2M} \#(\mathcal{L}_{j_1} \cap \mathcal{L}_{j_2}^c) \left(\frac{b}{4}\right)^2 + \frac{1}{2M} \#(\mathcal{L}_{j_2} \cap \mathcal{L}_{j_1}^c) \left(\frac{b}{4}\right)^2$$

$$= \frac{1}{4} \left(\frac{b}{4}\right)^2 + \frac{1}{4} \left(\frac{b}{4}\right)^2 = \frac{1}{2} \left(\frac{b}{4}\right)^2.$$

Hence, for all $j_1 \neq j_2$, $\| \mu_{N_{j_1}} f - \mu_{N_{j_2}} f \|_2 \geq b/4\sqrt{2}$. Since $j_i, j_2 = 1, \ldots, M$ is allowed here and $M \geq 1$ is arbitrary, this shows that the entropy of (μ_N) on $(\mathbb{T}, \Sigma, m, \sigma)$ is infinite, completing the proof.   □

Remark. Clearly D need only be dense in some $E \subset \mathbb{T}$ with $\lambda(E) > 0$.

2. Application of Entropy to Constructing Bad Sequences

Theorem 6.10 will allow us to show that any lacunary sequence is δ sweeping out and L_∞ inherently universally bad. First,

Exercise 6.11. Given $\alpha > 0$, there exists $r = r(\alpha)$ such that if $\inf_k n_{k+1}/n_k \geq r$, then for any choice of $\sigma_k = \pm 1$, there is a dense set $D \subset [0,1]$ such that for all $\theta \in D$ there is $k = k(\theta) \geq 1$ such that if $k \geq k(\theta)$, $|\exp(2\pi i n_k \theta) - \sigma_k| \leq \alpha$.

Exercise 6.12. 1) The above construction can be carried out to guarantee D consists only of irrationals. If one chooses all $\sigma_k = 1$ and $\alpha = 1/2$, this shows that for suitably lacunary (n_k) there exists $D \subset \mathbb{T}$, D dense, such that for all dissipative sequences supported in (n_k), if $\gamma \in \mathbb{T}$ there is $N_\gamma \geq 1$ such that for $N \geq N_\gamma$, $|\hat{\mu}_N(\gamma)| \geq 1/4$.
  2) If (n_k) is lacunary, does there exist $\theta \in (0,1)$ such that the limit

$$\lim_{N \to \infty} \frac{1}{N} \sum_{k=1}^{N} e^{2\pi i n_k \theta} = 1?$$

Theorem 6.13. *Suppose (n_k) is a sequence in $\mathbb{Z}$ with $\inf_{k \geq 1} n_{k+1}/n_k > 1$. Then any dissipative sequence (μ_N) supported in (n_k) is δ sweeping out for some $\delta > 0$; so (n_k) is L_∞ inherently universally bad and δ sweeping out for some $\delta > 0$.*

Proof. Suppose (μ_N) is a dissipative sequence supported in (n_k). Let $\gamma = \inf_k n_{k+1}/n_k$, and choose $R \geq 1$ such that $\gamma^R \geq r\left(\frac{1}{2}\right)$, where $r(\alpha)$ is as in Exercise 6.11. Each subsequence $(\eta_k^t) = (n_{kR+t} : k \geq 1)$, where

$t = 0, \ldots, R - 1$, satisfies $\inf\limits_{k} \eta^t_{k+1}/\eta^t_k \geq \gamma^R \geq r\left(\frac{1}{2}\right)$. Each $\mu_N = \sum\limits_{t=0}^{R-1} \nu_{Nt}$,
where ν_{Nt} are positive measures supported in (η^t_k). There is some $t_0 = 0, \ldots, R - 1, c > 0$, and an infinite sequence (N_s) such that $\| \nu_{N_s t_0} \|_1 \geq c$ for all $s \geq 1$. If $\nu_s = \nu_{N_s t_0}/ \| \nu_{N_s t_0} \|_1$, then (ν_s) is a dissipative sequence and the remark above shows that $\liminf\limits_{s \to \infty} |\hat\nu_s(\gamma)| \geq \frac{1}{4}$ for all γ in a dense subset D of $\mathbb{T}$. Theorem 6.10 then shows $(\nu_s : s \geq 1)$ is δ sweeping out, and thus (μ_N) is $c\delta$ sweeping out, because $\mu_{N_s} \geq c\nu_s$ for all $s \geq 1$, for some $\delta > 0$ $\qquad\qquad\qquad\qquad\qquad\qquad\qquad\qquad\qquad\qquad\qquad\qquad\quad \square$

Remark. Lacunary sequences are the prototype of what are now called Sidon sets. It is not known if every Sidon set is universally bad.

The technique of Theorem 6.10 also allows one to construct non-lacunary sequences which are still pervasively universally bad. Specifically, consider the following.

Exercise 6.15. There exists an L_∞ inherently universally bad sequence (n_k) such that $\lim\limits_{k \to \infty} n_{k+1}/n_k = 1$.

Hint: Consider $\{2^m 3^n : m, n \geq 1\}$ ordered as an increasing sequence (n_k). It is easy to see that $\lim\limits_{k \to \infty} n_{k+1}/n_k = 1$.

Remark. More is proved above than is stated: it is shown that any dissipative sequence (μ_N) supported in (n_k) is δ sweeping out.

Roger Jones pointed out another example of a non-lacunary L_∞ universally bad sequence which will be useful here. Let $\llbracket \cdot \rrbracket$ denote the greatest integer function.

Proposition 6.16. Let $n_k = k\llbracket \log_2 k \rrbracket$ for $k \geq 1$. Then (n_k) is δ sweeping out for some $\delta > 0$.

Proof. Given an ergodic dynamical system (X, Σ, m, τ), the averages
$$\mu_N 1_E(x) = \frac{1}{N} \sum_{k=1}^{N} 1_E(\tau^{n_k} x) \text{ satisfy } \mu_{2N-1} 1_E(x) \geq \frac{1}{2N-1} \sum_{k=N}^{2N-1} 1_E(\tau^{n_k} x) \geq$$
$\frac{1}{2} d_N 1_E(x)$, where $d_N = \frac{1}{N} \sum\limits_{k=N}^{2N-1} \delta_{n_k}$. So it suffices to show (d_N) is δ sweeping out. But if $N = 2^{2^M}$, $d_N = \frac{1}{N} \sum\limits_{k=N}^{2N-1} \delta_{k2^M}$. Hence for all dyadic rational θ, $\lim\limits_{M \to \infty} \hat d_{2^{2^M}}(\exp(2\pi i \theta)) = 1$. By Theorem 6.10, (d_N) is δ sweeping out for some $\delta > 0$. $\qquad\qquad\qquad\qquad\qquad\qquad\qquad\qquad\qquad\qquad\quad \square$

Remark. Here $(n_k) = (k\llbracket \log_2 k \rrbracket)$ is strictly increasing and $\lim\limits_{k \to \infty} n_{k+1}/n_k = 1$; so the above proof that (n_k) is δ sweeping out, and therefore L_∞ universally bad, gives another example like that of Theorem 6.12. However, it is not clear if this (n_k) is L_∞ persistently bad.

Generally, δ sweeping out does not imply strongly sweeping out, as the following two examples will show.

Example 6.17. a) To see this for sequences of probability measures, let
$\nu_N = \frac{1}{N} \sum_{k=1}^{N} \delta_k$ and let $\nu_N = \frac{1}{N} \sum_{k=1}^{N} \delta_{2^k}$ for $N \geq 1$. Let $\omega_N = \frac{1}{2}\nu_N + \frac{1}{2}\mu_N$.
Then because $\omega_N \geq \frac{1}{2}\nu_N$, Theorem 6.13 shows that (ω_N) is δ sweeping
out for some $\delta > 0$. However, such a δ cannot be larger than $\frac{1}{2}$; so (ω_N)
is not strongly sweeping out. Indeed, for all ergodic dynamical systems, if
$f \in L_1(X)$, $\lim_{N \to \infty} \mu_N f(x) = \int f\, dm$ a.e. So for ergodic dynamical systems,
$\limsup_{N \to \infty} \omega_N 1_E(x) = m(E) + \limsup_{N \to \infty} \frac{1}{2}\nu_N 1_E \leq m(E) + \frac{1}{2}$. This same method,
with $\omega_N = (1 - \varepsilon)\mu_N + \varepsilon\nu_N$ for some $\varepsilon > 0$, gives an example of (ω_N) which
is δ sweeping out, but only for $\delta \leq \varepsilon$.

b) The essence of Example a) can be captured for sequences (n_k) using
the sequence $(q_k) = (k \llbracket \log_2 k \rrbracket : k \geq 4)$ in Proposition 6.16. Let (p_k) be the
prime numbers. Bourgain [1988] (see also [Wierdl, 1988]) showed that for all
dynamical systems, if $f \in L_2(X)$, $\lim_{N \to \infty} \frac{1}{N} \sum_{k=1}^{N} f(\tau^{p_k} x)$ exists a.e. Now each
q_k is composite, so $\{p_k\} \cap \{q_k\} = \phi$. Also, it is easy to see that $\#\{q_k \leq N\} \sim \frac{N}{\log_2 N}$ as $N \to \infty$. But the prime number theorem says that $\#\{p_k \leq N\} \sim \frac{CN}{\log_2 N}$ as $N \to \infty$, where $C = \log_2 e$. Let $\mathcal{N} = \{p_k\} \cup \{q_k\}$, and emunerate
$\mathcal{N} = (n_k)$ as an increasing sequence. This (n_k) is δ sweeping out for some
$\delta > 0$, but it is not strongly sweeping out. To see this, write $\frac{1}{N} \sum_{k=1}^{N} \delta_{n_k} = \frac{N_1}{N} \frac{1}{N_1} \sum_{k=1}^{N_1} \delta_{p_k} + \frac{N_2}{N} \frac{1}{N_2} \sum_{h=1}^{N_2} \delta_{q_k}$, where $N_1 = \#\{p_k \leq n_N\}$ and $N_2 = \#\{q_k \leq n_N\}$. Then $N_1 + N_2 = N$, and so $\lim_{N \to \infty} \frac{N}{N_2} = 1 + \lim_{N \to \infty} \frac{N_1}{N_2} = 1 + C$. But
then for any ergodic dynamical system, $\lim_{N \to \infty} \frac{1}{N} \sum_{k=0}^{N} 1_E(\tau^{p_k} x) = f^*(x)$ a.e.
with $\int f^* dm = m(E)$. Thus, $\limsup_{N \to \infty} \frac{1}{N} \sum_{k=1}^{N} 1_E(\tau^{n_k} x) \leq \frac{C}{1+C} f^*(x) + \frac{1}{1+C}$.
Because $\int f^* dm = m(E)$, this shows (n_k) is at most $\frac{1}{1+C}$ sweeping out. It
also shows that $\limsup_{N \to \infty} \frac{1}{N} \sum_{k=1}^{N} 1_E(\tau^{n_k} x) \geq \frac{1}{1+C} \limsup_{N \to \infty} \frac{1}{N_1} \sum_{k=1}^{N_1} 1_E(\tau^{q_k} x) = \frac{1}{1+C} \limsup_{N \to \infty} \frac{1}{N} \sum_{k=1}^{N} 1_E(\tau^{q_k} x)$. So by Proposition 6.16, (n_k) is δ sweeping out
for some $\delta > 0$.

The examples of Exercise 6.15 and Proposition 6.16 show that rapidity
of growth of (n_k) is not particularly necessary for universally bad behavior.
In this same direction, consider the following two theorems.

Theorem 6.18. *Given (d_N) with $\lim_{N \to \infty} d_N = \infty$ and any (s_N), let $\mu_N = \frac{1}{N} \sum_{k=1}^{N} \delta_{s_N + k d_N}$ for $N \geq 1$. Then (μ_N) is δ sweeping out for some $\delta > 0$.*

Proof. By Theorem 6.10, it suffices to show there is a dense set $D \subset \mathbb{T}$, $b > 0$, and a sequence (N_s) such that $\liminf_{s \to \infty} |\hat{\mu}_{N_s}(\gamma)| \geq b$ for all $\gamma \in D$.

Since $|\hat{\mu}_N(\gamma)| = \frac{1}{N} | \sum_{k=1}^{N} \gamma^{kd_N}|$, without loss of generality all s_N can be taken to be 0.

Now the geometric sum $A_N = \frac{1}{N} \sum_{k=1}^{N} z^{kd}$, $z = \exp(2\pi i \theta)$, for $z^d \neq 1$, has

$|A_N| = \frac{1}{N} \frac{|z^{Nd}-1|}{|z^d-1|}$. Let $D_N = \{\theta : \text{for some } p = 0, \ldots, d-1, \frac{1}{3dN} \leq |\theta - \frac{p}{d}| \leq \frac{1}{2dN}\}$. If $\theta \in D_N$, then $|d\theta - p| \leq \frac{1}{2N}$ for some p, and so $|\exp(2\pi i d\theta) - 1| \leq \frac{\pi}{N}$. That is, $\frac{1}{N|z^d-1|} \geq \frac{1}{\pi}$. But $dN\theta$ cannot be too close to $\mathbb{Z}$. Indeed if there is $P \in \mathbb{Z}$ such that $|dN\theta - P| \leq \frac{1}{8}$, then $|d\theta - \frac{P}{N}| \leq \frac{1}{8N}$, and so, using $|d\theta - p| \leq \frac{1}{2N}$, it follows that $|p - \frac{P}{N}| \leq \frac{5}{8N}$. Hence, $|Np - P| \leq \frac{5}{8}$, and so $Np = P$. But then we can use $\frac{1}{3dN} \leq |\theta - \frac{p}{d}|$ for the same p as above, giving $\frac{1}{3} \leq |dN\theta - Np| = |dN\theta - P| \leq \frac{1}{8}$, a contradiction. Hence, $|dN\theta - P| > \frac{1}{8}$ for all $P \in \mathbb{Z}$, and thus $|\exp(2\pi i dN\theta) - 1| = |z^{dN} - 1| \geq C$ for some universal constant ($C = 2\sin\left(\frac{1}{16}\right)$ will do). Thus, for $\theta \in D_N, |A_N(\theta)| \geq \frac{C}{\pi} \geq \frac{3}{100}$.

Because $\lim_{N \to \infty} d_N = \infty$, there is a subsequence (d_{N_s}) which is sufficiently lacunary that $\mathcal{D} = \bigcup_{S=1}^{\infty} \bigcap_{s=S}^{\infty} D_s$ is dense in $[0,1]$. Let $D = \{\exp(2\pi i \theta) : \theta \in \mathcal{D}\}$. Then $\liminf_{s \to \infty} |\hat{\mu}_{N_s}(\gamma)| \geq \frac{3}{100}$ for all $\gamma \in D$. $\square$

Remark. a) The proof above can be carried out in any subsequence (N_t). Hence, actually any (μ_{N_t}) is δ sweeping out, and so (μ_N) is L_∞ persistently universally bad.

b) One corollary of this result is that there cannot be any pointwise ergodic theorem for averages of the form $\frac{1}{N} \sum_{k=1}^{N} f(\tau^{kN} x)$.

Theorem 6.19. *Given any (g_k) with $g_k \geq 1$ and $\lim_{k \to \infty} g_k = \infty$, there exists (n_k) which is δ sweeping out, and thus L_∞ universally bad, such that $n_{k+1} - n_k \leq g_k$ for all $k \geq 1$.*

Proof. Choose any (d_N), $d_N \geq 1$, and then choose (s_N) so that the blocks of integers $B_N = (s_N + d_N, \ldots, s_N + M_N d_N)$ are disjoint, but $s_N + M_N d_N + 1 = s_{N+1} + d_{N+1}$. By choosing $M_{N+1}/M_N \geq 2$ for all $N \geq 1$, $\# \bigcup_{N=1}^{N_0} B_N \leq \# B_{N_0+1}$. Let (n_k) be an enumeration of $\bigcup_{N=1}^{\infty} B_N$ in increasing order. Then $\frac{1}{M} \sum_{k=1}^{M} \delta_{n_k} \geq \frac{1}{2} \frac{1}{M_L} \sum_{k \in B_L} \delta_k$ when $\{n_1, \ldots, n_M\} = \bigcup_{N=1}^{L} B_N$. If $\mu_L = \frac{1}{M_L} \sum_{k \in B_L} \delta_k$, $L \geq 1$, then (μ_L) is δ sweeping out by Theorem 6.16 if $\lim_{N \to \infty} d_N = \infty$. Also, if (d_N) increases slowly enough, then

$n_{k+1} - n_k \leq g_k$ for all $k \geq 1$. Thus, it is possible to choose (d_N) so that both conditions are met.   □

Remark. It is probably possible to construct (n_k) with $n_{k+1} - n_k$ growing as slowly as one likes, but such that (n_k) is L_∞ inherently universally bad. However, if $n_{k+1} - n_k$ remains bounded, then (n_k) is never L_1 universally bad. Indeed, this bound shows $\liminf\limits_{N \to \infty} \frac{\#\{n_k \leq N\}}{N} > 0$. Hence, by the usual ergodic theorem, there is a weak maximal estimate $m\{x :$
$\sup\limits_{N \geq 1} \frac{1}{N} | \sum\limits_{k=1}^{N} f(\tau^{n_k} x)| > \lambda\} \leq \frac{C}{\lambda}\|f\|_1$, for some constant C, for all $\lambda > 0$.
But then for a.e. irrational θ, $(n_k \theta)$ is uniformly distributed mod 1. Letting $\tau(x) = (x + \theta)$ mod 1, for all continuous $f : [0, 1] \to \mathbb{R}$, $f(0) = f(1)$, $\lim\limits_{N \to \infty} \frac{1}{N} \sum\limits_{k=1}^{N} f(\tau^{n_k} x) = \int_0^1 f(s)\, ds$ uniformly in x. Thus, for all $f \in L_1[0, 1]$, $\lim\limits_{N \to \infty} \frac{1}{N} \sum\limits_{k=1}^{N} f(\tau^{n_k} x)$ exists a.e. Hence, for sequences with bounded gaps, the right question is whether the sequence is L_1 universally good. The example in [Bourgain, 1989] of a sequence (n_k) which is universally good for mean convergence, and has a positive density, but is not L_1 universally good, can be used to some purpose here. Indeed, let $\{\eta_k\} = 2\mathbb{Z} \cup \{2n_k + 1 : k \geq 1\}$. Then as in Example 6.15 b), (η_k) is not universally good, but it does have gaps $\eta_{k+1} - \eta_k \leq 2$. It would be very interesting to obtain a characterization of sequences (n_k) with bounded gaps which are L_1 universally good.

3. Good and Bad Behavior of Powers

There have been many interesting applications of Bourgain's entropy method to proving failure of almost everywhere convergence. The one that we want to present in this section is for convolution powers μ^n of one measure μ.

Theorems due to Bellow, Jones and Rosenblatt give a complete analysis of the pointwise behavior of $(\mu^n f)$ for μ with finite second moment. To best understand these results, both the positive and negative ones presented here, we need some terminology. A probability measure μ on $\mathbb{Z}$ is *strictly aperiodic* if $\operatorname{supp}(\mu) = \{k \in \mathbb{Z} : \mu\{k\} \neq 0\}$ is not contained in a proper arithmetic progression, i.e., $|\hat{\mu}(\gamma)| < 1$ if $\gamma \neq 1$, $|\gamma| = 1$.

Definition 6.20. A probability measure μ on $\mathbb{Z}$ has *bounded angular ratio* if μ is strictly aperiodic and there exists $\epsilon > 0$, $K < \infty$ such that if $\lambda \in \mathbb{C}$, $|\lambda| = 1$, $\lambda \neq 1$, $|\lambda - 1| < \epsilon$, then $\frac{|\hat{\mu}(\gamma)-1|}{1-|\hat{\mu}(\gamma)|} \leq K$.

That is, μ has bounded angular ratio if and only if $\sup\limits_{|\gamma|=1,\ \gamma \neq 1} \frac{|\hat{\mu}(\gamma)-1|}{1-|\hat{\mu}(\gamma)|} < \infty$. Another characterization is that μ has bounded angular ratio if and only if $\hat{\mu}\{\gamma \in \mathbb{C} : |\gamma| = 1\}$ is contained in some proper Stolz region.

Theorem 6.21. *If μ has bounded angular ratio and $1 < p \leq \infty$, then there exists a constant $K_p < \infty$ such that for all $f \in L_p(X)$, $\| \sup_n |\mu^n f| \|_p \leq K_p \|f\|_p$. Moreover, given $f \in L_p(X)$, there exists a unique τ-invariant function $f^* \in L_p(X)$ such that $\lim_{n \to \infty} \mu^n f(x) = f^*(x)$ for a.e. x. If τ is ergodic, then $f^*(x) = \int f \, dm$ for a.e. x.*

Remark. The constant K_p does not depend on (X, Σ, m, τ).

Before proving this theorem, let us consider some of the examples of probability measures μ to which it applies. For instance, if μ is symmetric, then $\hat{\mu}(\gamma)$ is always real, and so $\frac{|1-\hat{\mu}(\gamma)|}{1-|\hat{\mu}(\gamma)|} = \frac{1-\hat{\mu}(\gamma)}{1-\hat{\mu}(\gamma)} = 1$ for γ sufficiently close to 1. So if μ is symmetric and strictly aperiodic, then μ has bounded angular ratio.

Other examples are given by the following. First, we extend the definition given before Theorem 5.8.

Definition 6.22. For $p > 0$, the *pth moment* of μ is $\sum_{k=-\infty}^{\infty} |k|^p \mu(k)$ and is denoted by $m_p(\mu)$.

Exercise 6.23. If $m_2(\mu) < \infty$ and $E(\mu) = 0$, and if μ is strictly aperiodic, then μ has bounded angular ratio.

Remark. If $m_2(\mu) < \infty$, the proof shows that $\frac{|1-\hat{\mu}(e^{ix})|}{1-|\hat{\mu}(e^{ix})|}$ is bounded for x near 0 if and only if $E(\mu) = 0$. Hence, if μ is strictly aperiodic and $m_2(\mu) < \infty$, then μ has bounded angular ratio if and only if $E(\mu) = 0$. However, if $m_2(\mu) = \infty$, μ can be strictly aperiodic, $E(\mu)$ can exist and equal 0, but μ fails to have bounded angular ratio. See [Bellow, Jones, and Rosenblatt, 1994] for further details.

The remarks above show that μ can have bounded angular ratio and not have $m_p(\mu) < \infty$ for any $p > 0$. Indeed, for a suitable constant $C > 0$, the measure $\mu = C \sum_{n=1}^{\infty} \frac{1}{n(\log^2 n + 1)} (\delta_n + \delta_{-n})$ is a probability measure. Since it is symmetric and strictly aperiodic, it has bounded angular ratio, yet $m_p(\mu) = \infty$ for any $p > 0$. In addition, there are probability measures that have bounded angular ratio but are far from being symmetric. An example like $\mu = \frac{1}{5}(3\delta_{-1} + \delta_1 + \delta_2)$ shows this in one sense, by Exercise 6.23. But here is a more extreme case.

Example 6.24. Consider $h(x) = 1 - (1 - x)^{1/2}$ for $0 \leq x \leq 1$. We can expand $h(x)$ as a power series $\sum_{n=1}^{\infty} c_n x^n$ which converges to $h(x)$ for $0 \leq x \leq 1$. For all $n \geq 1$, $c_n > 0$ and $\sum_{n=1}^{\infty} c_n = 1$. But then $\mu_0 = \sum_{n=1}^{\infty} c_n \delta_n$ is a probability measure which is strictly aperiodic. It also has bounded angular ratio, as we see below, and yet its support is entirely in $\mathbb{Z}^+$. This example shows in a second sense how far from symmetry μ can be and still have $(\mu^n f)$ satisfying maximal inequalities.

These examples should help clarify the hypotheses in Theorem 6.21. Another form of the transfer principle, from Chapter III, after Lemma 3.5, will be needed here.

Exercise 6.25. Let (d_n) be a sequence of finite measures on $\mathbb{Z}$. If $1 < p < \infty$ and there is a constant $K < \infty$ such that $\|(\sum\limits_{n=1}^{\infty} |d_n * \varphi|^2)^{\frac{1}{2}}\|_2 \le K\|\varphi\|_2$ for all $\varphi \in \ell_2(\mathbb{Z})$, then for any dynamical system (X, Σ, m, τ), $\|(\sum\limits_{n=1}^{\infty} |d_n f|^2)^{\frac{1}{2}}\|_2 \le K\|f\|_2$ for all $f \in L_2(X)$.

Proof of Theorem 6.21. Since μ is strictly aperiodic, and so $|\hat{\mu}(\gamma)| < 1$ for all $\gamma \in \mathbb{C}$, $|\gamma| = 1$, $\gamma \ne 1$, it is well-known that $\lim\limits_{n \to \infty} \|\mu^{n+1} * \delta_1 - \mu^n\|_1 = 0$. See [Rosenblatt, 1981] for references. So

$$\limsup_{n \to \infty} \|\mu^n(f\circ\tau - f)\|_\infty \le \limsup_{n \to \infty} \|\mu^n * \delta_1 - \mu^n\|_1 \, \|f\|_\infty = 0$$

for all $f \in L_\infty(X)$. Also, if f is τ-invariant, then $\mu^n f = f$ for all $n \ge 1$. But the subspace $\{f_1 + (f_2\circ\tau - f_2) : f_2 \in L_\infty(X), f_1 \text{ is } \tau\text{-invariant, and } f_1 \in L_\infty(X)\}$ is dense in $L_p(X)$, $1 < p < \infty$. Thus, to prove the theorem only requires a maximal estimate.

Let $Mf(x) = \sup\limits_{n \ge 1} |\mu^n f(x)|$ and $f^\#(x) = \sup\limits_{N \ge 1} |\frac{1}{N} \sum\limits_{k=1}^{N} \mu^k f(x)|$. By the Dunford-Schwartz ergodic theorem [Dunford, Schwartz, 1986], if $1 < p < \infty$, there is a constant $K_p < \infty$ such that $\|f^\#\|_p \le K_p\|f\|_p$. By partial summation,

$$\mu^{N+1} = \frac{1}{N} \sum_{k=1}^{N} k(\mu^{k+1} - \mu^k) + \frac{1}{N} \sum_{k=1}^{N} \mu^k.$$

But the Cauchy-Schwartz inequality says

$$|\frac{1}{N} \sum_{k=1}^{N} k(\mu^{k+1} - \mu^k f)| \le (\sum_{k=1}^{N} k|(\mu^{k+1} - \mu^k)f|^2)^{\frac{1}{2}}$$

$$\le (\sum_{k=1}^{\infty} k|(\mu^{k+1} - \mu^k)f|^2)^{\frac{1}{2}}.$$

Thus,

$$\sup_{N \ge 1} |\mu^{N+1}f| \le f^\# + (\sum_{k=1}^{\infty} k|(\mu^{k+1} - \mu^k)f|^2)^{\frac{1}{2}}$$

and

$$\|\sup_{N \ge 1} |\mu^{N+1}f|\|_2 \le \|f^\#\|_2 + \|\sum_{k=1}^{\infty} k|(\mu^{k+1} - \mu^k)f|^2)^{\frac{1}{2}}\|_2.$$

Hence, if we show $\|(\sum\limits_{k=1}^{\infty} k|(\mu^{k+1}-\mu^k)f|^2)^{\frac{1}{2}}\|_2 \le C\|f\|_2$ for some constant C independent of $f \in L_2(X)$, we would have $\|\sup\limits_{N\ge 1} |\mu^{n+1}f|\|_2 \le (K_2+C)\|f\|_2$, giving the maximal inequality needed to prove the theorem for $p = 2$.

By Exercise 6.25, it suffices to show that $\|(\sum\limits_{k=1}^{\infty} k|(\mu^{k+1}-\mu^k)*\varphi|^2)^{\frac{1}{2}}\|_2 \le C\|\varphi\|_2$ for all $\varphi \in \ell_2(\mathbb{Z})$. But the Plancherel Theorem allows the computation

$$\|(\sum_{k=1}^{\infty} k|(\mu^{k+1}-\mu^k)*\varphi|^2)^{\frac{1}{2}}\|_2^2$$

$$= \sum_{s=-\infty}^{\infty}(\sum_{k=1}^{\infty} k|(\mu^{k+1}-\mu^k)*\varphi(s)|^2)$$

$$= \sum_{k=1}^{\infty} k\int_{\{|\gamma|=1\}} |\hat{\mu}^{k+1}(\gamma)-\hat{\mu}^k(\gamma)|^2|\hat{\varphi}(\gamma)|^2 d\gamma$$

$$\le (\sup_{|\gamma|=1}\sum_{k=1}^{\infty} k|\hat{\mu}^{k+1}(\gamma)-\hat{\mu}^k(\gamma)|^2)\int_{\{|\gamma|=1\}} |\hat{\varphi}(\gamma)|^2 d\gamma$$

$$= (\sup_{|\gamma|=1}\sum_{k=1}^{\infty} k|\hat{\mu}^{k+1}(\gamma)-\hat{\mu}^k(\gamma)|^2)\|\varphi\|_2^2.$$

Thus, it suffices to bound $\sum\limits_{k=1}^{\infty} k|\hat{\mu}^{k+1}(\gamma)-\hat{\mu}^k(\gamma)|^2$ independently of $\gamma, |\gamma| = 1$. But this series is 0 if $\gamma = 1$, and if $|\gamma| = 1$, $\gamma \ne 1$, then $|\hat{\mu}(\gamma)| < 1$. So

$$\sum_{k=1}^{\infty} k|\hat{\mu}^{k+1}(\gamma)-\hat{\mu}^k(\gamma)|^2 = |\hat{\mu}(\gamma)-1|^2\sum_{k=1}^{\infty} k|\hat{\mu}^k(\gamma)|^2$$

$$= |\hat{\mu}(\gamma)-1|^2|\hat{\mu}(\gamma)|^2\sum_{k=1}^{\infty} k(|\hat{\mu}(\gamma)|^2)^{k-1}$$

$$= |\hat{\mu}(\gamma)-1|^2|\hat{\mu}(\gamma)|^2\frac{1}{(1-|\hat{\mu}(\gamma)|^2)^2}$$

$$\le \left(\frac{|\hat{\mu}(\gamma)-1|}{1-|\hat{\mu}(\gamma)|}\right)^2.$$

Thus, because μ has bounded angular ratio, $\sup\limits_{|\gamma|=1}\sum\limits_{k=1}^{\infty} k|\hat{\mu}^{k+1}(\gamma)-\hat{\mu}^k(\gamma)|^2 < \infty$.

This gives the maximal inequality in $L_2(X)$. A similar calculation gives the maximal inequality $\|\sup\limits_{n\ge r} n^r \Delta^r \mu^n f\|_2 \le C_r\|f\|_2$ for the rth difference operator $\Delta^r\mu^n$. The rest of the argument giving a strong maximal inequality in $L_p(X)$, $1 < p < \infty$, proceeds by using this estimate on the difference operator and the Dunford-Schwartz Theorem. The proof is identical to the proof in [Stein, 1961], employing his complex interpolation theorem. □

Remark. The positive result of Theorem 6.21 does not as stated include $L_1(X)$, but proving an a.e. convergence result in $L_1(X)$ is of central importance. In Reinhold-Larsson's thesis, see [Reinhold-Larsson, 1991] and [Reinhold-Larsson, 1993], an $L_1(X)$ result is proved for μ with $m_3(\mu) < \infty$ and $E(\mu) = 0$. Recently, A. Bellow and A. Calderón have extended the $L_1(X)$ result to all μ with $m_2(\mu) < \infty$ and $E(\mu) = 0$, via methods from the theory of singular integrals.

The positive result of Theorem 6.21, with Exercise 6.23, should be contrasted with the negative result of Theorem 5.8. Indeed, these results suggest trying to show that if the angular ratio is unbounded, then (μ^n) will not be good for a.e. convergence on $L_2(X)$. While this is not yet known to be true, a stronger assumption of this type, together with Bourgain's entropy theorem, does give negative results.

First, for each arc $A \subset \{\gamma \in \mathbb{C} : |\gamma| = 1\}$, define $r_A = \inf\{|\hat{\mu}(\gamma)| : \gamma \in A\}$. Also, for $A \subset \{\gamma \in \mathbb{C} : |\gamma| = 1\}$, let $|A|$ denote the normalized Lebesgue measure of A. We also need to measure the number of rays meeting $\hat{\mu}(A)$. So given $B \subset \{z \in \mathbb{C} : |z| \leq 1\}$, let $B_\rho = \{e^{i\theta} :$ for some r, $0 \leq r \leq 1$, $re^{i\theta} \in B\}$. The measure $|B_\rho|$ will be denoted also as $|B|_\rho$.

Theorem 6.26. *Suppose μ has the property that there is a sequence of open arcs $(A(k))$ such that $\lim\limits_{k \to \infty} \frac{|\hat{\mu}(A(k))|_\rho}{1 - r_{A(k)}} = \infty$. Then there exists $\delta > 0$ such that for all ergodic dynamical systems and all $\varepsilon > 0$, there is $E \in \Sigma$, $m(E) < \varepsilon$, such that $\limsup\limits_{n \to \infty} \mu^n 1_E \geq \delta$ for a.e. x.*

Remark. Although the conclusion of Theorem 6.26 is not strong sweeping out, it still suffices to show that a.e. convergence of $(\mu^n f)$ fails even for some characteristic function f.

To see the scope of Theorem 6.26, in contrast to Theorem 5.8, we use largely this proposition. The proof of this proposition is left as an exercise.

Proposition 6.27. *If μ is a probability measure on $\mathbb{Z}$ with*

$$\lim_{\gamma \to 1} \frac{|\hat{\mu}(\gamma) - 1|}{1 - |\hat{\mu}(\gamma)|} = \infty \,,$$

then there exists a sequence of open arcs $(A(k))$ such that

$$\lim_{k \to \infty} \frac{|\hat{\mu}(A(k))|_\rho}{1 - r_{A(k)}} = \infty \,.$$

Here is an example where Theorem 6.26 applies, but Theorem 5.8 does not.

Example 6.28. Let $\mu = \sum_{k=1}^{\infty} \frac{1}{k(k+1)} \delta_{-k}$. Then $m_s(\mu) < \infty$ only for $s < 1$, and $E(\mu)$ does not exist. So Theorem 5.8 does not apply. However, we can explicitly calculate $\hat{\mu}(e^{i\theta})$ as follows. If $f(z) = \sum_{k=1}^{\infty} \frac{1}{k(k+1)} z^k$, then f is analytic for $|z| < 1$, f is continuous on $|z| = 1$, and $f(e^{i\theta}) = \hat{\mu}(e^{i\theta})$. For $|z| < 1$, $f(z) = 1 - \frac{(z-1)\log(1-z)}{z}$ by partial fractions, where $\log(1-z)$ is the usual branch of the logarithm. Thus, f can be analytically continued by this formula to $\mathbb{C}\backslash\{r : r \geq 1\}$. So for any z, $|z| = 1$, $z \neq 1$,

$$\frac{1-\overline{\hat{\mu}(z)}}{1-\hat{\mu}(z)} = \frac{1-f\left(\frac{1}{z}\right)}{1-f(z)} = \frac{z^2(\frac{1}{z}-1)\log(1-\frac{1}{z})}{(z-1)\log(1-z)}.$$

Hence, letting $z = e^{i\theta}$, as $\theta \to 0$,

$$\frac{1-\overline{\hat{\mu}(z)}}{1-\hat{\mu}(z)} = \frac{z\log(1-\frac{1}{z})}{-\log(1-z)} \sim -\frac{\log|\sin\theta| - i\frac{\pi}{2}}{\log|\sin\theta| + i\frac{\pi}{2}} \longrightarrow -1.$$

It is now easy to see that μ satisfies the conditions of Proposition 6.27, and so Theorem 6.28 applies.

Proof of Theorem 6.26. We have seen in 6.6 that the conclusion of the theorem holds if the L_2 entropy of (μ^n) is unbounded for some ergodic dynamical system. We will show actually that for any $\gamma_0 \in \mathbb{C}$, $|\gamma_0| = 1$, γ_0 of infinite order, the dynamical system on the circle $\mathbb{T}$ with the usual Lebesgue measure and with $\tau(\gamma) = \gamma_0\gamma$ has unbounded L_2 entropy. Specifically, we will show that for all $M \geq 1$ there exist distinct $n_1, \ldots, n_M$ and $f \in L_2(\mathbb{T}), \|f\|_2 \leq 1$, such that for any $i \neq j$, $\|\mu^{n_i}f - \mu^{n_j}f\|_2 \geq \frac{1}{4}$.

Given $M \geq 1$, we can show that there are distinct $n_1 \ldots, n_M$ such that for each pair (n_i, n_j), $i \neq j$, we can associate a set of 2^{M-1} integers, denoted by $S(i,j)$, selected from $\{1, \ldots, 2^M\}$; and there will be a sequence of pairwise disjoint non-empty open sets $I_1, \ldots, I_{2^M} \subset \{\gamma \in \mathbb{C} : |\gamma| = 1\}$, such that $|\hat{\mu}^{n_i}(\gamma) - \hat{\mu}^{n_j}(\gamma)| \geq \frac{1}{2}$ for $\gamma \in I_s$ if $s \in S(i,j)$. See [Bellow, Jones, and Rosenblatt, 1994] for the technical details.

This construction will finish the proof. Indeed, since $\{\overline{\gamma}_0^\ell : \ell \geq 1\}$ is dense in $\{\gamma \in \mathbb{C}: |\gamma| = 1\}$, we can choose distinct $\ell(s)$ such that $\gamma_0^{-\ell(s)} \in I_s$. Define $f(\gamma) = \frac{1}{\sqrt{2^M}} \sum_{s=1}^{2^M} \gamma^{-\ell(s)}$. Notice that

$$\mu^n f(\gamma) = \sum_{k=-\infty}^{\infty} \mu^n(k) f(\gamma_0^k \gamma)$$

$$= \frac{1}{\sqrt{2^M}} \sum_{k=-\infty}^{\infty} \mu^n(k) \sum_{s=1}^{2^M} \gamma_0^{-k\ell(s)} \gamma^{-\ell(s)}$$

$$= \frac{1}{\sqrt{2^M}} \sum_{s=1}^{2^M} \hat{\mu}^n(\gamma_0^{\ell(s)}) \gamma^{-\ell(s)}.$$

Thus, for $i \neq j$,

$$\|\mu^{n_i} f - \mu^{n_j} f\|_2^2 = \frac{1}{2^M} \sum_{s=1}^{2^M} |\hat{\mu}^{n_i}(\gamma_0^{\ell(s)}) - \hat{\mu}^{n_j}(\gamma_0^{\ell(s)})|^2$$

$$\geq \frac{1}{2^M} \sum_{s \in S(i,j)} \frac{1}{4} = \frac{1}{8}.$$

Hence, $\|\mu^{n_i} f - \mu^{n_j} f\|_2 \geq \frac{1}{2\sqrt{2}}$ for all $i \neq j$, and the theorem is proved. □

The two unresolved questions related to Theorems 5.8 and 6.28 here are essentially these. First, if μ fails to have bounded angular ratio, are the conclusions of Theorem 6.28 true (or is it at least the case that for some, or all, ergodic dynamical systems, $\mu^n f$ can fail to converge a.e. for some $f \in L_\infty(X)$)? Second, when μ fails to satisfy any almost everywhere convergence theorem, does (μ^n) have strong sweeping out? Since there are examples of measures μ which have unbounded angular ratio but do not satisfy the hypotheses of Theorem 6.18, the first question seems to be a technically difficult one. Also, since failure of almost everywhere convergence does not entail strong sweeping out even in non-trivial situations, the second question also seems to be difficult. Recently, G. Chistyakov has used Theorem 6.28 to construct probability measures μ with $m_s(\mu) < \infty$ for all $s < 2$, and $E(\mu) = 0$, but for which $\mu^n f$ fails to converge a.e. for some $f \in L_\infty(X)$.

Bourgain's entropy method has also been successful in proving bad behaviors of certain discrete averages for measure-preserving flows. Here is one such case for powers of one measure.

Theorem 6.29. *Let a, b, c be real numbers such that $1, a, b$ and c are rationally independent. Let $\{T_t : t \in \mathbb{R}\}$ be an aperiodic ergodic flow on a probability space (X, Σ, m). Let $\mu = \frac{1}{3}(\delta_a + \delta_b + \delta_c)$. Then for any increasing sequence (n_s), there is $f \in L_\infty(X, \Sigma, m)$ such that $\lim_{s \to \infty} \mu^{n_s} f(x)$ does not exist a.e..*

The results in [Jones, Rosenblatt, and Tempelman, 1994] show that because μ is strictly aperiodic (i.e., supp μ is not contained in a proper coset in $\mathbb{Z}$), $\mu^n f$ converges in norm in $L_p(X, \Sigma, m)$, for $1 \leq p < \infty$; the limit is $\int f\, dm$ for all $f \in L_p(X, \Sigma, m)$. But then also $\frac{1}{N} \sum_{k=1}^{N} \mu^k f$ converge a.e. and in L_p-norm to $P_I f$ for all $f \in L_p(X, \Sigma, m)$, $1 \leq p < \infty$. Nonetheless, the chaotic behavior of $sp(\mu)$ causes there to be no subsequence theorem for a.e. convergence of iterates of μ in this case.

Proof of Theorem 6.29. Because the flow is ergodic, the averages $A_N f = \frac{1}{N} \sum_{k=1}^{N} \mu^k f$ converge in L_2 norm to $\int f\, dm$ for all $f \in L_2(X, \Sigma, m)$.

Hence, to prove the result only requires denying the finiteness of the L_2-metric entropy for $(\mu^{n_s} : n \geq 1)$.

Choose $M \geq 1$ and (σ_{ij}), $i = 1, \ldots, 2^M$, $j = 1, \ldots, M$, with each $\sigma_{ij} = \pm 1$, such that column $(\sigma_{ij} : i \geq 1)$ consists of half $+1$ and half -1, and the columns $(\sigma_{ij} : i \geq 1)$ are independent of each other for distinct j. Using Exercise 6.11, choose a subsequence $(n_{m_k} : k \geq 1)$ with $n_{m_{k+1}}/n_{m_k} \geq r(\alpha)$ for all $k \geq 1$. Choose γ_i, $|\gamma_i| = 1$, $|\gamma_i^{n_{m_j}} - \sigma_{ij}| \leq \alpha$ for $i = 1, \ldots, 2^M$, $j = 1, \ldots, M$. Now choose real numbers r_i, $r_{i+1} \geq r_i + 1$, $i = 1, \ldots, M$, with $|\hat{\mu}(r_i) - \gamma_i| \leq \varepsilon_0$. This is possible by the rational independence of $1, a, b, c$. By suitable choices of α and ε_0, we see that we can arrange $|\hat{\mu}^{n_{m_j}}(r_i) - \sigma_{ij}| \leq \varepsilon$ for $i = 1, \ldots, 2^M$, $j = 1, \ldots, M$.

Using [Lind, 1975], the aperiodicity of $(T_t : t \in \mathbb{R})$ shows that for all $\varepsilon > 0$, there is a measure-theoretic copy of $(F \times [L, L], \mu_F \times \lambda_{\mathbb{R}})$ in (X, m), where $\lambda_{\mathbb{R}}$ is Lebesgue measure and μ_F is a positive measure on F with $m(X \backslash (F \times [-L, L])) < \varepsilon$, so that $T_t(x, r) = (x, r+t)$ for $(x, r) \in F \times [-L, L]$ with $t + r \in [-L, L]$ too. It follows that $\frac{1}{2L} \geq \mu_F(F) \geq \frac{1-\varepsilon}{2L}$.

Define $1_r(x, t) = \exp(-2\pi i r t)$ for $(x, t) \in F \times [-L, L]$, and $1_r(x, t) = 0$ otherwise. Here $1_r(t)$ will also denote $\exp(-2\pi i r t)$, for any $t \in \mathbb{R}$. Let

$$f(x, t) = \frac{1}{\sqrt{2^M}} \sum_{\ell=1}^{2^M} 1_{r_\ell}(x, t).$$

This f has L_2 norm which can be estimated by

$$\|f\|_2^2 = m(F \times [-L, L]) + \frac{1}{2^M} \sum_{\substack{\ell, \ell'=1 \\ \ell \neq \ell'}}^{2^M} \int 1_{r_\ell} \overline{1_{r_{\ell'}}} \, dm \leq 1 + \frac{2^M - 1}{2\pi L},$$

because of the approximate orthogonality of the functions $\{1_{r_\ell}\}$.

Also, for $(x, t) \in F \times [-L, L]$ such that $s + t \in [-L, L]$ for all $s \in supp(\mu^n)$,

$$\mu^n f(x, t) = \frac{1}{\sqrt{2^M}} \sum_{\ell=1}^{2^M} \sum_s \mu^n(s) 1_{r_\ell}(x, s+t)$$

$$= \frac{1}{\sqrt{2^M}} \sum_{\ell=1}^{2^M} \sum_s \mu^n(s) \exp(-2\pi i r_\ell(s+t))$$

$$= \frac{1}{\sqrt{2^M}} \sum_{\ell=1}^{2^M} \hat{\mu}^n(r_\ell) 1_{r_\ell}(t).$$

Hence, there is a constant K, depending only on M, such that if $|t| \leq L - K$, $x \in F$, and $k = 1, \ldots, M$, then $\mu^{n_{m_k}} f(x, t) = \frac{1}{\sqrt{2^M}} \sum_{\ell=1}^{2^M} \hat{\mu}^{n_{m_k}}(r_\ell) 1_{r_\ell}(t)$. But then for $k \neq k'$,

$$\|\mu^{n_m k} f - \mu^{n_m k'} f\|_2^2$$

$$\geq \int_F \int_{-L+K}^{L-K} |(\mu^{n_m k} f - \mu^{n_m k'} f)(x,t)|^2 dt d\mu_F(x)$$

$$= \int_F \int_{-L+K}^{L-K} \frac{1}{2^M} |\sum_{\ell=1}^{2^M} (\hat{\mu}^{n_m k}(r_\ell) - \hat{\mu}^{n_m k'}(r_\ell)) 1_{r_\ell}(t)|^2 dt d\mu_F(x)$$

$$\geq \frac{1-\varepsilon}{2L} \int_{-L+K}^{L-K} \frac{1}{2^M} |\sum_{\ell=1}^{2^M} (\hat{\mu}^{n_m k}(r_\ell) - \hat{\mu}^{n_m k}(r_\ell)) 1_{r_\ell}(t)|^2 dt .$$

It is not hard then to get an estimate that if L is large enough and $\varepsilon > 0$ small enough, for $k \neq k'$,

$$\|\mu^{n_m k} f - \mu^{n_m k'} f\|_2^2 \geq \frac{4}{5} .$$

At the same time, if L is large enough and $\varepsilon > 0$ is small enough, we can have $\frac{3}{2} \geq \|f\|_2^2$.

The conclusion is that for any $M \geq 1$ there exists $f \in L_2(X, \Sigma, m)$, $\|f\|_2 = 1$, and some $n_{m_1}, \ldots, n_{m_M}$ such that $\|\mu^{n_m k} f - \mu^{n_m k'} f\|_2 \geq \sqrt{\frac{8}{15}}$ for all $k, k' = 1, \ldots, M$, $k \neq k'$. This shows that the metric entropy of μ^n is infinite. $\qquad\square$

Remark. a) The same conclusion as 2.29 is true if we just know that some non-trivial arc in $\{z : |z| = 1\}$ is a subset of $cl_{\mathbb{C}}\{\hat{\mu}(r) : r \in \mathbb{R}\}$.

b) The example in 6.29 is important because all previously studied cases of the use of convolution powers to get ergodic theorems had the property that if there were an L_p-norm theorem, then there was at least an a.e. convergence theorem along a subsequence (n_s). See [Bellow, Jones, and Rosenblatt, 1988].

c) There should be an analogous theorem for a normal operator T with $sp(T)$ containing an arc in the unit circle.

d) An inspection of the proof of 6.29 shows that the constant δ in the entropy calculation can be, instead of $\sqrt{\frac{8}{15}}$, as close to $\sqrt{2}$ as we like. For geometrical reasons, this is certainly optimal. It follows, analogously to [Rosenblatt, 1988] that the powers μ^r are δ_0-sweeping out for some δ_0. Actually, an inspection of Bourgain's proof in [Bourgain, 1988b] shows that there is a universal constant C such that $\delta_0 \geq C\delta$, where δ is as above.

This example shows that behavior of discrete measures $\mu = \sum_{j=1}^{m} x_j \delta_{x_j}$ on $\mathbb{R}$ is different spectrally than the same in $\mathbb{Z}$. Let us consider this a little further. We assume μ is a probability measure, $\sum_{j=1}^{m} \alpha_j = 1$, $\alpha_j \geq 0$. From

the case of $\mathbb{Z}$, we would expect that centering, $E(\mu) = \sum_{j=1}^{m} x_j \alpha_j = 0$, would
imply good behavior of the μ averages for ergodic actions on a Hilbert
space. It does imply recurrence; see [Chung, Fuchs, 1951].

However, we have the following example. Choose $x < 0$, $y, z > 0$ such
that $\{1, x, y, z\}$ are rationally independent, e.g., $x = -\sqrt{2}$, $y = \sqrt{3}$, $z = \sqrt{5}$
will do. Then choose $\alpha, \beta, \gamma \geq 0$, $\alpha+\beta+\gamma = 1$, and let $\mu = \alpha\delta_x + \beta\delta_y + \gamma\delta_z$.
We can arrange for $E(\mu) = 0$ too, e.g., let $a = \frac{1}{\sqrt{2}}$, $b = \frac{1}{2\sqrt{3}}$, $c = \frac{1}{2\sqrt{5}}$. Let
$s = a+b+c$. Let $\alpha = a/s$, $\beta = b/s$, $\gamma = c/s$; so $\alpha, \beta, \gamma \geq 0$ and $\alpha+\beta+\gamma = 1$.
But also,

$$E(\mu) = \frac{a}{s}x + \frac{b}{s}y + \frac{c}{s}z = \frac{1}{s}\left(\frac{1}{\sqrt{2}}(-\sqrt{2}) + \frac{1}{2\sqrt{3}}(\sqrt{3}) + \frac{1}{2\sqrt{5}}(\sqrt{5})\right)$$
$$= \frac{1}{s}\left(-1 + \frac{1}{2} + \frac{1}{2}\right) = 0.$$

But since $\{1, x, y, z\}$ are rationally independent, for all $\varepsilon > 0$, $|\gamma_i| = 1$,
there exists $r \in \mathbb{R}$, $|e^{ixr} - \gamma_1| < \varepsilon$, $|e^{iyr} - \gamma_2| < \varepsilon$ and $|e^{izr} - \gamma_3| < \varepsilon$,
simultaneously. It follows easily that $cl_{\mathbb{C}}\{\hat{\mu}(r): r \in \mathbb{R}\} = \{z: |z| \leq 1\}$. The
method of Theorem 6.18 gives this result.

Theorem 6.30. *With μ as above, and any sequence $n_1 < n_2 < n_3 < \cdots$, and any ergodic strictly aperiodic flow $t \mapsto T_t$ on a probability space (X, Σ, m), there exists $E \in \Sigma$ such that $\mu^{n_*} 1_E$ fails to converge a.e..*

We see the problem here is that $E(\mu) = 0$ is not enough to guarantee
that $\mathrm{spec}(\mu)$, with μ as an operator on $L_2(X)$, only meets $\{z: |z| = 1\}$ at
$z = 1$, let alone that $\mathrm{spec}(\mu)$ is contained in a Stolz region.

The situation is somewhat better if $\mu = \frac{1}{m}\sum_{j=1}^{m}\delta_{x_j}$ for some $x_1, \ldots x_m$,
when $\sum_{j=1}^{m} x_j = 0$ so that $E(\mu) = 0$. First, consider the example $\mu = \frac{1}{2}\delta_{-1} + \frac{1}{2}\delta_1$. Although this is not strictly aperiodic, it does have $E(\mu) = 0$,
and choosing $e^{irs} \to -1$, we see $\hat{\mu}(rs) = \frac{1}{2}e^{+irs} + \frac{1}{2}e^{-irs} \to -1$ as $s \to \infty$.
So $\mathrm{Spec}(\mu)$ contains more of $\{z: |z| = 1\}$ than $\{1\}$. However, there is some
control. Let R_m denote the mth roots of unity in $\mathbb{C}$.

Proposition 6.31. *If $\mu = \frac{1}{m}\sum_{j=1}^{m}\delta_{x_j}$ and $\sum_{j=1}^{m} x_j = 0$, then $\mathrm{spec}(\mu) \cap \{z: |z| = 1\}$ is contained in R_m.*

Proof. It suffices to show that if $|\hat{\mu}(r_s)| \to 1$, then the limit points of
$\hat{\mu}(r_s)$ are in R_m. But

$$|\hat{\mu}(r_s)| \to 1 \quad \text{implies} \quad |e^{ix_j r_s} - e^{ix_\gamma r_s}| \to 0 \tag{1}$$

for all j, $j = 1, \ldots, m$. Indeed, to see (1) notice

$$\iint |e^{ixr_s} - e^{iyr_s}|^2 d\mu(x)d\mu(y) = 2 - \widehat{\mu}(-r_s)\widehat{\mu}(r_s) - \widehat{\mu}(-r_s)\widehat{\mu}(r_s)$$

$$= 2 - 2|\widehat{\mu}(r_s)|^2$$

goes to zero as s goes to ∞. Hence, for a.e. (x,y) $[\mu \times \mu]$, $|e^{ixr_s} - e^{iyr_s}| \to 0$. In the case that μ is discrete with finite support, this gives (1).

Without loss of generality, assume $\lim\limits_s e^{ix_1r_s} = \gamma_0$. Then the above says $e^{ix_jr_s} \to \gamma_0$ as $s \to \infty$ for all j. Hence,

$$1 = e^{i\sum\limits_{j=1}^{m} x_jr_s} = \prod_{j=1}^{m} e^{ix_jr_s} \to \gamma_0^m,$$

and $\gamma_0 \in R_m$. It follows that μ has the spectral property above. □

But this means that $\nu = \mu^m$ has the property that if $|\widehat{\nu}(r_s)| \to 1$, then $\widehat{\nu}(r_s) \to 1$. So $\operatorname{spec}(\nu) \cap \{z : |z| = 1\} = \{1\}$.

Proposition 6.32. *If $\mu = \frac{1}{m}\sum\limits_{i=1}^{m}\delta_{x_i}$ and $E(\mu) = 0$, then there exists a subsequence of powers (μ^{n_i}) such that for any flow $t \mapsto T_t$, $\mu^{n_i}f(x)$ converges a.e. for all $f \in L_2(X)$.*

Note. Actually (n_i) can be chosen inside any previously fixed sequence of powers.

Proof. See the result of [Bellow, Jones, and Rosenblatt, 1988]. □

Indeed, more is true. We have $\mu = \frac{1}{m}\sum\limits_{j=1}^{m}\delta_{x_j}$, $\sum\limits_{j=1}^{m} x_j = 0$, given. Then if $|\widehat{\mu}(r_s)| \to 1$, we see that $\widehat{\nu}(r_s) \to 1$, and actually if $\nu = \frac{1}{M}\sum\limits_{\ell=1}^{M}\delta_{y_\ell}$, then $E(\nu) = \frac{1}{M}\sum\limits_{\ell=1}^{M} y_\ell = 0$ too, and $e^{iy_\ell r_s} \to 1$ for all $\ell = 1, \ldots, M$.

Proposition 6.33. *The set $\{\widehat{\nu}(r) : r \in \mathbb{R}\}$ is contained in a Stolz angle.*

Corollary 6.34. *For any flow $t \mapsto T_t$, the maximal function $\sup |\nu^n f|$ is strong (p,p), $1 < p < \infty$.*

Corollary 6.35. *For any flow $t \mapsto T_t$, the maximal function $\sup |\mu^n f|$ is strong (p,p), $1 < p < \infty$.*

Corollary 6.36. *If μ is strictly aperiodic, $\mu = \frac{1}{m}\sum\limits_{j=1}^{m}\delta_{x_j}$, $E(\mu) = 0$, then for any flow and $f \in L_p(X)$, $1 < p < \infty$, $\mu^n f(x)$ converges a.e.*

Corollary 6.34 is immediate from [Bellow, Jones, and Rosenblatt, 1988]. This gives Corollary 6.35, since $\nu = \mu^m$. Corollary 6.36 follows because of the existence of a dense class on which a.e. convergence occurs. We see from Corollary 6.36 that for these special discrete measures, $\mu^n * f(r) \to 0$ a.e. r for $f \in L_2(\mathbb{R})$ too, because the flow can be on any σ-finite space. It is only the *negative* results on a.e. convergence which to date require a probability space.

Proof of Proposition 6.33. The spectral behavior of ν shows that it suffices to consider the function $F(r) = \frac{1}{M} \sum_{\ell=1}^{M} e^{-iy_\ell r}$ and show that there is a constant C such that if $e^{-iy_\ell r_s} \to 1$ for all $\ell \geq 1$, then $\frac{|1-F(r_s)|}{1-|F(r_s)|} \leq C$ for all $s \geq 1$. Since $e^{-iy_\ell r_s} \to 1$, we can take fractional powers defined in $\mathbb{C}$ with $\{z : z \text{ real}, z \leq 0\}$ as the cut in $\mathbb{C}$. Without loss of generality, assume $y_1 \neq 0$. Then

$$
\begin{aligned}
F(r) &= \frac{1}{M}(e^{-iy_1 r} + e^{-iy_1 r})^{y_2/y_1} + \cdots + (e^{-iy_1 r})^{y_m/y_1} \\
&= \frac{1}{M}(\gamma + \gamma^{y_1/y_1} + \cdots + \gamma^{y_M/y_1}),
\end{aligned}
$$

where $\gamma = e^{-iy_1 r}$. Consider the function $G(\gamma) = \frac{1}{M}(\gamma + \gamma^{y_2/y_1} + \cdots + \gamma^{y_M/y_1})$, where $|\gamma| = 1$. Then near 1 in $\mathbb{C}$, G is analytic. Also, $G(1) = 1$, and $G'(\gamma) = \frac{1}{M}\left(1 + \frac{y_2}{y_1}\gamma^{\frac{y_2-y_1}{y_1}} + \cdots + \frac{y_M}{y_1}\gamma^{\frac{y_M-y_1}{y_1}}\right)$. So

$$
G'(1) = \frac{1}{M}\left(1 + \sum_{j=2}^{M} y_j/y_1\right) = \frac{1}{M}\left(1 + \frac{-y_1}{y_1}\right) = 0.
$$

Also then,

$$
\begin{aligned}
G''(1) &= \frac{1}{M} \sum_{j=2}^{M} \frac{y_j(y_j - y_1)}{y_1^2} = \frac{1}{M} \frac{\sum_{j=2}^{M} y_j^2 - y_1(-y_1)}{y_1^2} \\
&= \frac{1}{M} \frac{\sum_{j=1}^{M} y_j^2}{y_1^2} = \sigma^2 > 0.
\end{aligned}
$$

Hence, as $\gamma \to 1$, $G(\gamma) = 1 + \frac{\sigma^2}{2}(\gamma - 1)^2 + o((\gamma - 1)^2)$. So, $|G(\gamma)|^2 = 1 + \sigma^2 Re(\gamma - 1)^2 + o((\gamma - 1)^2)$ and $|1 - G(\gamma)| = \frac{\sigma^2}{2}|\gamma - 1|^2 + o((\gamma - 1)^2)$. So $\frac{|1-G(\gamma)|}{1-|G(\gamma)|^2} \leq \frac{2|1-G(\gamma)|}{1-|G(\gamma)|^2} \sim \frac{2\frac{\sigma^2}{2}|\gamma-1|^2}{-\sigma^2 Re(\gamma-1)^2}$ as $\gamma \to 1$. But $|\gamma| = 1$, and so as $\gamma \to 1$, $\frac{|\gamma-1|^2}{-Re(\gamma-1)^2} \to 1$. Thus, for γ near 1, $\frac{|1-G(\gamma)|}{1-|G(\gamma)|} \leq 2$. But this, together with $|F(r_s)| \to 1$ only as $\gamma \to 1$, gives the bound needed. $\qquad\square$

Notes to Chapter VI

1. Theorem 6.6 and its derivation from Bourgain's entropy theorem are from [Rosenblatt, 1991].
2. In [Bourgain, 1988], the entropy theorem is used to prove several interesting results. The most significant ones are these two:

   a) if $x_k \to 0$ in $\mathbb{R}$ is a sequence with $x_k \neq 0$ for all k, then there is some $f \in L_1(\mathbb{R}) \cap L_\infty(\mathbb{R})$ such that $\lim\limits_{n\to\infty} \frac{1}{n} \sum\limits_{k=1}^{n} f(x + x_k)$ fails to exist a.e.;

   b) for any method of summation of the form $\frac{1}{\sum_n} \sum\limits_{k=1}^{n} \sigma_k f\{kx\}$, where $\sum_n = \sum\limits_{k=1}^{n} \sigma_k \to \infty$ as $n \to \infty$, there exists $f \in L_\infty[0,1]$ for which $\lim\limits_{n\to\infty} \frac{1}{\sum_n} \sum\limits_{k=1}^{n} \sigma_k f\{kx\}$ fails to exist a.e..

3. Theorem 6.10 and its corollaries are from [Rosenblatt, 1988].
4. Theorems 6.21 and 6.26 appear in [Bellow, Jones, and Rosenblatt, 1994].
5. Theorem 6.29 and related results for group actions are in [Jones, Rosenblatt, and Tempelman, 1994].
6. Many of the results in Chapter VI have been improved to include strong sweeping out in [Akcoglu, et al, 1994]. In that article some of the criteria are in terms of Fourier transforms and some are not, but the techniques of proof are generally combinatorial and measure-theoretic. Comparison of [Akcoglu, et al, 1994] with Chapter VI here will give the reader better insight into the material under consideration.

CHAPTER VII

SEQUENCES THAT ARE GOOD ONLY FOR SOME L^p

In this short chapter we discuss results of Bellow and Reinhold-Larsson concerning sequences that are L^p good for certain values of p only.

Definition 7.1. For a strictly increasing sequence $A = (a_n)$ of positive integers and a dynamical system (X, Σ, m, τ), we continue with the notation

$$M_t f(x) = M_t(A, f)(x) = \frac{1}{A(t)} \sum_{\substack{a \leq t \\ a \in A}} f(\tau^a x).$$

Let $1 \leq p < \infty$. We say that A is ∞-*sweeping out for* L^p in (X, Σ, m, τ) if and only if there is $f \in L^p(X)$ such that

$$\sup_t M_t(A, f)(x) = \infty$$

for a.e. $x \in X$. We say that A is *universally* ∞-*sweeping out for* L^p if and only if it is ∞-sweeping out for L^p in every *aperiodic (free)* dynamical system.

The strictly increasing sequence of positive integers B is called a *perturbation* of A if and only if

$$\lim_{t \to \infty} \frac{\#\{n \mid n \in (B \setminus A) \cup (A \setminus B), n \leq t\}}{A(t)} = 0.$$

Note that if B is a perturbation of A, then

$$\lim_{t \to \infty} \frac{B(t)}{A(t)} = 1. \tag{7.1}$$

In the sequel C will denote a "generic" positive constant, which is independent of those quantities it should be independent of, but it can have different values even in the same set of inequalities.

It was shown in [Bellow, 1989] that there exists a sequence of integers which is universally good for, say, L^2 but not good for L^p, $p < 2$. The existence of a sequence that is universally good for L^p, $p > 1$, but not good for L^1 was proved in [Reinhold-Larsson, 1991, 1994]. The method used in [Bellow, 1989] and [Reinhold-Larsson, 1991, 1994] is perturbation. The perturbation method has its origin in the paper of Emerson [1974]. Here we describe a technically simpler version of the perturbation method, and we prove a result, Theorem 7.2 below, that cannot be improved in the sense of the first remark after the enunciation of the theorem.

Theorem 7.2. *Let A be a strictly increasing sequence of positive integers with $\underline{d}(A) = 0$.*
(I) *Let $1 < q \leq \infty$. Suppose that A is universally good for L^q. Then there is a perturbation B of A which is also universally good for L^q, but which is universally ∞-sweeping out for L^p if $1 \leq p < q$.*
(II) *Let $1 \leq q < \infty$. Suppose that A is universally good for L^p for each $p > q$. Then there is a perturbation B of A which is also universally good for L^p for each $p > q$, but which is universally ∞-sweeping out for L^q.*

Remarks.

(1) Note that if the sequence A has positive lower density then any perturbation B of A also has positive lower density. It follows from Exercise 4 for Example 3.7 that the sequence B is good for some irrational rotation of the interval $[0, 1)$.

(2) We get interesting instances of Theorem 7.2 if we perturb the sequence of squares, the primes, or the sequence $([n^{3/2}])$. As we see, our theorem applies to such irregular sequences as the primes, while the perturbation used in [Bellow, 1989] and [Reinhold-Larsson, 1991, 1994] does not seem to be effective enough to handle these sequences.

(3) It is possible, using our method, to construct a sequence of integers that is universally good for $L \log L$ (of course, for finite measure spaces), but is ∞-sweeping out for L^1. Specific examples could be perturbations of the sequences $([n^{3/2}])$ or $([n \log n])$, since they are known to be universally good for $L \log L$ (cf. [Wierdl, 1989]).

Proof of Theorem 7.1. Here we just prove part **(II)**; the reader will have no difficulty proving the other part. The idea of the proof is the following: the new sequence B is formed by adding segments of "bad" sequences to A, and we shall do this so that the cardinality of these "perturbations" is not big enough to affect the good behaviour in L^p, $p > q$, but the perturbation is strong enough to destroy the maximal inequality in L^q. We will make this more quantitative in a minute, but first let us indicate what we mean by a "bad" sequence.

There are numerous ways to construct a bad sequence, but probably the easiest is to give a sequence which is not uniformly distributed among residue classes for infinitely many moduli. This means that fixing a modulus Q and a residue v, at one point sufficiently many elements of our sequence will be congruent to $v \mod Q$.

Below we shall define integers t_k, $k = 1, 2, \ldots$. For $u = 0, 1, \ldots$ set

$$A_u = \{k \mid k = 2^u, 2^u + 1, \ldots, 2^{u+1} - 1\}.$$

The new sequence B will contain A and is formed by adding to A a certain number of integers from the interval $[t_k, 2t_k)$ so that these added integers will be congruent to k mod 2^u if $k \in A_u$. To be specific, the cardinality of these numbers will be

$$\left(\frac{u}{2^u}\right)^{1/q} \cdot A(t_k), \qquad k \in A_u. \tag{7.2}$$

In order to be able to find this many integers in $[t_k, 2t_k)$, each congruent to k mod 2^u, we need that t_k is large enough: $t_k > u^{1/q} \cdot 2^{u(1-1/q)} \cdot A(t_k)$. We certainly achieve this if

$$t_k > u2^u \cdot A(t_k). \tag{7.3}$$

We also have to make sure that these new numbers are numerous enough to destroy the weak maximal inequality in L^q, hence we have to make sure that A does not have many elements in $[t_k, 2t_k)$. Indeed, we will choose t_k so that the number of elements of A in $[t_k, 2t_k)$ will not exceed $2A(t_k)$.

To sum up, the t_k will satisfy

(i) $t_k > 2n_{k-1}$;
(ii) $\frac{A(t_k)}{t_k} < \frac{1}{u2^u}, \quad k \in A_u$;
(iii) $A(2t_k) \leq 3A(t_k)$.

As a last requirement on the t_k, so B becomes a perturbation of A, we will have

(iv) $\left(\frac{u}{2^u}\right)^{1/q} \cdot A(t_k) > \sum_{i=1}^{k-1} A(t_i), \quad k \in A_u$.

The recursive construction of the t_k satisfying the above four properties is quite simple. Since $\underline{d}(A) = 0$, there is a sequence $\{m_j\}$ of positive integers such that

(v) $\lim_{j \to \infty} \frac{A(m_j)}{m_j} = 0$;
(vi) $\frac{A(m_j)}{m_j} \leq \frac{A(m)}{m}$ for $m \leq m_j$.

Having constructed $t_1, \ldots, t_{k-1}$, we just take $t_k = [m_j/2]$ for large enough j. It is clear that we can choose j so that (i) and (iv) are satisfied. To have (ii) we use (v) and the estimate

$$\frac{A(t_k)}{t_k} = \frac{A([m_j/2])}{[m_j/2]} \leq 3\frac{A(m_j)}{m_j}.$$

Finally, to see that we can choose j to have (iii), let us use (vi) and estimate

$$A(2t_k) \leq A(m_j) = \frac{A(m_j)}{m_j} \cdot m_j \leq \frac{A([m_j/2])}{[m_j/2]} \cdot m_j \leq$$
$$\leq 3A([m_j/2]) = 3A(t_k).$$

Proof that B is a perturbation of A.

Since B is formed by *adding* new terms to A we need to prove that

$$\lim_{n \to \infty} \frac{B(t) - A(t)}{A(t)} = 0. \tag{7.4}$$

Let n be arbitrary. Then for some k and u, $k \in A_u$, we have $t_k \leq n < t_{k+1}$. By property (iv), we can estimate (recall that the cardinality of the new numbers in the interval $[t_k, t_{k+1})$ is given in (7.2))

$$B(t) - A(t) \leq 2 \left(\frac{u}{2^u} \right)^{1/q} A(t_k) \,.$$

Since $A(t) \geq A(t_k)$, we have

$$B(t) - A(t) \leq 2 \left(\frac{u}{2^u} \right)^{1/q} A(t) \,.$$

This implies (7.4), for as $n \to \infty$ so do k and u. □

Proof that B is universally good for L^p, $p > q$.

Fix $p > q$ and the measure-preserving system (X, B, m, τ). Since we have property (7.1), we just need to prove that for $f \in L^p$

$$\frac{B(t)}{A(t)} \cdot M_t(B, f)(x) = \frac{1}{A(t)} \sum_{\substack{b \leq n \\ b \in B}} f(\tau^b x)$$

converges a.e.. Without loss of generality we can assume $f \geq 0$. Since A is a good sequence for L^p, it is enough to prove that

$$\limsup_{n \to \infty} \frac{1}{A(t)} \sum_{\substack{b \leq n \\ b \in \overline{B} \backslash A}} f(\tau^b x) = 0 \,.$$

Let t be arbitrary. Then for some k and u, $k \in A_u$, we have $t_k \leq t < t_{k+1}$. Noting that the extra elements of $B \cap [t_k, t_{k+1})$ are taken from the interval $[t_k, 2t_k)$, we can estimate

$$\frac{1}{A(t)} \sum_{\substack{b \leq t \\ b \in \overline{B} \backslash A}} f(\tau^b x) \leq \frac{1}{A(t_k)} \sum_{\substack{b \leq 2t_k \\ b \in \overline{B} \backslash A}} f(\tau^b x) \overset{\text{def}}{=} R_k f(x) \,.$$

So we just need to prove that

$$R_k f(x) \to 0 \quad \text{a.e..}$$

This will follow if we prove

$$\int_X \left(\sum_{k=1}^{\infty} (R_k f(x))^p \right) dm(x) = \sum_{k=1}^{\infty} \| R_k f(x) \|_{L^p}^p < \infty \,. \tag{7.5}$$

By the triangle inequality and by (iii), we can estimate

$$\sum_{k=1}^{\infty} \|R_k f(x)\|_{L^p}^p \leq \sum_{k=1}^{\infty} \left(\frac{1}{A(t_k)} \sum_{\substack{b \leq 2t_k \\ b \in B \backslash A}} \|f(\tau^b x)\|_{L^p} \right)^p$$

$$= \|f\|_{L^p}^p \sum_{k=1}^{\infty} \left(\frac{1}{A(t_k)} \sum_{\substack{b \leq 2t_k \\ b \in B \backslash A}} 1 \right)^p$$

$$= \|f\|_{L^p}^p \sum_{u=1}^{\infty} \sum_{k \in A_u} \left(\frac{1}{A(t_k)} \sum_{\substack{b \leq 2t_k \\ b \in B \backslash A}} 1 \right)^p \qquad \text{(by } (iii)\text{)}$$

$$\leq \|f\|_{L^p}^p \sum_{u=1}^{\infty} \sum_{k \in A_u} \left(\frac{\left(\frac{u}{2^u}\right)^{1/q} \cdot 3A(t_k)}{A(t_k)} \right)^p$$

$$= \|f\|_{L^p}^p \sum_{u=1}^{\infty} 2^u \cdot 3^p \left(\frac{u}{2^u} \right)^{p/q}$$

$$= C_p \|f\|_{L^p}^p < \infty,$$

since $p > q$. Therefore we proved (7.5). $\square$

Proof that B is universally ∞-sweeping out for L^q.

By the lemma below, we just need to disprove the existence of a maximal inequality on $\mathbb{Z}$.

In the rest of the proof, for $f : \mathbb{Z} \to \mathbb{R}$ and integer sequence B we use the notation

$$M_t f(x) = M_t(B, f)(x) = \frac{1}{B(t)} \sum_{\substack{b \leq n \\ b \in B}} f(x+b).$$

We also introduce the following definition for $f : \mathbb{Z} \to \mathbb{R}$:

$$D(f) = \limsup_{L \to \infty} \frac{1}{2L+1} \sum_{x=-L}^{L} f(x).$$

For a set A of integers, $D(A)$ will mean $D(1_A)$.

Lemma 7.3. *Let $0 < q < \infty$, and let B be a strictly increasing sequence of positive integers. Suppose that for every positive K and ϵ there is $f : \mathbb{Z} \to \mathbb{R}$, $D(|f|^q) \leq 1$, and a finite set of integers Λ with*

$$D\left\{ x \mid \max_{n \in \Lambda} M_t(B, f)(x) \geq K \right\} \geq 1 - \epsilon.$$

Then B is universally ∞-sweeping out for L^q.

Proof of Lemma 7.3. Let (X, B, m, τ) be an aperiodic, probability measure-preserving system. By the assumption of the lemma, and using Rokhlin's tower construction, we conclude that for each positive K and ϵ there is $\bar{f} : X \to \mathbb{R}$, $\|\bar{f}\|^q_{L^q(X)} \leq 1$ with

$$m\left\{x \mid \sup_t \bar{M}_t(B, \bar{f})(x) \geq K\right\} \geq 1 - \epsilon,$$

where

$$\bar{M}_t(B, \bar{f})(x) = \frac{1}{B(t)} \sum_{\substack{b \leq n \\ b \in B}} \bar{f}(\tau^b x).$$

It then follows that for each positive integer N there is $\bar{g} = \bar{g}_N : X \to \mathbb{R}$ with

$$\|\bar{g}\|^q_{L^q(X)} \leq 2^{-N},$$

and

$$m\left\{x \mid \sup_t \bar{M}_t(B, \bar{g})(x) \geq N\right\} \geq 1 - \frac{1}{N}.$$

Let us set

$$\bar{g}_0(x) = \sup_N \bar{g}_N(x).$$

Then

$$\|\bar{g}_0\|^q_{L^q(X)} \leq 1,$$

so $\bar{g}_0 \in L^q(X)$. Let us denote

$$E_N = \left\{x \mid \sup_t \bar{M}_t(B, \bar{g})(x) \geq N\right\}$$

and set

$$E = \bigcap_{J=1}^{\infty} \bigcup_{N=J}^{\infty} E_N.$$

It is clear that $m(E) = 1$, and also that if $x \in E$ then

$$\sup_t \bar{M}_t(B, \bar{g}_0)(x) = \infty.$$

$\square$

Let us now fix u, and assume it is large — large enough to satisfy (7.9) below. Define $f : \mathbb{Z} \to \mathbb{R}$ as follows:

$$f(x) = \begin{cases} 2^{u/q} & \text{if } 2^u \mid x; \\ 0 & \text{otherwise.} \end{cases}$$

Clearly,

$$D(|f|^q) = 1 .$$

We are going to show

$$\left\{ x \mid \max_{k \in A_u} M_{2t_k}(B, f)(x) \geq \frac{1}{4} u^{1/q} \right\} = \mathbb{Z} , \qquad (7.6)$$

which, by the lemma, would finish the proof.

Let $x \in \mathbb{Z}$. Then for some $k \in A_u$ we have $x \equiv -k \mod 2^u$. Recall —
(7.2) — that there are

$$\left(\frac{u}{2^u} \right)^{1/q} \cdot A(t_k)$$

numbers in $B \cap [t_k, 2t_k)$ that are congruent to $k \mod 2^u$. Let us denote
the set of these numbers by G. So we have

$$\#G = \left(\frac{u}{2^u} \right)^{1/q} \cdot A(t_k) , \qquad (7.7)$$

and

$$f(x + b) = 2^{u/q} \quad \text{for} \quad b \in G , \qquad (7.8)$$

since $2^u \mid x + b$. By property (iii), and since B is a perturbation of A, we
have, for large enough u,

$$B(2t_k) \leq 4A(t_k) . \qquad (7.9)$$

We can now estimate

$$M_{2t_k}(B, f)(x) = \frac{1}{B(2t_k)} \sum_{\substack{b \leq 2t_k \\ b \in B}} f(x + b) \qquad \text{(by (7.9))}$$

$$\geq \frac{1}{4A(t_k)} \sum_{b \in G} f(x + b) \qquad \text{(by (7.8))}$$

$$= \frac{1}{4A(t_k)} \sum_{b \in G} 2^{u/q} \qquad \text{(by (7.7))}$$

$$= 2^{u/q} \cdot \frac{\left(\frac{u}{2^u} \right)^{1/q} \cdot A(t_k)}{4A(t_k)} = \frac{1}{4} \cdot u^{1/q} ,$$

which proves (7.6). □

REFERENCES

1. M. Akcoglu, A. del Junco (1975), *Convergence of averages of point transformations*, Proc. Amer. Math. Soc. **49**, 265–266.
2. M. Akcoglu, A. Bellow, R. Jones, V. Losert, K. Reinhold-Larsson, M. Wierdl (1994), *Strong sweeping out of ergodic averages, Riemann sums and related matters*, preprint.
3. I. Assani (1992), *A Wiener-Wintner property for the helical transform of the shift on $[0,1]^Z$.*, Ergodic Th. Dynamical Syst. **12**, 659–672.
4. I. Assani (1993), *The helical transform and the a.e. convergence of Fourier series.*, Illinois J. Math. **37**, 123–146.
5. I. Assani, K. Petersen (1992), *The helical transform as a connection between ergodic theory and harmonic analysis*, Trans. Amer. Math. Soc. **331**, 131–142.
6. I. Assani, K. Petersen, H. White (1991), *Some connections between ergodic theory and harmonic analysis*, Almost everywhere convergence II (A. Bellow, R. L. Jones, eds.), Academic Press, San Diego, pp. 17-40.
7. A. Bellow (1983), *On "bad universal" sequences in ergodic theory* II, Lecture Notes in Mathematics, vol. 1033, Springer-Verlag.
8. A. Bellow (1989), *Perturbation of a sequence*, Adv. Math. **78**, no. 2, 131–139.
9. A. Bellow, R. L. Jones, J. Rosenblatt (1988), *Harmonic analysis and ergodic theory*, Almost everywhere convergence: Proceedings of the International Conference on Almost Everywhere Convergence in Probability and Ergodic Theory; Columbus, Ohio (G. Edgar, L. Sucheston, eds.), Academic Press, pp. 73–98.
10. _____ (1990), *Convergence for moving averages*, Ergodic Th. Dynamical Syst. **10**, 43–62.
11. _____ (1994), *Almost everywhere convergence of convolution powers*, to appear in Ergodic Theory and Dynamical Systems, preprint, 19 pages.
12. A. Bellow, V. Losert (1984), *On sequences of density zero in ergodic theory*, Conference in Modern Analysis, in "Contemporary Mathematics", vol. 26, Amer. Math. Soc., Providence, RI.
13. _____ (1985), *The weighted pointwise ergodic theorem and the individual ergodic theorem along subsequences*, Trans. Amer. Math. Soc. **288**, 307–345.
14. G.D. Birkhoff (1931), *Proof of the ergodic theorem*, Proc. Natl. Acad. Sci. USA **17**, 656–660.
15. J. R. Blum, R. Cogburn (1975), *On ergodic sequences of measures,*, Proc. Amer. Math. Soc. **51**, 359-365.
16. P. Bohl (1909), *Über ein in der Theorie der säkutaren Störungen vorkommendes Problem*, J. reine angew. Math **135**, 189-283.
17. M. Boshernitzan (1983), *Homogeneously distributed sequences and Poincaré sequences of integers of sublacunary growth*, Monatsh. Math. **96**, no. 3, 173–181.
18. J. Bourgain (1988), *On the maximal ergodic theorem for certain subsets of the integers*, Israel J. Math. **61**, 39–72.
19. _____ (1988a), *On the pointwise ergodic theorem on L^p for arithmetic sets*, Israel J. Math. **61**, 73–84.
20. _____ (1988b), *Almost sure convergence and bounded entropy*, Israel J. Math. **63**, 79–97.
21. _____ (1989), *Pointwise ergodic theorems for arithmetic sets (Appendix: The return time theorem)*, Publ. Math. IHES **69**, 5–45.
22. _____ (1990), *Problems of almost everywhere convergence related to harmonic analysis and number theory*, Israel J. Math. **71**, no. 1, 97–127.
23. _____ (1990a), *Double recurrence and almost sure convergence*, J. reine angew. Math. **404**, 140–161.
24. A. Calderón (1968), *Ergodic theory and translation invariant operators*, Proc. Natl. Acad. Sci. USA **59**, 349–353.

25. J. Campbell, K. Petersen (1989), *The spectral measure and Hilbert transform of a measure-preserving transformation*, Trans. Amer. Math. Soc. **313**, 121–129.

26. A. Carbery, M. Christ, J. Vance, S. Wainger, D.K. Watson (1989), *Operators associated to flat plane curves: L^p-estimates via dilation methods*, Duke Math. J. **59**, no. 3, 675–700.

27. K.L. Chung, P. Erdős (1951), *Probability limit theorems assuming only the first moment*, Memoirs Amer. Math. Soc. **6**, 13–19.

28. K. L. Chung, W. H. J. Fuchs (1951), *On the distribution of values of sums of random variables*, Memoirs Amer. Math. Soc. **6**, 1–6.

29. J. P. Conze (1973), *Convergence des moyennes ergodiques pour des sous-suites*, Bull. Soc. Math. France **35**, 7–15.

30. J. P. Conze, E. Lesigne (1984), *Théorèmes ergodiques pur des mésures diagonales*, Bull. Soc. Math. France **112**, 143–145.

31. M. Cotlar (1955), *A unified theory of Hilbert transforms and ergodic theorems*, Rev. Mat. Cuyana **1**, 105–167.

32. N. Dunford, J. Schwartz (1986), *Linear operators, Vol. I*, John Wiley and Sons, New York.

33. R. Durrett (1991), *Probability: Theory and Examples*, Wadsworth & Cole, Pacific Grove, California.

34. F. and W. Ellison (1985), *Prime Numbers*, John Wiley & Sons.

35. M. Ellis, N. Friedmann (1978), *Sweeping out on a set of integers*, Proc. Amer. Math. Soc. **72 (3)**, 509–512.

36. W.R. Emerson (1974), *The pointwise ergodic theorem for amenable groups*, Amer. J. Math. **96 (3)**, 472–487.

37. H. Furstenberg (1990), *Nonconventional ergodic averages*, Proc. of Symp. Pure Math. **50**, 43–56.

38. A. Garsia (1970), *Topics in Almost Everywhere Convergence*, Chicago, Markham Publ. Co., 1970.

39. M. de Guzmán (1981), *Real variable methods in Fourier analysis*, Math. Studies, North Holland, New York.

40. P. R. Halmos (1956), *Lectures on Ergodic Theory*, Math. Soc. of Japan.

41. H. Helson (1991), *Harmonic Analysis*, Wadsworth & Cole, Pacific Grove, California.

42. E. Hopf (1937), *Ergodentheorie*, Ergebnisse d. Math., V/2, Berlin.

43. Y. Huang (1992), *Random sets for the pointwise ergodic theorem*, Ergodic Th. Dynamical Syst. **12**, 85–94.

44. R. Jones, I. Ostrovskii, J. Rosenblatt, *Square functions in ergodic theory*, preprint, 49 pages.

45. R. Jones, J. Olsen, M. Wierdl (1992), *Subsequence ergodic theorems for L^p-contractions*, Trans. Amer. Math. Soc. **331**, 837–850.

46. R. Jones, J. Rosenblatt, A. Tempelman, *Convergence of probability measures on groups in ergodic theory*, to appear in Ill. J. Math., preprint, 30 pages.

47. A. del Junco, J. Rosenblatt (1979), *Counterexamples in ergodic theory and number theory*, Math. Ann. **245**, 185–197.

48. J. P. Kahane (1985), *Some Random Series of Functions*, 2nd ed., Cambridge Univ. Press.

49. S. Kakutani, K. Yosida (1939), *Birkhoff's ergodic theorem and the maximal ergodic theorem*, Proc. Imp. Acad. Tokyo **15**, 165–168.

50. A. Ya. Khintchin (1933), *Zur Birkhoff's Lösung des Ergodensproblems*, Math. Ann. **107**, 485–488.

51. J. F. Koksma (1952), *A diophantine property of some summable functions*, J. Indian Math Soc. **15**, 87–96.

52. U. Krengel (1971), *On the individual ergodic theorem for subsequences*, Ann. Math. Statist. **42**, 1091–1095.

53. ———— (1985), *Ergodic Theorems*, vol. 6, de Gruyter Studies in Mathematics.

54. M. Lacey, *The Carleson - Hunt theorem, Littlewood - Paley inequalities and return-time theorems in ergodic theory*, preprint.
55. M. Lacey, K. Petersen, D. Rudolph, M Wierdl, *Random ergodic theorems with universally representative sequences*, Ann. Inst. H. Poincaré (to appear).
56. E. Lesigne, Personal communication.
57. D. Lind (1975), *Locally compact measure preserving flows*, Advances in Math **15**, 175-193.
58. M. Marcus, G. Pisier (1981), *Random Fourier Series with Applications to Harmonic Analysis*, Ann. Math. Studies, vol. 101, Princeton Univ. Press, Princeton.
59. J. M. Marstrand (1970), *On Khintchine's conjecture about strong uniform distribution*, Proc. London Math. Soc. **21**, 540-556.
60. A. Nagel, E. M. Stein (1984), *On certain maximal functions and approach regions*, Adv. Math. **54**, 83-106.
61. R. Nair (1991), *On polynomials in primes and J. Bourgain's circle method approach to ergodic theorems*, Ergodic Th. Dynamical Syst. **11**, 485-499.
62. J. von Neumann (1932), *Proof of the quasi ergodic hypothesis*, Proc. Natl. Acad. Sci. USA **18**, 70-82.
63. K. Petersen (1989), *Ergodic Theory*, Cambridge Univ. Press; Cambridge and New York.
64. K. Reinhold-Larsson (1991), *Almost everywhere convergence of weighted averages, Ph.D Thesis*, The Ohio State University.
65. K. Reinhold-Larsson (1993), *Almost everywhere convergence of convolution powers*, Illinois Journal of Math **37**, 666-679.
66. K. Reinhold-Larsson (1994), *Discrepancy of behavior in L^p spaces of perturbed sequences*, to appear in AMS Proceedings, 16 pages.
67. F. Riesz (1938), *Sur un théorème de maximum de MM. Hardy et Littlewood*, J. London Math. Soc. **7**, 10-13.
68. F. Riesz, B. Sz.-Nagy (1990), *Functional Analysis*, Dover Publications, Inc., New York.
69. V. A. Rokhlin (1962), *On the fundamental ideas of measure theory*, Amer. Math. Soc. Trans., vol. 10, pp. 1-54.
70. J. Rosenblatt (1981), *Ergodic and mixing random walks on locally compact groups*, Math Annalen **257**, 31-42.
71. _____(1988), *Almost everywhere convergence of series*, Math. Ann. **280**, 565-577.
72. _____(1991), *Universally bad sequences in ergodic theory*, Almost everywhere convergence (II) (A. Bellow, R. Jones, eds.), Academic Press, pp. 227-245.
73. _____(1994), *Norm convergence in ergodic theory and the behavior of Fourier transforms*, Can. Journal of Math **46**, 184-199.
74. J. Rosenblatt, M. Wierdl (1992), *A new maximal inequality and its applications*, Ergodic Th. Dynamical Syst. **12**, 509-558.
75. H. Royden (1988), *Real Analysis, 3rd ed.*, Macmillan Publishing Co., New York.
76. D. Rudolph (1994), *A joinings proof of Bourgain's return time theorem*, Ergodic Th. Dynamical Syst. **14**, 197-203.
77. R. Salem, A. Zygmund (1954), *Trigonometric series whose terms have random sign*, Acta Math. **91**, 245-301.
78. S. Sawyer (1966), *Maximal inequalities of weak type*, Ann. Math. **84**, 157-174.
79. W. Sierpiński (1910), *Sur la valeur asymptotique d'une certaine somme*, Bull. Intl. Acad. Polonaise des Sci. et des Lettres (Cracovie) **series A**, 9-11.
80. E. M. Stein (1961a), *On limits of sequences of operators*, Ann. Math. **74**, 140-170.
81. E. M. Stein (1961b), *On the maximal ergodic theorem*, Proc. Nat. Acad. Sci. **47**, 1894-1897.
82. E. M. Stein, S. Wainger (1978), *Problems in harmonic analysis related to curvature*, Bull. Amer. Math. Soc. **84**, 1239-1295.

83. ———(1990), *Discrete analogues of singular Radon transforms*, Bull. Amer. Math. Soc. (New Series) **23**, no. 2, 537–544.

84. A. Tempelman (1967), *Ergodic theorems for general dynamical systems*, Dokl. Akad. Nauk SSSR **176**, 790-793 (English translation: Soviet Math Doklady 8 (1967)1213-1216.

85. J. P. Thouvenot (1990), *La convergence presque sûre des moyennes ergodiques suivant certaines sous-suites d'entières (d'après J. Bourgain)*, no. 189-190, Exp. No. 719, Seminaire Bourbaki, pp. 133–153.

86. R. C. Vaughan (1981), *The Hardy-Littlewood Method*, Cambridge Univ. Press.

87. A. Weil (1979), *Number Theory for Beginners*, Springer-Verlag, New York.

88. H. Weyl (1910), *Über die Gibbs'sce Erscheinung und verwandte Konvergenzphänomene*, Rendiconti del Circolo Matematico di Palermo **330**, 377-407.

89. H. Weyl (1916), *Über die Gleichverteilung von Zâhlen mod. Eins*, Math. Ann. **77**, 313-352.

90. R. Wheeden, A. Zygmund (1977), *Measure and Integral (An introduction to real analysis)*, Marcel Dekker, Inc., New York and Basel.

91. N. Wiener (1939), *The ergodic theorem*, Duke Math. J. **5**, 1-8.

92. N. Wiener, A. Wintner (1941), *Harmonic analysis and ergodic theory*, Amer. J. Math. **63**, 415-426.

93. M. Wierdl (1988), *Pointwise ergodic theorem along the prime numbers*, Israel J. Math. **64**, no. 2, 315-336.

94. ———(1989), *Almost everywhere convergence and recurrence along subsequences in ergodic theory*, Ph.D Thesis, The Ohio State University.

95. R. Wittmann, *On a maximal inequality of Rosenblatt and Wierdl*, preprint.

96. A. Zygmund (1968), *Trigonometric Series*, Cambridge Univ. Press, London.

DEPARTMENT OF MATHEMATICS, THE OHIO STATE UNIVERSITY, 231 W. 18TH AVE., COLUMBUS, OHIO 43210
E-mail address: jrsnblatt@math.ohio-state.edu

DEPARTMENT OF MATHEMATICS, LUNT HALL, NORTHWESTERN UNIVERSITY, EVANSTON, ILLINOIS 60208
E-mail address: mate@math.nwu.edu

HARMONIC ANALYSIS IN RIGIDITY THEORY

R. J. SPATZIER

CONTENTS

Partially supported by the NSF

1. INTRODUCTION

This article surveys the use of harmonic analysis in the study of rigidity properties of discrete subgroups of Lie groups, actions of such on manifolds and related phenomena in geometry and dynamics. Let me call this circle of ideas *rigidity theory* for short. Harmonic analysis has most often come into play in the guise of the representation theory of a group, such as the automorphism group of a system. I will concentrate on this avenue in this survey. There are certainly other ways in which harmonic analysis enters the subject. For example, harmonic functions on manifolds of nonpositive curvature, the harmonic measures on boundaries of such spaces and the theory of harmonic maps play an important role in rigidity theory. I will mention some of these developments.

Rigidity theory became established as an important field of research during the last three decades. The first rigidity results date back to about 1960 when A. Selberg, E. Calabi and A. Vesentini and later A. Weil discovered various deformation, infinitesimal and perturbation rigidity theorems for certain discrete subgroups of Lie groups. At about the same time, M. Berger proved his purely geometric 1/4-pinching rigidity theorem for positively curved manifolds [184, 28, 27, 205, 206]. But the most important and influential early result was achieved by G. D. Mostow in 1968. In proving his celebrated Strong Rigidity Theorem, Mostow not only provided a global version of the earlier local results, but also introduced a battery of novel ideas and tools from topology, differential and conformal geometry, group theory, ergodic theory, and harmonic analysis. Mostow's results were the catalyst for a host of diverse developments in the ensuing years. Mostow himself generalized his strong rigidity theorem to locally symmetric spaces in 1973 [146]. In 1974, a second major breakthrough occurred when G. A. Margulis discovered his ingenious superrigidity and arithmeticity theorems for higher rank locally symmetric spaces [127, 129]. All along, H. Furstenberg had been developing his probabilistic approach to rigidity and introduced the idea of boundaries of groups [66, 67, 69]. R. J. Zimmer has been building his important deep program of studying actions of "large" groups on manifolds since 1979 [220, 223, 225]. M. Ratner's work over the last decade has provided a deep and fundamental analysis of the rigidity of horocycle flows and unipotent actions. [169, 170, 172, 173]. Rigidity theory blossomed throughout the 1980s and early 1990s. Further major contributors were: W. Ballmann, Y. Benoist, M. Brin, K. Burns, K. Corlette, P. Eberlein, P. Foulon, E. Ghys, M. Gromov, A. Katok, U. Hamenstadt, S. Hurder, F. Labourie, N. Mok, P. Pansu, R. Schoen, Y.T. Siu, R. Spatzier and others.

Let me briefly outline the paper. Section 2 presents a synopsis of rigidity theory. At the heart of this section lies Mostow's theorem and a summary of its proof. I also discuss Margulis' superrigidity and arithmeticity theorems and outline further major developments in rigidity theory. In the next three sections, I describe various tools from harmonic analysis and how they are used in rigidity theory. Each of these sections addresses a particular tool. In section 3, I explain Mautner's phenomenon, the ergodicity of homogeneous flows and its application to Mostow's theorem. Mautner's phenomenon is closely related to the vanishing of matrix coefficients and their rate of decay. This as well as two applications of the rate of decay to rigidity constitute the remainder of Section 3. In Section 4, I introduce amenable actions, and use them to set up the first step of the proof of Margulis' superrigidity theorem. Some other applications to the rigidity of group actions are given as well. Section 5 introduces Kazhdan's property (T). It has been enormously successful, especially in the study of group actions. I include a selection of these topics. In Section 6, I report on a variety of other applications of harmonic analysis to rigidity theory.

Finally, let me thank S. Adams and my two referees for their many valuable comments and suggestions.

2. A SYNOPSIS OF RIGIDITY THEORY

2.1. Early results. It is probably futile to attempt a general definition of rigidity. However, a common feature of a rigidity theorem is that some fairly weak conditions suddenly and surprisingly force strong consequences. Rigidity theorems usually come in one of the following forms:

   a): a deformation or perturbation of a system is equivalent to the original system,

   b): a system preserving a weak structure is forced to preserve a strong structure,

   c): a weak isomorphism between two objects implies a strong isomorphism.

Already the early history of rigidity theory has theorems of all three types. To begin with, A. Selberg proved in 1960 that a discrete subgroup Γ of $SL(n, \mathbb{R})$ with $n \geq 3$ and $SL(n, \mathbb{R})/\Gamma$ compact cannot be continuously deformed except by inner automorphisms of $SL(n, \mathbb{R})$ [184]. Selberg's method extended to the other classical groups of real rank at least 2. Around the same time, E. Calabi and A. Vesentini proved that the complex structure of a compact quotient of a bounded symmetric domain is rigid under deformations [28]. Slightly later, E. Calabi proved deformation rigidity of compact hyperbolic n-spaceforms for $n \geq 3$ [27]. Remarkably, there are smooth non-isometric families of compact hyperbolic 2-space forms, as was long known at the time. A. Weil generalized Selberg's and Calabi's results to all semisimple groups without compact or three-dimensional factors in 1962 [205, 206].

In 1961, M. Berger proved that a Riemannian manifold of curvature between 1 and 4 which is not homeomorphic to a sphere is isometric to a symmetric space. While this is a classical example of a rigidity theorem (of the second type), its proof is entirely differential geometric. More

importantly, the source of rigidity in this case is ellipticity rather than
hyperbolicity, unlike most of the examples we will discuss (cf. [79] for a
discussion of non-hyperbolic rigidity phenomena).

There are some other early rigidity results. J. Wolf found in 1962 that
if G/Γ is compact, then the rank of G is determined by Γ [208]. This was
later generalized to non-uniform lattices by G. Prasad and M. S. Raghunathan in [160]. Using probabilistic considerations, H. Furstenberg showed
in 1967 that a lattice in $SL(n, \mathbb{R})$ cannot be a lattice in $SO(n, 1)$ [66]. Y.
Matsushima and S. Murakami as well as M. S. Raghunathan obtained vanishing theorems for the cohomology of representations of uniform lattices
[136, 166, 168, 167].

2.2. Mostow's strong rigidity theorems. In 1968, Mostow proved his
celebrated strong rigidity theorem for hyperbolic space forms. It is the
prime example of a rigidity theorem of the third type. We will discuss it
in detail in the remainder of this section. Let us first fix some notations.
We will always call compact manifolds without boundary *closed*. Given a
Riemannian manifold M, we denote its universal cover by $\tilde{M}$, its tangent
bundle by TM and its unit tangent bundle by SM. Denote by γ_v the
unique unit speed geodesic tangent to v at 0. Recall that the *geodesic flow*
$g_t : SM \to SM$ takes a unit vector v to $\gamma'_v(t)$, the vector tangent to γ_v
at time t. Recall that the sectional curvature function K is a real valued
function on the set of 2-planes in TM.

Theorem 2.1. [Strong Rigidity Theorem, Mostow, 1968] *Suppose
M and N are closed manifolds of constant sectional curvature -1. Assume
M has dimension at least 3. If there is an isomorphism $\psi : \pi_1(M) \to \pi_1(N)$
then M and N are isometric.*

For surfaces, this theorem fails completely. In fact, there is a $6g - 6$-
dimensional space, the so-called *moduli space*, that parametrizes all metrics
of constant curvature up to diffeomorphism on a surface of genus $g \geq 2$.
Here I will just outline the so-called "pair of pants" construction to exhibit
a non-trivial continuous family of constant curvature metrics on a closed
surface of genus $g \geq 2$. This construction can be used to describe the
moduli space completely [26].

Elementary considerations in hyperbolic geometry show that, given any positive real numbers
l_1, l_2 and l_3, there exists a hexagon in the hyperbolic plane $\mathcal{H}^2$ with geodesic edges $e_1, \ldots, e_6$ such
that the length of e_{2i} is l_i for $i = 1, 2$ and 3 [26].
By gluing two such hexagons along e_1, e_3 and e_5,
we obtain a metric of constant curvature -1 on a
two-sphere minus three open disjoint disks, a "pair
of pants". Note that the boundary curves of this
pair of pants are geodesics.

Gluing two such pairs of pants along edges of equal length, we obtain a surface Σ of genus 2 with a metric of constant curvature -1 which has closed geodesics of length $2\,l_1, 2\,l_2$ and $2\,l_3$. Recall that the lengths of closed geodesics with respect to a fixed metric form a countable set of numbers. Hence, varying the l_i continuously, we obtain non-isometric metrics on a surface of genus 2.

Similar constructions can be made for any closed surface of genus g at least 2. In fact, such a surface is obtained by gluing $2\,(g-1)$ pairs of pants. Note that the rotation by which two boundary geodesics of two pairs of pants are glued together gives another parameter of the construction. Careful analysis shows that these $6g - 6$ length and rotation parameters actually parametrize the set of all metrics of constant curvature -1 on Σ [26].

Theorem 2.1 is completely equivalent to a theorem about subgroups of Lie groups. Given any locally compact group G, we call a discrete subgroup Γ of G a *lattice* if the Haar measure μ on G/Γ is finite and G-invariant. We call a lattice *uniform* or *cocompact* if G/Γ is compact. Otherwise we call Γ *non-uniform*.

Let $SO(n,1)$ be the group of $n+1 \times n+1$ real matrices which preserve the bilinear form

$$\langle x,y \rangle = x_0 y_0 - \sum_{i=1}^{n} x_i y_i$$

on $\mathbb{R}^{n+1}$. Let $SO_0(n,1)$ denote the connected component of the identity of $SO(n,1)$. Then $SO_0(n,1)$ acts transitively and effectively on the sheet of the hyperboloid

$$\mathcal{H}^n \overset{def}{=} \{x \in \mathbb{R}^{n+1} \mid x_0 > 0, \langle x,x \rangle = 1\}$$

through $(1,0,\dots,0)$. The restriction of the quadratic form $\langle \, , \, \rangle$ to the tangent bundle of $\mathcal{H}^n$ is positive definite, and the resulting Riemannian metric has constant curvature -1. We call $\mathcal{H}^n$ the *real hyperbolic space* of dimension n. Note that the isotropy group of $(1,0,\dots,0)$ consists of the matrices of the form

$$\begin{pmatrix} 1 & 0\dots0 \\ 0 & \\ . & \\ . & A \\ . & \\ 0 & \end{pmatrix}$$

where A is in $SO(n)$, i.e. A is an orthogonal $n \times n$ matrix matrix of determinant 1. Thus $\mathcal{H}^n$ is the homogeneous space $SO(n) \setminus SO_0(n,1)$, and the metric $\langle \, , \, \rangle$ on $\mathcal{H}^n$ can be described in terms of the unique $SO_0(n,1)$-invariant quadratic form on $SO_0(n,1)$, the so-called Cartan-Killing form. It is well-known that any complete simply connected manifold of constant curvature -1 is isometric to the real hyperbolic space $\mathcal{H}^n$. Thus the universal cover of any closed manifold M of constant negative curvature -1 is $\mathcal{H}^n$. Therefore we can write M as $\mathcal{H}^n/\Gamma$ or $SO(n) \setminus SO_0(n,1) / \Gamma$ where Γ is a group of isometries of $\mathcal{H}^n$. It is elementary that $\mathcal{H}^n/\Gamma$ is isometric to $\mathcal{H}^n/\Gamma'$ for some discrete subgroup Γ' of $SO_0(n,1)$ if and only if Γ and Γ' are conjugate in $SO_0(n,1)$. It follows that the geometric version of Mostow's rigidity theorem above is completely equivalent to the following group theoretic rigidity theorem. These remarks also explain the relationship between Selberg's, Calabi's and Weil's results mentioned above.

Theorem 2.2. [Strong Rigidity Theorem, Algebraic Form, Mostow, 1968] *Let Γ be a cocompact lattice in $SO_0(n,1)$, and $n \geq 3$. Let $\psi : \Gamma \to SO_0(m,1)$ for some m be an injective homomorphism such that $\psi(\Gamma)$ is a cocompact lattice in $SO_0(m,1)$. Then $m = n$ and ψ is the restriction of an inner automorphism of $SO_0(n,1)$.*

G. Prasad generalized this theorem to non-uniform lattices in 1973 [159]. M. Gromov proved the theorem in a totally different and completely geometric way in 1979 [80, 196]. Mostow's original proof has also been streamlined and extended to other discrete subgroups of $SO_0(m,1)$, latest by Ivanov in 1993 [198, 8, 103].

Let us give an outline of Mostow's proof. For more details we refer to [196, 81]. Harmonic analysis enters in step 3 below. As we point out there, while historically significant, one can easily substitute a geometric argument (e.g. Hopf's argument) for the harmonic analysis used here. However, the same ideas from harmonic analysis have since proved useful for other rigidity theorems.

Outline of Mostow's proof: Set $\Gamma = \pi_1(M)$. The argument proceeds in several steps.

Step 1: Call a map ϕ between two metric spaces (X, d) and (X', d') a *quasi-isometry* if there are constants $C > 0$ and $E > 0$ such that

$$\frac{1}{C}d(x,y) - E < d'(\phi(x), \phi(y)) < C\,d(x,y) + E.$$

The isomorphism $\psi : \Gamma \to \pi_1(N)$ gives rise to a quasi-isometry $\phi : \tilde{M} \to \tilde{N}$ between the universal covers of M and N. Moreover, ϕ is Γ-equivariant. This means that for all $\gamma \in \Gamma$ and all $x \in \tilde{M}$, we have

$$\phi(\gamma\,x) = \psi(\gamma)\,\phi(x)$$

where γ and $\psi(\gamma)$ act on $\tilde{M}$ and $\tilde{N}$ by deck transformations.

Essentially one constructs ϕ as follows. First one maps a fundamental domain of Γ in $\tilde{M}$ to a fundamental domain of $\pi_1(N)$ in $\tilde{N}$ in an arbitrary way. This map extends uniquely to a Γ-equivariant map on $\tilde{M}$. Note that ϕ need not be continuous, as there maybe discontinuities along the boundary of the fundamental domain. Note that Mostow in his original argument showed how to make Φ into a homeomorphism.

Step 2: Compactify $\mathcal{H}^n$ as follows. Call two geodesic rays *asymptotic* if they are a finite distance apart. The set of asymptote classes is called the *sphere at infinity* S^{n-1}. One can topologize S^{n-1} as well as $\mathcal{H}^n \cup S^{n-1}$ by the so-called "cone topology" [14]. The sphere at infinity then is homeomorphic to a standard sphere.

In the unit disk model of hyperbolic space, this amounts to compactifying the open ball $\mathcal{H}^n$ in Euclidean space to the closed ball. While the above construction works for any simply-connected complete manifold of nonpositive curvature, the sphere at infinity of $\mathcal{H}^n$ has a differentiable and a conformal structure, as one can see from the unit disk model. Standard Lebesgue measure induces a measure class on S^{n-1}. One can abstractly describe this as a *visual measure*. This means, that the measure of a set X in the sphere at infinity is given by the measure of the unit tangent vectors v at a fixed point $p \in \mathcal{H}^n$ such that the asymptote class of the geodesic ray determined by v lies in X.

Any action on $\mathcal{H}^n$ by isometries naturally extends to a continuous action on S^{n-1}. Furthermore, the extension is conformal and preserves the measure class on S^{n-1}.

Thus both $\tilde{M}$ and $\tilde{N}$ are compactified by their spheres at infinity S^{n-1} and S^{m-1}. The quasi-isometric image of a geodesic ray in $\tilde{M}$ lies a finite distance from a (unique) geodesic ray. This crucial fact follows from the the the negative curvature of $\tilde{N}$. Thus the quasi-isometry $\phi : \tilde{M} \to \tilde{N}$ extends to a map of the spheres at infinity $\bar{\phi} : S^{n-1} \to S^{m-1}$. By the Γ-equivariance of ϕ, $\bar{\phi}$ is also Γ-equivariant.

One also shows that $\bar{\phi}$ is *quasi-conformal*. This means that the images of all small spheres have bounded *distortion*. Here the distortion of a set is the infimum of the fractions of the radii of circumscribed and inscribed balls. It follows from analysis, that $\bar{\phi}$ is differentiable almost everywhere, and the derivatives and their inverses have uniformly bounded norm. Hence we see that $n \leq m$, and by symmetry that $m = n$. Thus we can identify $\tilde{M}$ and $\tilde{N}$ with $\mathcal{H}^n$.

Step 3: The action of Γ on S^{n-1} is ergodic. Mostow deduced this directly from the Mautner phenomenon in representation theory (cf. Section 3.1). Ergodicity of the Γ-action also follows from the ergodicity of the geodesic flow. The latter can be seen either again using the Mautner phenomenon or via the Hopf argument (cf. Section 3.1).

Using the quasi-conformality and the ergodicity of Γ on S^{m-1}, one can show that the distortion of $\bar{\phi}$ is constant. Then one argues that the distortion is 1. Therefore $\bar{\phi}$ is conformal.

Step 4: Any conformal map of S^{n-1} extends to a unique isometry of $\mathcal{H}^n$. By the Γ-equivariance of $\bar{\phi}$, the isometric extension $\tilde{\phi}$ of $\bar{\phi}$ is Γ-equivariant. Hence $\tilde{\phi}$ descends to an isometry of M to N. This finishes the proof of Mostow's theorem. ◇

Step 2 above established the existence of a conjugacy $\bar{\phi}$ between the two actions of Γ on two spheres at infinity, and that $\bar{\phi}$ is quasi-conformal. The remainder of the proof then concludes that a quasi-conformal conjugacy is conformal, and the two embeddings of Γ into $SO(n,1)$ are conjugate. We can interpret this last part of Mostow's proof as a dynamical rigidity theorem, about conjugacy of actions of groups of isometries of hyperbolic space on the sphere at infinity. More generally one can ask what kind of conjugacies between two actions of a group on the sphere at infinity of hyperbolic space imply conjugacy of the groups? Continuity alone of course is not sufficient. R. Bowen and D. Sullivan showed around 1980 that absolute continuity of the conjugacy suffices [23, 193]. In 1985, P. Tukia further minimized the amount of differentiability needed [198]. For simplicity, let us state a version of his theorem for cocompact lattices in $PSL(2,\mathbb{R}) = SL(2,\mathbb{R})/\{ \overset{+}{-} 1 \}$.

Theorem 2.3. [Tukia, 1985] *Assume* Γ_1 *and* Γ_2 *are cocompact lattices in* $PSL(2,\mathbb{R})$. *Suppose* Γ_1 *and* Γ_2 *are isomorphic and that the actions of* Γ_1 *and* Γ_2 *on the circle at infinity are conjugate by a homeomorphism which has a non-zero derivative in at least one point. Then* Γ_1 *and* Γ_2 *are conjugate in* $PSL(2,\mathbb{R})$.

This avenue was further pursued in [6, 7, 8]. In a recent preprint, N. Ivanov gave a remarkably simple proof of Tukia's theorem and generalized it to higher derivatives and dimensions [103].

Mostow himself generalized his theorem from hyperbolic spaces to compact locally symmetric spaces of the noncompact type in 1973 [146]. To exclude two-dimensional hyperbolic space, he assumed that the space does not have closed one or two dimensional geodesic subspaces which are direct factors locally. Any locally symmetric space can be written as a double quotient $K \backslash G/\Gamma$ where G is the isometry group of the universal cover, K is a maximal compact subgroup of G and Γ is a torsion-free uniform lattice in G. Specific examples are hyperbolic spaces or $SO(n) \backslash SL(n,\mathbb{R})/\Gamma$ where Γ is a uniform lattice in $SL(n,\mathbb{R})$. As for hyperbolic spaces, there is a geometric and a group theoretic version of Mostow's strong rigidity theorem for locally symmetric spaces [146]. Let us only present the algebraic version in detail.

Theorem 2.4. [**Strong Rigidity Theorem, Mostow, 1973**] *Let* G *and* H *be connected semisimple real Lie groups without compact factors and trivial center. Let* Γ *be a lattice in* G, *and assume that there is no factor* G' *of* G, *isomorphic to* $PSL(2,\mathbb{R})$ *which is closed modulo* Γ, *i.e. such that* $\Gamma G'$ *is closed in* G.

Suppose $\psi : \Gamma \to H$ *is an injective homomorphism such that* $\psi(\Gamma)$ *is a lattice in* H. *Then* ψ *extends to a smooth isomorphism* $G \to H$.

Mostow proved this theorem for uniform lattices. Strong rigidity for certain non-uniform arithmetic lattices had been obtained by algebraic and arithmetic means in 1967 by H. Bass, J. Milnor and J. P. Serre and also

by M. S. Raghunathan [15, 168]. For general non-uniform lattices, strong
rigidity was established by G. Prasad in [159]. While substantially more
complicated, Mostow's ideas in the proof for locally symmetric spaces are
similar to the constant curvature case. Again one argues that there is a
quasi-isometry between the universal covers which extends to the sphere at
infinity. For symmetric spaces with negative curvature, there is a generalized conformal structure on the sphere at infinity, and one can proceed as
in the constant curvature case. If the symmetric space has some 0 curvature, the sphere at infinity carries the structure of a so-called *spherical Tits
building* [146, 197]. They are generalizations of classical projective geometry. Tits had invented them to discuss all semisimple algebraic groups from
a geometric point of view in a uniform way. Mostow then used Tits' extension of the fundamental theorem of projective geometry to Tits geometries
to see that the map induced by the quasi-isometry on the sphere at infinity
is induced by an isomorphism of the ambient Lie groups.

2.3. Margulis' superrigidity theorem. The assumption in Mostow's
strong rigidity theorem that the image $\psi(\Gamma)$ be a lattice is quite restrictive.
G. A. Margulis achieved a breakthrough in 1974 when he determined the
homomorphisms of lattices in higher rank semisimple Lie groups G into Lie
groups H under only mild assumptions [129, 127]. Higher rank refers to the
real rank of G. Geometrically, this is the maximal dimension of a totally
geodesic flat subspace of the symmetric space $K \setminus G$. Algebraically, one
can define it in terms of the maximal dimension of a so-called *split Cartan*
subgroup (cf. Section 3.2). Let us call a lattice Γ in G *reducible* if there
are connected infinite normal subgroups G' and G'' in G such that $G' \cap G''$
is central in G, $G' \cdot G'' = G$, and the subgroup $(\Gamma \cap G') \cdot (\Gamma \cap G'')$ has
finite index in Γ. Otherwise we call Γ *irreducible*. Note that any lattice
in a simple Lie group is irreducible. Let us note here that any connected
simple real algebraic group always contains both uniform and non-uniform
lattices [22, 21, 165]. While Margulis originally only considered semisimple
range groups, he later developed the following much refined version of his
superrigidity theorem [127, 129].

Theorem 2.5. [Superrigidity Theorem, Margulis, 1974] *Let G be
a connected semisimple Lie group with finite center, real rank at least 2,
and without compact factors. Let Γ be an irreducible lattice in G. Let
$\psi : \Gamma \to H$ be a homomorphism of Γ into a real algebraic group H. Then
the Zariski closure of the image $\psi(\Gamma)$ is semisimple. Suppose that $\psi(\Gamma)$ is
not relatively compact, that the Zariski closure of $\psi(\Gamma)$ has trivial center
and does not have non-trivial compact factors. Then ψ extends uniquely to
a continuous homomorphism from G to H.*

Again, the main point in the proof is to construct a Γ-equivariant map
between certain "boundaries" of G and H, and to show that such maps
are automatically smooth, and give rise to an extension of ψ to G. While

the original proof did not directly involve harmonic analysis, R. J. Zimmer in 1979 used amenability to construct such a boundary map [216, 220]. This also allowed him to generalize superrigidity to cocycles of actions of semisimple groups. We will discuss these developments in Section 4.

Margulis refined his theorem to homomorphisms taking values in algebraic groups over arbitrary and in particular p-adic fields [127, Ch. VII, Theorem 6.5]. As a corollary, Margulis obtained his famed arithmeticity theorem [127, 129]. Briefly, an *arithmetic group* is any group commensurable with the points of integers of an algebraic group defined over the integers. Arithmetic groups are always lattices [165]. A typical example is $SL(n, \mathbb{Z})$ in $SL(n, \mathbb{R})$. Note though that $SL(n, \mathbb{Z})$ is a non-uniform lattice.

Theorem 2.6. [Arithmeticity Theorem, Margulis, 1974] *Let Γ and G be as in Theorem 2.5. Then Γ is arithmetic.*

Superrigidity and arithmeticity theorems fail for lattices in many real rank 1 groups [126, 201, 82, 147, 148, 150, 45]. Using differential geometric methods from the theory of harmonic maps, K. Corlette in the Archimedean case, and M. Gromov and R. Schoen in the non-Archimedean case extended Margulis' results to certain rank one spaces.

Theorem 2.7. [Corlette, 1990, Gromov-Schoen, 1992] *Let Γ be a lattice in $G = Sp(n, 1)$, $n \geq 2$, and the exceptional real rank 1 group F_4^{-20}. Then any homomorphism $\Gamma \to GL_n$ into the general linear group over a local field extends unless the image of Γ is precompact. Furthermore, Γ is arithmetic.*

The idea in the proof is very different from Mostow's scheme. One first finds a Γ-equivariant harmonic map between the associated globally symmetric spaces, using a general existence theorem originally due to J. Eells and J. Sampson [49]. Then one shows that this map is totally geodesic using a "Bochner formula" for a certain 4-tensor. This idea goes back to Y.-T. Siu's strong rigidity theorem for Kähler manifolds of negative bisectional curvature [186]. In the quaternionic case, the Bochner formula is so strong that one obtains superrigidity rather than just strong rigidity.

N. Mok, Y.-T. Siu and S.-K. Yeung recently found Bochner formulas for suitable 4-tensors for all higher rank globally symmetric spaces, quaternionic hyperbolic spaces and the Cayley plane [141, 140, 139]. Thus they reprove and extend both Margulis' and Corlette's superrigidity theorems using harmonic maps in a fairly unified geometric way. In part, this work is a nonlinear version of the ideas of Y. Matsushima and S. Murakami [136]. Similar results were obtained by J. Jost and S.-T. Yau [104]. These methods also yield a fair amount of information for complex hyperbolic spaces and other Kähler manifolds, as was shown by W. Goldman and J. Millson and later J. Carlson and D. Toledo [74, 29].

Another high point is Margulis' finiteness theorem. It asserts that all normal subgroups of an irreducible lattice Γ in a semisimple Lie group of the noncompact type and higher rank are either central or have finite index. In the course of the proof, Margulis also determined the measurable quotients of the natural Γ-action on the "boundaries" of G, e.g. projective spaces [127].

2.4. Further developments.

We will now give a brief outline of further important developments in rigidity theory during the last two decades. In later sections we will discuss those advances that are connected with harmonic analysis in more detail.

2.4.1. *Actions of semisimple groups and their lattices.* Due to the Margulis' superrigidity theorems, one essentially understands the finite dimensional representations of an irreducible lattice in a semisimple Lie group of higher rank. Naturally one asks if other representations of such a group are similarly restricted.

Infinite dimensional unitary representations of such a lattice in general are quite unwieldy. In fact, E. Thoma showed in 1964 that a discrete group is type I if and only if it has an abelian subgroup of finite index [195]. On the positive side however, such representations satisfy some restrictions, such as Kazhdan's property (cf. Section 5) or the recent results of M. Cowling and T. Steger on restrictions of unitary representations of the ambient semisimple group to the lattice (cf. Section 3.4) [119, 36].

The "finite dimensional nonlinear representation theory" of such groups, especially the study of smooth actions on manifolds, lies in between and is much more restricted. R. J. Zimmer started their study in 1979, when he generalized Margulis' superrigidity theorems to cocycles over finite measure preserving actions of semisimple groups and their lattices.

Given an action of a group G on a measure space M with measure μ and another group H, a measurable map $\beta : G \times M \to H$ is called a *measurable cocycle* if it satisfies the cocycle identity $\beta(g_1 g_2, m) = \beta(g_1, g_2 m) \beta(g_2, m)$ for μ-a.e. $m \in M$ [220]. The simplest cocycles are the constant cocycles, i.e. those μ-a.e. constant in M. They correspond to homomorphisms $G \to H$. As another example, suppose G acts differentiably on a manifold M by α. Let μ be a Lebesgue measure on M and choose a measurable framing of M. Then the derivatives of $g \in G$ acting on M determine elements in $GL(n, \mathbb{R})$ at every point of M. Due to the chain rule, this defines a cocycle, the so-called *derivative cocycle* of α.

Call two cocycles β and β^* *measurably cohomologous* if there exists a measurable function $P : M \to H$, called a *measurable coboundary*, such that $\beta^*(a, x) = P(ax)^{-1} \beta(a, x) P(x)$ for all $a \in G$ and a.e. $x \in M$. For example, a change of the measurable framing of a manifold determines a cohomologous derivative cocycle. Thus often we are only interested in the cohomology class of the cocycle.

Let us call a finite measure preserving ergodic action *irreducible* if it remains ergodic when restricted to any non-trivial normal subgroup of G.

Theorem 2.8. [**Cocycle Superrigidity, Zimmer, 1980**] *Let G be a connected semisimple real algebraic group without compact factors and of rank at least 2. Let α be an irreducible ergodic finite measure preserving action of G on a measure space M. Let H be a connected non-compact simple real algebraic group. Suppose $\beta : G \times M \to H$ is a measurable cocycle which is not cohomologous to a cocycle taking values in an algebraic subgroup L of H. Then β is cohomologous to a constant cocycle.*

Amenability plays an essential role in the proof of this theorem (cf. Section 4). There are also versions of this theorem for p-adic and complex semisimple groups as range groups. It is not known if Theorem 2.8 holds for range groups which are not semisimple. However, suppose G acts by principal bundle automorphisms on a principal H-bundle $P \to M$ over a compact manifold M. Trivialize the bundle measurably. Then the bundle automorphisms give rise to a cocycle $\beta : G \times M \to H$. Zimmer extended Theorem 2.8 to these cocycles in 1990 [226]. This extends earlier work by G. Stuck and Zimmer himself and gives a full generalization of Margulis' superrigidity theorem for these cocycles [191, 225].

Using harmonic maps, K. Corlette and R. Zimmer found a quaternionic version of the cocycle superrigidity theorem under somewhat stronger assumptions [33]. M. Cowling and R. Zimmer had earlier developed certain rigidity statements for lattices in $Sp(n,1)$ and their actions using von Neumann algebra techniques [37] [37].

The cocycle superrigidity theorems impose severe restrictions on the derivative cocycle of a smooth volume preserving action, especially when the action preserves further geometric structures. This was shown by Zimmer and others in a remarkable series of papers (cf. e.g. [225, 223]). By now the study of such actions is an important area in rigidity theory. We will be discussing various aspects of this below.

2.4.2. *Riemannian geometry.* The last decade has brought several important developments on the differential geometric side of rigidity. M. Gromov showed in 1981 that any compact non-positively curved manifold with fundamental group isomorphic to that of a compact locally symmetric space of the non-compact type of higher rank is isometric to the symmetric space [14]. This was also shown for locally reducible spaces by P. Eberlein in [47]. Gromov's proof explores the Tits geometry of such a space, and is quite close to Mostow's ideas. W. Ballmann, M. Brin, P. Eberlein and I introduced a purely differential geometric notion of rank in 1984 [12, 13]. Given a complete Riemannian manifold M, its *rank* is the minimal dimension of the space of parallel Jacobi fields along any bi-infinite geodesic. If M has nonpositive sectional curvature, M has higher rank if and only if every geodesic is contained in a totally geodesic flat 2-plane. We established a structure theory for such spaces in [12, 13]. These developments culminated in the rank-rigidity theorem by W. Ballmann and independently K. Burns and myself [11, 25, 14].

Theorem 2.9. [Ballmann, Burns-Spatzier, 1985] *A closed locally irreducible higher rank Riemannian manifold of non-positive curvature is locally symmetric.*

The proof by Burns and myself is again inspired by Mostow's approach. The use of the dynamics of the geodesic flow proved to be a major ingredient in both our and Ballmann's proofs.

P. Eberlein and J. Heber generalized the rank rigidity theorem to certain noncompact manifolds, in particular spaces of finite volume [48]. Heber established the theorem for homogeneous manifolds of nonpositive curvature [91]. S. Adams extended the theorem to leaves of foliations of closed manifolds by manifolds of nonpositive curvature and higher rank [4]. The general case of a noncompact manifold remains open,

The quaternionic hyperbolic spaces and the Cayley upper half plane play a special role within manifolds of strictly negative curvature. For these spaces, P. Pansu extended Mostow's work on quasi-isometries in a remarkable way in 1989 [156].

Theorem 2.10. [Pansu, 1989] *Any quasi-isometry of a quaternionic hyperbolic space or the Cayley upper half plane is a finite distance from an isometry.*

The real and complex hyperbolic cases are very different. For $\mathcal{H}^n$ for example, any diffeomorphism of S^{n-1} extends to a quasi-isometry of $\mathcal{H}^n$. It is not known if Pansu's theorem extends to the irreducible higher rank symmetric spaces of nonpositive curvature.

Berger's 1/4-pinching theorem found its counterpart for compact quotients of rank one non-constant curvature symmetric spaces in the works of U. Hamenstadt, L. Hernandez, and S. T. Yau, F. Zheng [87, 95, 209].

Theorem 2.11. [Hamenstadt, Hernandez, Yau-Zheng, 1990's] *Any 1/4-pinched metric on a closed locally symmetric space of negative, non-constant curvature is symmetric.*

This theorem was first proved in dimension 4 by M. Ville via an inequality between the signature and the Euler characteristic [200].

The latter papers use harmonic maps in an essential way. They had been introduced to rigidity theory by Y. T. Siu in his generalization of Mostow's rigidity theorem to certain Kähler manifolds in 1980 [186]. J. Sampson improved Siu's argument to show that harmonic maps from a Kähler manifold to a manifold with nonpositive curvature operator are pluriharmonic [181]. J. Carlson and D. Toledo used these results systematically to study harmonic maps from Kähler manifolds into locally symmetric spaces [29]. This approach to rigidity culminated in the extensions of the superrigidity and arithmeticity theorems to quaternionic spaces by Corlette and Gromov and

Schoen [32, 83], and the general geometric proof by Mok, Siu and Yeung [141].

T. Farrell and L. Jones proved topological analogues of Mostow's rigidity theorem for manifolds of variable negative and nonpositive curvature over the last decade. While metric rigidity clearly fails in negative curvature, they established rigidity of the homeomorphism type [52, 53, 54, 55].

Theorem 2.12. [**Farrell-Jones, 1980's**] *Any two closed non-positively curved manifolds of dimension bigger than 5 with isomorphic fundamental group are homeomorphic.*

Farrell and Jones obtained similar rigidity results for manifolds of nonpositive curvature [54]. Further generalizations to polyhedra of negative curvature were obtained by B. Hu [100]. Farrell and Jones also disproved rigidity of the diffeomorphism type.

Theorem 2.13. [**Farrell-Jones, 1990's**] *In dimension 7 and up, there are closed manifolds of negative sectional curvature which are homeomorphic but not diffeomorphic to a real (or complex) hyperbolic space.*

The idea of their construction is to glue an exotic sphere into a closed hyperbolic space M. The new space is homeomorphic but not diffeomorphic to M. Then they explicitly construct a metric of negative sectional curvature on the new space. In the complex hyperbolic space, the metrics obtained have sectional curvatures between -1 and $-4 - \delta$ for arbitrarily small $\delta > 0$. This complements Theorem 2.11. P. Ontaneda recently obtained 6-dimensional closed manifolds which are homeomorphic but not diffeomorphic to a closed real hyperbolic manifold by a totally different technique [152]. Note that none of these examples is diffeomorphic to a locally symmetric space by Mostow's rigidity theorem. There are also negatively curved closed manifolds which are not even homotopy equivalent to a locally symmetric space. G. Mostow and Y. T. Siu constructed a certain 4-dimensional Kähler manifold of this type [151]. Later M. Gromov and W. Thurston found a rather flexible construction of such spaces using ramified coverings of closed real hyperbolic spaces [84]. In particular, they found such spaces with sectional curvatures arbitrarily close to -1.

2.4.3. *Dynamics of amenable groups.* One of the most profound developments in dynamical rigidity is M. Ratner's work on the horocycle flow. Recall that the horocycle flow on the unit tangent bundle of a surface is the flow along the stable manifolds parametrized by arc length. Group theoretically, it is the flow of the one parameter group generated by the matrix $\left(\begin{smallmatrix} 0 & 1 \\ 0 & 0 \end{smallmatrix} \right)$. Ratner showed in 1980 that if the horocycle flows of two surfaces of constant curvature -1 are measurably isomorphic, then the surfaces are isometric [169]. She later determined the factors and joinings of the horocycle flows. Various generalizations to higher dimensional locally symmetric spaces and variable curvature were established by J. Feldman and D. Ornstein, C. Croke, P. Otal, D. Witte, L. Flaminio and myself [56, 59, 207, 61].

In a breakthrough in 1990, M. Ratner proved a very general theorem about invariant sets and measures for unipotent flows on homogeneous spaces which subsumes her previous works mentioned above. Recall that a subgroup U of a connected Lie group G is called *unipotent* if for each $u \in U$, Ad (u) is a unipotent automorphism of the Lie algebra of G. The horocycle flow is an example of a unipotent subgroup of $SL(2, \mathbb{R})$. M. S. Raghunathan had conjectured that all closed invariant subsets of a unipotent flow on a homogeneous space G/Γ, Γ a lattice in G, are algebraic. Ratner showed this via the following measure theoretic generalization of Raghunathan's conjecture.

Theorem 2.14. [Ratner, 1990] *Let U be a unipotent subgroup of G, and Γ a lattice in G. Then any invariant probability measure μ for the action of U on G/Γ is a Haar measure on a closed homogeneous subspace of G/Γ.*

Special cases of this theorem had been obtained by H. Furstenberg, W. Parry, and S. G. Dani [64, 68, 158, 43, 44]. G. A. Margulis had shown Raghunathan's topological conjecture for $SL(3, \mathbb{R})$ and used it to prove the Davenport conjecture in number theory [133]. The theorem was generalized to p-adic groups by M. Ratner and independently by G. Margulis and G. Tomanov [175, 174, 134, 135].

Ratner's theorem has strong implications for actions of semisimple groups. R. J. Zimmer and later Zimmer and A. Lubotzky used it to find topological obstructions to the existence of actions of semisimple groups on a manifold [229, 123]. The following is a typical example of their work.

Theorem 2.15. [Lubotzky-Zimmer, 1993] *Suppose a connected simple non-compact Lie group G of $\mathbb{R}$-rank at least 2 acts real analytically on a closed manifold M, preserving a real analytic connection and a volume. Then any faithful linear representation of $\pi_1(M)$ in $GL(m, \mathbb{C})$ contains a lattice in a Lie group which locally contains G.*

A. Katok and the author used Ratner's theorem to give a partial classification of invariant measures for homogeneous Anosov $\mathbb{R}^k$-actions for $k \geq 2$ [117].

The classification of Anosov flows and diffeomorphisms with smooth stable foliation was another major theme in the last decade. E. Ghys started this investigation in the three-dimensional case in 1987 [71]. Recall the relevant definitions. Let M be a closed manifold with a fixed Riemannian norm$\| \; \|$. Then a C^∞-flow ϕ_t on M is called *Anosov* if there is a splitting of the tangent bundle $TM = E^s \oplus E^u \oplus \frac{d}{dt}\phi_t$ of M into invariant subbundles E^s and E^u and the flow direction $\frac{d}{dt}\phi_t$ and there exist constants $C > 0$ and

$m > 0$ such that for all $v \in E^s$ and $t > 0$ ($(v \in E^u$ and $t < 0$ respectively)

$$\| d\phi_t(v) \| \leq C\, e^{-m|t|}.$$

The distributions E^s and E^u are called the *stable* and *unstable* distributions respectively. They are integrable, and define the *stable* and *unstable foliations* of ϕ_t. While the individual leaves of the stable and unstable foliations are C^∞, the dependence of the leaves on the initial point in general is only Hölder. Y. Benoist, P. Foulon and F. Labourie obtained the following description of Anosov flows with smooth stable foliations [16].

Theorem 2.16. [Benoist-Foulon-Labourie, 1990] *Suppose a C^∞ Anosov flow ϕ_t preserves a contact structure. If the stable foliation of ϕ_t is C^∞, then a finite cover of ϕ_t is C^∞-conjugate to a time change of a geodesic flow of a locally symmetric space of negative curvature.*

They also determined which time changes can occur. Furthermore, if ϕ_t is the geodesic flow of a manifold of negative curvature, then ϕ_t itself is C^∞-conjugate to a geodesic flow of a locally symmetric space of negative curvature. This generalizes earlier work of M. Kanai, R. Feres and A. Katok who prove this theorem for geodesic flows of manifolds with sufficiently pinched negative curvature [106, 57, 58]. There are similar theorems for Anosov diffeomorphisms [17].

P. Foulon and F. Labourie applied the techniques used in Theorem 2.16 to negatively curved asymptotically harmonic manifolds, i.e. manifolds whose horospheres have constant mean curvature [63]. C. Yue then combined these techniques with his work on the Margulis' function to resolve the Green conjecture in odd dimensions [211].

Theorem 2.17. [Foulon-Labourie, Yue, 92] *Let M be a closed Riemannian manifold of negative curvature such that the mean curvature of a horosphere at v only depends on the footpoint of v. Then M is asymptotically harmonic. Its geodesic flow is C^∞-conjugate to that of a locally symmetric space of negative curvature. Furthermore, if $\dim M$ is odd, then M has constant curvature.*

Earlier C. Yue had shown that closed Riemannian manifolds of negative curvature for which the harmonic measure equals the Liouville measure are asymptotically harmonic [210]. Combining this with Theorem 2.17 shows that odd dimensional manifolds of this kind have constant negative curvature. This affirms a particular instance of the Sullivan conjecture.

Quite recently, rigidity phenomena associated with higher rank have been observed in the dynamics of small groups. These phenomena first surfaced in the study of the rigidity of the standard action of $SL(n, \mathbb{Z})$ on the n-torus by S. Hurder, A. Katok, J. Lewis and R. J. Zimmer where the local C^∞-rigidity of the action of $n - 1$ commuting Anosov automorphisms of T^n proved crucial [101, 110, 111, 112]. Recall that an action α of a discrete group G is called *locally C^k-rigid* if any perturbation of the action which is C^1-close on a generating set is C^k-conjugate to the original action. This definition does not generalize well to Lie groups. Indeed, we may always compose α with an automorphism ρ of G, close to the identity, to get a perturbation of α. Thus let us call an action of an arbitrary group G *locally C^k-rigid* if any perturbation of the action which is C^1-close on a generating set is C^k-conjugate to α composed with an automorphism of G.

Call an action of a group *Anosov* if one element of the group acts normally hyperbolically to the orbit foliation (cf Section 3.3 for more details). In 1992, A. Katok and the author introduced a certain class of homogeneous Anosov actions of $\mathbb{R}^k$ and $\mathbb{Z}^k$, called *standard Anosov actions* [114]. A typical example is the action of the diagonal group in $SL(n, \mathbb{R})$ on $SL(n, \mathbb{R})/\Gamma$ by left translations where Γ is a uniform lattice in $SL(n, \mathbb{R})$ and $n > 2$. Other examples are generated by commuting Anosov toral automorphisms. All the standard actions are actions by $\mathbb{R}^k$ or $\mathbb{Z}^k$ for $k \geq 2$. Conjecturally, all homogeneous actions of $\mathbb{R}^k$ are standard unless they have a rank one factor, i.e. a factor on which a hyperplane in $\mathbb{R}^k$ acts trivially. We established a local rigidity theorem for such actions [114, 115, 116, 118].

Theorem 2.18. [Katok-Spatzier, 1992] *The standard Anosov $\mathbb{R}^k$- or $\mathbb{Z}^k$-actions are C^∞-rigid if $k \geq 2$.*

Much earlier, in 1976, R. Sacksteder had established infinitesimal rigidity for expanding toral endomorphisms [180].

In the proof of Theorem 2.18 we need to show that certain cocycles are cohomologous to constant cocycles. Unlike in Zimmer's superrigidity theorem however, all our cocycles and coboundaries are C^∞ or Hölder and take values in abelian groups.

Theorem 2.19. [Katok-Spatzier, 1992] *Every $\mathbb{R}$-valued C^∞ (Hölder)-cocycle over a standard $\mathbb{R}^k$-Anosov action is C^∞ (Hölder)-conjugate to a constant cocycle if $k \geq 2$.*

This last theorem uses harmonic analysis in an essential way, as I will explain in detail in Section 3.3. Later we applied our techniques to obtain the local C^∞-rigidity of projective actions of lattices [118]. We also classified a certain class of invariant measures for the standard Anosov actions [115]. A. Katok and K. Schmidt extended Theorem 2.19 to automorphisms of compact abelian groups other than tori [113]. A. and S. Katok obtained vanishing results for higher cohomologies of higher rank abelian automorphism groups of the torus [109].

There is very little known about general Anosov actions by higher rank abelian groups. They seem to be quite rare. J. Palis and J. C. Yoccoz have shown that generically, Anosov diffeomorphisms on tori only commute with their own powers [154, 155].

3. MAUTNER'S PHENOMENON AND ASYMPTOTICS OF MATRIX COEFFICIENTS

Rigidity problems in geometry, group theory and dynamics are often closely related, as our discussion of Mostow's theorem has shown. It is natural to try to apply harmonic analysis in the guise of representation theory as a tool. I will now describe various techniques and ideas from harmonic analysis that have proved useful in the study of rigidity.

E. Hopf discovered the ergodicity of the geodesic flow of a manifold of constant negative curvature in 1939 by a geometric argument [97]. The relation between such geodesic flows and certain one-parameter subgroups of $SL(2, \mathbb{R})$ was only discovered in S. V. Fomin's and I. M. Gelfand's article [62] in 1952. There, Fomin and Gelfand identified the geodesic flow of a compact surface $\mathcal{H}/\Gamma$ of constant negative curvature with the homogeneous flow induced by the one-parameter subgroup $\begin{pmatrix} e^t & 0 \\ 0 & e^{-t} \end{pmatrix}$ on $SL(2, \mathbb{R})/\Gamma$. Then they used the representation theory of $SL(2, \mathbb{R})$ to determine the spectrum of the geodesic flow. In the same paper, they also initiated the general study of homogeneous flows. O. Parasyuk used these ideas in 1953 to discuss the horocycle flow of a compact surface. In 1957, F. Mautner determined the ergodic components of the geodesic flow of a compact or finite volume locally symmetric space M of the non-compact type [137]. Like Fomin and Gelfand, he modeled the geodesic flow of such a space by certain homogeneous flows. He then proved their ergodicity by analyzing isotropy groups of unitary representations. C. C. Moore discovered a beautiful and simple criterion for the ergodicity of a homogeneous flow on a homogeneous space of a semisimple group in 1966 [142]. His analysis was based on Mautner's ideas on isotropy groups. Homogeneous flows and their ergodicity were further investigated by L. Auslander, L. Green, F. Hahn, C. C. Moore, J. Brezin and others [10, 9, 142, 190, 41, 42, 143, 24].

In this section we will first describe Mautner's and Moore's results and give a proof in a simple case. Then I will introduce matrix coefficients, discuss their rough asymptotics and derive Moore's theorem from them. In the remainder of this section, I will describe precise results on the decay of matrix coefficients of smooth vectors. This fine asymptotics of matrix coefficients has been applied to rigidity in dynamics in three ways.

(1) R. J. Zimmer showed that higher rank lattices cannot act ergodically on sufficiently low dimensional manifolds, preserving a volume, provided the action is "N-distal" [225];

(2) A. Katok and the author showed that the first cohomology of certain hyperbolic abelian actions is trivial;

(3) M. Cowling and T. Steger investigated restrictions of unitary representations of the ambient Lie group to lattices. A. Iozzi applied their results to equivariant maps between lattice actions.

I will discuss applications 2) and 3) in some detail. For the first application I refer to Zimmer's survey [225].

3.1. The Mautner-Moore results. Let us start with some definitions. Let G denote a connected semisimple real Lie group without compact factors. The prime example of such a group is $SL(n, \mathbb{R})$. In most of our discussion, the reader may substitute $SL(n, \mathbb{R})$ for a general G. I will illustrate the majority of the results and definitions by that example. If Γ is a lattice in G and $\{a_t\} \subset G$ a one-parameter subgroup, call the flow by the left translations of $\{a_t\}$ on G/Γ *homogeneous*. If G is unimodular, then $\{a_t\}$ leaves Haar measure μ on G/Γ invariant. If μ is ergodic for $\{a_t\}$, we call the homogeneous flow ergodic.

The next theorem is Moore's extension of Mautner's fundamental work on the ergodicity of the geodesic flow of a symmetric space [137].

Theorem 3.1. [Moore, 1966] *Let Γ be a lattice in G, and $\{a_t\} \subset G$ a one-parameter subgroup. Then the homogeneous flow by $\{a_t\}$ on G/Γ is ergodic if and only if $\{a_t\}$ is not precompact in G.*

Note that these homogeneous flows encompass geodesic and horocycle flows as well as various frame flows and other extensions. These flows exhibit a variety of geometric behaviors which make a geometric treatment of the ergodicity of such flows rather difficult.

Mautner's discovery was based on a fixed point phenomenon for unitary representations of G. For simplicity, let us consider the case when $G = SL(2, \mathbb{R})$ and

$$g_t = \begin{pmatrix} e^t & 0 \\ 0 & e^{-t} \end{pmatrix}.$$

Let

$$h_s^+ = \begin{pmatrix} 1 & s \\ 0 & 1 \end{pmatrix}.$$

Then g_t and h_s^+ satisfy the commutation relation

$$g_{-t}\, h_s^+\, g_t = h_{s\,e^{-2t}}^+.$$

Geometrically, one can interpret g_t as a geodesic flow and h_s^+ as its horocycle flow. Then the commutation relation above just expresses the standard exponential contraction of the stable horospheres by the geodesic flow.

Now suppose that ρ is a continuous unitary representation of $SL(2, \mathbb{R})$ on a Hilbert space V, and that g_t fixes a vector $v \in V$. The commutation relation above implies for fixed s that $g_{-t}\, h_s^+\, g_t \to 1$ as $t \to \infty$. Hence we see that

$$\| \rho(h_s^+)v - v \| = \| \rho(h_s^+)\rho(g_t)v - v \| = \| \rho(g_{-t}\, h_s^+\, g_t)v - v \| \to 0$$

as $t \to \infty$. Thus v is fixed by h_t.

A similar argument applies to the opposite horocycle flow given by the matrices

$$h_s^- = \begin{pmatrix} 1 & 0 \\ s & 1 \end{pmatrix}.$$

Since g_t, h_s^+ and h_s^- generate $SL(2, \mathbb{R})$, we see that v is fixed by all of $SL(2, \mathbb{R})$. Thus we have shown that a vector of a unitary representation of $SL(2, \mathbb{R})$ fixed by g_t is already fixed by all of $SL(2, \mathbb{R})$. This is the so-called *Mautner phenomenon*. It extends to various one-parameter subgroups of more general groups G [127, Chapter II, 3].

Now let us apply Moore's theorem to the unitary representation on $L^2(G/\Gamma)$, induced from the action by left translations on a homogeneous space G/Γ. As the one-parameter group g_t is not precompact, we see that a g_t-invariant L^2-function is already G-invariant, and hence constant. Thus the homogeneous flow induced by g_t is ergodic.

Let M be the surface $\mathcal{H}^2/\Gamma$. Then its unit tangent bundle can be identified with $SL(2, \mathbb{R})/\Gamma$ as $SL(2, \mathbb{R})$ acts transitively on points and directions of $\mathcal{H}^2$. The geodesic flow of M becomes the homogeneous flow of g_t on $SL(2, \mathbb{R})/\Gamma$. Thus the argument above proves the ergodicity of the geodesic flow of a surface of constant curvature. E. Hopf gave a much more geometric argument for the ergodicity of the geodesic flow in constant negative curvature [97]. The spirit of it however is the same. Hopf's main idea is to show that a function invariant under the geodesic flow is constant along stable and also unstable manifolds. The latter are the orbits of h_s^+ and h_s^-. The stable manifolds are contracted exponentially fast by the geodesic flow. This is the geometric interpretation of the commutation relation between g_t and h_s^+, and also the key to Hopf's argument.

For $G = SO(n, 1)$ with $n > 2$, matters get more complicated, since the geodesic flow itself is not homogeneous anymore. However, it is covered by the frame flow, which itself can be viewed as a homogeneous flow on $SO(n, 1)/\Gamma$ for some lattice $\Gamma \subset SO(n, 1)$. Thus Mautner's and Moore's work proves the ergodicity of the frame and geodesic flow of any compact manifold of constant negative curvature. This justifies the third step in the proof of Mostow's rigidity theorem. The extensions of this to other symmetric spaces are similar in nature. In the next section we will derive Theorem 3.1 from asymptotic information about matrix coefficients.

3.2. Asymptotics of matrix coefficients. Let G be a connected semisimple Lie group. We will consider irreducible unitary representations π of G on a Hilbert space $\mathcal{H}$. Define the *matrix coefficient* or *correlation function* of v and $w \in \mathcal{H}$ as the function $\phi_{v,w} : G \to \mathbb{R}$ given by

$$g \to \langle \pi(g)v, w \rangle.$$

Harish-Chandra used matrix coefficients and their asymptotic properties as an essential tool in his seminal work on the representation theory of semisimple Lie groups (cf. e.g. [89]). His results were refined and extended by a number of people during the last two decades [185, 212, 99, 34, 98, 30, 144, 171]. While I will describe the best asymptotic results further below, their proofs lie outside the scope of this article. Instead we will now discuss a vanishing result and its proof.

Let X be a locally compact topological space. We say that a function $f : X \to \mathbb{R}$ *vanishes at* ∞ if $f(x) \to 0$ as x leaves compact subsets of X. We also say that a sequence $x_n \in X$ *tends to* ∞ if x_n leaves any compact subset of X. The following result appeared in papers by T. Sherman, R. Howe and C. C. Moore, and R. J. Zimmer [185, 99, 212].

Theorem 3.2. *Let G be a connected semisimple real Lie group without compact factors. Suppose π is a unitary representation of G in a Hilbert space V for which no non-trivial normal subgroup G' has invariant vectors. Then the matrix coefficients of π vanish at ∞.*

Note that Theorem 3.1 follows immediately in case G is simple. Indeed, decompose $L^2(G/\Gamma)$ into unitary irreducible representations of G. Suppose some function f orthogonal to the constants is invariant under a homogeneous flow $\{a_t\}$. Then some component f' of f in some non-trivial irreducible subrepresentation π of $L^2(G/\Gamma)$ is non-zero and $\{a_t\}$-invariant. In particular, the matrix coefficient of f' does not vanish at ∞. Since G is simple and π is non-trivial, no non-trivial normal subgroup of G has an invariant vector. Thus Theorem 3.2 gives a contradiction. The general case in Theorem 3.1 requires some technicalities about irreducible lattices.

Next we will outline the proof of this decay result for the case $G = SL(2, \mathbb{R})$ (for a detailed treatment of this and the general case see [220]). The main idea of the proof is similar to that of Mautner's phenomenon. There, we used the commutation relation between geodesic and horospherical flow applied to the representation π itself. Here, we will exploit it in a more sophisticated way, by looking at the dual action of the geodesic flow on the set of characters for the horocyclic flow it defines.

Outline of proof : Recall that any element $g \in SL(2, \mathbb{R})$ can be written as a product $g = k \, a \, l$ where a is a diagonal matrix and k and l are orthogonal matrices. Clearly, a sequence g_n tends to ∞ if and only if the diagonal factor a_n in the above decomposition tends to ∞. Thus it suffices to prove that matrix coefficients vanish at infinity when restricted to the diagonal subgroup A.

Let N be the group of strictly upper triangular matrices. As N is isomorphic to $\mathbb{R}$, the irreducible unitary representations $\hat{N} = \hat{\mathbb{R}}$ of N are precisely the one-dimensional unitary representations π_θ given by multiplication by $e^{i\theta t}$ for $\theta \in \mathbb{R}$. Furthermore, any unitary representation of N decomposes as a direct integral of irreducible unitary representations. For $m \in \{\infty, 0, 1, 2, \ldots\}$, let $m \, \pi_\theta$ denote the direct sum of m copies of π_θ. We call m the *multiplicity* of π_θ. Thus, restricting π to N, there is a measure ν on $\hat{\mathbb{R}}$ and multiplicities $m_\theta \in \{\infty, 0, 1, 2, \ldots\}$ such that

$$\pi \mid_N = \int_{\hat{\mathbb{R}}} m_\theta \pi_\theta d\nu(\theta).$$

As A normalizes N, $a \in A$ also acts on the unitary dual $\hat{N}$ via $(a\,\pi_\theta)(t) = \pi_\theta(\operatorname{Ad} a^{-1}t)$. One calculates that $a\,\pi_\theta = \pi_{a^2\,\theta}$.

Now there is a dichotomy, either $\nu(0) = 0$ or $\nu(0) > 0$. In the first case, the support of $a^2 \nu$ moves to infinity in $\mathring{\mathbb{R}}$ as $a \to \infty$. This implies the vanishing of the matrix coefficients of π restricted to A. In the second case, when $\nu(0) = 0$, we get A- and N-invariant vectors in π. Note that this argument also shows that any A-invariant vector is N-invariant. By a similar argument, any A-invariant vector is fixed by the the strictly lower triangular subgroup N^-. Since A, N and N^- generate $SL(2, \mathbb{R})$, we conclude that either all matrix coefficients vanish at ∞ or there are $SL(2, \mathbb{R})$-invariant vectors. ◇

While the last theorem always insures the vanishing of the matrix coefficients, the rate of decay can be very slow in general. This can be seen as follows. As first shown by Fomin and Gelfand [62], the geodesic flow of a surface with constant negative curvature has countable Lebesgue spectrum (indeed, it is K and even Bernoulli). Thus there is an L^2-function f orthogonal to all its translates by g_T^n for some fixed T and arbitrary n. Considering functions of the form $\sum_{i=0}^{\infty} \beta_i (g_{Ti} f)$, we see that the decay can be as slow as the the decay of the l_2-norm of the tail of an l_2-sequence.

Harish-Chandra on the other hand found explicit exponential decay estimates for the matrix coefficients of sufficiently nice vectors [89]. These estimates were recently reproved and refined by M. Cowling, W. Casselman and D. Miličić and R. Howe [34, 30, 98] by various methods.

Recall that any locally compact group has maximal compact subgroups. For a Lie group, they are unique up to conjugacy. As an example, the group $SO(n)$ of orthogonal matrices of determinant 1 is a maximal compact subgroup for $SL(n, \mathbb{R})$.

Let G be a connected semisimple Lie group with finite center as above. Fix a maximal compact subgroup K of G. A vector $v \in \mathcal{H}$ is called K-finite if the K-orbit of v spans a finite dimensional vector space. Let $\hat{K}$ denote the unitary dual of K. One can then decompose

$$\mathcal{H} = \oplus_{\mu \in \hat{K}} \mathcal{H}_\mu$$

where $\mathcal{H}_\mu$ is $\pi(K)$-invariant and the action of K on $\mathcal{H}_\mu$ is equivalent to $n\mu$ where n is an integer or $+\infty$, called the *multiplicity* of μ in $\mathcal{H}$. The K-finite vectors form a dense subset of $\mathcal{H}$. One calls $\mathcal{H}_\mu$ the *μ-isotypic component* of π.

Recall that a *Cartan subgroup* is a maximal abelian subgroup of G consisting of semisimple elements. It decomposes into a product of a torus and an $\mathbb{R}^k$ for some k. The $\mathbb{R}^k$-factor is called a *split Cartan subgroup* of G. We call it *maximal* if its dimension is maximal amongst all split Cartan subgroups.

Let A be a maximal split Cartan subgroup of G, and $\mathcal{A}$ its Lie algebra. For $G = SL(n, \mathbb{R})$ for example, we may take A to be the subgroup of diagonal matrices of determinant 1. Then $\mathcal{A}$ becomes the Lie subalgebra of trace 0 diagonal matrices. Recall that the *(restricted) roots* α are linear functions on $\mathcal{A}$ such that the eigenvalues of the adjoint representation of $x \in \mathcal{A}$ are given by the $\alpha(x)$ as α ranges over the roots. For $G = SL(n, \mathbb{R})$ and $\mathcal{A}$ the traceless diagonal matrices, the roots are just the functionals $e_i - e_j$ on $\mathcal{A}$, where e_i denotes the i-th diagonal entry of $x \in \mathcal{A}$.

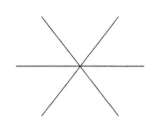

In general, the kernels of the roots determine finitely many hyperplanes in $\mathcal{A}$. Removing these hyperplanes, we obtain finitely many connected components, called the *Weyl chambers*. Let us fix such a Weyl chamber $\mathcal{C}$, and call it the *positive* Weyl chamber. The roots whose hyperplanes bound $\mathcal{C}$ are a basis of $\mathcal{A}$. Call them the *elementary* roots. Any other root can be expressed as an integral linear combination of the elementary roots with either all coefficients nonnegative or nonpositive. Thus we can speak of the *positive* and *negative* roots. In fact, picking a basis of roots for $\mathcal{A}$ with this property, is equivalent to picking a positive Weyl chamber. In our basic example of $G = SL(n, \mathbb{R})$, we can choose $e_{i+1} - e_i$, $i = 2, \ldots, n$ as a set of elementary roots. For $n = 3$, the Weyl chambers are represented by the familiar picture on the left.

Call π *strongly L^p* if there is a dense subset of $\mathcal{H}$ such that for v, w in this subspace, $\phi_{v,w} \in L^p(G)$. Let ρ be half the sum of the positive roots on $\mathcal{A}$. R. Howe obtained the following estimate for matrix coefficients of K-finite vectors in 1980 [98, Corollary 7.2 and §7].

Theorem 3.3. *Let π be a strongly L^p-representation of G on $\mathcal{H}$. Let μ and ν be in $\hat{K}$. Then the matrix coefficients of $v \in \mathcal{H}_\mu$ and $w \in \mathcal{H}_\nu$ satisfy the estimate:*

$$| \phi_{v,w}(\exp tA) | \leq D \, \| v \| \, \| w \| \, \dim \mu \; \dim \nu \, e^{-\frac{1}{2p} \rho(A)}$$

where $A \in \bar{\mathcal{C}}$ and $D > 0$ is a universal constant.

Howe's version of the decay estimates is particularly strong, since it gives explicit and uniform control in terms of the norms of the K-finite vectors and some universal constants. The question remains which representations are strongly L^p. Fortunately, M. Cowling had already resolved it in 1979 [34].

Theorem 3.4. *Every irreducible unitary representation of G with discrete kernel is strongly L^p for some p. Furthermore, if $\mathcal{G}$ does not have factors isomorphic to $so(n, 1)$ or $su(n, 1)$ then p can be chosen independently of π.*

For applications in dynamics, it is important to obtain decay results for more general vectors, for example C^∞-functions on a manifold. These are easy corollaries of Howe's and Cowling's results. Here the uniformity in Howe's estimates becomes particularly important.

A vector $v \in \mathcal{H}$ is called C^∞ or *smooth* if the map $g \in G \to \pi(g)v$ is C^∞. Let $m = \dim K$ and $X_1, \ldots, X_m$ be an orthonormal basis of the Lie algebra $\mathcal{K}$ of K. Set $\Omega = 1 - \sum_{i=1}^m X_i^2$. Then Ω belongs to the center of the universal enveloping algebra of $\mathcal{K}$, and acts on the K-finite vectors in $\mathcal{H}$ since K-finite vectors are smooth. Now A. Katok and the author obtained the following estimate [114]:

Corollary 3.5. *Let v and w be C^∞-vectors in an irreducible unitary representation π of G with discrete kernel. Then there is a universal constant $E > 0$ and an integer $p > 0$ such that for all $A \in \bar{C}$ and large enough l*

$$|\langle \exp(tA)v, w \rangle| \leq E\, e^{-\frac{t}{2p}\rho(A)} \parallel \Omega^l(v) \parallel \parallel \Omega^l(w) \parallel .$$

In fact, p can be any number for which π is strongly L^p. Furthermore, if $\mathcal{G}$ does not have factors isomorphic to $so(n,1)$ or $su(n,1)$, p only depends on G.

Note that v and w only need to be C^k with respect to K for some large k. Combining this with Moore's Theorem 3.1 as in [114], we obtain:

Corollary 3.6. *Let G be a semisimple connected Lie group with finite center. Let Γ be an irreducible cocompact lattice in G. Assume that $\mathcal{G}$ does not have factors isomorphic to $so(n,1)$ or $su(n,1)$. Let $f_1, f_2 \in L^2(G/\Gamma)$ be C^∞-functions orthogonal to the constants. Let C be a positive Weyl chamber in a maximal split Cartan $\mathcal{A}$. Then there is an integer $p > 0$ which only depends on G and a constant $E > 0$ such that for all $A \in \bar{C}$ and all large l*

$$\langle (\exp tA)_*(f_1), f_2 \rangle \leq E\, e^{-\frac{t}{2p}\rho(A)} \parallel f_1 \parallel_l \parallel f_2 \parallel_l$$

Here $\parallel f \parallel_l$ denotes the l'th Sobolev norm of f.

This corollary extends to G locally isomorphic to $SO(n,1)$ or $SU(n,1)$ if we allow p to depend on the lattice Γ as well as on G. In fact, such a p is roughly inversely proportional to the bottom of the spectrum of the Laplacian on non-constant functions on the locally symmetric space $K \backslash G / \Gamma$. As $K \backslash G / \Gamma$ is compact, the latter is not 0. Thus one might hope for a positive resolution of the following question.

Problem 3.7. *Does the corollary extend to all G for a p that depends only on the lattice?*

Already for $G = SL(2, \mathbb{R}) \times SL(2, \mathbb{R})$, this seems to be a very subtle problem.

Our results for the exponential decay of matrix coefficients for smooth functions extend to Hölder vectors and functions. This was first shown for representations of $SL(2, \mathbb{R})$ by M. Ratner [171]. C. C. Moore gave an alternative proof for arbitrary rank one groups in [144]. However, he had to assume that the vectors had a Hölder exponent bigger than $\dim K/2$ where K is the maximal compact subgroup. In a private communication, G. A. Margulis outlined an argument for arbitrary Hölder vectors for general G.

3.3. Higher rank hyperbolic abelian actions. The second application of the decay of matrix coefficients concerns the the rigidity of certain homogeneous actions of higher rank abelian groups. This is joint work of A. Katok and myself [114]. While our theorems hold for fairly general hyperbolic homogeneous actions, the so-called *standard partially hyperbolic actions*, I will restrict the outline of the proof here to the semisimple case. I already summarized our results for the Anosov case in Section 2.4.

Let me first introduce the notion of a partially hyperbolic action.

Definition 3.8. *Let A be $\mathbb{R}^k$ or $\mathbb{Z}^k$. Suppose A acts C^∞ and locally freely on a manifold M with a Riemannian norm $\|\ \|$. Call an element $g \in A$ partially hyperbolic if there exist real numbers $\lambda > \mu > 0$, $C, C' > 0$ and a continuous splitting of the tangent bundle*

$$TM = E_g^+ + E_g^0 + E_g^-$$

such that for all $p \in M$, for all $v \in E_g^+(p)$ ($v \in E_g^-(p)$ respectively) and $n > 0$ ($n < 0$ respectively) we have for the differential $g_* : TM \to TM$

$$\| g_*^n(v) \| \le Ce^{-\lambda |n|} \| v \|$$

and for all $n \in \mathbb{Z}$ and $v \in E_g^0$ we have

$$\| g_*^n(v) \| \ge C' e^{-\mu |n|} \| v \| .$$

Furthermore, we assume that the distribution E_g^0 is uniquely integrable.

Call an A-action partially hyperbolic if it contains a partially hyperbolic element. We also say that the A acts normally hyperbolically with respect to the foliation defined by E_g^0. We call E_g^+ and E_g^- its stable and unstable distribution respectively.

Note that every Anosov flow and diffeomorphism is partially hyperbolic. More generally, we call a partially hyperbolic action *Anosov* if E_g^0 equals the tangent distribution of the orbit foliation of A for some partially hyperbolic element $g \in A$.

Our prime examples of such actions are algebraic. Let G be a connected semisimple Lie group of the non-compact type, A a maximal split Cartan and Γ a uniform lattice in G. Then the homogeneous action of A on G/Γ is partially hyperbolic. Let M be the compact part of the centralizer of A in G. Then the homogeneous action of A on G/Γ descends to an action of A on $M \setminus G/\Gamma$. The latter action is always an Anosov action, called the *Weyl chamber flow* on G/Γ. If we further assume that G does not have local factors isomorphic to $SO(n,1)$ or $SU(n,1)$, then all of these actions are standard partially hyperbolic actions.

Let us first describe how one reduces local rigidity of the standard Anosov actions to a cocycle theorem. By structural stability, any perturbation of an Anosov action is still Anosov, and C^0-conjugate to the original action α. By a geometric argument, one can see that the homeomorphism is a C^∞-diffeomorphism. Thus, up to a conjugacy, the perturbed action is a C^∞-*time change* of α, i.e. a C^∞-action whose orbits coincide with those of α. While this time change involves only small changes of time, we can actually exclude global time changes even for the standard partially hyperbolic $\mathbb{R}^k$-actions.

Theorem 3.9. [Katok-Spatzier, 1992] *All C^∞-time changes of a standard partially hyperbolic $\mathbb{R}^k$-action are C^∞-conjugate to the original action up to an automorphism.*

Every C^∞-time-change α^* of an action α determines a C^∞-cocycle β : $\mathbb{R}^k \times M \to \mathbb{R}^k$ via the equation

$$\alpha(a, x) = \alpha^*(\beta(a, x), x).$$

This is clear on orbits without isotropy. From this and the fact that most orbits of Anosov actions do not have isotropy, one can extend the cocycle everywhere.

If β is C^∞-cohomologous to a constant cocycle given by an automorphism $\phi : \mathbb{R}^k \to \mathbb{R}^k$, then α^* is C^∞-conjugate to $\alpha \circ \phi$. Thus it suffices to prove the following extension of Theorem 2.19.

Theorem 3.10. [Katok-Spatzier, 1992] *Any C^∞-cocycle $\beta : A \times M \to \mathbb{R}^m$ over a standard partially hyperbolic A-action is C^∞-cohomologous to a constant cocycle.*

I will now illustrate the proof of this theorem in the semisimple case. Thus let G be a connected semisimple Lie group of the non-compact type and real rank at least 2, A a maximal split Cartan and Γ a uniform lattice in G. Further assume that G does not have factors locally isomorphic to $SO(n, 1)$ or $SU(n, 1)$. Let A act on G/Γ by left translations. Pick a regular element $a \in A$, i.e. a does not lie on the wall of a Weyl chamber. I will show that β is cohomologous to the homomorphism $\rho(b) = \int_M \beta(b, x)dx$, or that $\beta - \rho$ is cohomologous to 0. Thus we may assume that β has 0 averages.

Let $f : G/\Gamma \to \mathbb{R}^k$ be the function given by $f(x) = \beta(a, x)$. Denote by $a\,f$ the function $a\,f(x) = f(ax)$. Define formal solutions of the cohomology equation by

$$P_a^+ = \sum_{k=0}^{\infty} a^k f \quad \text{and} \quad P_a^- = -\sum_{k=-\infty}^{-1} a^k f.$$

By our assumptions on G, we can apply Corollary 3.6 on the exponential decay of matrix coefficients of smooth functions on G/Γ. It implies that P_a^+ and P_a^- are not just formal solutions, but define distributions on G/Γ.

The key argument is to show that $P_a^+ = P_a^-$. Again this uses the exponential decay of matrix coefficients crucially. Since A has rank at least 2, we can pick $b \in A$ independent of a. Let $f^*(x) = \beta(b, x)$. Since β is a cocycle, it follows that

$$\sum_{k=-l}^{l} a^k bf - \sum_{k=-l}^{l} a^k f = \sum_{k=-l}^{l} a^{k+1} f^* - a^k f^* = a^{l+1} f^* - a_1^{-l} f^*.$$

Since the matrix coefficients decay, it follows that $P_a^+ - P_a^-$ is b-invariant. By exponential decay, the sum

$$\sum_{m=-\infty}^{\infty} \sum_{k=-\infty}^{\infty} \langle a^k f, b^m g \rangle = \lim_{m \to \infty} 2m \sum_{k=-\infty}^{\infty} \langle a^k f, g \rangle$$

converges absolutely. It follows that $P_a^+ - P_a^- = 0$ as desired.

Note that P_a^+ is C^∞ along the stable manifold. For the first derivative for example, the derivatives $\frac{d}{dv}(f \circ a^k)$ decay exponentially for any vector v tangent to the stable manifold, and hence the sum $\sum_{k=0}^{\infty} \frac{d}{dv}(a^k \circ f)$ is absolutely convergent. Similarly, P_a^- is C^∞ along the unstable manifold. Since $P_a^+ = P_a^-$, we see that P_a^+ is C^∞ along both stable and unstable directions. The stable and unstable distribution together with their Lie brackets generate the whole tangent bundle. Using subelliptic estimates for sums of even powers of vectorfields in these directions, due to Rothschild, Nourrigat, Helffer and others, one can show that P_a^- is C^∞ on M. These estimates are very strong generalizations of the famed Hoermander square theorem. Let us note here that these subelliptic estimates themselves were developed using the harmonic analysis of nilpotent groups [178, 93].

While we used non-commutative harmonic analysis in the above, ordinary Fourier series arguments can be used to prove Theorem 3.9 for standard actions on tori. However, the superpolynomial decay of Fourier coefficients of smooth functions replaces the exponential decay estimates. Thus while the result depends on harmonic analysis in any of the cases, there is no uniform method of proof at this point of time.

3.4. Restrictions of representations to lattices and equivariant maps. M. Cowling and T. Steger used the fine decay estimates on matrix coefficients to see when restrictions of irreducible unitary representations of semisimple groups to lattices are irreducible and isomorphic [36]. This generalized earlier work of Steger for the case of $SL(2, \mathbb{R})$. For simplicity, I will state their theorem for irreducible lattices.

Let G be a semisimple group without compact factors. The irreducible subrepresentations of the left regular representation of G on $L^2(G)$ are called the *discrete series* representations of G. One can characterize the discrete series representations as the irreducible unitary representations with a matrix coefficient which belongs to $L^2(G)$.

Theorem 3.11. [Cowling-Steger, 1991] *Suppose Γ is an irreducible lattice in a connected semisimple group G with finite center and without compact factors. Suppose π and $\hat{\pi}$ are unitary representations of G. Then*

(1) *if π is a discrete series representation of G, then $\pi \mid_\Gamma$ is reducible;*
(2) *if π is not a discrete series representation, then $\pi \mid_\Gamma$ is irreducible;*
(3) *if $\pi \mid_\Gamma$ and $\hat{\pi} \mid_\Gamma$ are irreducible and unitarily equivalent, then π and $\hat{\pi}$ are unitarily equivalent.*

A. Iozzi applied this result to actions of connected semisimple Lie groups G with finite center and without compact factors. She called an action on a space (X, μ), μ a finite G-invariant measure, *purely atomic* if the representation of G on $L^2(X, \mu)$ is a direct sum of irreducible representations [102].

Theorem 3.12. [Iozzi, 1992] *Let G be as above and Γ a lattice in G. Suppose G acts on X_1 and X_2 preserving finite invariant measures μ_1 and μ_2. Suppose the actions have purely atomic spectra and are either essentially free or essentially transitive. Then any measure preserving measurable Γ-equivariant map $\phi : X_1 \to X_2$ is G-equivariant.*

Iozzi first showed that the spectrum of such an action cannot be a sum of discrete series representations of G. Let K is a maximal compact subgroup of G. The decomposition of the discrete series representations into irreducible subrepresentations of K is given by the so-called *Blattner formula*, which was established by H. Hecht and W. Schmid [182, 92]. In particular, one sees that discrete series representations do not contain K-invariant vectors. Suppose now that the spectrum of the G-action on X_i consists only of discrete series representations. Then there are no K-invariant nonconstant functions on the X_i. Thus K acts transitively, and G cannot act with discrete stabilizer. Since ϕ is measure preserving, ϕ gives rise to a unitary intertwining operator T of the restrictions of the representations to the lattice. By Theorem 3.11, T intertwines the non-discrete series parts $\mathcal{H}_0^i$ of the G-representations. Now consider the factor Y_i of X_i with Borel algebra generated by the $\mathcal{H}_0^i$. Then the Y_i have G-actions. Since T intertwines the G-actions, there is a G-equivariant map $Y_1 \to Y_2$. To complete the proof, Iozzi showed that K acts transitively on the fibers of the X_i over the Y_i. This uses the pure atomicity of the spectrum.

4. AMENABILITY

4.1. Definitions and basic results. Let us begin with the representation theoretic definition of amenability. Call an action of a (locally compact topological) group G on a compact convex subset W of a locally convex topological vector space *affine* if for all $w_1, w_2 \in W$, all $0 \le t \le 1$ and all $g \in G$, we have $g(t w_1 + (1 - t) w_2) = t(g w_1) + (1 - t)(g w_2)$. Call G *amenable* if every continuous affine action of G on a compact convex subset W of a locally convex topological vector space fixes some $w \in W$, i.e. $g w = w$ for all $g \in G$. In fact it is sufficient to require such fixed points for affine actions on subsets of duals of separable Banach spaces W.

There are several other conditions equivalent to amenability such as the existence of an invariant mean on G or the existence of an invariant probability measure for any action of G on a compact non-empty topological space X. A direct characterization of amenability in terms of the group can be given using Følner sets [75]. There are also other representation theoretic criteria. The *regular representation* of G is the representation of G on $L^2_\mu(G)$, where μ is Haar measure. A unitary representation π is called *weakly contained* in a unitary representation σ if every matrix coefficient of π is a uniform limit on compact subsets of matrix coefficients of σ. Then amenability is equivalent to the trivial one-dimensional representation (and in fact any irreducible unitary representation) being weakly contained in the regular representation of G [75]. Various people have also considered an extension of amenability, K-amenability, a property of the K-theory of the group (cf. e.g. [105]). While every amenable group is K-amenable there are also certain rank 1 groups such as $SL(2, \mathbb{R})$ and $SU(1, 1)$ that are K-amenable. The dynamic impact of this property has not been studied.

Any abelian or solvable group is amenable. Conversely, amenable connected Lie groups are always compact extensions of solvable Lie groups [72]. On the other hand, there are very complicated amenable discrete groups.

The representation theoretic notion of amenability as a fixed point property quite easily generalizes to group actions. This was first done by R. J. Zimmer in 1978 [213]. Recall the notion of a measurable cocycle from Section 2.4, before Theorem 2.8. Suppose we are given an action of G on X with quasi-invariant measure ν, a measurable cocycle $\beta : G \times X \to H$ and an action of H on a measure space Y. Then a measurable map $s : X \to Y$ is called a *section* of α if for ν-a.e. $x \in X$ and all $g \in G$ we have $s(g x) = \beta(g, x) s(x)$. Let E be a separable Banach space. Consider cocycles β taking values in the isometry group $Iso(E)$. Let β^* denote the dual cocycle taking values in $Iso(E^*)$. Let E_1^* denote the unit ball of E^*. A *(Borel) field of convex convex sets* A_x, $x \in X$, is a family of convex compact subsets of E_1^* such that $\{x, e) \mid e \in A_x\}$ is Borel. Call such a field A_x β-*invariant* if $A_{g x} = \beta(g, x)^*(A_x)$.

Definition 4.1. *An action α of G on a measure space (X, ν) is* amenable *if for any separable Banach space E, every cocycle $\beta : G \times X \to Iso(E)$ and every β-invariant field of convex sets A_x, there is a section s of β^* such that $s(x) \in A_x$ for ν-a.e. $x \in X$.*

It is easy to rephrase this condition as an ordinary fixed point condition in terms of the representation of G on $L^1(X, E)$ skewed by β [220, Section

4.3].

Any action of an amenable group is amenable. Furthermore, a homogeneous action of G on G/H, H a closed subgroup of G, is amenable if and only if H is amenable [220]. This gives us many examples of amenable actions. The next example lies at the heart of many applications of amenability to rigidity theory.

Example 4.2. Let Γ be a lattice in $SL(2, \mathbb{R})$. The action of Γ on the sphere at infinity S^1 of $\mathcal{H}^2$ is equivalent to the action of Γ on $SL(2, \mathbb{R})/H$ where H is the subgroup of upper triangular matrices. For any closed subgroup L of $SL(2, \mathbb{R})$, the action of Γ on $SL(2, \mathbb{R})/L$ is amenable if and only if the action of L on $SL(2, \mathbb{R})/\Gamma$ is amenable. Thus the action of Γ on S^1 is amenable (w.r.t. Lebesgue measure on S^1) since H is solvable and hence amenable.

More generally, R. J. Zimmer and the author showed in 1991 that the action of any discrete group of isometries on the sphere at infinity of a complete simply-connected Riemannian manifold with sectional curvature $-b^2 \leq K \leq -a^2 < 0$ is amenable with respect to any quasi-invariant measure [187, 189]. One says that the action is *universally amenable*. This was further generalized to actions of Gromov-hyperbolic groups on their spheres at infinity by S. Adams [1].

While the above shows the amenability of the action of a lattice on the sphere at infinity of a globally symmetric space of negative curvature, the situation is somewhat different for a semisimple group G of the noncompact type of real rank at least 2. For such G, it is more suitable to consider the so-called *Furstenberg boundaries*. These are the homogeneous spaces G/P where P is a parabolic in G (i.e. an algebraic subgroup P such that G/P is compact). There are 2^k such parabolics up to conjugacy where k is the real rank of G (including $P = G$). These boundaries actually appear as orbits of the G-action on the sphere at infinity of the globally symmetric space G/K, K a maximal compact subgroup of G. Up to conjugacy, there is exactly one parabolic of smallest dimension, the so-called *minimal parabolic*. It is always a compact extension of a solvable group. Hence we see as above that the action of a lattice on the maximal boundary G/P, P a minimal parabolic, is amenable.

4.2. Amenability, superrigidity and other applications. The following lemma lies at the heart of most of the applications of amenability to rigidity. For the special case of G-actions on maximal boundaries, it is due to Furstenberg [69]. Denote by $\mathcal{P}(Y)$ the set of probability measures on a space Y.

Lemma 4.3. [Furstenberg, 1973, Zimmer, 1980] *Let G act amenably on a space X with quasi-invariant measure μ. Suppose $\beta : G \times X \to H$ is a cocycle into a group H. If H acts on a compact metric space Y, then there exists a β-invariant section $X \to \mathcal{P}(Y)$.*

The lemma readily follows from the definition of amenability, applied to the Banach space of continuous functions on Y.

As a first application, let me now explain the first step in the proof of Margulis' superrigidity theorem. Let Γ be a lattice in a semisimple group G without compact factors and of rank at least 2. Suppose $\psi : \Gamma \to H$ is a homomorphism into a connected Lie group H. For simplicity, we will assume that H is also semisimple without compact factors. As in Mostow's proof of strong rigidity, the first goal is to find a Γ-equivariant map ϕ between boundaries of G and H. Margulis originally used the multiplicative ergodic theorem to find such a map [129]. In his extension of Margulis' superrigidity theorem to cocycles, Zimmer later modified Margulis' approach. He used only the amenability of the Γ-action on the maximal boundary G/P of G to construct ϕ [216]. In fact, let B be a maximal boundary of H. By Lemma 4.3 there is a β^*-invariant section $s : G/P \to \mathcal{P}(B)$ with values in the space of probability measures $\mathcal{P}(B)$. Zimmer then showed, using a method of Furstenberg, that the action of H on $\mathcal{P}(B)$ is smooth, i.e. all orbits are locally closed [65, 70, 214]. Let $\pi : \mathcal{P}(B) \to \mathcal{P}(B)/H$ be the projection. Then the composition $\pi \circ s : G/P \to \mathcal{P}(B)/H$ is a Γ-invariant map into a standard probability space. Note that P acts ergodically on G/Γ by the Mautner-Moore phenomenon. Hence Γ acts ergodically on G/P, and $\pi \circ s$ is a.e. constant. This means that s essentially takes values in a single orbit H/L of H on $\mathcal{P}(B)$. One can show that L is a parabolic. Therefore we get the desired measurable Γ-equivariant mapping. Note that we get such a map even in real rank one. However, to show that this map is an algebraic map $G/P \to H/L$, Margulis used the higher rank condition. To get a homomorphism $G \to H$ then is easy.

There are quite a few other applications of this idea. To start with, Zimmer proved his superrigidity theorem for cocycles along the same lines [216]. He later used it, in conjunction with an extension of Margulis' theorem about equivariant measurable quotients of boundaries, to prove the following theorem about *Riemannian foliations* [218]. These are foliations of a measure space (X, ν) such that a.e. leaf has a Riemannian structure. Note that the notions of cocycle and thus amenability naturally extend to equivalence relations, and in particular to foliations.

Theorem 4.4. [Zimmer, 1982] *Let $\mathcal{F}_1$ and $\mathcal{F}_2$ be Riemannian ergodic foliations with transversely invariant measure and finite total volume. Suppose that a.e. leaf of $\mathcal{F}_1$ is simply-connected and complete, and that the sectional curvatures are negative and uniformly bounded away from 0. Further suppose that $\mathcal{F}_2$ is irreducible and that a.e. leaf of $\mathcal{F}_2$ is isometric to a symmetric space of the noncompact type of rank at least 2. Then $\mathcal{F}_1$ and $\mathcal{F}_2$ are not transversely equivalent. Furthermore, $\mathcal{F}_1$ and $\mathcal{F}_2$ are not amenable, and thus not transversely equivalent to the orbit foliation of an action of an amenable group.*

The idea of an amenable foliation had already proved useful when Zimmer extended the Gromoll-Wolf and Lawson-Yau result on solvable fundamental groups of manifolds of non-positive curvature to foliations [219]. The general philosophy is that assumptions on the fundamental group of a manifold can be replaced by suitable hypotheses about foliations.

Theorem 4.5. [Zimmer, 1982] *Let $\mathcal{F}$ be an amenable Riemannian measurable foliation with transversely invariant measure and finite total volume. Suppose a.e. leaf is simply-connected and complete, and has non-positive sectional curvature. Then a.e. leaf is flat.*

Again one main idea is to find a section of probability measures on the spheres at infinity of the leaves invariant under the "holonomy" of $\mathcal{F}$.

Zimmer and I obtained restrictions on the fundamental group of spaces on which a higher rank semisimple group can act [189].

Theorem 4.6. [Spatzier-Zimmer, 1991] *Let G be a connected simple Lie group with finite center, finite fundamental group, and real rank at least 2. Suppose G acts on a closed manifold M preserving a real analytic connection and a finite measure. Then $\pi_1(M)$ cannot be isomorphic to the fundamental group Λ of a complete Riemannian manifold N with sectional curvature $-a^2 \leq K \leq -b^2 < 0$ for some real numbers a and b.*

S. Adams generalized this theorem to Gromov hyperbolic groups and spaces [1, 3]. The proof of Theorem 4.6 is is inspired by that of Theorem 4.4. However, we use the amenability of the boundary actions twice, in very different ways.

Outline of Proof: For simplicity, assume G is simply-connected. Suppose $\pi_1(M)$ is isomorphic to some Λ as in Theorem 4.6. View the universal cover $\tilde{M}$ as a Λ-principal bundle over M. Then the lift of the G-action to $\tilde{M}$ acts by bundle automorphisms, and gives rise to a cocycle $\beta : G \times M \to \Lambda$. Let P be a minimal parabolic of G. Then the action of G on $M \times G/P$ is amenable. Let $\tilde{\beta}$ be the lift of β to $M \times G/P$. By Lemma 4.3, there is $\tilde{\beta}$-invariant section $\phi : M \times G/P \to \mathcal{P}(N_\infty)$, where N_∞ is the sphere at infinity of N, and $\mathcal{P}(N_\infty)$ is the set of probability measures on N_∞. Let $\psi : M \times G/P \to M \times \mathcal{P}(N_\infty)$ be the map $\psi(m, x) = (m, \phi(m, x))$. Then project the product measure class on $M \times G/P$ to $M \times \mathcal{P}(N_\infty)$. Since we can project to the first factor of $M \times \mathcal{P}(N_\infty)$, we see that $M \times \mathcal{P}(N_\infty)$ lies in between $M \times G/P$ and M. Zimmer's generalization of Margulis' measurable quotient theorem asserts that $M \times \mathcal{P}(N_\infty)$ is of the form $M \times G/P^*$ for some parabolic P^* of G.

Under the geometric assumptions of Theorem 4.6, M. Gromov showed that the G-action on $\tilde{M}$ is proper on a set of full measure. Hence β does not take values in a finite subgroup of Λ. In fact, no restriction of β to a noncompact closed subgroup of G is cohomologous to a cocycle taking values in a finite subgroup.

On the other hand, the section $\phi : M \times G/P \to \mathcal{P}(N_\infty)$ gives rise to a $\beta \mid_{P \times M}$-invariant map $\phi_0 : M \to \mathcal{P}(N_\infty)$. A probability measure on $P(N_\infty)$ invariant under an infinite group of isometries is necessarily atomic with one or two atoms, due to the negative curvature on N. As a cocycle analogue, ϕ_0 as above is supported on one or two points for a.e. $m \in M$ unless $\beta \mid_{P \times M}$ is cohomologous to a cocycle taking values in a finite subgroup of Λ. The latter is not possible by the last paragraph. Let $Q = (N_\infty \times N_\infty)/S_2$ where S_2 is the permutation group in two letters. By the above, $M \times P(N_\infty) = M \times G/P^*$ is isomorphic to the skew product $M \times_\beta Q$.

If $P^* \neq G$ there is a non-compact closed abelian subgroup $A \subset P^*$ that fixes a non-atomic probability measure μ on G/P^*. Under the above isomorphism, μ corresponds to a $\beta \mid_{A \times M}$-invariant map $\theta : M \to P(Q)$. Lifting θ to a map into $P(N_\infty \times N_\infty)$ we see that the essential range of θ is supported on atomic measures on Q. This contradicts the fact that μ is non-atomic.

We conclude that $P^* = G$. This implies that there is a β-invariant map $F : M \to P(N_\infty)$. Again the essential range of F lies in $Q = (N_\infty \times N_\infty)/S_2$. Since N is negatively curved, N_∞ is universally amenable, and so is Q (cf. Example 4.2). Endowing Q with the image of the measure on M, we get a map from the G-action on M into an amenable action. Since G is Kazhdan (cf. Section 5), almost the opposite of amenability, this leads to a contradiction, unless β is cohomologous to a cocycle into a finite subgroup. As above, the latter is impossible by Gromov's result. $\diamond$

R. Zimmer used the amenability of the boundary action in a novel way to distinguish actions of product groups and certain irreducible lattices. Furstenberg's Lemma 4.3 also plays an essential role in the proof.

Theorem 4.7. [Zimmer, 1983] *Let Γ be a lattice in a connected simple Lie group with finite center or a fundamental group of a closed manifold of negative curvature. Suppose Γ acts essentially freely and ergodically on S, preserving a finite measure. If this action is orbit equivalent to a product action of two discrete groups Γ_1 and Γ_2 on $S_1 \times S_2$, then either S_1 or S_2 is essentially finite.*

S. Adams generalized this theorem to hyperbolic groups in Gromov's sense in [2]. This is significant as the class of Gromov hyperbolic groups is much larger than fundamental groups of negatively curved closed Riemannian manifolds.

Let us illustrate the idea of the proof very roughly in case Γ is the fundamental group of a manifold M of negative curvature. First pick amenable subrelations of the actions of Γ_i on S_i. By Lemma 4.3, we get sections $S \to P(M_\infty)$, invariant under the subrelations. Since the subrelations commute with the full action on the other factor, one can conclude that these sections are actually invariant under the full Γ-action. Thus again, this section takes values in atomic measures and hence the Γ-action is an extension of an amenable action, and hence amenable. Since Γ preserves a finite measure, this implies that Γ is amenable. This is impossible.

5. KAZHDAN'S PROPERTY

5.1. Definitions and basic results. Rigidity properties typically are strongest for higher rank semisimple Lie groups and fail for $SL(2, \mathbb{R})$. For the other rank one groups, they may or may not fail. For example, Mostow's strong rigidity theorem holds for all of them. Superrigidity and arithmeticity are only known to hold for $Sp(n, 1)$ and F_4^{-20}. In this section we will introduce a representation theoretic property, called Kazhdan's property (T), which falls right in between higher rank and just excluding $SL(2, \mathbb{R})$. D. Kazhdan discovered it in 1967. He realized that for $SL(3, \mathbb{R})$, the trivial representation is isolated within the unitary representations [119]. For $SL(2, \mathbb{R})$ on the other hand, it is well known that the trivial representation is not isolated. He used this property to show that lattices in higher rank simple Lie groups of the non-compact type are finitely generated, and their first Betti number vanishes. Kazhdan's property (T) has proven amazingly successful in the study of the dynamics of semisimple groups. Interestingly, its range of validity within semisimple groups coincides exactly with that of of the superrigidity and arithmeticity theorems. I will now introduce this property in detail, and discuss a handful of representative applications to rigidity theory. I refer to the survey of P. de la Harpe and A. Valette for a more complete discussion of this property [90]. They also discuss some other exciting applications, such as telephone networks, that lie outside our presentation.

Let G be a locally compact second countable group. Let π be a unitary representation of G on a Hilbert space V. For any $\varepsilon > 0$ and compact subset K of G, we call a unit vector $v \in V$ (ε, K)-*invariant* if $\| \pi(k)\, v - v \| < \varepsilon$ for all $k \in K$.

Definition 5.1. [**Kazhdan, 1967**] *Call G a Kazhdan group if any unitary representation of G which has (ε, K)-invariant vectors for all compact subsets K and $\varepsilon > 0$, has G-invariant vectors.*

Any compact group is Kazhdan, as follows easily from a standard averaging argument. On the other hand, amenable Kazhdan groups are necessarily compact. Any connected semisimple real Lie group is Kazhdan provided that it does not have $SO(n, 1)$ or $SU(n, 1)$ as a local factor. This was shown by D. Kazhdan for the higher rank groups and by B. Kostant for the rank one groups [119, 120]. P. de la Harpe and A. Valette gave a streamlined proof for the rank one groups in [90]. In the p-adic case, only the higher rank or compact semisimple groups are Kazhdan [90]. S. P. Wang showed in 1975 that certain skew products such as $SL(n, \mathbb{R}) \ltimes \mathbb{R}^n$ are Kazhdan. To get examples of discrete groups, D. Kazhdan observed that a lattice in a group G is Kazhdan if and only if G is Kazhdan. Finally note that any quotient group of a Kazhdan group is Kazhdan.

Let me note here that all known examples of Kazhdan groups are obtained via the above procedures, thus eventually through the (more or less) explicit representation theory of a Lie group. The geometry of Kazhdan groups is not well understood. E. Ghys for example asked:

Problem 5.2. [Quasi-isometric Invariance] *Suppose two finitely generated groups Γ_1 and Γ_2 are quasi-isometric with respect to their word metrics. Suppose Γ_1 is Kazhdan. Is Γ_2 necessarily Kazhdan?*

Note that the negatively curved locally symmetric spaces with Kazhdan fundamental groups all achieve sectional curvatures -1 and -4. Little is known for variable curvature.

Problem 5.3. [Pinching] *Let M be a closed Riemannian manifold with sectional curvatures $-4 < K < -1$. Can $\pi_1(M)$ be Kazhdan?*

Fundamental groups of closed manifolds with sectional curvature very close to -1, diameter bounded above and volume bounded below, are never Kazhdan. This is an immediate consequence of Gromov's compactness theorem for the space of geometrically bounded Riemannian metrics.

D. Kazhdan's original motivation for considering Kazhdan groups lay in the following result.

Theorem 5.4. [Kazhdan, 1967] *Let Γ be a discrete Kazhdan group. Then Γ is finitely generated and its abelianization is finite.*

Proof. List the elements of Γ by $\gamma_1, \ldots, \gamma_n, \ldots$. Let Γ_n denote the subgroup generated by $\gamma_1 \ldots \gamma_n$. Set $\pi = \oplus L^2(\Gamma/\Gamma_n)$. Since for any n, $\gamma_1 \ldots \gamma_n$ fix a vector, π has almost invariant vectors for any ε and finite subset of Γ. Since Γ is Kazhdan, there is a fixed unit vector v in π. Then the component v_n of v in any $L^2(\Gamma/\Gamma_n)$ is Γ-invariant. Thus for some n, v_n is a non-zero constant L^2-function on Γ/Γ_n. This implies that Γ/Γ_n is finite, and hence that Γ is finitely generated.

Since Γ is finitely generated, so is its abelianization $\Gamma/[\Gamma, \Gamma]$. It $\Gamma/[\Gamma, \Gamma]$ is infinite, it factors through $\mathbb{Z}$. As quotient groups of Kazhdan groups are Kazhdan, this is a contradiction. $\diamond$

I will now discuss some of the many applications of Kazhdan's property to rigidity theory.

5.2. Lorentz actions. In 1984, R. J. Zimmer obtained the following remarkable result about actions of Kazhdan groups on Lorentz manifolds [221]. His analysis rides on an understanding of the continuous homomorphisms from Kazhdan groups into non-Kazhdan Lie groups.

Lemma 5.5. *Let Γ be Kazhdan, G a connected simple Lie group which is not Kazhdan. Then the image of any homomorphism $\rho : \Gamma \to G$ is contained in a compact subgroup of G.*

Proof. Let π be a unitary representation of G which does not contain a G-fixed vector. By Theorem 3.2 on the vanishing of matrix coefficients for π and G, we conclude that $\rho(\Gamma)$ is precompact. ◇

The lemma actually extends to real algebraic groups G with all simple factors locally isomorphic to $SO(n,1)$ and $SU(n,1)$ [221, Corollary 20]. The cocycle analogue of the last lemma holds true as well [183, 217, 221]. The proof uses two more ingredients from harmonic analysis. For one, unitary representations of G as above have locally closed orbits provided that the π-image of the projective kernel of π is closed. Also, simple groups with faithful finite-dimensional unitary representations are compact.

Theorem 5.6. [Zimmer, 1984] *Let a Kazhdan group Γ act ergodically on a standard probability space S, preserving a measure μ. Let H be a real algebraic group. Then any cocycle $\beta : S \times \Gamma \to H$ is cohomologous to a cocycle taking values in an algebraic Kazhdan subgroup of H.*

For amenable ranges, this is due to K. Schmidt and R. J. Zimmer [183, 217].

Now suppose Γ acts on a compact Lorentz manifold preserving the Lorentz structure. Then the derivative cocycle takes values in $O(n,1)$. By Theorem 5.6, there is a measurable framing of the tangent bundle such that the derivative cocycle takes values in a compact group. Averaging a Riemannian structure for each such compact group, we see that there is a measurable Riemannian metric invariant under Γ. Using higher order jet bundles and Sobolev theory, Zimmer improved this conclusion as follows [221, 222].

Theorem 5.7. [Zimmer, 1984] *Let a Kazhdan group Γ act isometrically on a closed Lorentz manifold M. Then Γ preserves a C^∞-Riemannian metric on M. Thus the action is C^∞-equivalent to an action of Γ on a homogeneous space K/K_0 of a compact group K via a homomorphism $\Gamma \to K$.*

Lie group actions by Lorentz transformations are even more restricted. Zimmer showed that then either the Lie group is locally isomorphic to a product of $SL(2, \mathbb{R})$ and a compact group or it is amenable with nilradical at most of step 2 [224]. Lorentz actions have been further analyzed by G. D'Ambra and M. Gromov, using geometric means [78, 40].

5.3. Infinitesimal and local rigidity of actions. Recall A. Weil's local rigidity result. Namely, any small deformation of a uniform lattice Γ in a semisimple group G is conjugate to the lattice, provided that G does not have $SL(2, \mathbb{R})$ as a local factor [205, 206]. He actually showed "infinitesimal rigidity", namely that the cohomology $H^1(\Gamma, \text{Ad})$ vanishes. Applying the implicit function theorem, he then deduced local rigidity from that.

This motivates the following definition of infinitesimal rigidity for group actions. Suppose a group Γ act smoothly on a manifold M. Denote by

$Vect(M)$ the space of smooth vector fields on M. Call the Γ-action *infinitesimally rigid* if the first cohomology $H^1(\Gamma, Vect(M))$ vanishes. Due to the delicate nature of the implicit function theorem in infinite dimensions, there is no clear connection between infinitesimal and local rigidity.

Let G be a connected semisimple Lie group G with finite center and without compact factors, and Γ a uniform lattice in G. Given a homomorphism $\pi : G \to H$ of G into another Lie group H and a uniform lattice $\Lambda \subset H$, then Γ acts naturally on H/Λ. R. J. Zimmer proved infinitesimal rigidity for such actions in [227].

Theorem 5.8. [Zimmer, 1990] *Assume that G does not have factors locally isomorphic to $SO(n, 1)$ or $SU(n, 1)$ and that $\pi(\Gamma)$ is dense in H. Then the Γ-action on H/Λ is infinitesimally rigid.*

Let me indicate how Kazhdan's property enters the proof. Let $Vect_2(M)$ denote the space of L^2-vector fields on M. Zimmer first showed that the action is L^2-*infinitesimally rigid*, i.e. the canonical map

$$H^1(\Gamma, Vect(M)) \to H^1(\Gamma, Vect_2(M))$$

is zero. Thus there always is an L^2-coboundary, and Theorem 5.8 becomes a regularity theorem. To show L^2-*infinitesimal rigidity*, note that for $M = H/\Lambda$, the Γ-action on $Vect(M)$ is isomorphic with the Γ-action on the space of functions from M to the Lie algebra $\mathfrak{h}$ of H where Γ acts via $Ad \circ \pi$. Generalizing the calculations of Y. Matsushima and S. Murakami, Zimmer showed L^2-*infinitesimal rigidity* for all the non-trivial irreducible components of $Ad \circ \pi$ [136, 227]. For the trivial representations contained in $Ad \circ \pi$, the first cohomology in the L^2-functions already vanishes. The latter follows from the following characterization of Kazhdan's property due to J. P. Serre (in a letter to A. Guichardet) [90].

Theorem 5.9. [Serre] *A locally compact group G is Kazhdan if and only if for all unitary representations ρ of G, $H^1(G, \rho) = 0$.*

Note that these arguments apply to other finite dimensional representations besides $Ad \circ \pi$.

In his thesis in 1989, J. Lewis managed to adapt Zimmer's arguments to certain actions of non-uniform lattices [121].

Theorem 5.10. [Lewis, 1989] *The action of $SL(n, \mathbb{Z})$ on the n-torus T^n is infinitesimally rigid for all $n \geq 7$.*

Deformation rigidity for this and other toral actions of lattices was obtained by S. Hurder [101]. A. Katok, J. Lewis and R. J. Zimmer later obtained local and semi-global rigidity results [110, 111, 112].

Theorem 5.11. [Hurder, Katok, Lewis, Zimmer, 1990's] *The standard action of $SL(n, \mathbb{Z})$ or any subgroup of finite index on the n-torus is locally C^∞-rigid.*

The Kazhdan property of $SL(n, \mathbb{Z})$ is used here to see that any small perturbation of the action preserves an absolutely continuous probability measure. N. Qian recently announced the deformation rigidity for "most" actions of irreducible higher rank lattices Γ on tori by automorphisms [161, 162, 163, 164].

R. J. Zimmer also obtained a local rigidity result for isometric actions of Kazhdan groups in [225].

Theorem 5.12. [Zimmer, 1987] *Let a Kazhdan group Γ act on a closed manifold preserving a smooth Riemannian metric. Then any small enough volume preserving ergodic perturbation of the action leaves a smooth Riemannian metric invariant.*

5.4. Discrete spectrum. Let a locally compact group G act on a measure space X preserving a probability measure μ. We say that the action has *discrete spectrum* if $L^2(X, \mu)$ decomposes into a direct sum of finite dimensional representations. Furthermore call an action of G *measurably isometric* if it is measurably conjugate to an action of G on a homogeneous space K/K_0 of a compact group K via a homomorphism $G \to K$. Measurably isometric actions always have discrete spectrum by the Peter-Weyl theorem. J. von Neumann in the commutative case and G. W. Mackey in general proved the converse [204, 125].

Note that measurably isometric actions always leave a measurable Riemannian metric invariant. The converse however is not true in general. A. Katok constructed an example of a volume preserving weak mixing diffeomorphism of a closed manifold which preserves a measurable Riemannian metric. For Kazhdan groups on the other hand, R. J. Zimmer obtained the complete equivalence in [228].

Theorem 5.13. [Zimmer, 1991] *Let a discrete Kazhdan group Γ act smoothly on a closed manifold preserving a smooth volume. Then the action has discrete spectrum if and only if Γ preserves a measurable Riemannian metric. Furthermore, if Γ is an irreducible lattice in a higher rank semisimple group, then discrete spectrum is also equivalent to the vanishing of the metric entropy for every element $\gamma \in \Gamma$.*

It is fairly easy to see that the Γ-action has some discrete spectrum. For every point $m \in M$, approximate the Γ-invariant measurable metric by a smooth metric ω_m in a neighborhood of m. Let $f_{r,m}$ be the normalized characteristic function of a ball of size r for ω_m about m. Set $F_r(m, x) = f_{r,m}(x)$ for $(x, m) \in M \times M$. For any finite set K in Γ and $\varepsilon > 0$, there is a small enough r such that F_r is (ε, K)-invariant. This follows since the original measurable metric is Γ-invariant. Since Γ is Kazhdan, F_r is close to a Γ-invariant function in $L^2(M \times M)$. This gives rise to a non-trivial Γ-invariant finite dimensional subspace of $L^2(M)$.

5.5. Ruziewicz' problem. By uniqueness of Haar measure, there is a unique countably additive rotation-invariant measure on any sphere S^n. S. Ruziewicz asked if the finitely additive rotation-invariant measures, defined on all Lebesgue measurable sets, are also unique. For the circle, S. Banach found other such measures in 1923. The problem remained open for the higher dimensional spheres until the early 80's. Then G. A. Margulis and D. Sullivan showed independently, using a partial result of J. Rosenblatt, that such measures do not exist on S^n for $n > 3$ [132, 194]. Their main idea is that $SO(n + 1)$ for $n > 3$ contains discrete Kazhdan groups Γ dense in $SO(n + 1)$. J. Rosenblatt on the other hand showed that a finitely additive rotation-invariant measure distinct from Lebesgue measure μ gives rise to Γ-almost invariant vectors in the orthogonal complement to the constants in $L^2(S^n, \mu)$. By Kazhdan's property, there are Γ- and hence $SO(n + 1)$-invariant non-constant functions, a contradiction. The remaining cases, S^2 and S^3, were resolved by V. G. Drinfel'd in 1984 [46]. Though $SO(3)$ and $SO(4)$ do not contain discrete Kazhdan groups, he was able to exhibit discrete subgroups for which certain unitary representations do not contain almost invariant vectors. Drinfel'd's approach used deep theorems in number theory. To summarize, we have the complete resolution of Ruziewicz' problem.

Theorem 5.14. [Rosenblatt,Margulis,Sullivan,Drinfeld, 1980's] *For $n > 1$, there is a unique rotation invariant finitely additive measure on S^n, defined on all Lebesgue measurable sets.*

G. A. Margulis also resolved the analogous problem for Euclidean spaces [132]. There we have uniqueness on $\mathbb{R}^n$ exactly when $n > 2$. K. Schmidt has further analyzed the connection between actions of Kazhdan groups and unique invariant means [183]. I refer to [90] for a detailed exposition of the Ruziewicz' problem as well as other interesting applications of Kazhdan's property outside rigidity theory.

5.6. Gaps in the Hausdorff dimension of limit sets. Let $\mathcal{H}^n$ denote either a quaternionic hyperbolic space or the Cayley plane. Compactify $\mathcal{H}^n$ by a sphere S as usual. The *limit set* $L(\Gamma)$ of a discrete group of isometries Γ of $\mathcal{H}^n$ is the set of accumulation points of a Γ-orbit of a point $x \in \mathcal{H}^n$ in S. This is independent of the choice of initial point x. The *ordinary set* $O(\Gamma)$ is the complement $S \setminus L(\Gamma)$ of $L(\Gamma)$ in S. Then Γ acts properly discontinuously on $\mathcal{H}^n \cup O(\Gamma)$. Call Γ *geometrically cocompact* if $\mathcal{H}^n \cup O(\Gamma)/\Gamma$ is compact. This is a generalization of convex cocompact groups on real hyperbolic space. K. Corlette used Kazhdan's property to estimate a gap in the Hausdorff dimension of the limit set of such groups in [31].

Theorem 5.15. [Corlette, 1990] *A geometrically cocompact discrete subgroup of $Sp(n, 1)$ (or F_4^{-20}) is either a lattice or its limit set has Hausdorff* *codimension at least 2 (respectively 6). Furthermore, $\mathcal{H}^n/\Gamma$ has at most one end.*

Corlette proved similar results for higher rank semisimple groups. However, it is not clear if there are any non-trivial examples of geometrically cocompact groups in higher rank.

To outline the proof of Theorem 5.15, Corlette first established a connection between the bottom of the spectrum λ_0 of the Laplacian on $\mathcal{H}^n/\Gamma$ and the Hausdorff dimension δ_Γ of the limit set, following work of Akaza, Beardon, Bowen, Elstrodt, Patterson and Sullivan in the real hyperbolic case. He calculated that $\lambda_0 = \delta_\Gamma(N - \delta_\Gamma)$ where N is the Hausdorff dimension of S. On the other hand, the Laplacian on $L^2(\mathcal{H}^n/\Gamma)$ can be determined via the unitary representation of $Sp(n,1)$, say, on $L^2(Sp(n,1)/\Gamma)$. B. Kostant had determined the unitary dual of $Sp(n,1)$ and established Kazhdan's property for them with an estimate of the isolation of the trivial representation. The trivial representation of $Sp(n,1)$ is only present in $L^2(Sp(n,1)/\Gamma)$ if Γ is a lattice. Thus Corlette could apply Kostant's work to get a lower bound on λ_0 and thus an upper bound on δ_Γ, in case Γ is not a lattice. The second claim follows easily from the first, since the limit set of Γ cannot disconnect the ordinary set.

5.7. Variants of Kazhdan's property. A group G is Kazhdan if the trivial representation of G is isolated in the space of unitary representations of G. By changing the class of representations under consideration, one obtains many variants of Kazhdan's property.

Around 1980 already, M. Cowling established a stronger version of Kazhdan's property by enlarging the class of representations [34, 35]. Consider representations of a group G on a Hilbert space $\mathcal{H}$ which are not necessarily unitary but uniformly bounded in G. If G is amenable, then any uniformly bounded representation is unitarizable. For general G however, e.g. $G = SL(2, \mathbb{R})$, the two classes of representations differ.

Theorem 5.16. [Cowling, 1980] *Let G be a connected real simple Lie group with finite center. Then the trivial one-dimensional representation of G is isolated within the uniformly bounded irreducible representations precisely when the real rank of G is at least 2.*

A. Lubotzky and R. J. Zimmer investigated various weakenings of Kazhdan's property for discrete groups G by decreasing the class of representations [122]. For example, consider $G = SL(n, \mathbb{Q})$ for $n > 2$. Then G is not Kazhdan since it is not finitely generated. On the other hand, Lubotzky and Zimmer showed that the trivial representation is isolated both within the class of finite dimensional unitary representations as well as the class of unitary representations whose matrix coefficients vanish at infinity. More generally, this holds for irreducible lattices Γ in the product of a noncompact simple group with a semisimple Kazhdan group. They obtained the following geometric consequence.

Theorem 5.17. [Lubotzky-Zimmer, 1989] *Isometric ergodic actions on closed manifolds by lattices* Γ *as above are infinitesimally rigid.*

6. MISCELLANEOUS APPLICATIONS

6.1. Isospectral rigidity. There are now many examples of Riemannian manifolds, both in positive and negative curvature whose Laplacians have the same spectrum but which are not isometric. Negatively curved manifolds however exhibit somewhat more isospectral rigidity . To begin with, V. Guillemin and D. Kazhdan established deformation rigidity [85]. A Riemannian manifold has *simple length spectrum* if the ratio of the lengths of any two distinct closed geodesics are irrational.

Theorem 6.1. [Guillemin-Kazhdan, 1980] *Let* (M, g_0) *be a negatively curved closed surface with simple length spectrum. Then any deformation* g_t *of* g_0 *such the spectrum of the Laplace operator is independent of* t *is trivial.*

Guillemin and Kazhdan generalized this theorem later to higher dimensional manifolds with sufficiently pinched negative curvature [86]. The proof involves analyzing the representations of $O(n)$ on the tangent bundle and A. N. Livshitz' theorem on the cohomology of the geodesic flow.

Analysis of the length spectrum of a Riemannian manifold M, i.e. the set of lengths of closed geodesics is closely related with the spectrum of the Laplacian. If M is negatively curved, then each free homotopy class of loops contains precisely one closed geodesic. Assigning its length to the homotopy class defines a function from $\pi_1(M) \to \mathbb{R}$, the so-called *marked length spectrum* of M. While the length spectrum itself is not rigid, J. P. Otal and independently C. Croke showed in [153, 38]

Theorem 6.2. [Otal, Croke, 90] *The marked length spectrum determines a closed surface of negative curvature up to isometry.*

This has been generalized to nonpositively curved surfaces in [39]. Little is known in higher dimension.

Let me note here that the L^2-spectrum of an action of a semisimple group does not determine the action even measurably [188]. The construction of the counterexamples is based on Sunada's ideas on the spectral non-rigidity of the Laplace operator on a Riemannian manifold.

Theorem 6.3. [Spatzier, 1989] *Let* G *be a noncompact almost simple connected classical group of real rank at least 27. Then* G *has properly ergodic actions which are not measurably conjugate and have the same* L^2-*spectrum. Moreover, the actions can be chosen such that the* L^2 *of the spaces decomposes into a countable direct sum of irreducible representations of* G.

Let Γ be a lattice in G. It follows from Theorem 3.12 that not even the restrictions of the above purely atomic actions to Γ are measurably isomorphic.

Guillemin's and Kazhdan's work suggests the following problem.

Problem 6.4. *Let G be a semisimple group without compact and $PSL(2, \mathbb{R})$-factors. Are volume preserving actions of G and its lattices locally isospectrally rigid (or deformation rigid)?*

At least for the natural homogeneous actions, Guillemin's and Kazhdan's techniques may be helpful.

6.2. Entropy rigidity. Given a closed Riemannian manifold M, there are two fundamental measures of the complexity of its geodesic flow, namely its topological entropy h_{top} and its metric entropy h_λ with respect to the Liouville measure λ. If the sectional curvature of M is non-positive, the topological entropy can be interpreted as the exponential rate of growth of the volume of balls in the universal cover. Furthermore, for negatively curved M, there is a unique measure, called the *Bowen-Margulis measure*, whose metric entropy coincides with the topological entropy [128]. In general, the topological entropy always majorizes any metric entropy. Naturally, one asks when $h_{top} = h_\lambda$. A. Katok showed in 1982 that the metric and topological entropy of the geodesic flow of a closed surface with negative curvature coincide precisely when the surface has constant curvature [107, 108]. For higher dimensions, he conjectured that the entropies are equal if and only if the manifold is locally symmetric. He showed that this holds within the conformal class of a locally symmetric metric.

Extremal properties of the entropies are closely related. Given a closed locally symmetric space M with maximal sectional curvature -1, P. Pansu showed in 1989, using quasi-conformal methods, that any other metric on M with sectional curvature bounded above by -1 is at least as big as the topological entropy of the locally symmetric metric [157]. In 1990, U. Hamenstaedt characterized the extremal such metrics as the locally symmetric metrics amongst these [87].

Theorem 6.5. [Hamenstadt, 1990] *Let M be a closed locally symmetric space with maximal sectional curvature -1. Then the locally symmetric metrics on M are precisely the metrics which minimize the topological entropy amongst all metrics with upper bound -1 for the sectional curvature.*

Normalizing the volume rather than the maximal sectional curvature, M. Gromov conjectured that the locally symmetric metrics again are precisely the metrics which minimize the topological entropy [77]. A partial answer was found by G. Besson, G. Courtois and S. Gallot [20, 19, 20]. [1] They

---

[1]They have recently shown that on a closed real hyperbolic manifold, the topological entropy on the space of metrics of equal volume is minimized by the constant curvature metrics.

endow the space of metrics on a manifold with the Sobolev topologies H^s, $s > 0$.

Theorem 6.6. [**Besson-Courtois-Gallot, 1991**] *Let g_0 be a metric of constant curvature -1 on a closed manifold M of dimension n. Then for all $s > n/2$, there is an H^s-neighborhood $\mathcal{U}$ of g_0 in the space of metrics with volume equal to that of g_0 on which the topological entropy is minimal at g_0. Furthermore, any other metric in $\mathcal{U}$ with minimal entropy has constant curvature -1.*

In fact, they can choose $\mathcal{U}$ to be a neighborhood of the conformal class of g_0, saturated by conformal classes.

L. Flaminio established a version of this theorem for C^2-deformations of g_0 transversal to the orbit of g_0 under the diffeomorphism group, using representation theory [60]. This allows him to get explicit estimates of the second derivative of the topological entropy at the constant curvature metric in terms of the L^2-norm of the variation of the metric. Interestingly, the metric entropy h_λ is neither maximized nor minimized at the constant curvature metric [60]. However, Flaminio showed that the difference $h_{top} - h_\lambda$ is convex near the constant curvature metric. He thus obtained a local entropy rigidity theorem resolving Katok's conjecture affirmatively locally [60].

Theorem 6.7. [**Flaminio, 1992**] *Let (M, g_0) be a closed manifold of constant negative curvature. Then along any sufficiently short path of C^∞-metrics g_t starting at g_0 and transverse to the orbit of g_0 under the diffeomorphism group, equality of topological and metric entropy implies constant curvature.*

In the proof, Flaminio calculated the derivatives of the topological and metric entropies at g_0 using the representation theory of $SO(n, 1)$. This is based on Guillemin's and Kazhdan's work on isospectral rigidity (cf. Section 6.1). To obtain precise quantitative results on the size of the derivatives, Flaminio used the full knowledge of the unitary dual of $SO(n, 1)$.

Finally let us mention a third canonical measure on a closed manifold of negative curvature, the so-called *harmonic measure*. D. Sullivan conjectured that such a space is locally symmetric provided the harmonic measure coincides with the Liouville measure. C. Yue made substantial progress in this direction [211]. Let M be a closed manifold of negative sectional curvature. Let ν_x be the harmonic measure on the ideal boundary $\tilde{M}(\infty)$ of the universal cover $\tilde{M}$ of M, i.e. the hitting probability measure for Brownian motion starting at $x \in \tilde{M}$. Let m_x be the push forward of Lebesgue measure on the unit tangent sphere S_x at x using the canonical projection from S_x to $\tilde{M}(\infty)$. Similarly, one can project the Bowen-Margulis measure of maximal entropy to measures μ_x on $\tilde{M}(\infty)$. Combining Yue's principal result with Theorem 2.17, we have

Theorem 6.8. [Yue, 1992] *The horospheres in $\tilde{M}$ have constant mean curvature provided that either $m_x = \nu_x$ for all $x \in \tilde{M}(\infty)$ or $\nu_x = \mu_x$ for all $x \in \tilde{M}(\infty)$. In either case, the geodesic flow is C^∞-conjugate to that of a locally symmetric space. Furthermore, M has constant curvature if its dimension is odd.*

6.3. Unitary representations with locally closed orbits. An action of a topological group G on a Borel space X is called *smooth* or *tame* if the quotient space X/G is countably separated. By work of J. Glimm and E. G. Effros, an action on a complete separable metrizable space X is tame if and only if all orbits are locally closed [73, 50]. The tameness of certain actions has proved very useful in rigidity theory. In the proof of Margulis' superrigidity theorem for example, the tameness of the action of a real semisimple group G on the space of probability measures on the maximal boundary H/P_H of a second group H allowed us to construct an equivariant map from the maximal boundary G/P of G to a homogeneous space of H.

R. J. Zimmer showed that unitary representations π of real algebraic groups are typically tame [215]. Define the *projective kernel, P_π*, of π by

$$P_\pi = \{g \in G \mid \pi(g) \text{ is a scalar multiple of } 1\}.$$

Theorem 6.9. [Zimmer, 1978] *If π is an irreducible unitary representation of a real algebraic group G, then π is tame provided that $\pi(P_\pi)$ is closed.*

As explained in Section 5.2, Zimmer applied this theorem to Lorentz actions.

Notice that the regular representation of a discrete group is always tame. This observation motivated the proof of the following theorem.

Theorem 6.10. [Adams-Spatzier, 1990] *Let a discrete Kazhdan group G act ergodically on a measure space S preserving a finite measure. Suppose $\beta : G \times S \to H_1 *_{H_3} H_2$ is a cocycle into an amalgamated product. Then β is cohomologous to a cocycle into H_1 or H_2.*

Amalgamated products act on trees. One can associate unitary representations of G to these trees which have almost invariant vectors and are tame. Combining this with Kazhdan's property and Zimmer's techniques from his proof of the cocycle superrigidity theorem yields the proof. The theorem easily generalizes to automorphism groups of real trees.

REFERENCES

1. S. Adams, *Boundary amenability for word hyperbolic groups and an application to smooth dynamics of simple groups*, Topology, to appear.
2. S. Adams, *Indecomposability of equivalence relations generated by word hyperbolic groups*, Topology, to appear.
3. S. Adams, *Reduction of cocycles with hyperbolic targets*, preprint.

4. S. Adams, *Rank rigidity for foliations by manifolds of nonpositive curvature*, preprint.
5. S. Adams and R. J. Spatzier, *Kazhdan groups, cocycles and trees*, Amer. J. of Math. **112** (1990), 271-287.
6. S. Agard, *A geometric proof of Mostow's rigidity theorem for groups of divergence type*, Acta Math. **151** (1983). 321-252.
7. S. Agard, *Mostow rigidity on the line: A survey*, in: *Holomorphic functions and moduli, II*, ed. by D. Drasin, C. Earle, F. Gehring, I. Kra, A. Marden, MSRI publications **11** (1988), 1-12.
8. K. Astala and M. Zinsmeister, *Mostow rigidity and Fuchsian groups*, C. R. Acad. Sci. Paris, **311**, Série I (1990), 301-306.
9. L. Auslander, *An exposition of the structure of solvmanifolds*, Part I: Algebraic Theory, Bull. AMS **79** (1973), 227-261; Part II: G-induced flows, ibid, 262-285.
10. L. Auslander, L. Green and F. Hahn, *Flows on homogeneous spaces*, Princeton University Press, Princeton, New Jersey, 1963.
11. W. Ballmann, *Nonpositively curved manifolds of higher rank*, Annals of Math. 122 (1985), 597-609.
12. W. Ballmann, M. Brin and P. Eberlein, *Structure of manifolds of non-positive curvature I*, Ann. of Math. **122** (1985), 171-203.
13. W. Ballmann, M. Brin and R. J.Spatzier, *Structure of manifolds of non-positive curvature II*, Ann. of Math. **122** (1985), 205-235.
14. W. Ballmann, M. Gromov and V. Schroeder, *Manifolds of nonpositive curvature*, Progress in Mathematics, Birkhauser, Boston, 1985.
15. H. Bass, J. Milnor and J. P. Serre, *Solution of the congruence subgroup problem for SL_n and Sp_2n*, Publ. Math. IHES **33** (1967), 59-137.
16. Y. Benoist, P. Foulon and F. Labourie, *Flots d'Anosov à distributions stable et instable différentiables*, J.A.M.S. **5** (1992), 33-74.
17. Y. Benoist and P. Foulon, *Sur les difféomorphismes d'Anosov affines a feuilletages stable et instable differentiable*, Invent. Math. **111** (1993), 285-308.
18. M. Berger, *Sur les variétés à courbure positive de diamètre minimum*, Comment. Math. Helv. **35** (1961), 28-34.
19. G. Besson, G. Courtois and S. Gallot, *Volume et entropie minimale des espaces localement symetriques*, Invent. Math. **103** (1991), 417-445.
20. G. Besson, G. Courtois and S. Gallot, *Sur les extrema locaux de l'entropie topologique*, preprint.
21. A. Borel, *Compact Clifford-Klein forms of symmetric spaces*, Topology **2** (1963), 111-122.
22. A. Borel and Harish-Chandra, *Arithmetic subgroups of algebraic groups*, Ann. of Math. **75**, (1962), 485-535.
23. R. Bowen, *Hausdorff dimension of quasicircles*, Publ. Math. IHES **50** (1979), 11-25.
24. J. Brezin and C. C. Moore, *Flows on homogeneous spaces: A new look*, Amer. J. Math. **103** (1981), 571-613.
25. K. Burns and R. J. Spatzier, *Manifolds of nonpositive curvature and their buildings*, Publ. Math. IHES **65** (1987), 35-59.
26. P. Buser, *Geometry and spectra of compact Riemannian surfaces*, Boston, Birkhäuser 1992.
27. E. Calabi, *On compact Riemannian manifolds with constant curvature I*, Differential geometry, Proc. Symp. Pure Math. vol. 3, Amer. Math. Soc., Providence, R.I. 1961, 155-180.
28. E. Calabi and E. Vesentini, *On compact locally symmetric Kähler manifolds*, Ann. of Math. **71** (1960), 472-507.
29. J. Carlson and D. Toledo, *Harmonic mappings of Kähler manifolds into locally*

symmetric spaces, Publ. Math. **69** (1989), 359–372.

30. W. Casselman and D. Miličić, *Asymptotic behavior of matrix coefficients of admissible representations*, Duke J. of Math. **49** (1982), 869–930.

31. K. Corlette, *Hausdorff dimension of limit sets I*, Invent. Math. **102** (1990), 521–542.

32. K. Corlette, *Archimedean superrigidity and hyperbolic geometry*, Ann. of Math. **135** (1992), 165–182.

33. K. Corlette and R. J. Zimmer, *Superrigidity for cocycles and hyperbolic geometry*, preprint.

34. M. Cowling, *Sur les coefficients des représentations unitaires des groupes de Lie simple*, Lecture Notes in Mathematics **739**, 1979, 132–178, Springer Verlag.

35. M. Cowling, *Unitary and uniformly bounded representations of some simple Lie groups*, in: Harmonic Analysis and group representations, C.I.M.E. 1980, Liguori Napoli, 1982, 49–128.

36. M. Cowling and T. Steger, *The irreducibility of restrictions of unitary representations to lattices*, J. Reine Angew. Math. **420** (1991), 85–98.

37. M. Cowling and R. J. Zimmer, *Actions of lattices in $Sp(n,1)$*, Erg. Th. and Dynam. Syst. **9** (1980), 221–237.

38. C. Croke, *Rigidity for surfaces of nonpositive curvature*, Comment. Math. Helv. **65** (1990), 150–169.

39. C. Croke, A. Fathi and J. Feldman, *The marked length-spectrum of a surface of nonpositive curvature*, Topology **31** (1992), 847–855.

40. G. D'Ambra and M. Gromov, *Lectures on Transformation groups*, in: Surveys in differential geometry, Cambridge, MA, 1990.

41. S. G. Dani, *Kolmogorov automorphisms on homogeneous spaces*, Amer. J. Math **98** (1976), 119–163.

42. S. G. Dani, *Spectrum of an affine transformation*, Duke J. of Math. **98** (1977), 129–155.

43. S. G. Dani, *Invariant measures of horospherical flows on noncompact homogeneous spaces*, Invent. Math. **47** (1978), 101–138.

44. S. G. Dani, *Invariant measures and minimal sets of horospherical flows*, Invent. Math. **64** (1981), 357–385.

45. P. Deligne and G. D. Mostow, *Monodromy of hypergeometric functions and non-lattice integral monodromy*, Publ. Math. IHES **63** (1986), 5–89.

46. V. G. Drinfel'd, *Finitely additive measures on S^2 and S^3 invariant with respect to rotations*, Funct. Anal. Appl. **18** (1984), 245–246.

47. P. Eberlein, *Rigidity of lattices of nonpositive curvature*, Ergod. Th. and Dynam. Syst. **3** (1983), 47–85.

48. P. Eberlein and J. Heber, *A differential geometric characterization of symmetric spaces of higher rank*, Publ. Math. IHES **71** (1990), 33–44.

49. J. Eels and J. Sampson, *Harmonic maps of Riemannian manifolds* Am. J. Math. **86** (1964), 109–160.

50. E. G. Effros, *Transformation groups and C^*-algebras*, Ann. of Math. **81** (1965), 38–55.

51. J. Faraut and K. Harzallah, *Distances Hilbertiennes invariantes sur un espace homogène*, Ann. Inst. Fourier, Grenoble, **24** (1974), 171–217.

52. T. Farrell and L. Jones, *A topological analogue of Mostow's rigidity theorem*, J. Amer. Math. Soc. **2** (1989), 257–370.

53. T. Farrell and L. Jones, *Negatively curved manifolds with exotic smooth structures*, J. Amer. Math. Soc. **2** (1989), 899–908.

54. T. Farrell and L. Jones, *Rigidity in geometry and topology*, Proceedings of the International Congress of Mathematicians, Vol. I, II (Kyoto, 1990), 653–663.

55. T. Farrell and L. Jones, *Complex hyperbolic manifolds and exotic smooth structures*, preprint.

56. J. Feldman and D. Ornstein, *Semirigidity of horocycle flows over compact surfaces of variable negative curvature*, Erg. Th. and Dynam. Syst. **7** (1987), 49–72.

57. R. Feres, *Geodesic flows on manifolds of negative curvature with smooth horospheric foliations*, Erg. Th. and Dyn. Syst. **11** (1991), 653–686.

58. R. Feres and A. Katok, *Anosov flows with smooth foliations and rigidity of geodesic flows on three-dimensional manifolds of negative curvature*, Erg. Th. and Dyn. Syst. **10** (1990), 657–670.

59. L. Flaminio, *An extension of Ratner's rigidity theorem to n-dimensional hyperbolic space*, Erg. Th. and Dyn. Syst. **7** (1987), 73–92.

60. L. Flaminio, *Local entropy rigidity*, preprint.

61. L. Flaminio and R. J. Spatzier, *Geometrically finite groups, Patterson-Sullivan measures and Ratner's rigidity theorem*, Invent. Math. **99** (1990), 601–626.

62. S. V. Fomin and I. M. Gelfand, *Geodesic flows on manifolds of constant negative curvature*, Uspehi Mat. Nauk **7**, no. 1, 118–137 (1952).

63. P. Foulon and F. Labourie, *Sur les varietes compactes asymptotiquement harmoniques*, Invent. Math. **109** (1992), 97–111.

64. H. Furstenberg, *Strict ergodicity and transformations on the torus*, Amer. J. Math. **83** (1961), 573–601.

65. H. Furstenberg, *A Poisson Formula for semi-simple Lie groups*, Ann. of Math. **77** (1963), 335–383.

66. H. Furstenberg, *Poisson boundaries and envelopes of discrete groups*, Bull. AMS **73** (1967), 350–356.

67. H. Furstenberg, *Random walks and discrete subgroups of Lie groups*, in: *Advances in Probability and Related Topics*, 1–63, vol. 1, Dekker, New York, 1971.

68. H. Furstenberg, *The unique ergodicity of the horocycle flow*, in *Recent Advances in Topological Dynamics*, Springer, 1972, 95–115.

69. H. Furstenberg, *Boundary theory and stochastic processes on homogeneous spaces*, in: *Harmonic analysis on homogeneous spaces*, Proc. of Symp. in Pure and Applied Math., **26** (1973), 193–29.

70. H. Furstenberg, *A note on Borel's density theorem*, Proc. AMS **55**, (1976), 209–212.

71. E. Ghys, *Flots d'Anosov dont les feuilletages stables sont differentiables*, Ann. Sci. Ecole Norm. Sup. **20** (1987), 251–270.

72. S. Glasner, *Proximal flows*, Lect. Notes in Math. **517**, Springer-Verlag, Berlin-New York, 1976.

73. J. Glimm, *Locally compact transformation groups*, TAMS **101** (1961), 124–138.

74. W. Goldman and J. J. Millson, *Local rigidity of discrete groups acting on complex hyperbolic space*, Invent. Math. **88** (1987), 495–520.

75. F. Greenleaf, *Invariant means on topological groups*, van Nostrand, New York, 1969.

76. M. Gromov, *Hyperbolic manifolds according to Thurston and Jorgensen*, Sém. Bourbaki, 32 année, 1979/80, 40–53.

77. M. Gromov, *Filling Riemannian manifolds*, J. Diff. Geom. **18** (1983). 1–47.

78. M. Gromov, *Rigid transformation groups*, D. Bernard, Choquet-Bruhat (eds.), Géométrie differentielle, Travaux en course, Hermann, Paris 1988.

79. M. Gromov, *Stability and pinching*, preprint IHES (1990).

80. M. Gromov, *Hyperbolic manifolds according to Thurston and Jorgensen*, Sém. Bourbaki, 32 année, 1979/80, 40–53.

81. M. Gromov and P. Pansu, *Rigidity of lattices: an introduction*, in: Geometric topology: recent developments, Montecatini Terme, 1990, 39–137.

82. M. Gromov and I. Piatetski-Shapiro, *Non-arithmetic groups in Lobachevsky spaces*, Publ. Math. IHES **66** (1988), 93–103.

83. M. Gromov and R. Schoen, *Harmonic maps into singular spaces and p-adic superrigidity for lattices in groups of rank one*, Publ. Math. IHES **76** (1992), 165–246.

84. M. Gromov and W. P. Thurston, *Pinching constants for hyperbolic manifolds*, In-

vent. Math. **89** (1987), 1–12.

85. V. Guillemin and D. Kazhdan, *Some inverse spectral results for negatively curved 2-manifolds*, Topology **19** (1980), 301–312.

86. V. Guillemin and D. Kazhdan, *Some inverse spectral results for negatively curved n-manifolds*, in : *Geometry of the Laplace operator*, Proc. Sympos. Pure Math. **36**, 1980, AMS, Providence.

87. U. Hamenstadt, *Entropy-rigidity of locally symmetric spaces of negative curvature*, Ann. of Math. **131** (1990), 35–51.

88. Harish Chandra, *Spherical functions on a semisimple Lie group*, *I*, Amer. J. of Math. **80** (1958), 241–310.

89. Harish Chandra, *Discrete series for semisimple groups*, II, Acta Math. **116** (1966), 1–111.

90. P. de la Harpe and A. Valette, *La propriété (T) de Kazhdan pour les groupes localement compacts*, Astérisques **175** (1989), Soc. Math. de France.

91. J. Heber, *On the geometric rank of homogeneous spaces of nonpositive curvature*, Invent. Math. **112** (1993), 151–170.

92. H. Hecht and W. Schmid, *A proof of Blattner's formula* Invent. Math. **31** (1975), 129–154.

93. B. Helffer and F. Nourrigat, *Caracterisation des operateurs hypoelliptiques homogenes invariants à gauche sur un groupe nilpotent gradué*, Comm. in P. D. E. **4** (8) (1979), 899–958.

94. J. Heber, *On the geometric rank of homogeneous spaces of nonpositive curvature*, Invent. Math. **112** (1993), 151–170.

95. L. Hernandez, *Kahler manifolds and 1/4-pinching*, Duke Math. J. **62** (1991), 601–611.

96. L. Hörmander, *The analysis of linear partial differential operators III,IV*, Springer Verlag, Berlin 1985.

97. E. Hopf, *Statistik der geodätischen Linien in Mannigfaltigkeiten negativer Krümmung*, Leipzig Ber. Verhandl. Sächs. Akad. Wiss. **91** (1939), 261–304.

98. R. Howe, *A notion of rank for unitary representations of the classical groups*, in: A. Figà Talamanca (ed.), Harmonic analysis and group representations, CIME 1980.

99. R. Howe and C. C. Moore, *Asymptotic properties of unitary representations*, J. Funct. Anal. **32** (1979), 72–96.

100. B. Hu, *A PL geometric study of algebraic K theory*, TAMS **334** (1992), 783–808.

101. S. Hurder, *Rigidity for Anosov actions of higher rank lattices*, Ann. of Math. **135** (1992), 361–410.

102. A. Iozzi, *Equivariant maps and purely atomic spectrum*, preprint.

103. N. V. Ivanov, *Action of Möbius transformations on homeomorphisms: stability and rigidity*, preprint 1993.

104. J. Jost and S.-T. Yau, *Harmonic maps and superrigidity*, Proc. Symp. Pure Math. **54** (1993), 245–280.

105. P. Julg and A. Valette, *K-theoretic amenability for $SL_2(\mathbb{Q}_p)$, and the action on the associated tree*, J. Funct. Anal. **58** (1984), 194–215.

106. M. Kanai, *Geodesic flows of negatively curved manifolds with smooth stable and unstable foliations*, Erg. Th. and Dyn. Syst. **8** (1988), 215–239.

107. A. Katok, *Entropy and closed geodesics*, Ergod. Th. and Dynam. Syst. **2** (1982), 339–365.

108. A. Katok, *Four applications of conformal equivalence to geometry and dynamics*, Erg. Th. and Dynam. Syst. **8** (19988), 139–152.

109. A. Katok and S. Katok, *Higher cohomology for abelian groups of toral automorphisms*, preprint.

110. A. Katok and J. Lewis, *Local rigidity for certain groups of toral automorphism*, Isr. J. of Math. **75** (1991), 203–241.

111. A. Katok and J. Lewis, *Global rigidity for lattice actions on tori, and new examples*

of volume-preserving actions, preprint.

112. A. Katok, J. Lewis and R. J. Zimmer, *Cocycle superrigidity and rigidity for lattice actions on tori*, preprint.

113. A. Katok and K. Schmidt, *Cohomology of expanding $\mathbb{Z}^d$-actions by compact automorphisms*, preprint.

114. A. Katok and R. J. Spatzier, *First cohomology of Anosov actions of higher rank Abelian groups and applications to rigidity*, to apear in Publ. Math. IHES.

115. A. Katok and R. J. Spatzier, *Differential Rigidity of Hyperbolic Abelian Actions*, MSRI preprint 1992.

116. A. Katok and R. Spatzier, *Subelliptic estimates of polynomial differential operators and applications to rigidity of abelian actions*, Math. Res. Letters **1** (1994), 1–11.

117. A. Katok and R. J. Spatzier, *Invariant measures for higher rank hyperbolic Abelian actions*, preprint.

118. A. Katok and R. J. Spatzier, *Local differential rigidity of hyperbolic abelian actions, their orbit foliations and projective actions of lattices*, preprint in preparation.

119. D. Kazhdan, *Connection of the dual space of a group with the structure of its closed subgroups*, Func. Anal. and Appl. **1** (1967), 63–65.

120. B. Kostant, *On the existence and irreducibility of certain series of representations*, Bull. AMS **75** (1969), 627–642.

121. J. Lewis, *Infinitesimal rigidity for the action of $SL(n, \mathbb{Z})$ on T^n*, TAMS **324** (1991), 421–445.

122. A. Lubotzky and R. J. Zimmer, *Variants of Kazhdan's property for subgroups of semisimple groups*, Israel J. Math. **66** (1989), 289–299.

123. A. Lubotzky and R. J. Zimmer, *Arithmetic structure of fundamental groups and actions of semisimple Lie groups*, preprint.

124. R. Lyons, *On measures simultaneously 2- and 3- invariant*, Israel J. of Math. (1988), 219–224.

125. G. W. Mackey, *Ergodic transformations groups with a pure point spectrum*, Ill. J. Math. **6** (1962), 327–335.

126. V. Makarov, *On a certain class of discrete Lobachevsky space groups with infinite fundamental domain of finite measure*, Soviet Math. Dokl. **7** (1966), 328–331.

127. G. A. Margulis, *Discrete subgroups of semisimple Lie groups*, Springer Verlag, Berlin 1991.

128. G. A. Margulis, *Applications of ergodic theory to the investigation of manifolds of negative curvature*, Funct. Anal. Appl. **3** (1969), 335–336.

129. G. A. Margulis, *Discrete groups of motions of manifolds of non-positive curvature*, Proc. Int. Congress Math. (Vancouver 1974), AMS Transl. **109** (1977), 33–45.

130. G. A. Margulis, *Explicit construction of expanders*, Probl. Inform. Transmission (1975), 325–332.

131. G. A. Margulis, *Arithmeticity of irreducible lattices in semisimple groups of rank greater than 1*, appendix to the Russian translation of M. Raghunathan, *Discrete subgroups of Lie groups*, Mir, Moscow 1977 (in Russian). Invent. Math. **76** (1984), 93–120.

132. G. A. Margulis, *Finitely-additive invariant measures on Euclidean spaces*, Ergod. Theory and Dynam. Syst. **2** (1983), 383–396.

133. G. A. Margulis, *Discrete subgroups and ergodic theory*, Proc. Conf. in honor of A. Selberg, 1989, Oslo.

134. G. A. Margulis and G. M. Tomanov, *Measure rigidity for algebraic groups over local fields*, C. R. Acad. Sci. Paris **315**, Sér. 1 (1992), 1221–1226.

135. G. A. Margulis and G. M. Tomanov, *Invariant measures for actions of unipotent groups over local fields on homogeneous spaces*, Invent. Math. **116** (1994), 347–392.

136. Y. Matsushima and S. Murakami, *On vector bundle-valued harmonic forms and automorphic forms on symmetric Riemannian manifolds*, Ann. of Math. **78** (1963), 361–416.

202   R. J. SPATZIER

137. F. I. Mautner, *Geodesic flows on symmetric Riemannian spaces*, Ann. of Math. **65** (1957), 416–431.
138. G. Metivier, *Fonction spectrale et valeurs propres d'une classe d'operateurs non elliptiques*, Comm. in P. D. E. **1** (5) (1976), 467–519.
139. N. Mok, *Topics in complex differential geometry*, in: *Recent topics in differential and analytic geometry*, 1–141, Academic Press, Boston, MA, 1990.
140. N. Mok, *Aspects of Kahler geometry on arithmetic varieties*, in: *Several complex variables and complex geometry*, Part 2, 335–396, Santa Cruz, California, 1989.
141. N. Mok, Y.-T. Siu and S.-K. Yeung, *Geometric superrigidity*, Invent. Math. **113** (1993), 57–83.
142. C. C. Moore, *Ergodicity of flows on homogeneous spaces*, Amer. J. Math. **88**, (1966), 154–178.
143. C. C. Moore, *The Mautner phenomenon for general unitary representations*, Pacific J. Math. **86** (1980), 155–169.
144. C. C. Moore, *Exponential decay of correlation coefficients for geodesic flows*, in: C. C. Moore (ed.), Group representations, ergodic Theory, operator algebras, and mathematical physics, Proceedings of a conference in Honor of George Mackey, MSRI publications, Springer Verlag 1987.
145. G. D. Mostow, *Quasi-conformal mappings in n-space and the rigidity of the hyperbolic space forms*, Publ. Math. IHES **34** (1968), 53–104.
146. G. D. Mostow, *Strong rigidity of locally symmetric spaces*, Ann. Math. Studies, No. 78, Princeton University Press, Princeton, N.J. 1973.
147. G. D. Mostow, *Existence of a non-arithmetic lattice in SU(2,1)*, Proc. Nat. Acad. Sci. USA **75** (1978), 3029–3033.
148. G. D. Mostow, *On a remarkable class of polyhedra in complex hyperbolic space*, Pac. J. of Math. **86** (1980), 171–276.
149. G. D. Mostow, *A compact Kähler surface of negative curvature not covered by the ball*, Ann. Math. **112** (1980), 321–360.
150. G. D. Mostow, *On discontinuous action of monodromy groups on the complex n-ball*, JAMS **1** (1988), 555–586.
151. G. D. Mostow and Y. T. Siu, *A compact Kähler surface of negative curvature not covered by the ball*, Ann. Math. **112** (1980), 321–360.
152. P. Ontaneda, thesis, Stony Brook 1994.
153. J.-P. Otal, *Le spectre marque des longueurs des surfaces a courbure negative*, Ann. of Math. **131** (1990), 151–162.
154. J. Palis and J. C. Yoccoz, Rigidity of centralizers of diffeomorphisms, Ann. scient. Éc. Norm. Sup. **22** (1989), 81–98.
155. J. Palis and J. C. Yoccoz, Centralizers of Anosov diffeomorphisms on tori, Ann. scient. Éc. Norm. Sup. **22** (1989), 99–108.
156. P. Pansu, *Metriques de Carnot-Caratheodory et quasiisometries des espaces symetriques de rang un*, Ann. of Math. **129** (1989), 1–60.
157. P. Pansu, *Dimension conforme et sphere a l'infini des variétés a courbure negative*, Ann. Acad. Sci. Fenn. Ser. A I Math. **14** (1989), 177–212.
158. W. Parry, *Ergodic properties of affine transformations and flows on nilmanifolds*, Amer. J. Math. **91** (1969), 757–771.
159. G. Prasad, *Strong rigidity of Q-rank 1 lattices*, Invent. Math. **21** (1973), 255–286.
160. G. Prasad and M. S. Rahunathan, *Cartan subgroups and lattices in semi-simple groups*, Ann. of Math. **96** (1972), 296–317.
161. N. Qian, *Anosov automorphisms for nilmanifolds and rigidity of group actions*, preprint.
162. N. Qian, *Topological deformation rigidity of actions of higher rank lattices on tori*, preprint.
163. N. Qian, *Local rigidity of Anosov SL(n, ℤ)-actions on tori*, peprint.
164. N. Qian, personal communication

165. M. S. Raghunathan, *Discrete subgroups of Lie groups*, Springer Verlag, New York, 1972.
166. M. S. Raghunathan, *On the first cohomology of discrete subgroups of semisimple Lie groups*, Amer. J. Math. **87** (1965), 102–139.
167. M. S. Raghunathan, *Vanishing theorems for cohomology groups associated to discrete subgroups of semi-simple Lie groups*, Osaka J. Math. **3** (1966), 243–256. Corrections, Osaka J. Math. **16** (1979), 295–299.
168. M. S. Raghunathan, *Cohomology of arithmetic subgroups of algebraic groups* I, II, Ann. of Math. **86** (1967), 409–424 and **87** (1968), 279–304.
169. M. Ratner, *Rigidity of horocycle flows*, Ann. of Math. **115** (1982), 587–614.
170. M. Ratner, *Horocycle flows, joinings and rigidity of products*, Ann. of Math. **118** (1983), 277–313.
171. M. Ratner, *The rate of mixing for geodesic and horocycle flows*, Ergod. Th. and Dyn. Syst. **7** (1987), 267–288.
172. M. Ratner, *On Raghunathan's measure conjecture*, Ann. of Math. **134** (1991), 545–607.
173. M. Ratner, *Raghunathan's topological conjecture and distribution of unipotent flows*, Duke Math. J. **63** (1991), 235–280.
174. M. Ratner, *Invariant measures and orbit closures for unipotent actions on homogeneous spaces*, Publ. Math. Sci. Res. Inst.), Berlin, Heidelberg, New York: Springer (to appear).
175. M. Ratner, *Raghunathan's conjectures for p-adic Lie groups*, Int. Math. Research Notices **5** (1993), 141–146 (in Duke Math. J. 70:2).
176. M. Rees, *Some $\mathbb{R}^2$-Anosov flows*, preprint.
177. C. Rockland, *Hypoellipticity on the Heisenberg group: representation-theoretic criteria*, T.A.M.S. **240** (517) (1978), 1–52.
178. L. P. Rothschild, *A criterion for Hypoellipticity of operators constructed from vectorfields*, Comm. in P. D. E. **4** (6) (1979), 645–699.
179. L. P. Rothschild and E. M. Stein, *Hypoelliptic differential operators and nilpotent groups*, Acta Math. **137** (1976), 247–320.
180. R. Sacksteder, *Semigroups of expanding maps*, TAMS 221 (1976), 281–288.
181. J. Sampson, Applications of harmonic maps to Kähler geometry, Contemp. Math. **49** (1986), 125–133.
182. W. Schmid, *Some properties of square-integrable representations of semisimple Lie groups*, Ann. of Math. **102** (1975), 535–564.
183. K. Schmidt, *Amenability, Kazhdan's property (T), strong ergodicity and invariant means for ergodic group actions*, Erg. Th. Dynam. Syst. 1 (1981), 223-246.
184. A. Selberg, *On discontinuous groups in higher dimensional symmetric spaces*, International Colloquium on Function Theory, Tata Inst. of Fund. Research (Bombay) 1960, 147–164.
185. T. Sherman, *A weight theory for unitary representations*, Canadian J. Math. **18** (1966), 159–168.
186. Y. T. Siu, *The complex-analyticity of harmonic maps and the strong rigidity of compact Kähler manifolds*, Ann. of Math. **112** (1980), 73–112.
187. R. J. Spatzier, *An example of an amenable action from geometry*, Erg. Th. and Dynam. Syst. **7** (1987), 289–293.
188. R. J. Spatzier, *On isospectral locally symmetric spaces and a theorem of von Neumann*, Duke J. of Mathematics **59** (1989), 289–294, Errata, **60** (1990), 561.
189. R. J. Spatzier and R. J. Zimmer, *Fundamental groups of negatively curved manifolds and actions of semisimple groups*, Topology **30** (1991), 591–601.
190. A. M. Stepin, *Dynamical systems on homogeneous spaces of semi-simple Lie groups*, Izv. Akad. Nauk, SSSR **37** (1973), 1091–1107.
191. G. Stuck, *Cocycles of ergodic group actions and vanishing of first cohomology for S-arithmetic groups*, Amer. J. Math. **113** (1991), 1–23.

192. D. Sullivan, *On the ergodic theory at infinity of an arbitrary discrete group of hyperbolic motions*, Annals of Math. Studies, No. 91, 1978, 465-496.
193. D. Sullivan, *Discrete conformal groups and measurable dynamics*, Bull. AMS **6** (1) (1982), 57-73.
194. D. Sullivan, *For n > 3 there is only one finitely additive rotationally invariant measure on the n-sphere defined on all Lebesgue measurable subsets*, Bull. AMS **4** (1981), 121-123.
195. E. Thoma, *Über unitäre Darstellungen abzählbarer, diskreter Gruppen*, Math. Ann. **153** (1964), 111-138.
196. W. Thurston, *The geometry and topology of 3-manifolds*, mimeographed notes, Princeton Univeristy, 1977.
197. J. Tits, *Buildings of spherical type and finite BN-pairs*, Springer Lect. Notes in Math. **386** (1970).
198. P. Tukia, *Differentiability and rigidity of Möbius groups*, Invent. Math. **82** (1985), 557-578.
199. W. Veech, *Periodic points and invariant pseudomeasures for toral endomorphisms*, Erg. Th. and Dynam. Syst. **6** (3) (1986), 449-473.
200. M. Ville, *On $\frac{1}{4}$-pinched 4-dimensional Riemannian manifolds of negative curvature*, Ann. Global Anal. Geom. **3** (1985), 329-336.
201. E. Vinberg, *Discrete groups generated by reflections in Lobachevsky spaces*, Math. USSR-Sb. **1** (1967), 429-444.
202. S. P. Wang, *On isolated points in the dual spaces of locally compact groups*, Math. Ann. **218** (1975), 19-34.
203. G. Warner, *Harmonic Analysis on semisimple Lie groups I*, Springer Verlag, Berlin 1972.
204. J. von Neumann, *Zur Operatorenmethode in der klassischen Mechanik*, Ann. Math. **33** (1932), 587-642.
205. A. Weil, *On discrete subgroups of Lie groups I*, Ann. Math. **72** (1960), 369-384.
206. A. Weil, *On discrete subgroups of Lie groups II*, Ann. Math. **75** (1962), 578-602.
207. D. Witte, *Rigidity of horospherical foliations*, Erg. Th. and Dyn. Syst. **9** (1989), 191-205.
208. J. Wolf, *Discrete groups, symmetric spaces, and global holonomy*, Amer. J. Math. **84** (1962), 527-542.
209. S. T. Yau, F. Zheng, *Negatively $\frac{1}{4}$-pinched Riemannian metric on a compact Kähler manifold*, Invent. Math. **103** (1991), 527-535.
210. C. Yue, *On the Sullivan conjecture*, Random Comput. Dynamics **1** (1992/93), 131-145.
211. C. Yue, *On Green's conjecture*, preprint.
212. R. J. Zimmer, *Orbit spaces of unitary representations, ergodic theory, and simple Lie groups*, Ann. of Math. **106**, (1977), 573-588.
213. R. J. Zimmer, *Amenable ergodic group actions and an application to Poisson boundaries of random walks*, J. Func. Anal. **27** (1978), 350-372.
214. R. J. Zimmer, *Induced and amenable actions of Lie groups*, Ann. Sci. Ec. Norm. Sup.**11** (1978), 407-428.
215. R. J. Zimmer, *Uniform subgroups and ergodic actions of exponential Lie groups*, Pac. J. Math. **78** (1978), 267-272.
216. R. J. Zimmer, *Strong rigidity for ergodic actions of semisimple Lie groups*, Ann. of Math. **112** (1980), 511-529.
217. R. J. Zimmer, *On the cohomology of ergodic actions of semisimple Lie groups and discrete subgroups*, Amer. J. Math. **103** (1981), 937-950.
218. R. J. Zimmer, *Ergodic theory, semisimple groups and foliations by manifolds of negative curvature*, Publ. Math. IHES **55** (1982), 37-62.
219. R. J. Zimmer, *Curvature of leaves in amenable foliations*, Amer. J. Math. **105** (1983), 1011-1022.

220. R. J. Zimmer, *Ergodic theory and semisimple groups*, Boston: Birkhäuser 1984
221. R. J. Zimmer, *Kazhdan groups acting on compact manifolds*, Invent. Math. **75** (1984), 425–436.
222. R. J. Zimmer, *Lattices in semisimple groups and distal geometric structures*, Invent. Math. **80** (1985), 123–137.
223. R. J. Zimmer, *Actions of semisimple groups and discrete subgroups*, Proceedings of the ICM, (1986: Berkeley, CA), A.M.S., Providence, RI, 1987, 1247–1258.
224. R. J. Zimmer, *On the automorphism group of a compact Lorentz manifold and other geometric manifolds*, Invent. Math. **83** (1986), 411–424.
225. R. J. Zimmer, *Lattices in semisimple groups and invariant geometric structures on compact manifolds*, in Discrete groups in geometry and analysis, R. Howe (ed.), Birkhäuser (1987), 152–210.
226. R. J. Zimmer, *On the algebraic hull of an automorphism group of a principal bundle*, Comment. Math. Helvetici **65** (1990), 375-387.
227. R. J. Zimmer, *Infinitesimal rigidity of discrete subgroups of Lie groups*, J. Diff, Geom. **31** (1990), 301–322.
228. R. J. Zimmer, *Spectrum, entropy, and geometric structures for smooth actions of Kazhdan groups*, Israel J. Math. **75** (1991), 65–80.
229. R. J. Zimmer, *Superrigidity, Ratner's theorem, and fundamental groups*, Israel J. Math. **74** (1991), 199–207.

DEPARTMENT OF MATHEMATICS, UNIVERSITY OF MICHIGAN, ANN ARBOR, MI 48103
E-mail address: spatzier@math.lsa.umich.edu

SOME PROPERTIES AND APPLICATIONS
OF JOININGS IN ERGODIC THEORY

J.-P. THOUVENOT

0. INTRODUCTION

We will be dealing with actions of groups on Lebesgue spaces. That is, given a locally compact group G and a Lebesgue space $(X, \mathcal{A}, m)$, a G-action is a measurable mapping $G \times X \to X$, $(g, x) \to g(x)$, such that for all g, $h \in G$, $g(h(x)) = gh(x)$ and $ex = x$ for almost all $x \in X$ (where e is the identity in G). Furthermore, $x \to g(x)$ is measure-preserving for every $g \in G$. In case the group G is continuous we will assume the following: for every $A \in \mathcal{A}$, $\forall \epsilon > 0$ there exists $V(e)$ (a neighborhood of the identity in G) such that $m(A \triangle gA) < \epsilon$ for every g in $V(e)$. We will be mainly concerned with $G = \mathbb{Z}$ or $G = \mathbb{R}$.

A *joining* of two actions of the same group $((X_1, \mathcal{A}_1, m_1, G)$ and $(X_2, \mathcal{A}_2, m_2, G))$ is a probability measure λ on $(X_1 \times X_2, \mathcal{A}_1 \times \mathcal{A}_2)$ which is invariant under the diagonal action of G $(g(x_1, x_2) = (g(x_1), g(x_2)))$ and whose marginals on $\mathcal{A}_1 \times X_2$ and $X_1 \times \mathcal{A}_2$ are m_1 and m_2 respectively (i.e., if $A_1 \in \mathcal{A}_1$, $\lambda(A_1 \times X_2) = m_1(A_1)$; and if $A_2 \in \mathcal{A}_2$, $\lambda(X_1 \times A_2) = m_2(A_2)$). The set of joinings is never empty. (Take $\lambda = m_1 \otimes m_2$.) If we call $\mathcal{V} = \{A \times X_2, A \in \mathcal{A}_1\}$ and $\mathcal{H} = \{X_1 \times A, A \in \mathcal{A}_2\}$, we say that $\mathcal{V} \subset \mathcal{H}$ (λ) if for every set $\tilde{A}$ in $\mathcal{V}$ there is a set $\tilde{B}$ in $\mathcal{H}$ such that $\lambda(\tilde{A} \triangle \tilde{B}) = 0$. As $(X_1, \mathcal{A}_1, m_1)$ and $(X_2, \mathcal{A}_2, m_2)$ are Lebesgue spaces, a joining λ of two actions such that $\mathcal{V} = \mathcal{H}$ (λ) defines an isomorphism of the two actions. If $(X_i, \mathcal{A}_i, m_i, G)$, $1 \leq i \leq 2$, are two copies of the same action, a joining λ such that $\mathcal{V} = \mathcal{H}$ (λ) is exactly (Id $\times S)\Delta$, where Δ is the diagonal joining $\Delta(A_1 \times A_2) = m(A_1 \cap A_2)$ and S is an invertible element of the centralizer (the set of measure-preserving transformations commuting with the action of every g in G). As we deal with Lebesgue spaces, a joining λ of two ergodic actions $(X_1, \mathcal{A}_1, m_1, G)$ and $(X_2, \mathcal{A}_2, m_2, G)$ has the property that there exists a Lebesgue space Ω and a probability P on Ω such that $\lambda = \int_{\Omega} \lambda(\omega) dP(\omega)$, where every $\lambda(\omega)$ is ergodic. (This is just the ergodic decompostion of λ, and as the marginals of λ are ergodic a.e., λ_ω is a joining). Therefore the set of *ergodic* joinings is never empty.

As a consequence of the previous observations, let us describe the proof (due to Mariusz Lemanczyk) of the fact that two ergodic dynamical systems with discrete spectrum which are spectrally isomorphic are isomorphic. In

fact, let us take an ergodic joining λ of two such systems. We know by ergodicity that every proper value (in the joining) has only (up to a constant)
one eigenfunction (which is therefore both $\mathcal{V}$ and $\mathcal{H}$ measurable). As the
linear span of the eigenfunctions is the whole L^2 for both actions, we have
that $\mathcal{V} = \mathcal{H}$ and λ is an isomorphism.

Given two actions $(X_i, \mathcal{A}_i, m_i, G)$, $1 \leq i \leq 2$, the set of joinings is
topologized as follows: Choose A_n, $n \geq 1$ in $\mathcal{A}_1$ and B_m, $m \geq 1$ in $\mathcal{A}_2$, two
countable families of sets which are dense in $\mathcal{A}_1$ and $\mathcal{A}_2$ respectively for the
metrics $d_i(A, B) = m_i(A \triangle B)$, $1 \leq i \leq 2$, $A, B \in \mathcal{A}_i$. λ and λ' being two
joinings, we define

$$d(\lambda, \lambda') = \sum_{n,m \geq 0} \frac{1}{2^{n+m}} |\lambda(A_n \times B_m) - \lambda'(A_n \times B_m)|.$$

Endowed with this topology, to which we will refer as the weak topology,
the set of joinings is compact. We shall frequently be using the relatively
independent joining, which we define here for $\mathbb{Z}$ actions. Assume that two
dynamical systems $(X, \mathcal{A}, m, T)$ and $(X', \mathcal{A}', m', T')$ both have as a factor
the system $(Y, \mathcal{B}, \mu, S)$. (That is, there are given two maps $\varphi : X \to Y$ and
$\varphi' : X' \to Y'$ such that $\varphi^* m = \mu$, $\varphi'^* m' = \mu$, $S\varphi = \varphi T$ a.e., $S\varphi' = \varphi' T'$
a.e.) For $\mathcal{E} = \varphi^{-1}(\mathcal{B})$ we identify every $\mathcal{E}$-measurable function g on X to a
$\mathcal{B}$-measurable function $\tilde{g}$ on Y by $\tilde{g}(y) = g(\varphi^{-1}(y))$. Let $\mathcal{E}' = \varphi'^{-1}(\mathcal{B})$. The
relatively independent joining λ of $(X, \mathcal{A}, m, T)$ with $(X', \mathcal{A}', m', T')$ above
$(Y, \mathcal{B}, \mu, S)$, given φ and φ', is defined as follows:

$$\lambda(A \times B) = \int \tilde{E}^{\mathcal{E}} 1_A \tilde{E}^{\mathcal{E}'} 1_B d\mu, \qquad A \in \mathcal{A}, \ B \in \mathcal{B}.$$

It will be very frequent that $(X, \mathcal{A}, m, T)$ and $(X', \mathcal{A}', m', T')$ will be given
as products

$$((X, \mathcal{A}, m, T) = (X_1, \mathcal{A}_1, m_1, T_1) \times (Y, \mathcal{B}, \mu, S);$$
$$(X', \mathcal{A}', m', T') = (X_2, \mathcal{A}_2, m_2, T_2) \times (Y, \mathcal{B}, \mu, S)).$$

In that case the relatively independent joining of (X, T) and (X', T') above
(Y, S) will mean that in the above definition φ and φ' are taken to be the
projections $\varphi(x_1, y) = y$ and $\varphi'(x_2, y) = y$.

Historically, joinings were introduced by H. Furstenberg in his paper
[Furstenberg, 1967] on disjointness. In particular, he defined the important
notion of disjointness in the following way: T and S are *disjoint* if the
only joining between them is the product joining. We are going to give an
account of (some of) those results in ergodic theory in which joinings occur
in a rigid, algebraic way, of which disjointness is an example.

At the opposite end, joinings can be used to describe the isomorphism
theory of D. Ornstein. Most convenient and attractive in this framework

is their flexibility. The Ornstein isomorphism theorem can be stated in the following way: Given two Bernoulli shifts with the same entropy, the set of joinings between them which are isomorphisms is a dense G_δ.

All proofs of results alluded to in this Introduction can be found in [Rudolph, 1990].

I am very much indebted to the unknown referees for fruitful criticisms, suggestions, and comments, and to Karl Petersen for his most communicative enthusiasm and energy, without the help of which this would never have been written.

I. GENERAL STUDY OF JOININGS

Lemma 1. *If* $(X, \mathcal{A}, m, T)$ *and* $(Y, \mathcal{B}, \mu, S)$ *are two measure-preserving transformations with singular spectral measures, they are disjoint.*

Proof. In $X \times Y$ endowed with a joining measure λ, consider $f \in L^2(X)$ and $g \in L^2(Y)$ and consider H_f the L^2 closure of the linear span of the functions $T^n f - \int f dm$, $n \in \mathbb{Z}$. The projection of g on H_f (within $L^2(X \times Y, d\lambda)$) will have a spectral measure absolutely continuous with respect to the spectral type of T and thus has to be 0. Therefore $g \perp (f - \int f dm)$ in $L^2(d\lambda)$, and $\int f g d\lambda = \int f dm \int g d\mu$.

A nice and important application, which will be used in the sequel, is the following.

Corollary 1. *The identity transformation is disjoint from any ergodic transformation.*

As another nice application to Lemma 1, the reader may prove that if $(X, \mathcal{A}, m, T)$ is an ergodic transformation with simple spectrum, its factor algebras are reticulated, that is, if $\mathcal{B}_1$ and $\mathcal{B}_2$ are any two factor algebras $\mathcal{B}_1 \overset{\mathcal{B}_1 \wedge \mathcal{B}_2}{\perp} \mathcal{B}_2$. (In particular, if $\mathcal{B}_1$ and $\mathcal{B}_2$ have no sets in common, $\mathcal{B}_1 \perp \mathcal{B}_2$.)

The second example is the following.

Lemma 2. *If* T *and* S *are such that the entropy* $E(T) = 0$ *and* S *is a* K-*automorphism, then* S *and* T *are disjoint.*

We will not give the proof of this Lemma, as this lecture is doomed to be entropy-free (In fact all the results which we are going to mention, all the examples which we are going to consider, with only one exception, will remain inside the class of zero-entropy transformations.) We are now going to give two examples of results which are easily obtained using joinings.

Lemma 3. *Let* $(X, \mathcal{A}, m, T)$ *be a mixing transformation and let* $(\widehat{X}, \widehat{\mathcal{A}}, \widehat{m}, \widehat{T})$ *be a finite extension of it (i.e., there exists* $k \in \mathbb{N}$ *and a measurable mapping* Φ *from* X *to* S_k *((the group of permutations of* $[1, k]$*)) such that* $\widehat{X} = X \times [1, k]$, $\widehat{T}(x, i) = (Tx, \Phi(x)i)$, $\widehat{m} = m \otimes \delta$ *where* δ, *is the normalized counting measure on* $[1, k]$*).*

If $\widehat{T}$ *is weakly mixing, then* $\widehat{T}$ *is mixing.*

Proof. We consider the joining of $(\widehat{X}, \widehat{\mathcal{A}}, \widehat{m}, \widehat{T})$ with itself which is the product measure $\widehat{m} \otimes \widehat{m}$. $\widehat{T} \times \widehat{T}$ is ergodic for this measure, as $\widehat{T}$ was assumed weakly mixing. An interesting measure when we join, as here, two copies of the same transformation, is the diagonal measure Δ, defined by $\Delta(A \times B) = \widehat{m}(A \cap B)$. It is also clear that $\mathrm{Id} \times \widehat{T}^n(\Delta) = \Delta_n$ is a joining for all $n \in \mathbb{Z}$. If $\widehat{T}$ were not mixing, we could find a subsequence n_i such that $\Delta_{n_i} \to \lambda$ in the weak sense, with λ a joining different from $\widehat{m} \otimes \widehat{m}$. However, for all sets A, B in $\widehat{X}$,

$$\varlimsup_{n \to \infty} \widehat{m}(A \cap \widehat{T}^n B) < k\widehat{m}(A)\widehat{m}(B).$$

Therefore, λ has to be absolutely continuous with respect to $\widehat{m} \otimes \widehat{m}$ (with Radon-Nikodym derivative bounded by k), and since this last measure is ergodic, $\lambda = \widehat{m} \otimes \widehat{m}$, a contradiction.

The original proof of this fact is taken from [Ornstein, 1967]. The interested reader may try to prove the generalization of this result to isometric extensions (see [Rudolph, 1985]) and to a proof of the fact that mixing of order n in the base lifts to $\widehat{T}$.

As a second example of an application of joinings, we shall give a proof of the fact that the horocycle flow is mixing of all orders. In fact the result is more general and deals with actions of $\mathrm{SL}(2, \mathbb{R})$.

Theorem 1. *Let us consider an action of the group* $\mathrm{SL}(2, \mathbb{R})$ *by measure-preserving transformations of the space* $(X, \mathcal{A}, m)$. *If this action is ergodic, then the actions of the group* G_t, *induced by the elements* $\begin{pmatrix} e^t & 0 \\ 0 & e^{-t} \end{pmatrix}$, $t \in \mathbb{R}$, *and* H_s, *induced by the elements* $\begin{pmatrix} 1 & s \\ 0 & 1 \end{pmatrix}$, $s \in \mathbb{R}$, *are mixing. The action of* H_s *is mixing of all orders.*

Proof. The action of G_t alone is ergodic, because if $x \in L^2(X, \mathcal{A}, m)$ is G_t invariant ($G_t x = x$ a.e. $\forall t \in \mathbb{R}$), it is also H_s invariant. In fact, fixing $s \in \mathbb{R}$,

$$\int H_s x \cdot x \, dm = (H_s x | x) = (H_s G_t x | G_t x) = (G_{-t} H_s G_t x | x)$$

$$= (H_{se^{-2t}} x | x) \qquad \forall t \in \mathbb{R}.$$

(We used the commutation relation $\begin{pmatrix} e^t & 0 \\ 0 & e^{-t} \end{pmatrix} \begin{pmatrix} 1 & s \\ 0 & 1 \end{pmatrix} \begin{pmatrix} e^{-t} & 0 \\ 0 & e^t \end{pmatrix} = \begin{pmatrix} 1 & se^{2t} \\ 0 & 1 \end{pmatrix}$.)

When $t \to +\infty$, $(H_{se^{-2t}} x | x)$ goes to $\|x\|^2$, and therefore $(H_s x | x) = \|x\|^2$, hence $H_s x = x$; the same will be true for the action of $\widetilde{H}_s$ induced by the elements $\begin{pmatrix} 1 & 0 \\ s & 1 \end{pmatrix}$. The relation

$$\begin{pmatrix} 1 & s \\ 0 & 1 \end{pmatrix} \begin{pmatrix} 1 & 0 \\ a & 1 \end{pmatrix} \begin{pmatrix} 1 & t \\ 0 & 1 \end{pmatrix} = \begin{pmatrix} 1+as & t+s(at+1) \\ a & at+1 \end{pmatrix} \qquad (*)$$

implies (letting $as = -2$, $at = -2$, where $a \to 0$) that if $\tilde{I}$ is the action of $\begin{pmatrix} -1 & 0 \\ 0 & -1 \end{pmatrix}$, $\tilde{I}x = x$. This proves the ergodicity of G_t. To prove that H_s is ergodic, we now only need to show that if x is H_s invariant, it is also G_t invariant.

Let $\epsilon > 0$ be given, and fix $t_0 \in \mathbb{R}$.

$$(\tilde{H}_a x | x) = (\tilde{H}_a H_t x | H_{-s} x) = (H_s \tilde{H}_a H_t x | x), \qquad \text{for all } a, s, t \in \mathbb{R}^3.$$

Choose a small enough so that

$$\left| (\tilde{H}_a x | x) - \|x\|^2 \right| < \epsilon \quad \text{and} \quad \left\| \begin{pmatrix} e^{t_0} & 0 \\ a & e^{-t_0} \end{pmatrix} x - G_{t_0} x \right\| < \epsilon.$$

(By $\begin{pmatrix} e^{t_0} & 0 \\ a & e^{-t_0} \end{pmatrix} x$ we mean the image of x under the action induced by the element $\begin{pmatrix} e^{t_0} & 0 \\ a & e^{-t_0} \end{pmatrix}$.) We can find, using the relation $*$, s and t such that $H_s \tilde{H}_a H_t = \begin{pmatrix} e^{t_0} & 0 \\ a & e^{-t_0} \end{pmatrix}$, and therefore

$$\left| (G_{t_0} x | x) - \|x\|^2 \right| < 2\epsilon,$$

and x is G_t invariant.

We now prove, using an argument in Guivarch [1992], that G_t is mixing. Let $s \neq 0$ be fixed. For all x, y in $L^2(X, \mathcal{A}, m)$,

$$(G_t(H_s x - x)|y) = (G_t H_s G_{-t} G_t x - G_t x | y) = (G_t x | G_t H_{-s} G_t y) - (G_t x | y)$$

and $\lim t \to -\infty$ $(G_t(H_s x - x)|y) = 0$ (as $\|y - G_t H_{-s} G_t y\| \to 0$ when $t \to -\infty$). Therefore, as H_s is ergodic, the linear combinations of vectors $H_s x - x$ are dense in $L_0^2(X, \mathcal{A}, m)$ (the subspace of functions with 0 integral), and for all $z \in L_0^2(X, \mathcal{A}, m)$, $\lim_{t \to -\infty} (G_t z | y) = 0$. This gives the announced mixing.

The original proof of the fact that the horocycle flow is mixing of all orders is due to B. Marcus [1978]. Simpler proofs using joinings have been given independently by S. Mozes [1992] (where a more general result is shown) and V. Ryzhikov [1991]. We follow here Ryzhikov.

Let I be an interval and $A_0, A_1, \ldots, A_k$, $k > 1$, measurable sets in $\mathcal{A}$. Let τ_i^r, $i \in \mathbb{N}$, $1 \leq r \leq k$, be positive reals such that

$$\lim_{i \to +\infty} |\tau_i^r - \tau_i^s| = +\infty \text{ when } r \neq s.$$

Then

$$\lim_{i \to +\infty} \frac{1}{|I|} \int_I m \left(A_0 \cap H_{t\tau_i^1} A_1 \cap H_{t\tau_i^2} A_2 \cap \cdots \cap H_{t\tau_i^k} A_k \right) dt$$

$$= \prod_{i=0}^{k} m(A_i). \quad (**)$$

We proceed to prove (∗∗). Let $\tau_i = \max_r \tau_i^r$, $\lambda_i^r = \tau_i^r / \tau_i$. We can assume, by taking a subsequence if necessary, that $\lambda_i^r \to \bar{\lambda}^r$, $1 \le r \le k$. For every $i \in \mathbf{N}$, the integral on the left of the relation (∗∗) gives rise to a joining ν_i of $(k+1)$ copies of the action of H_s on $(X, \mathcal{A}, m)$. Using another subsequence, we may assume that ν_i converges to ν, which has to be invariant under the action of $\mathrm{Id} \times H_{\bar{\lambda}^1} \times H_{\bar{\lambda}^2} \times \cdots \times H_{\bar{\lambda}^k}$. (Writing

$$\nu_i(A_0 \times H_{\bar{\lambda}^1} A_1 \cdots \times H_{\bar{\lambda}_k} A_k)$$
$$= \frac{1}{|I|} \int_I m(A_0 \cap H_{\tau_i^1(t + \frac{\bar{\lambda}^1}{\tau_i^1})}) A_1 \cdots \cap H_{\tau_i^k(t + \frac{\bar{\lambda}^k}{\tau_i^k})} A_k) dt,$$

this differs as $i \to +\infty$ by less and less from

$$\frac{1}{|I|} \int_I m(A_0 \cap H_{\tau_i^1(t + \frac{\bar{\lambda}^1}{\tau_i^1})}) A_1 \cdots \cap H_{\tau_i^k(t + \frac{1}{\tau_i})} A_k) dt,$$

which in turn converges to $\nu(A_0 \times A_1 \times \cdots \times A_k)$.)

(A) In the case when $k = 1$, since $H_{\bar{\lambda}^1}$ is ergodic (if H_{s_0} is ergodic for some s_0, H_s has to be ergodic for all s, as H_s is conjugate to H_{s_0} and the global ergodicity of the flow forces the existence of ergodic times), $\nu = m \otimes m$ as an application of Corollary 1. The mixing of the horocycle flow follows now by contradiction: If it were not mixing, one could find sets A_0, A_1 and a sequence $\tau_n \to +\infty$ and $\epsilon > 0$ such that

$$(\mathbf{1}_{A_0} \mid H_{\tau_n} \mathbf{1}_{A_1}) > m(A_0) m(A_1) + \epsilon.$$

For $|u| < \eta$, $(G_{-u} \mathbf{1}_{A_0} \mid H_{\tau_n} G_{-u} \mathbf{1}_{A_1}) > m(A_0) m(A_1) + \frac{\epsilon}{2}$, and therefore, for $I = [e^{-2\eta}, e^{2\eta}]$, we would have $\frac{1}{|I|} \int_I m(A_0 \cap H_{\tau_n t} A_1) dt > m(A_0) m(A_1) + \frac{\epsilon}{2}$, yielding the announced contradiction.

(B) An induction argument, based on the fact that H_t is mixing, hence weakly mixing (whence ergodicity of the direct product actions) and repeated applications of Corollary 1, give that for all $k \ge 1$ the limit ν satisfies

$$\nu(A_0 \times A_1 \cdots \times A_k) = \prod_{i=0}^{k} m(A_i).$$

(In fact (∗∗) is true for all weakly mixing actions.)

As before, if H_t were not mixing of order k, one could find sets A_i, $0 \le i \le k$, and positive reals τ_n^i, $0 \le i \le k$, and $\epsilon > 0$ such that $|\tau_n^i - \tau_n^j| \to +\infty$, $i \le j$, and $m(A_0 \cap H_{\tau_n^1} A_1 \cap \ldots H_{\tau_n^k} A_k) - \prod_{i=0}^{k} m(A_i) > \epsilon$ for all n (for instance). For $|u| < \eta$,

$$m(G_{-u} A_0 \cap H_{\tau_n^1} G_{-u} A_1 \cap \cdots \cap H_{\tau_n^k} G_{-u} A_k) > \frac{\epsilon}{2} + \prod_{i=0}^{k} m(A_i),$$

and for $I = [e^{-2\eta}, e^{2\eta}]$,

$$\frac{1}{|I|} \int_I m\left(A_0 \cap H_{\tau_n^1 t} A_1 \cdots \cap H_{\tau_n^k t} A_k\right) dt > \frac{\epsilon}{2} + \prod_{i=0}^{k} m(A_i) \quad \text{for all } n,$$

a contradiction.

It was Rudolph [1979] who discovered that it could be of interest to study the joinings of a transformation with itself. He first defined a property, *minimal self joinings*, which basically says that the transformation has no other joinings than the obvious ones. He proved the existence of transformations satisfying this property, out of which many striking examples can be produced. Weaker notions were introduced later on (*simplicity* by Veech [1982], *finite joining rank* by King [1988], *semi-simplicity* by del Junco, Lemanczyk and Mentzen [1993]) as well as extensions to flows by Ratner [1982 a and b], which could be applied to several classes of transformations. All these notions are based on the fact that when the self joinings of a transformation are not too abundant and are well enough described, substantial information can be given about the transformation, especially about the structure of its factors, its centralizer, and also on the way it can be joined to another transformation. On the other hand, the joinings of a transformation with strictly positive entropy are extremely abundant; in fact, every transformation of strictly positive entropy can be realized as the joining of three Bernoulli shifts (see [Smorodinsky-Thouvenot, 1979]).

Definition 1. An ergodic transformation $(X, \mathcal{A}, m, T)$ has *minimal self joinings* (M.S.J.) if the following is true:

For all $n \geq 2$, an ergodic joining λ of the product of n copies of $(X, \mathcal{A}, m, T)$ (λ is a probability measure on $\prod_{i=1}^{n}(X_i, \mathcal{A}_i)$ invariant under $\prod_{i=1}^{n} T_i$, which satisfies $\lambda(A_i \times \prod_{j \neq i} X_j) = m(A_i)$ for all $1 \leq i \leq n$, for all $A_i \in \mathcal{A}_i$; $(X_i, \mathcal{A}_i, m_i, T_i)$, $1 \leq i \leq n$, is a copy of $(X, \mathcal{A}, m, T)$) satisfies the following: The set $[1, n]$ can be decomposed into a disjoint union of subsets E_k, $1 \leq k \leq r$, for which the following properties are true:

(1) The algebras $\mathcal{B}_k = \prod_{i \in E_k} \mathcal{A}_i \times \prod_{j \in E_k^c} X_j$, $1 \leq k \leq r$, are λ independent.

(2) Let $K =$ the cardinality of E_k. For every $1 \leq k \leq r$, there exists integers $n_{i_1}, n_{i_2}, \ldots, n_{i_{K-1}}$ such that λ restricted to $\mathcal{B}_k$ is exactly $(\mathrm{Id} \times T_{n_{i_1}}, T_{n_{i_2}} \times \cdots \times T_{n_{i_{K-1}}})\Delta$, where Δ denotes the diagonal measure

$$\Delta(A_1 \times \cdots \times A_K) = m(A_1 \cap A_2 \cdots \cap A_K).$$

Definition 2. If the previous properties only hold for a joining of $n = 2$ copies of $(X, \mathcal{A}, m, T)$, we say that this transformation has *2-fold minimal self joinings*. (In the same manner one can define k-fold M.S.J.)

An important generalization of the notion of minimal self joinings is the notion of simplicity.

Definition 3. An ergodic transformation $(X, \mathcal{A}, m, T)$ is *simple* if the following is true: For all $n \geq 2$, an ergodic joining λ of the product of n copies of $(X, \mathcal{A}, m, T)$ satisfies the following: The set $[1, n]$ can be decomposed into a disjoint union of subsets E_k, $1 \leq k \leq r$, for which the following properties are true:

(1) The r algebras $\mathcal{B}_k = \prod_{i \in E_k} \mathcal{A}_i \times \prod_{j \in E_k^c} X_j$ are λ-independent.

(2) For every $1 \leq k \leq r$, λ restricted to $\mathcal{B}_k$ satisfies the property that the $|E_k|$ algebras

$$\mathcal{A}_i^{(k)} = \mathcal{A}_i \times \prod_{\substack{j \in E_k \\ j \neq i}} X_j, \qquad i \in E_k,$$

are λ-identical. (That is, for every pair $(i, j) \in E_k$, for every set A_i in $\mathcal{A}_i$, there exists a set A_j in $\mathcal{A}_j$ such that $\lambda(A_i \times X_j \triangle X_i \times A_j) = 0$.) An equivalent way to describe λ restricted to $\mathcal{B}_k$ is to say that there exist elements $S_{i_1}, S_{i_2}, \ldots, S_{i_{K-1}}$ $(K = |E_k|)$ in $C(T)$ (the group of invertible measure-preserving transformations that commute with T), the centralizer of T, such that

$$\lambda = (\mathrm{Id} \times S_{i_1} \times \cdots \times S_{i_{K-1}}) \Delta.$$

Let us remark that T has M.S.J. is equivalent to T simple and $C(T)$ is reduced to the powers of T.

Definition 4. If the previous properties only hold for a joining of $n = 2$ copies of $(X, \mathcal{A}, m, T)$ we say that the transformation is *2-simple*. In the same way we can define *k-simple*.

Historically Veech [1982] introduced 2-simplicity, which was later extended to simplicity by del Junco and Rudolph [1987]. The previous definitions of M.S.J. and simplicity entail the verification of infinitely many conditions. However, this is not necessary. It was J. King [1992] who first noticed that 4-fold M.S.J. or 4-simple implied M.S.J. and simplicity respectively. This was improved by Glasner, Host, and Rudolph [1992] to the following:

Theorem 2. *If $(X, \mathcal{A}, m, T)$ is weakly mixing and 3-simple, it is simple.*

(In particular, if $(X, \mathcal{A}, m, T))$ has 3-fold M.S.J. it has M.S.J.)

Proof. (1) Let λ be an ergodic joining of two copies of $(X, \mathcal{A}, m, T)$ with an ergodic $(Y, \mathcal{B}, \mu, S)$ which is pairwise independent, not globally independent. We denote this joining in a condensed way by $(X_1 \times X_2 \times Y, \lambda)$. The hypothesis is therefore, in the same condensed way, that

$$(X_1) \perp (X_2), \quad (X_1) \perp (Y), \quad (X_2) \perp (Y) \qquad (\text{for } \lambda).$$

Let $n > 0$. Call λ_n the image of λ under $(\mathrm{Id} \times T^n \times \mathrm{Id})$. Note that λ_n is again a pairwise-independent joining. Consider $(X_1 \times X_2 \times Y \times X_1', \tilde{\lambda}_n)$, the relatively independent joining of $(X_1 \times X_2 \times Y, \lambda)$ with $(X_1 \times X_2 \times Y, \lambda_n)$ above $(X_2 \times Y)$. Then

$$(X_1') \perp (X_1) \qquad \text{(for } \tilde{\lambda}_n), \qquad\qquad \textbf{(A)}$$

because, if in the ergodic decomposition of $\tilde{\lambda}_n$ into ergodic joinings $\tilde{\lambda}_n$ such that $\tilde{\lambda}_n = \int \tilde{\lambda}_n(\omega) dP(\omega)$ we have that P almost surely $(X_1') \perp (X_1)$ (for $\tilde{\lambda}_n(\omega)$), then $(X_1') \perp (X_1)$ (for $\tilde{\lambda}_n$). Otherwise there would exist $\tilde{\lambda}_n(\omega_0)$ for which, as a consequence of 2-simplicity, $\tilde{\lambda}_n(\omega_0)$ restricted to $X_1' \times X_1$ equals $(\text{Id} \times \sigma)\Delta$ (σ in $C(T)$, Δ diagonal measure); let $f_1 \in L^2(X_1)$, $f_2 \in L^2(X_2)$ and $g \in L^2(Y)$. We can compute $\int f_1 f_2 g d\lambda$ which if we read it in the algebra $(X_1' \times X_2 \times Y, \lambda_n)$ (which is $\tilde{\lambda}_n(\omega_0)$ identical to $(X_1 \times X_2 \times Y, \lambda)$), equals $\int (f_1 \circ \sigma)(f_2 \circ T^n) g d\lambda$, and hence λ is $(\sigma \times T^n \times \text{Id})$-invariant. As T^n is weakly mixing, $(X_1 \times X_2) \overset{\mathcal{I}}{\perp} (Y)$ (for λ) (where $\mathcal{I}$ is the sub-σ-algebra of $\mathcal{A}_1$ of σ-invariant sets). This is a consequence of Corollary 1 and of the fact that a direct product of a weakly mixing transformation with an ergodic transformation is ergodic. As $\mathcal{I} \perp (Y)$, we see that $(Y) \perp (X_1 \times X_2)$ (for λ), which contradicts the assumption that λ is not globally independent. Thus there is no "bad" $\tilde{\lambda}_n(\omega_0)$, and this proves (A).

(2) Let us prove 4-simplicity. Let λ be a joining of 4 copies of $(X, \mathcal{A}, m, T)$ which is not globally independent but for which all restrictions to the product of three factors are independent. We call this joining $(X_1 \times X_2 \times X_3 \times X_4, \lambda)$, and we consider the joining $(X_1 \times X_2 \times X_3 \times X_4 \times X_1', \tilde{\lambda}_n)$, $(n \in \mathbb{N})$ resulting from the relatively independent joining of $(X_1 \times X_2 \times X_3 \times X_4, \lambda_n)$ with $(X_1 \times X_2 \times X_3 \times X_4, \lambda)$ above $(X_2 \times X_3 \times X_4, \lambda)$. ($\lambda_n$ is here the image of λ by $(\text{Id} \times \text{Id} \times \text{Id} \times T^n)$). Because of (1)(A), $(X_1) \perp (X_1')$ (for $\tilde{\lambda}_n$). (The role of Y is played by $X_2 \times X_3$.) Therefore, by 3-simplicity, $\tilde{\lambda}_n$ restricted to $X_1 \times X_1' \times X_4$ (which are pairwise independent) must be globally independent. This says that if $f \in L^2(X_1)$, $f' \in L^2(X_1')$, $h \in L^2(X_4)$, $\int E_{234} f \times h \times (\text{Id} \times \text{Id} \times T^n) E_{234} f' d\lambda (= \int f h f' d\tilde{\lambda}_n) = \int f dm \times \int h dm \times \int f' dm$, where $E_{234} f$ is the conditional expectation of f with respect to the σ-algebra generated by $X_2 \times X_3 \times X_4$ in the joining $(X_1 \times X_2 \times X_3 \times X_4, \lambda)$. If in the previous identity we take $f = f'$, $h = h \times h T^n$ and $\int f dm = 0$, we then have

$$\int E_{234} f h \times (\text{Id} \times \text{Id} \times T^n) E_{234} f h = 0.$$

As $X_2 \times X_3$ is the algebra of the invariant sets of the action of $\text{Id} \times \text{Id} \times T$ on $X_2 \times X_3 \times X_4$, an application of the Ergodic Theorem gives

$$\int (E_{23} f h)^2 d\lambda = 0$$

whenever $\int f dm = 0$. (As before, E_{23} is the conditional expectation with respect to $(X_2 \times X_3)$ in the joining $(X_1 \times X_2 \times X_3 \times X_4, \lambda)$.) This forces

$(X_1 \times X_4) \perp (X_2 \times X_3)$ (for λ), and therefore global independence. A trivial induction completes the proof.

The interested reader may want to prove as an exercise that if T is weakly mixing 2-simple but not 3-simple, there exists in fact for every $k \geq 3$ an ergodic joining of k copies of T which is pairwise independent with no three copies globally independent. The result which has just been proved says that from now on we need only distinguish two categories of M.S.J. and simplicity: they are 2-M.S.J. and M.S.J. (the same with simplicity). It should also be emphasized that there are now no examples known of transformations which are 2-simple but not simple, and that this is the main open problem in this area. In connection with this problem, let us mention without proof an important result of Host [1991].

Theorem 3. *Let $(X, \mathcal{A}, m, T)$ be a weakly mixing transformation with singular spectral measure. Let $k \geq 3$ be an integer and $(X_i, \mathcal{A}_i, m_i, T_i)$, $1 \leq i \leq k$, be k copies of $(X, \mathcal{A}, m, T)$. If λ is a joining of $(\prod_1^n X_i, \prod_1^n \mathcal{A}_i, \prod_1^n T_i)$ such that $\lambda|\mathcal{A}_i = m_i, 1 \leq i \leq k$, and $\mathcal{A}_i$ and $\mathcal{A}_j$ are λ-independent whenever $i \neq j$, then λ is the product measure $\prod_1^k m_i$.*

There are two corollaries by Host [1991].

Corollary 2. *If $(X, \mathcal{A}, m, T)$ is mixing with singular spectral measure, it is mixing of all orders.*

Corollary 3. *If $(X, \mathcal{A}, m, T)$ is 2-simple with singular spectral measure, it is simple.*

We now give a theorem due to Veech [1982] concerning the structure of simple transformations.

Theorem 4. *Let $(X, \mathcal{A}, m, T)$ be a 2-simple weakly mixing transformation and $\mathcal{B}$ be a non-trivial invariant subalgebra of $\mathcal{A}$ (i.e., $T\mathcal{B} = \mathcal{B}$ and $\mathcal{B} \neq \{\phi, X\}$ and $\mathcal{B} \neq \mathcal{A}$). Then there exists a compact group K in $C(T)$ such that $\mathcal{B}$ is exactly the σ-algebra of invariant sets for the action of K (and T is then isomorphic to a compact group extension of its restriction to $\mathcal{B}$).*

Proof. We consider the relatively independent joining $(X_1 \times X_2, \lambda)$ of X with itself above $\mathcal{B}$. (X_1 and X_2 are two copies of $(X, \mathcal{A}, m, T)$).

In the ergodic decomposition $\lambda = \int \lambda(\omega) dP(\omega)$, we know from 2-simplicity that either $\lambda(\omega) = m \otimes m$ or $\lambda(\omega) = (\mathrm{Id} \times S_\omega)\Delta$ (Δ the diagonal measure, $S_\omega \in C(T)$). Let $E = \{\omega \mid \lambda(\omega) = m \otimes m\}$. Then $P(E) = 0$, as for $B \in \mathcal{B}$, $0 < m(B) < 1$,

$$\lambda(B \times B^C) = 0 = P(E)m(B)m(B^C) + \int_{E^C} m(B \cap S_\omega B^C)\, dP(\omega).$$

We have also shown that, P almost everywhere, $S_\omega B = B$. Therefore, if we call K the subgroup of $C(T)$ consisting of those elements S in $C(T)$ such that $SB = B$ m almost everywhere for every $B \in \mathcal{B}$, we can write

$$\lambda = \int_K (\mathrm{Id} \times S) \Delta dP,$$

where P becomes a probability measure on K (as $\mathcal{A}$ is countably generated). Considering the fact that for every $S_0 \in K$, $\lambda = (\mathrm{Id} \times S_0)\lambda$ (easily checked), and the uniqueness of the ergodic decomposition of λ, we see that P is a left-translation-invariant probability measure on K. Endowed with the weak topology $(d(S,T) = \sum'_{n \geq 1} m(SA_n \Delta TA_n)/2^n$, $\{A_n, \ n \geq 1\}$, a dense sequence of sets in $\mathcal{A})$, K is a separable metric topological group. Separability plus left invariance implies $P(U) > 0$ for every open set U in K. If K were not compact, there would exist a neighborhood V of the identity in K and elements S_i, $i \in \mathbb{N}$, in K such that $S_i V \cap S_j V = \emptyset$, for all $i \neq j$, a contradiction.

Let us remark that exactly the same proof yields the following.

Theorem 5. *Let $(X, \mathcal{A}, m, T)$ be an ergodic transformation and let $\mathcal{B}$ be a T-invariant subalgebra of $\mathcal{A}$. If in the ergodic decomposition of the relatively independent joining λ of T with itself above $\mathcal{B}$, almost every ergodic component λ_ω satisfies that the two algebras $(X \times \mathcal{A})$ and $(\mathcal{A} \times X)$ are λ_ω-identical, then T is a compact group extension of its restriction to $\mathcal{B}$. The converse is also true.*

(That the converse is true is left to the reader as an exercise.)

We now give a theorem, due to del Junco and Rudolph [1987], which describes the joinings of a simple transformation with an ergodic transformation. We first give a definition.

Definition 5. Let $(X, \mathcal{A}, m, T)$ be a simple weakly mixing transformation and $\mathcal{H}$ a T-invariant subalgebra of $\mathcal{A}$. An integer n being given, we denote by $\mathcal{W}_{\mathcal{H}}^{(n)}$ the factor of $(X^{(n)}, \mathcal{A}^{(n)}, m^{(n)}, T^{(n)})$ (the direct product of n copies of $(X, \mathcal{A}, m, T)$) which is defined as the subalgebra in $\mathcal{H} \times \mathcal{H} \times \cdots \times \mathcal{H}$ of sets invariant under the transformations $(x_1, x_2, \ldots, x_n) \to (x_{\sigma(1)}, x_{\sigma(2)}, \ldots, x_{\sigma(n)})$ for all $\sigma \in S_n$ (the symmetric group). Note that indeed these transformations commute with $T^{(n)}$ and preserve $m^{(n)}$.

Theorem 6. *Let $(X, \mathcal{A}, m, T)$ be a simple weakly mixing transformation and $(Y, \mathcal{B}, \mu, S)$ be an ergodic transformation. Let $(X \times Y, \mathcal{A} \times \mathcal{B}, \lambda, T \times S)$ be an ergodic joining of S and T which is not the product (i.e., $\lambda \neq m \times \mu$). Then there exist an integer $n \geq 1$ and a T-invariant subalgebra $\mathcal{H}$ of $\mathcal{A}$ such that λ extends to a joining λ_n of $(X^{(n)}, \mathcal{A}^{(n)}, m^{(n)}, T^{(n)})$ with $(Y, \mathcal{B}, \mu, S)$ for which $\mathcal{W}_{\mathcal{H}}^{(n)}$ is a factor of $(Y, \mathcal{B}, \mu, S)$, $\mathcal{B} \overset{\mathcal{W}_{\mathcal{H}}^{(n)}}{\perp} \mathcal{A}^{(n)}$, and the restriction of λ_n to $X_1 \times Y$ is λ.*

Proof. Given an integer n, a joining λ_n of X^n (the product of n copies of X) and Y is *good* if the restriction of λ_n to any component $X_i \times Y$ ($1 \leq i \leq n$) is exactly λ and if its restriction to X^n is the product joining. There exists a greatest integer n for which a good joining λ_n of X^n and Y exists — otherwise one could produce (by compactness) a joining $(X^{\mathbb{N}} \times Y, \lambda_\infty)$ where $(X^{\mathbb{N}}, \lambda_\infty)$ is the product joining of countably many copies of X and $(X_i \times Y, \lambda_\infty)$ coincides with $(X \times Y, \lambda)$ for all $i \in \mathbb{N}$. This is incompatible with the 0–1 law for independent and identically distributed processes, which forces $X \perp Y$ (for λ). Therefore, an application of Theorem 5 gives that for such a joining (which we may assume ergodic) with greatest n, $X^n \times Y$ is a compact group extension of Y (as every ergodic component of the relatively independent joining $(X^n \times Y \times X'^n, \bar{\lambda}_n)$ of $(X^n \times Y, \lambda_n)$ with itself above Y identifies X^n with X'^n, since otherwise we would get, by simplicity, a good joining of Y with X^{n_1} for some $n_1 > n$). Let K be this group. Since $X^{(n)}$ is invariant under the action of K, and Y is the algebra of invariant sets of the action of K, we have that $X^{(n)} \overset{Y \wedge X^{(n)}}{\perp} Y$ (for λ_n) (as a simple variant to Corollary 1). The action of K is transitive on the coordinates of $X^{(n)}$, otherwise we would get two subgroups K_1 and K_2 of K whose actions restricted to $X^{(n)}$ would have independent algebras of invariant sets, which is impossible. The algebra of invariants of the restriction of the action of K to $X^{(n)}$ is therefore exactly a $\mathcal{W}_{\mathcal{H}}^{(n)}$.

An interesting corollary to this result is the following (also proved in [del Junco-Rudolph, 1987]).

Corollary 4. *If $(X, \mathcal{A}, m, T)$ and $(Y, \mathcal{B}, \mu, S)$ are both weakly mixing and simple, every one of their joinings has the form given by Theorem 6, but with $n = 1$. (In particular, simple transformations with no isomorphic common factors are disjoint.)*

Proof. Consider the joining $(X^{(n)} \times Y, \lambda_n)$ given by the previous theorem. Assume $n > 1$, and consider the joining $(\tilde{Y} \times \tilde{X}^{n-1} \times X_1 \times X^{n-1} \times Y, \tilde{\lambda})$ obtained as the relatively independent joining of $(X^{(n)} \times Y, \lambda_n)$ with itself above X_1. If an ergodic component of $\tilde{\lambda}$ makes Y and $\tilde{Y}$ independent, we get that $\mathcal{W}_{\mathcal{H}}^{(n)} \perp (X_1)$ (by simplicity of X), which is impossible. Therefore (by simplicity of Y), Y and $\tilde{Y}$ are identified for almost every ergodic component $\tilde{\lambda}_\omega$ of $\tilde{\lambda}$ and $\mathcal{W}_{\mathcal{H}}^{(n)}$ and $\tilde{\mathcal{W}}_{\mathcal{H}}^{(n)}$ have to be factors of the same number of copies of X (since otherwise the maximality of n would be contradicted as the smallest number of copies of X necessary to define $\mathcal{W}_{\mathcal{H}}^{(n)} \vee \tilde{\mathcal{W}}_{\mathcal{H}}^{(n)}$ would be greater than n), which means that $X^{n-1} \times X_1$ is a compact group extension of X_1 (by Theorem 5), which is not possible either. Hence $n = 1$. (Note that we only used 2-simplicity of Y.)

We now are going to prove that a transformation which is the product of pairwise disjoint simple maps has the unique factorization property. First, we prove a lemma, also due to del Junco and Rudolph [1987], which is very useful and interesting in its own right.

Lemma 4. *Assume that $(X \times Y, \lambda)$ is an ergodic joining of $(X, \mathcal{A}, m, T)$ and $(Y, \mathcal{B}, \mu, S)$. If, when we consider the joining $(X \times Y \times X_1, \lambda_1)$ obtained as the relatively independent joining of $(X \times Y, \lambda)$ with itself above Y, the two algebras (X) and (X_1) are λ_1-independent, then (X) and (Y) are λ-independent.*

Proof. Let $f \in L^2(X)$. The hypothesis is therefore that $\int (E_Y f)^2 = (\int f)^2$. ($E_Y f$ is the conditional expectation of f with respect to the factor algebra Y in the joining λ.) Thus $E_Y f = \int f$, and the Lemma is proved.

Theorem 7. *Let $(X_i, \mathcal{A}_i, m_i, T_i), i \in \mathbb{N}$, be a family of pairwise disjoint weakly mixing simple systems. Let $(X, \mathcal{A}, m, T)$ be the product of these systems. Assume that $(X, \mathcal{A}, m, T)$ is isomorphic to the direct product of two systems $(Y_1, \mathcal{B}_1, \mu_1, S_1)$ and $(Y_2, \mathcal{B}_2, \mu_2, S_1)$. Then $\mathbb{N}$ can be split into a disjoint union $\mathbb{N}_1 \cup \mathbb{N}_2$ such that*

$$(Y_1, \mathcal{B}_1, \mu_1, S_1) = \prod_{i \in \mathbb{N}_1} (X_i, \mathcal{A}_i, m_i, T_i),$$

$$(Y_2, \mathcal{B}_2, \mu_2, S_2) = \prod_{i \in \mathbb{N}_2} (X_i, \mathcal{A}_i, m_i, T_i),$$

and this decomposition is unique.

Proof. If Z_1 and Z_2 are simple, weakly mixing, and disjoint, and if k and ℓ are integers, the only possible joining $(Z_1^k \times Z_2^\ell, \lambda)$ of Z_1^k and Z_2^ℓ is the product joining. (Consider first $(Z_1^k \times Z_2, \lambda)$; construct its relatively independent joining with itself above Z_2, and apply simplicity of Z_1 plus the preceding lemma to conclude that the restriction of λ to $Z_1^k \times Z_2$ is the product measure; then apply the same proof by forming the relatively independent joining of λ with itself above Z_1^k to get the result.) The same is true for any finite family $(Z_i, \Theta_i, \tau_i, R_i)$, $i \leq i \leq s$, of transformations which are simple, weakly mixing, and pairwise disjoint. If λ is a joining of $Z_1^{k_1} \times Z_2^{k_2} \times \cdots \times Z_s^{k_s}$, $(k_i \in \mathbb{N}, 1 \leq i \leq s)$, it must be the product joining. (This is seen by induction on s, applying the observation which has just been made to the relatively independent joining of λ with itself above $Z_s^{k_s}$, using the induction hypothesis and Lemma 4.)

We now form the relatively independent product of $(X, \mathcal{A}, m, T)$ with itself above $(Y_1, \mathcal{B}_1, \mu_1, S_1)$, which we denote $(\prod_{i \in \mathbb{N}} X_i \times \prod_{i \in \mathbb{N}} X_i', \tilde{\lambda})$. It is ergodic (isomorphic to $Y_1 \times Y_2 \times Y_2'$ with product measure and Y_2' a copy of Y_2). We can divide $\mathbb{N}$ into two sets, $\mathbb{N}_1 = \{i \in \mathbb{N}$ such that X_i is identified to X_i' (as an algebra) in the previous $\tilde{\lambda}$ joining (and therefore $X_i \subset \mathcal{B}_1$ by Theorem 5)$\}$ and $\mathbb{N}_2 = \{i \in \mathbb{N}$ such that $X_i \perp \mathcal{B}_1\}$.

By what has just been proved, $\prod_{i \in \mathbb{N}_2} X_i \times \prod_{i \in \mathbb{N}_2} X_i'$ is globally independent under λ, and by Lemma 4, $\mathcal{B}_1$ is independent of $\prod_{i \in \mathbb{N}_2} X_i$. As $\mathcal{B}_1$ is measurable with respect to $\prod_{i \in \mathbb{N}} X_i$, contains $\prod_{i \in \mathbb{N}_1} X_i$, is independent of $\prod_{i \in \mathbb{N}_2} X_2$ (and $\mathbb{N} = \mathbb{N}_1 \cup \mathbb{N}_2$), it is forced that $\mathcal{B}_1 = \prod_{i \in \mathbb{N}_1} X_i$. Working the same way with Y_2 finishes the proof.

We now describe a notion weaker than M.S.J., due to King [1988], which gives interesting properties of a transformation.

Definition 6. An ergodic system $(X, \mathcal{A}, m, T)$ is said to have the *finite joining rank* property if there exists an integer N such that if

$$\left(\prod_1^{N+1} X_i, \prod_1^{N+1} \mathcal{A}_i, \lambda, \prod_1^{N+1} T_i \right)$$

is an ergodic joining of the product of $N + 1$ copies of $(X, \mathcal{A}, m, T)$, then there exists $i \neq j$ such that either $\mathcal{A}_i \perp \mathcal{A}_j$ (for λ) or there exists an integer $k \in \mathbb{Z}$ such that $\lambda | X_i \times X_j$ is exactly $(\mathrm{Id} \times T_j^k) \Delta$ (where Δ is the diagonal measure). The smallest N for which this is true is of course called the *joining rank* of T.

Before we proceed to the next result we are going to state (without proof) a structure theorem of Zimmer [1976] and Furstenberg [1977], as we are going to use it in the proof.

Theorem 8. *Let $(Y, \mathcal{B}, \mu, S)$ be an ergodic dynamical system and $\mathcal{C}$ be an S-invariant factor algebra of $\mathcal{B}$. Then there exists a factor algebra $\widehat{\mathcal{C}} \supset \mathcal{C}$ relative to which S is weakly mixing (which is equivalent to the fact that the relatively independent product of S with itself above $\widehat{\mathcal{C}}$ is ergodic) and such that $\widehat{\mathcal{C}}$ is obtained from $\mathcal{C}$ in the following way: There exists a countable family of S-invariant factor algebras indexed by ordinals $\mathcal{C}_\eta$, $\eta \leq \eta_0$ such that $\mathcal{C}_1 = \mathcal{C}$, $\mathcal{C}_{\eta_0} = \widehat{\mathcal{C}}$, for every $\xi < \eta$, $\mathcal{C}_\xi < \mathcal{C}_\eta$, S restricted to $\mathcal{C}_{\eta+1}$ is an isometric extension of its restriction to $\mathcal{C}_\eta$, and if ξ is a limit ordinal $\mathcal{C}_\xi = \lim \uparrow \mathcal{C}_\eta (\eta \uparrow \xi)$. We call $\widehat{\mathcal{C}}$, which is canonical, the saturated distal extension of $\mathcal{C}$.*

The next theorem is in [King-Thouvenot, 1991].

Theorem 9. *If $(X, \mathcal{A}, m, T)$ has finite joining rank, then T is a finite extension of a power of a transformation S which has 2-M.S.J.*

Proof. Consider

$$\left(\prod_1^N X_i, \prod_1^N \mathcal{A}_i, \lambda, \prod_1^N T_i \right)$$

(which we condense as (X^N, T^N, λ)), a maximal ergodic joining with no two factors in trivial position. (Two factors X_i and X_j are in *trivial position* if either $\lambda | X_i \times X_j$ is $m \otimes m$ or $\lambda | X_i \times X_j$ is $(\mathrm{Id} \times T^k) \Delta$ for some $k \in \mathbb{Z}$).

In (X^N, T^N, λ), it must be that every component X_i satisfies $\widehat{X}_i = X^N$. (Pick $j \neq i$ and consider the relative product of $(N + 1)$ copies of X_j above

$\widehat{X}_i$ (in the joining λ), which is ergodic by Theorem 8; two of them must be in trivial position, but it cannot be independence in view of Lemma 4, therefore $\widehat{X}_i \supset X_j(\lambda)$).

We now use another lemma which is of disjointness type. Let $(X, \mathcal{A}, m, T)$ be an ergodic dynamical system, $\mathcal{B}_1$ and $\mathcal{B}_2$ be two factor algebras such that $\mathcal{B}_1 \perp \mathcal{B}_2$. If $T|\mathcal{B}_2$ is weakly mixing, and $\tilde{\mathcal{B}}_1$ is a factor containing $\mathcal{B}_1$ which is an isometric extension of $\mathcal{B}_1$, then $\tilde{\mathcal{B}}_1 \perp \mathcal{B}_2$ (and therefore $\hat{\mathcal{B}}_1 \perp \mathcal{B}_2$). (We leave the proof of this lemma which is a relative version of the fact that weakly mixing systems are disjoint from systems with pure point spectrum, a particular case of Lemma 1, as an exercise.) Let $(X^N, T^N, X'^N, T'^N, \mu)$ be an ergodic joining of two maximal joinings (X^N, T^N, λ) and $(X'^N T'^N, \lambda')$. By maximality, there exists i and j such that X_i and X'_j are in trivial position. In case $X_i \perp X'_j$ (for μ) the preceding observations and Theorem 8 imply that $X_i \perp X'^N$ (and this is true for any $1 \leq i \leq N$). Otherwise, there exists a substitution σ of $[1\ N]$ such that (X'^N, λ') is the image of (X^N, λ) by the map $(x_1, \ldots, x_N) \to (x_{\sigma(1)}, x_{\sigma(2)}, \ldots, x_{\sigma(N)})$ and the joining is the image of the diagonal joining by $\mathrm{Id} \times \prod_1^N (T'_i)^k$ for some $k \in \mathbb{Z}$. (This is a consequence of Corollary 1.) This implies that the maximal joining is unique (take an ergodic component of the relatively independent joining of (X^N, T^N, λ) with (X'^N, T'^N, λ') above their first coordinates and apply the preceding) and that if it were weakly mixing, it would be 2-simple. However, Theorem 5 applies and the maximal joining is a finite group extension of any of its nontrivial factors. We call this maximal joining (Y, J). It also follows from Theorem 5 that $C(J)$ (the centralizer of J) has the property that $C(J)/\mathbb{Z}$ is finite (and bounded by $N!$). We now prove that if K_1 and K_2 are two finite subgroups of $C(J)$, the group spanned by K_1 and K_2 is again finite. Let $x \in K_1$ and $y \in K_2$, and consider the set A of elements of $C(J)$ which are conjugate to x or to y. A is finite (as two distinct conjugates of x are in different classes of $C(J)$ mod its center). The group H generated by A is finite. To see that, let us write $A = \{a_1, \ldots, a_n\}$ and call r_i the order of a_i; let $\overline{N} = \sum_1^n r_i$. H is the set of finite products of elements in A (as every element in A is of finite order). In fact, any element in H can be written as the product of at most $\overline{N}$ elements in A. Let $z = a_{i_1} a_{i_2} \ldots a_{i_k}$, $k > \overline{N}$. Then there exists i_0 such that a_{i_0} appears at least r_{i_0} times in z. Given any element a_{i_j} in z, we can still write z as a product of k elements with a_{i_j} shifted by one to the right. $(a_{i_j} a_{i_{j+1}} a_{i_j}^{-1} = a_{i'} \in A)$, therefore $z = a_{i'_1} \ldots a_{i'_{k-1}} a_{i_0}^{r_{i_0}}$, and z has been rewritten as a product of $k - 1$ elements. Every element in H is of finite order and this finishes the proof. (This last proof was shown to me by M. Lazarus.) There is a biggest finite subgroup of $C(J)$ (the subgroup of elements with finite order) which corresponds to the smallest factor of J. This factor is a factor of every component of J, and hence a factor of $(X, \mathcal{B}, m, T)$. The preceding analysis shows that the action of T restricted to this factor is a power of an action with 2–M.S.J. This finishes the proof.

We introduce an important property, the R–property of Marina Ratner [1982a], which is the key to the understanding of the horocycle flows. The next definition is made for flows.

Definition 7. A measure-preserving flow T_t on the space $(X, \mathcal{A}, m)$ endowed with a metric d such that (X, d) is σ-compact is said to have the property R_p, $(p \neq 0)$, if the following is true: For every $\epsilon > 0$ and $N > 0$, there exists $\alpha(\epsilon)$, $\delta(\epsilon, N) > 0$ and a subset $A(\epsilon, N) \subset X$ such that $m(A) > 1 - \epsilon$ with the property that if $x, y \in A$ and $d(x, y) < \delta(\epsilon, N)$ and y is not on the T_t orbit of x, then there are $L = L(x, y)$ and $M = M(x, y) \geq N$ with $M/L \geq \alpha$ such that if we call

$$K^{\pm}(x, y) = \{n \in \mathbb{Z} \cap [L, L + M] \mid d(T_{np}(x), T_{(n\pm 1)p}(y)) < \epsilon\},$$

then either

$$\frac{|K^+|}{M} > 1 - \epsilon \quad \text{or} \quad \frac{|K^-|}{M} > 1 - \epsilon.$$

The next theorem is exactly Theorem III of Marina Ratner [1983].

Theorem 10. *Let (X, T_t, m, d) and $p \neq 0$ such that the flow has the property R_p and the transformation T_p is ergodic. Consider an ergodic dynamical system $(Y, \mathcal{B}, \mu, S)$ and λ an ergodic joining of T_p and S. Then either λ is the product joining, or the system $(X \times Y, \mathcal{A} \otimes \mathcal{B}, T_p \times S, \lambda)$ is a finite extension of its factor $(Y, \mathcal{B}, \mu, S)$.*

Proof. By Corollary 1, if λ is not the product joining, it is not invariant under the action of $T \times \mathrm{Id}$ (from now on in this proof, T means T_p). Therefore there are two open sets $A \subset X$ and $B \subset Y$ with boundaries of measure 0 and a number $\epsilon > 0$ such that $\lambda(A \times B)$ and $\lambda(TA \times B)$ differ by more than ϵ. Let ϵ_1 be such that $m(A \triangle V_{\epsilon_1}(A)) < \frac{\epsilon}{100}$ where $V_{\epsilon_1}(A))$ is the set of x' for which there exists $x \in A$ with $d(x, x') < \epsilon_1$. There exists, by the Ergodic Theorem, a set U in $X \times Y$, with $\lambda(U) > \frac{3}{4}$ and an integer N_1 such that for all $(x, y) \in U$, and all $n \geq N_1$,

$$\left| \frac{1}{n} \sum_{i=1}^{n} \mathbf{1}_{A \times B}(T^i \times S^i(x, y)) - \lambda(A \times B)) \right| < \frac{\epsilon \alpha(\epsilon_1)}{3}$$

and

$$\left| \frac{1}{n} \sum_{i=1}^{n} \mathbf{1}_{TA \times B}(T^i \times S^i(x, y)) - \lambda(TA \times B) \right| < \frac{\epsilon \alpha(\epsilon_1)}{3}.$$

The hypothesis implies that if $(x, y) \in U$ and $(x', y) \in U$, then either there exists $t > 0$ such that $T_t x = x'$ or $d(x, x') > \delta(\epsilon_1, N_1)$. Therefore, either there exists a fraction of Y of strictly positive measure on which the union of the atomic masses of the fiber measures is > 0 (and by ergodicity we have the conclusion of the theorem), or for almost every y the fiber is almost surely covered by pieces of countably many orbits of the flow T_t. But this would, by $T \times S$ invariance of the preceding decomposition and by ergodicity, imply that for every x in the fiber of y, $E_y(x) = \{t \mid T_t x$ is in the y fiber$\}$ is a constant set, an impossibility.

One may notice some analogies between this proof and the proof of del Junco, Rahe, and Swanson [1980] that the Chacon transformation has M.S.J. This result can be extended to the following, which is exactly Lemma 3.1 of [Ratner, 1983].

Theorem 11. *Let (X, T_t, m, d) and $p \neq 0$ be given such that the flow has the property R_p and the transformation T_p is weakly mixing. Assume that T_t acts continuously for the metric d. (This is really no restriction.) If λ is an ergodic joining*

$$\left(\prod_{i=1}^{3} X_i, \prod_{i=1}^{3} \mathcal{A}_i, \lambda, \prod_{i=1}^{3} T_t^i \right)$$

of three copies $(X_i, \mathcal{A}_i, m_i, T_i)$, $1 \leq i \leq 3$, of (X, T_t, m) such that $\mathcal{A}_i \perp \mathcal{A}_j$ (λ) for $i \neq j$ (and $\lambda | \mathcal{A}_i = m_i$, $1 \leq i \leq 3$), then λ is the product joining (i.e., $\lambda = m_1 \otimes m_2 \otimes m_3$).

Proof. As before we decide that T means T_p. If λ were not $(\text{Id} \times \text{Id} \times T)$-invariant there would exist $A_1 \times A_2 \in \mathcal{A}_1 \otimes \mathcal{A}_2$ and $B \in \mathcal{A}_3$ (open sets with boundaries of measure 0) and a strictly positive ϵ such that $|\lambda(A_1 \times A_2 \times B) - \lambda(A_1 \times A_2 \times TB)| > \epsilon$. Applying Theorem 10, we know that $(X_1 \times X_2 \times X_3, \lambda)$ is a finite extension of $(X_1 \times X_2, m_1 \otimes m_2)$. Therefore there exists a partition $\mathcal{P} = (B_1, B_2, \ldots, B_k)$ of $X_1 \times X_2 \times X_3$ such that $\lambda(B_i) = \frac{1}{k}$, $1 \leq i \leq k$, and $\mathcal{A}_1 \otimes \mathcal{A}_2 \vee \mathcal{P} = \mathcal{A}_1 \otimes \mathcal{A}_2 \otimes \mathcal{A}_3$ (λ). For every i, $1 \leq i \leq k$, we define a map $\varphi_i : X_1 \times X_2 \to X_3$ such that $(x_1, x_2, \varphi_i(x_1, x_2)) \in B_i$. There exists a set C in $X_1 \times X_2 \times X_3$ such that

(1) $\lambda(C) > \frac{9}{10}$;
(2) every point $(x_1, x_2, x_3) \in C$ is generic for λ;
(3) there exists N_1 such that for all $(x_1, x_2, x_3) \in C$, and all $n \geq N_1$,

$$\left| \frac{1}{n} \sum_{k=1}^{n} \mathbf{1}_{A_1 \times A_2 \times B}(T^k \times T^k \times T^k)(x_1, x_2, x_3) - \lambda(A_1 \times A_2 \times B)) \right| < \frac{\epsilon \alpha(\epsilon_1)}{3}$$

and

$$\left| \frac{1}{n} \sum_{k=1}^{n} \mathbf{1}_{A \times A_2 \times TB}(T^k \times T^k \times T^k)(x_1, x_2, x_3) - \lambda(A_1 \times A_2 \times TB)) \right| < \frac{\epsilon \alpha(\epsilon_1)}{3}$$

(ϵ_1 is defined in the same way as in the proof of Theorem 10);
(4) for $1 \leq i \leq k$, restricted to C (that is, whenever $(x_1, x_2, \varphi_i(x_1, x_2)) \in C$), φ_i is uniformly continuous;
(5) there exists an integer K such that when $(x_1, x_2, \varphi_i(x_1, x_2)) \in C$, then $(x_1, T_{\frac{1}{K}} x_2, \varphi_i(x_1, T_{\frac{1}{K}} x_2)) \in C, d(\varphi_i(x_1, x_2), \varphi_i(x_1 T_{\frac{1}{K}} x_2) < \delta(\epsilon_1, N_1)$, and furthermore $d(x_2, T_{\frac{1}{K}} x_2)$ is such that for all $n > N_1$,

$$\frac{1}{n} \sum_{k=1}^{n} 1_{A_2 \triangle T_{\frac{1}{K}} A_2}(T^k x_2) < \epsilon \frac{\alpha(\epsilon_1)}{3}.$$

((5) is in part a consequence of (4).)

The property R_p, (3) and (5) imply that if $(x_1, x_2, \varphi_i(x_1, x_2)) \in C$, there exists $\eta(x_1, x_2)$ such that $\varphi_i(x_1, T_{\frac{1}{K}} x_2) = T_{\eta(x_1, x_2)} \varphi_i(x_1, x_2)$. We can then find a point $(x_1, x_2, x_3) \in C$ such that $(x_1, T_{\frac{1}{K}} x_2, T_\eta x_3) \in C$. As every point in C is generic for λ, this implies that λ is (Id $\times T_{\frac{1}{K}} \times T_\eta$)-invariant, and, by Corollary 1, λ must be product measure. ($T_{\frac{1}{K}}$ is weakly mixing.)

This contradiction achieves the proof.

We now introduce some spectral properties which are sufficient to force a behaviour rigid enough on joinings to imply nice consequences on the structure of the action. The definition which comes is due to Stepin [1987]. See also [Oceledets, 1969].

Definition 8. Given a transformation $(X, \mathcal{A}, m, T)$ and a number $\alpha \in [0, 1]$, we say that T is α *weakly mixing* if there exists a subsequence n_i such that for every two sets A and B in $\mathcal{A}$,

$$m(T^{n_i} A \cap B) \to \alpha(A)m(B) + (1 - \alpha)m(A \cap B), \qquad (i \to +\infty).$$

The next theorem is a particular case of a more general result due to del Junco and Lemanczyk [1989]. Although it is proved by spectral methods, it has a strong "joining flavor".

Theorem 12. *Suppose that* $(X, \mathcal{A}, m, T)$ *is* α *weakly mixing for some* $\alpha \in (0, 1)$, *and consider the transformation* $S = \prod_{i \in \mathbb{N}} (X_i, \mathcal{A}_i, m_i, T_i)$ *(where* $(X_i, \mathcal{A}_i, m_i, T_i)$ *is a copy of* $(X, \mathcal{A}, m, T)$ *for each* $i \in \mathbb{N}$). *If* $\mathcal{B}$ *is an* S-*invariant subalgebra of* $\prod_{i \in \mathbb{N}} \mathcal{A}_i$ *restricted to which* S *acts isomorphically to a factor of* T, *then* $\mathcal{B}$ *is a factor of some* $\mathcal{A}_i$.

Proof. The Spectral Theorem together with a standard polarization imply that T α-mixing is equivalent to

$$\int_{\mathbb{T}} z^{n_i} f(z) d\sigma(z) \to \alpha \int_{\mathbb{T}} f(z) d\sigma(z)$$

for every $f \in L^1(d\sigma)$, where $d\sigma$ is the maximal spectral type of U_T (the unitary operator associated to T) acting on $L_2^0(X, m)$ (the subspace of functions with 0 integral.) This in turn implies $\sigma^{*n} \perp \sigma^{*m}$ whenever $n \neq m$. In fact, for every $f \in L^1(d\sigma^{*n})$, $\int z^{n_i} f d\sigma^{*n} \to \alpha^n \int f d\sigma^{*n}$. Therefore, if σ^{*n} and σ^{*m} have a common absolutely continuous part, this forces $\alpha^n = \alpha^m$ and $n = m$. As the spectral type of U_S is equivalent to $\sum_{n \geq 1} \frac{\sigma^{*n}}{2^n}$, where by the preceding observation, the subspace on which the restriction of U_S has spectral type equivalent to σ is exactly $\sum_{i \in \mathbb{N}} L_0^2(\mathcal{A}_i)$, if A is a set in $\mathcal{B}$, $1_A - m(A) = \sum f_i$ where $f_i \in L^2(\mathcal{A}_i)$. Since a two-valued function is indecomposable, and the f_i belong to independent algebras, it must be that the sum contains only one term f_{i_0} and $A \in \mathcal{A}_{i_0}$. If $A' \in \mathcal{B}$ belongs to $\mathcal{A}_{i_1}$ with $i_1 \neq i_0$, $A \cap A'$ is not in $\mathcal{A}_i$ for any i. This finishes the proof.

Del Junco and Lemanczyk [1989] have defined a strengthening of α-weak mixing, $(\alpha_1, \ldots, \alpha_k)$-weakly mixing, meaning that T^s is α_s-weakly mixing, $1 \leq s \leq k$, with a sequence which does not depend upon s, with the help of which they show the existence of transformations with properties close enough to M.S.J. to construct most of Rudolph's examples.

We now give a last statement about the "arithmetic" of transformations with M.S.J.

Definition 9. Let $(X, \mathcal{A}, m, T)$ be weakly mixing and have M.S.J. Let $n > 0$. In

$$\left(\prod_1^n X_i, \prod_1^n \mathcal{A}_i, \prod_1^n m_i, \prod_1^n T_i \right),$$

where $(X_i, \mathcal{A}_i, m_i, T_i)$ is a copy of $(X, \mathcal{A}, m, T)$ for all $1 \leq i \leq n$, consider the action U_σ of $S(n)$ (the group of substitutions of the set $[1, n]$) defined in the following way:

if $\sigma \in S(n)$, $U_\sigma(x_1, \ldots, x_n) = (x_{\sigma(1)}, x_{\sigma(2)}, \ldots, x_{\sigma(n)})$.

This action commutes with $\prod_i^n T_i$, and the algebra of the sets invariant under all U_σ, $\sigma \in S(n)$, defines a factor $\mathcal{S}_n$. (Such a factor occurred in Definition 5.)

Proposition 1. Let $(X, \mathcal{A}, m, T)$ be weakly mixing with M.S.J. Consider a product $(Y_1, \mathcal{B}_1, \mu_1, S_1) \times (Y_2, \mathcal{B}_2, \mu_2, S_2)$, and assume that there exists an integer n and a factor algebra $\mathcal{B}$ of $\mathcal{B}_1 \otimes \mathcal{B}_2$ restricted to which the action of $S_1 \times S_2$ is isomorphic to $\prod_1^n T_i$ acting on $\mathcal{S}_n$. Then there exists an integer $0 \leq k \leq n$ such that $\mathcal{B}_1$ contains a factor algebra restricted to which S_1 is isomorphic to $\prod_1^k T_i$ acting on $\mathcal{S}_k$, and $\mathcal{B}_2$ contains a factor algebra restricted to which S_2 is isomorphic to $\prod_1^{n-k} T_i$ acting on $\mathcal{S}_{n-k}$, and these two factors generate $\mathcal{B}$.

Proof. By an abuse of notation, we still call $\mathcal{S}_n$ the factor algebra of $\mathcal{B}_1 \otimes \mathcal{B}_2$ on which $S_1 \times S_2$ acts as $\prod_1^n T_i$ on $\mathcal{S}_n$. Consider, for $k \in \mathbb{Z}$, $(\text{Id} \times S_2^k)\mathcal{S}_n = \mathcal{S}_n^k$.

There exists a transformation S of which $S_1 \times S_2$ is a factor for which the following property is true: for every $k \in \mathbb{Z}$ there exists a set A_k of n

independent factor algebras of S, restricted to every one of which S acts isomorphically to T and such that the symmetric factor associate to this product action of S is exactly $\mathcal{S}_n^k$. (That this can be done is a consequence of the definition of $\mathcal{S}_n$ and of the M.S.J. property for T.) It is clear that for any ℓ, any $k_1, k_2, \ldots, k_\ell \in \mathbb{Z}$, any k,

$$\#\left(A_{k_1} \cap A_{k_2} \cap \cdots \cap A_{k_\ell} \right) = \#\left(A_{k_1+k} \cap A_{k_2+k} \cap \cdots \cap A_{k_\ell+k} \right).$$

Consider an atom of such an intersection $E = A_{k_1} \cap \cdots \cap A_{k_\ell}$. (That is, the intersection of E with any A_s is either empty or E itself.) Then the sequence $E_s = (A_{k_1+s} \cap \cdots \cap A_{k_\ell+s})$ is either made of pairwise disjoint sets or some s_0 is such that $E_{s_0} = E$. In fact this last case must be true for every atom: A transformation with M.S.J. has zero entropy (since it has no nontrivial factor algebras); $\mathcal{S}_n$ must be in the zero-entropy factor algebra (the Pinsker algebra) of $S_1 \times S_2$, which is the product of the Pinsker algebras of S_1 and S_2, respectively. Therefore we may assume that both S_1 and S_2 have zero entropy. If a sequence E_s is made of pairwise disjoint sets, $\#(\cup_{i=1}^m A_i) \geq \alpha m$ for all m, with a strictly positive α. This means that an isometric extension of S_2 has infinite entropy, which is impossible. It must then be that there exists an m for which $A_m = A_0$. So, there exists a finite extension $\hat{\mathcal{S}}_n$ of $\mathcal{S}_n$ (isomorphic to a factor of $\prod_1^n T_i$) which is invariant by $\text{Id} \times S_2^m$. As such finite extensions split into a direct product of transformations which only commute with their powers (due to the M.S.J. property of T), we must have that there exist an integer r, integers k and k_i, $1 \leq i \leq r$, and an independent family of factor algebras $\mathcal{A}_i$, $1 \leq i \leq r$, of $S_1 \times S_2$ such that

$$(\text{Id} \times S_2^{km}) \Big|\hat{\mathcal{S}}_n = \prod_{i=1}^r (S_1 \times S_2)^{k_i} \Big| \mathcal{A}_i.$$

(By this we mean that the restriction to the factor algebra $\mathcal{A}_i$ of the action of $\text{Id} \times S_2^{km}$ is the same as the restriction of $(S_1 \times S_2)^{k_i}$.) Hence, on $\mathcal{A}_i$, $S_1^{k_i} \times S_2^{km-k_i} = \text{Id}$. As the restriction of $(S_1 \times S_2)$ to $\mathcal{A}_i$ is weakly mixing, then either $k_i = 0$ and $\mathcal{A}_i \subset \mathcal{B}_1$ (the algebra of invariants of $\text{Id} \times S_2$), or $km = k_i$ and $\mathcal{A}_i \subset \mathcal{B}_2$. This finishes the proof.

As an application of this result and Theorem 6, the reader may prove that a transformation with M.S.J. is disjoint from any Gaussian process, and more generally from any infinitely divisible transformation.

II. EXAMPLES.

1. Locally rank one transformations. See Ferenczi [1984].

Definition 10. An ergodic transformation T on the space $(X, \mathcal{A}, m)$ is said to be *locally rank one* if the following is true: There exists a number $a > 0$ such that for any finite partition P of X, for every $\epsilon > 0$, there exists an integer $n > 0$, a set F_n such that

(α) the sets $T^i F_n$, $0 \le i \le n$, are disjoint;

(β) $m(\bigcup_{i=0}^n T^i F_n) > a - \epsilon$;

(γ) there exists a partition $\tilde{P}$ of $\bigcup_{i=0}^n T^i F_n = A_n$ with the same alphabet as P, measurable with respect to the partition of A_n into the sets $T^i F_n$, $0 \le i \le n$, such that $|\tilde{P} - P|A_n| < \epsilon$ (by $P|A_n$ we mean the partition of A_n into sets $p \cap A_n$, where p runs among the atoms of P; the distance between partitions P and $\tilde{P}$ with the same alphabet is the usual $|P - \tilde{P}| = \sum m(p_i \triangle \tilde{p}_i)$.)

We now indicate the proof to a theorem of J. King [1988].

Theorem 13. *A mixing locally rank-one transformation has finite joining rank.*

Proof. We keep the notations of Definition 10. We are going to show that the joining rank of T is bounded by $[\frac{1}{a}]$. Let $(x_1, \ldots, x_r)$ be a generic point of an ergodic joining λ of a product of $r = [\frac{1}{a}] + 1$ copies of $(X, \mathcal{A}, m, T)$.

An application of the Ergodic Theorem and a variant of the Weak Law of Large Numbers (we omit the details) give that for an infinite sequence of integers $n_i, i \ge 1$, the frequency of i's for which $x_j \in \overline{A}_{n_i} = \bigcup_{k=0}^{n_i(1-\epsilon)} T^k F_{n_i}$. ($F_{n_i}$ as in Definition 10, ϵ so that $r(a - \epsilon) > 1$, $1 \le j \le r$) is greater than $(1 - \epsilon)a$.

The choice of r forces that there exist two points x_1 and x_2 (for instance) which belong simultaneously to a positive proportion of the $\overline{A}_{n_i}$. We shall show that they are generic for a trivial joining (either product or (Id $\times$ $T^k)\Delta$). Call ρ the joining for which they are generic (λ restricted to $X_1 \times X_2$). If A and B are two cylinder sets, the assumption that x_1 and x_2 are in $\overline{A}_n$ implies that

$$x_1 \in T^{k_1} F_n, \quad 1 \le k_1 \le n(1 - \epsilon),$$
$$x_2 \in T^{k_2} F_n, \quad 1 \le k_2 \le n(1 - \epsilon).$$

Assume $k_1 \ge k_2$ and call $S_n = k_1 - k_2$.

$$(\text{Id} \times T^{S_n})\Delta(A \times B) = \alpha \rho_n(A \times B) + (1 - \alpha)\eta_n(A \times B),$$

where ρ_n is the frequency of $i \in [0, n - k_1]$ such that $T^i x_1 \in A$, $T^i x_2 \in B$, and $\alpha \ge \epsilon a$. (We used that $(x_1, T^{S_n} x_1)$ is generic for (Id $\times T^{S_n})\Delta$).

If the S_n are unbounded, taking limits and weak limits, mixing implies

$$m \times m = \alpha \rho + (1 - \alpha) \eta \qquad (\eta \text{ a joining}).$$

The ergodicity of $T \times T$ implies $\alpha = 1$. If the S_n are bounded, then the same reasoning implies that there exists an s such that $(\text{Id} \times T^s)\Delta = \rho$.

In the particular case where $a = 1$ (rank one transformations) the preceding implies that T has 2-M.S.J. In fact, in the case of rank one transformations, King [1988] proved that partial mixing suffices to imply finite joining rank (as a variant of the preceding); therefore, by Theorem 9 and his weak-closure theorem [King, 1986], it also implies 2-M.S.J. (this is Theorem 7.1 in [King-Thouvenot, 1991]). Coming back to the mixing case, a deep theorem of Kalikow [1984] (for rank one-transformations, two-fold mixing implies three-fold mixing), an extension of the previous proof and Theorem 2 imply that any mixing rank one has M.S.J. It is not known whether for transformations which are locally rank one, mixing implies two-fold mixing. Recently Adams [1993] and Friedman and Adams [1993] have given explicit constructions for mixing rank-one transformations. Historically, the first explicit transformation to have been shown to satisfy M.S.J. is the Chacon transformation (a particular rank one which is weakly mixing but not mixing). This was done by del Junco, Rahe, Swanson [1980]. A. del Junco [1983] has described an interesting family of interval-exchange transformations which have properties close to M.S.J. In his paper where he defines 2-simplicity, Veech asked the following: Is it true that almost every interval exchange on $k \geq 4$ intervals is simple? (In fact it is not known whether such transformations are almost surely weakly mixing—see [Nogueira and Rudolph, 1990].)

2. Horocycle flows.

In ergodic theory, no matter how pleasant and interesting to deal with a concept is, it really gets full justification when it meets the essential criterion of naturality. By "natural" is meant a property possessed by a transformation which is the model for a system whose evolution is governed by the laws of classical mechanics, or, more generally, a transformation modeled after a smooth diffeomorphism of a compact manifold leaving invariant an absolutely continuous measure.

It is remarkable that the theory of horocycle flows worked out by Marina Ratner [1982a, 1982b, 1983] gives rise to examples of simple maps and M.S.J. (in fact the extension of M.S.J. to $\mathbb{R}$ actions) coming from smooth actions (algebraic in nature).

In all that follows Γ will be a discrete subgroup with finite covolume in $\mathrm{SL}(2, \mathbb{R})$. Let H_t, $t \in \mathbb{R}$, be the action on $\mathrm{SL}(2, \mathbb{R})/\Gamma = X$ defined by $g\Gamma \to H_t g\Gamma$, where $H_t = \begin{pmatrix} 1 & t \\ 0 & 1 \end{pmatrix}$. H_t preserves the measure on X induced by the Haar volume on $\mathrm{SL}(2, \mathbb{R})$ and is mixing (by Theorem 1).

Theorem 14. [Ratner, 1982a]

   The horocycle action (X, m, H_t) has the property R_p for all $p \neq 0$.

Proof. We only give the proof for $p = 1$, the general case being treated exactly in the same way. We work on the universal cover $\mathrm{SL}(2, \mathbb{R})$ of X equipped with a right-invariant metric d. For an element $g = \begin{pmatrix} a & b \\ c & d \end{pmatrix} \in \mathrm{SL}(2, \mathbb{R})$, we define $\Delta(g) = \max(|1 - a|, |b|, |c|)$ and note that there is a constant $A > 1$ such that $A^{-1}\Delta(g) \leq d(e, g) \leq A\Delta(g)$ for all $g \in \mathrm{SL}(2, \mathbb{R})$ such that $d(e, g) \leq 1$. Since $d(H_s x, H_s y) = d(H_s xy^{-1} H_{-s}, e)$. $x, y \in \mathrm{SL}(2, \mathbb{R})$, $s \in \mathbb{R}$, $d(H_s x, H_{s\pm 1} y)$ will be small when the matrix $H_s xy^{-1} H_{-s}$ is close to the matrix $H_{\pm 1}$. If we denote

$$xy^{-1} = \begin{pmatrix} a & b \\ c & d \end{pmatrix}, \quad H_s xy^{-1} H_{-s} = \begin{pmatrix} a + cs & -cs^2 + (d - a)s + b \\ c & d - cs \end{pmatrix}.$$

We now notice that given $\epsilon > 0$ there exists α which only depends upon ϵ and $\delta(\epsilon, N)$ such that in the family of polynomials $P_{\epsilon_1 \epsilon_2 \epsilon_3} \equiv \epsilon_1 s^2 + \epsilon_2 s + \epsilon_3$ with $|\epsilon_i| < \delta$, $1 \leq i \leq 3$, the interval $I_\epsilon = [s_1, s_2]$ (with $s_1 > 0$) such that either $|P_{\epsilon_1 \epsilon_2 \epsilon_3}(s) - 1| \leq \epsilon$ or $|P_{\epsilon_1 \epsilon_2 \epsilon_3}(s) + 1| \leq \epsilon$ for all $s \in I_\epsilon$ satisfies both $s_2 - s_1 \geq N$ and $(s_2 - s_1)/s_1 > \alpha$ as well as $|\epsilon_1| s_2 < \epsilon$ as soon as ϵ_1 and ϵ_2 are not both 0. (This is easily seen: $\delta = (\epsilon/N)^2$, $\alpha = \epsilon/10$.) Note that here $A = X$. This finishes the proof.

We now make a list of observations: Again Γ is a discrete subgroup with finite covolume in $\mathrm{SL}(2, \mathbb{R})$; H_t is the action on $\mathrm{SL}(2, \mathbb{R})/\Gamma = X$ defined by $g\Gamma \to H_t g\Gamma$ where $H_t = \begin{pmatrix} 1 & t \\ 0 & 1 \end{pmatrix}$. Assume that there exists $\alpha \in \mathrm{SL}(2, \mathbb{R})$ such that $\Gamma_\alpha = \Gamma \cap \alpha^{-1}\Gamma\alpha$ is again a subgroup with finite covolume in $\mathrm{SL}(2, \mathbb{R})$. (Therefore $|\Gamma/\Gamma_\alpha| < +\infty$, $|\alpha^{-1}\Gamma\alpha/\Gamma_\alpha| < +\infty$.)

1) The action of H_t on X is isomorphic to the action of H_t on $\mathrm{SL}(2, \mathbb{R})/\alpha^{-1}\Gamma\alpha = X_\alpha$ by the isomorphism $g\Gamma \to g\alpha^{-1}\Gamma$.

2) The action of H_t on $\mathrm{SL}(2, \mathbb{R})/\Gamma_\alpha = \widetilde{X}_\alpha$ has naturally as factors the actions of H_t on X (by sending $g\Gamma_\alpha \to g\Gamma$) and the action of H_t on X_α (by sending $g\Gamma_\alpha \to g\alpha^{-1}\Gamma\alpha$).

3) Taking $s \in \mathbb{R}$, the image by $\mathrm{Id} \times H_s$ of the diagonal joining of two copies of H_t acting on $\widetilde{X}_\alpha$ gives rise, via the factor maps and the isomorphism which has been described in 2) and 1) to a joining of two actions of H_t on X. Let us call *quasi off-diagonal* a joining of the type which we have just described. Let us also notice that the two factors described in 2) span the whole Borel algebra of $\widetilde{X}_\alpha$, and that the joining is thus algebraic.

We list now without proof some of the results obtained by Marina Ratner [1983].

Theorem 15. *Let $X = SL(2,\mathbb{R})/\Gamma$ and H_t be as before. An ergodic joining of two copies of the action of H_t on X is either product measure or is quasi off-diagonal.*

We define now, for the sake of clarity a quasi off-diagonal joining of k copies of the action of H_t on X. Assume $\alpha_i \in SL(2,\mathbb{R})$, $1 \leq i \leq k$, have the property that

$$\bigcap_{i=1}^{k} \alpha_i^{-1} \Gamma \alpha_i = \Gamma_{\alpha_1,\dots,\alpha_k}$$

is again a group with finite covolume in $SL(2,\mathbb{R})$. Call

$$\widetilde{X}_{\alpha_1,\dots,\alpha_k} = SL(2,\mathbb{R})/\Gamma_{\alpha_1,\alpha_2,\dots,\alpha_k}.$$

The restriction of $(\mathrm{Id} \times H_{s_1} \times \cdots \times H_{s_{k-1}})\Delta$ ($s_i \in \mathbb{R}$, $1 \leq i \leq k-1$), where Δ is the diagonal measure on the product of k copies of the action of H_t on $\widetilde{X}_{\alpha_1,\alpha_2,\dots,\alpha_k}$, to the product of the k factors obtained by sending $g\Gamma_{\alpha_1\dots\alpha_k} \to g\alpha_i^{-1}\Gamma\alpha_i$ ($1 \leq i \leq k$) gives rise (via the k isomorphisms $\psi_i : g\alpha_i^{-1}\Gamma \to g\Gamma$) to a joining of k copies of the action of H_t on X, which is called a *quasi off-diagonal joining*.

As a consequence of Theorem 15 and Theorem 11, we can state

Theorem 16. *Any ergodic joining of the product of finitely many copies of the action of H_t on X splits as a product of quasi off-diagonals.*

As a consequence of a theorem of Margulis [1977], in the case where the subgroup Γ is maximal and not arithmetic, the horocycle flow on $SL(2,\mathbb{R})/\Gamma$ has M.S.J. (as an $\mathbb{R}$ action). In fact Theorem 15 is a particular case of a result of Ratner, who actually described joinings of different horocycle flows. Consider two discrete subgroups Γ_1 and Γ_2 of $SL(2,\mathbb{R})$ such that $X_1 = SL(2,\mathbb{R})/\Gamma_1$ and $X_2 = SL(2,\mathbb{R})/\Gamma_2$ have finite volume.

Theorem 17. *Consider an ergodic joining λ of the two actions of H_t on X_1 and H_t on X_2. If λ is not the product measure, there exists $\alpha \in SL(2,\mathbb{R})$ and $s \in \mathbb{R}$ such that $\alpha^{-1}\Gamma_2\alpha \cap \Gamma_1 = \Gamma_{3,\alpha}$ has finite covolume in $SL(2,\mathbb{R})$, and λ is obtained in the following way: Let $\widetilde{X}_\alpha = SL(2,\mathbb{R})/\Gamma_{3,\alpha}$. Let $\widetilde{\lambda}$ be the image by $\mathrm{Id} \times H_s$ of the diagonal joining of two copies of H_t acting on $\widetilde{X}_\alpha$. The action of H_t on $\widetilde{X}_\alpha$ has naturally as factors the actions of H_t on X_1 (by the mapping $\psi_1 : g\Gamma_{3,\alpha} \to g\Gamma_1$) and on $SL(2,\mathbb{R})/\alpha^{-1}\Gamma_2\alpha$ (by $\psi_2 : g\Gamma_{3,\alpha} \to g\alpha^{-1}\Gamma_2\alpha$).*

This last action is isomorphic to the action of H_t on X_2 (by $\psi_2 : g\alpha^{-1}\Gamma_2 \to g\Gamma_2$). λ is the image of $\widetilde{\lambda}$ by $(\psi_1, \psi_\alpha\psi_2)$. The H_t action defined by the joining λ is isomorphic to the action of H_t on $\widetilde{X}_\alpha$.

This theorem gives, with a small amount of work, as two corollaries that measure-theoretically isomorphic horocycle flows are canonically algebraically isomorphic, and that factors of horocycle flows are algebraic

(Corollary 4 and Corollary 5 in [Ratner, 1983]). Corollary 4 was first proved in [Ratner, 1982b] and Corollary 5 in [Ratner, 1982a].

Let us mention one more result (proved by E. Glasner and B. Weiss [1993] in the cocompact case).

Theorem 18. *Let* $\Gamma \subset SL(2, \mathbb{R})$ *be discrete with finite covolume. Let* H_t *act on* $SL(2, \mathbb{R})/\Gamma = X$. *Then it is a factor of a simple transformation (as an* $\mathbb{R}$ *action).*

Proof. Let $\overline{\Gamma} = \{\alpha \mid \alpha\Gamma\alpha^{-1} \cap \Gamma = \Gamma_\alpha$ has finite covolume in $SL(2, \mathbb{R})\}$. $\overline{\Gamma}$ is a countable group. If $\alpha_1, \ldots, \alpha_n \in \overline{\Gamma}$, the group $\bigcap_{i=1}^{n} \Gamma_{\alpha_i} = \Gamma_{\alpha_1, \ldots, \alpha_n}$ has again finite covolume in $SL(2, \mathbb{R})$, and the family of actions of H_t on $SL(2, \mathbb{R})/\Gamma_{\alpha_1, \alpha_2, \ldots, \alpha_n} = X_{\alpha_1, \ldots, \alpha_n}$ indexed by the directed set of finite subsets of $\overline{\Gamma}$ has an inverse limit $(\overline{X}, H_t)$. This inverse limit is in fact simple. To this end, consider an ergodic joining λ of $(\overline{X}, H_t)$ with (X, H_t). The result will be proved if we can show that either λ is the product joining or λ makes X a factor of $\overline{X}$. (For in that case X will be a factor of some $X_{\alpha_1 \ldots \alpha_n}$, hence some X_{α_i}, by Theorem 17; we also use the same observations as in the proof of Theorem 9). By Theorem 10, we know, assuming that λ is not the product joining, that $\overline{X} \times X$ is a finite extension, with fiber having cardinality k of $\overline{X}$ ($k \geq 2$). As $\overline{X}$ is an algebraic limit, there exist finitely many $\alpha_i \in \overline{\Gamma}$, $1 \leq i \leq n$, such that λ restricted to $X_{\alpha_1 \ldots \alpha_n} \times X$ is already an extension of cardinality k of $X_{\alpha_1 \ldots \alpha_n}$. By Theorem 17, there exists $\tilde{\alpha}_{n+1} \in SL(2, \mathbb{R})$ such that H_t acting on $X_{\alpha_1 \ldots \alpha_n} \times X$ equipped with λ is isomorphic to H_t acting on $SL(2, \mathbb{R})/\Gamma_{\alpha_1 \ldots \alpha_n} \cap \tilde{\alpha}_{n+1}^{-1}\Gamma\tilde{\alpha}_{n+1}$. Let us call this action of H_t on $X_{\alpha_1 \ldots \alpha_n} \times X$, $\tilde{X}_{\alpha_1, \ldots, \alpha_n, \tilde{\alpha}_{n+1}}$. By definition of $\overline{X}$, $X_{\alpha_1, \ldots, \alpha_n}$ has an $\overline{X}$-measurable extension $X_{\alpha_1, \ldots, \alpha_n, \tilde{\alpha}_{n+1}}$ which is isomorphic to $\tilde{X}_{\alpha_1 \ldots \alpha_n, \tilde{\alpha}_{n+1}}$. The two extensions are relatively independent, as the cardinality of the fiber of X in $\tilde{X}_{\alpha_1, \alpha_2, \ldots, \alpha_n, \tilde{\alpha}_{n+1}}$ is the same as the cardinality of the fiber of X in $X_{\alpha_1, \alpha_2, \ldots \tilde{\alpha}_{n+1}}$. Therefore, λ restricted to $X_{\alpha_1, \alpha_2 \ldots \tilde{\alpha}_{n+1}} \times \tilde{X}_{\alpha_1, \alpha_2, \ldots \alpha_n, \tilde{\alpha}_{n+1}}$ is isomorphic to the relatively independent joining of $X_{\alpha_1 \alpha_2 \ldots \tilde{\alpha}_{n+1}}$ with itself above $X_{\alpha_1 \alpha_2 \ldots \alpha_n}$. But this is never ergodic. Therefore $k = 1$. This finishes the proof as it implies 2-simplicity of $(\overline{X}, H_t)$ (which is mixing) and simplicity by Theorem 11 and Theorem 2.

This result has been used by Glasner and Weiss [1993] to prove the existence of simple maps with non-unique prime factors (using an appropriate Γ). It is remarkable to note that such an example seems to be out of reach by way of the cutting and stacking constructions. Let us remark that there is some analogy between $\overline{X}$ and the maximal joining in the proof of Theorem 9. Theorem 17 is proved by Marina Ratner by showing that the joining λ of the two actions of H_t on X_1 and X_2 has the property that for some $\sigma \in \mathbb{R}$, the joining $(\text{Id} \times H_\sigma)\lambda$ is in fact a joining of the whole actions of $SL(2, \mathbb{R})$ on X_1 and X_2. This in turn is a consequence of the R property, Theorem 10, and of the commutation relations. In fact, Marina Ratner

even proved (Theorem 4 in [Ratner, 1983]) that if λ is an ergodic joining of
the actions of H_1 on $X_1 = \mathrm{SL}(2, \mathbb{R})/\Gamma_1$ and $X_2 = \mathrm{SL}(2, \mathbb{R})/\Gamma_2$ (Γ_1, Γ_2 with
finite covolume), then λ is a joining of the actions of H_t, $t \in \mathbb{R}$, on X_1 and
X_2. (This implies that Theorem 18 extends to the following: *the time-one
map H_1 of every horocycle flow is a factor of a simple map*.)

It would be nice to find other examples where joinings of "small subgroups" of group actions are (up to the centralizer of the small subgroup)
automatically joinings of the whole group actions.

It would also be nice to know whether the R-property itself for a flow
suffices to imply that it is a factor of a simple flow.

3. Gaussian Processes.

A class of Gaussian processes has a property which is like a generalization
to the continuous case of the unique factorization property of a product of
pairwise disjoint simple systems (Theorem 7).

Definition 11. Let σ be a finite, symmetric, atomless positive measure on
S_1. Consider the stationary, centered Gaussian process, X_n, $n \in \mathbb{Z}$, whose
covariance is given by

$$E(X_n X_{n+m}) = \int_{S_1} e^{imx} d\sigma(x).$$

The shift transformation on this process gives rise to a measure-preserving
dynamical system $(X, \mathcal{A}, m, T_\sigma)$, which is weakly mixing.

Definition 12. A finite, positive, symmetric measure on S_1 is said to be
Kronecker if its support is $\Lambda \cup (-\Lambda)$, where Λ is a Kronecker set. (A
Kronecker set is one for which every continuous function of modulus one is
a uniform limit of characters.)

The following result was announced in [Thouvenot, 1986].

Proposition 2. *Let σ be an atomless Kronecker measure on S_1. If T_σ
is isomorphic to a direct product $T_1 \times T_2$, then there exist σ_1 and σ_2 such
that $\sigma_1 \perp \sigma_2$, $\sigma = \sigma_1 + \sigma_2$, $T_1 = T_{\sigma_1}, T_2 = T_{\sigma_2}$, and this decomposition is
canonical.*

Sketch of the proof. We shall use the theorem of Foïas and Stratila
[1967] which states the following: If $(Y, \mathcal{B}, \mu, S)$ is an ergodic dynamical
system, and if $f \in L_0^2(Y)$ has the property that the spectral measure μ_f is
a Kronecker measure, then the sequence $T^n f$, $n \in \mathbb{Z}$, defines a stationary
Gaussian process. We apply this to the following.

In $L^2(X, T_\sigma)$, call H_σ the Gaussian space which is the closure of the linear
span of the X_n, $n \in \mathbb{Z}$. If we consider an ergodic joining λ of (X, T_σ) with
(X', T_σ') ((X', T_σ') is a copy of (X, T_σ)), there correspond two closed subspaces, H_σ and its copy H_σ' in $L^2(X', T_\sigma')$. The result is that in $L_0^2(X', T_\sigma')$,
$(H_\sigma + H_\sigma')$ is still a Gaussian space (because the spectral measure of any

element in $(H_\sigma + H'_\sigma)$ is a Kronecker measure). The proof will be finished (using the relatively independent joining of T_σ with itself above the factor algebra generated by the factor T_1) once we notice that, in a Gaussian space H, if H_1 and H_2 are two closed subspaces, $\mathcal{B}(H_1 \cap H_2) = \mathcal{B}(H_1) \wedge \mathcal{B}(H_2)$ (where $\mathcal{B}(H_i)$ is the smallest σ-algebra that makes measurable all the variables in H_i, $1 \le i \le 2$). This is seen since in $L^2(\mathcal{B}(H))$ identified to $\sum_{n\ge 0} H^{n\odot}$ (the Wiener Chaos decomposition), the positive operator $\mathcal{W} = E^{\mathcal{B}(H_1)} E^{\mathcal{B}(H_2)} E^{\mathcal{B}(H_1)}$ ($E^{\mathcal{A}}$ the conditional expectation with respect to $\mathcal{A}$) commutes with the projections on $H^{n\odot}$, $n \in \mathbb{N}$; and restricted to $H^{n\odot}$, $\mathcal{W} = A^{n\odot}$, where A is the operator in H, $A = P_{H_1} P_{H_2} P_{H_1}$ (P_{H_i} the projection on H_i); $(L^2[\mathcal{B}(H_1) \wedge \mathcal{B}(H_2)]$ is the subspace corresponding to the eigenvalue 1 of $\mathcal{W}$; if $\tilde{H} = H_1 \cap H_2$, $\tilde{H}$ is the eigenspace corresponding to the eigenvalue 1 of A; $\tilde{H}^{n\odot}$ is the eigenspace of $A^{n\odot}$ in $H^{n\odot}$ corresponding to the proper value 1).

The same observations yield that a pairwise-independent joining of three copies of T_σ, σ Kronecker, must be globally independent.

Finally, let us mention the generalization of M.S.J. to transformations leaving a measure quasi-invariant by Rudolph and Silva [1989], with nice examples and also interesting questions; and the use of joinings of non-commutative actions (automorphisms of a C^*-algebra having an invariant state) with commutative ones in Sauvageot-Thouvenot [1992] for the purpose of giving an alternative presentation of the Connes-Narnhofer-Thirring entropy for noncommutative dynamical systems.

REFERENCES

1. Adams, T., *Smorodinsky's conjecture*, preprint (1993).
2. _____ and Friedman, W., *Staircase mixing*, preprint (1993).
3. del Junco, A., Rahe, M. and Swanson, L., *Chacon's automorphism has minimal self joinings*, J. d'Analyse Math. **37** (1980), 276–284.
4. del Junco, A., *A family of counterexamples in ergodic theory*, Israel J. Math. **44** (1983), 160–180.
5. _____ and Rudolph, D., *On ergodic actions whose self joinings are graphs*, Erg. Th. Dyn. Sys. **7** (1987), 531–557.
6. _____ and Lemańczyk, M., *Generic spectral properties of measure preserving maps and applications*, preprint (1989).
7. _____, Lemańczyk, M. and Mentzen, M., *Semisimplicity, joinings, and group extensions*, preprint (1993).
8. Ferenczi, S., *Systèmes localement de rang un*, Ann. Inst. Poincaré **20** (1984), 35–51.
9. Foïias, C. and Stratila, S., *Ensembles de Kronecker dans la théorie ergodique*, C. R. Acad. Sci. Paris **267, 20A** (1967), 166–168.
10. Furstenberg, H., *Disjointness in ergodic theory, minimal sets, and a problem in diophantine approximation*, Math. Sys. Theory **1** (1967), 1–49.
11. _____, *Ergodic behavior of diagonal measures and a theorem of Szemeredi on arithmetic progressions*, J. d'Analyse Math. **31** (1977), 204–256.
12. Glasner, E., Host, B. and Rudolph, D., *Simple systems and their higher order self-joinings*, Israel J. Math. **78** (1992), 131–142.

13. Glasner, E. and Weiss, B., *A simple weakly mixing transformation with non-unique prime factors*, preprint (1993).

14. Guivarch, Y., *Proprietés de mélange pour les groups à un parametre de $SL(d, \mathbb{R})$*, preprint (1992).

15. Host, B., *Mixing of all orders and pairwise independent joinings of systems with singular spectrum*, Israel J. Math. **76** (1991), 289–298.

16. Kalikow, S., *Twofold mixing implies threefold mixing for rank one transformations*, Erg. Th. Dyn. Sys. **4** (1984), 237–261.

17. King, J., *The commutant is the weak closure of the powers for rank one transformations*, Erg. Th. Dyn. Sys. **6** (1986), 363–384.

18. ———, *Joining rank and the structure of finite rank mixing transformations*, J. d'Analyse Math. **51** (1988), 182–227.

19. ———, *Ergodic properties where order 4 implies infinite order*, Israel J. Math. **80** (1992), 65–86.

20. King, J. and Thouvenot, J.-P., *A canonical structure theorem for finite joining rank maps*, J. d'Analyse Math. **56** (1991), 211–230.

21. Marcus, B., *The horocycle flow is mixing of all degrees*, Invent. Math. **46** (1978), 201–209.

22. Margulis, G., *Discrete groups of motions of manifolds of non-positive curvature*, Amer. Math. Soc. Translations **109** (1977), 33–45.

23. Mozes, S., *Mixing of all orders of Lie group actions*, Invent. Math. **107** (1992), 235–248.

24. Nogueira, A. and Rudolph, D., *Topological weak mixing of interval exchange maps*, preprint (1990).

25. Ornstein, D., *On the root problem in ergodic theory*, Proc. Sixth Berkeley Symp. Math. Stat. and Prob., Vol. II, 1967, pp. 347–356.

26. Oceledets, V. I., *An automorphism with simple and continuous spectrum and without the group property*, English transl. in Math. Notes **5** (1969), 196–198.

27. Ratner, M., *Factors of horocycle flows*, Erg. Th. Dyn. Sys. **2** (1982a), 465–489.

28. ———, *Rigidity of horocycle flows*, Ann. Math. **115** (1982b), 587–614.

29. ———, *Horocycle flows, joinings and rigidity of products*, Ann. Math. **118** (1983), 277–313.

30. Rudolph, D., *An example of a measure preserving map with minimal self-joinings and applications*, J. d'Analyse Math. **35** (1979), 97–122.

31. ———, *k-fold mixing lifts to weakly mixing isometric extensions*, Erg. Th. Dyn. Sys. **5** (1985), 445–449.

32. ——— and Silva, C., *Minimal self joinings for nonsingular transformations*, Erg. Th. Dyn. Sys. **9** (1989), 759–800.

33. ———, *Fundamentals of Measurable Dynamics*, Oxford Science Publications, 1990.

34. Ryzhikov, V. V., *Connection between the mixing properties of a flow and the isomorphicity of the transformations that compose it*, Math. Notes Acad. Sci. U.S.S.R. **49** (1991), 621–627.

35. Sauvageot, J.-L. and Thouvenot, J.-P., *Une nouvelle définition de l'entropie dynamique des systémes non commutifs*, Comm. Math. Phys. **145** (1992), 411–423.

36. Smorodinsky, Meir and Thouvenot, J.-P., *Bernoulli factors that span a transformation*, Israel J. Math. **32** (1979), 39–43.

37. Stepin, A. M., *Spectral properties of generic dynamical systems*, Math. U.S.S.R. Izv. **29** (1987), 159–192.

38. Thouvenot, J.-P., *The metrical structure of some Gaussian processes*, Proceedings On Ergodic Theory And Related Topics, II, Georgenthal, 1986, pp. 195–198.

39. Veech, W., *A criterion for a process to be prime*, Monats. Math. **94** (1982), 335–341.

40. Zimmer, R., *Ergodic group actions with generalized discrete spectrum*, Ill. J. Math. **20** (1976), 555–588.

UNIVERSITÉ PIERRE ET MARIE CURIE, LABORATOIRE DE PROBABILITÉS, 4 PLACE JUSSIEU - TOUR 56, 75252 PARIS CEDEX 05, FRANCE

PART II

RESEARCH PAPERS

ERGODIC BAKER'S TRANSFORMATIONS

C.J. Bose and P. Grzegorczyk

1. Introduction and the Main result

Let $(I, \mathcal{B}, m)$ denote the probability space consisting of the unit interval $I = [0,1]$ with Borel subsets $\mathcal{B}$ and Lebesgue measure m. Let $0 \leq f \leq 1$ be a measurable function on I. Define $a = \int_I f \, dm$, $I_0 = [0, a]$, $I_1 = [a, 1]$ and maps $\phi_i : I_i \to I$, $i = 0, 1$ by the formulas

$$\phi_0^*(x) = \int_0^x f \, dm \qquad \forall x \in I,$$

$$\phi_1^*(x) = 1 - \int_x^1 1 - f \, dm \qquad \forall x \in I,$$

$$\phi_0(x) = \inf\{t \mid \phi_0^*(t) \geq x\} \qquad \forall x \in I_0,$$

$$\phi_1(x) = \inf\{t \mid \phi_1^*(t) \geq x\} \qquad \forall x \in I_1.$$

By definition the ϕ_i are increasing and left continuous.

We define the generalized baker's transformation (associated to f) on the unit square $I \times I$ by

$$T(x,y) = \begin{cases} (\phi_0(x), \; f[\phi_0(x)] \, y), & (x,y) \in I_0 \times I \\ (\phi_1(x), \; 1 - (1 - f[\phi_1(x)]) \, (1 - y)), & (x,y) \in I_1 \times I. \end{cases}$$

It is easy to see that T is $\mathcal{B} \times \mathcal{B}$-measurable and preserves the measure $m \times m$.

f is called the *cut function* for the transformation T^{-1}. The action of T may be visualized as follows. The square is partitioned into two vertical columns over I_0 and I_1 respectively. The mass over I_0 is moved under the graph of f and the mass over I_1 is moved over the graph of f in such a way that vertical fibres are mapped into vertical fibres and measure is preserved. When $f \equiv \frac{1}{2}$, T is the (classical) baker's transformation (as described for example in [Halmos, 1956] or [Walters, 1981]). For the benefit of the reader we shall state without proof a number of basic facts about this class of transformations. Details may be found in [Bose, 1989]. We say that f has

Both authors supported under grant GPOO4658690 of the Natural Sciences and Engineering Research Council of Canada.

margins if there exists a constant $c > 0$ satisfying $c \leq f \leq 1 - c$. We also
define a two set partition of $I \times I$ as $P = \{P_0, P_1\}$ where

$$P_0 = \{(x,y) \in I \times I \mid 0 \leq y < f(x), \quad x \in I\},$$

$$P_1 = I \times I / P_0.$$

If f has margins, P generates under T. (Weaker sufficient conditions for P
to be a generator are known: see [Rahe, 1993] or [Bose, 1989].) When P generates, T is the *natural-extension* of the 2-1 Lebesgue-measure-preserving
endomorphism ϕ on I defined by

$$\phi(x) = \begin{cases} \phi_0(x) & x \in I_0 \\ \phi_1(x) & x \in I_1. \end{cases}$$

ϕ is an example of a piecewise monotone interval map. When f has margins,
ϕ is *expanding*, with $|\phi'| \geq \frac{1}{1-c}$. In general, the *entropy of T with respect
to P* is computed as

$$H(T, P) = \int_0^1 f \log_2 f \, dm + \int_0^1 (1 - f) \log_2 (1 - f) \, dm,$$

and this equals the *entropy of T, $H(T)$* when P generates, in particular
when f has margins. Although not necessary for the investigation at hand,
one can show that every ergodic automorphism S with $0 < H(S) < 1$ is
isomorphic to a generalized baker's transformation. We are interested in
the question: Can one determine ergodicity of T from the function f?

Extensive literature on piecewise monotone and expanding interval
maps provides motivation (and indeed, in many cases proof) for ergodic
statements about T. For example, it follows from the work of [Keller, 1985]
that if ϕ is expanding and piecewise $C^{1+\delta}$ then its natural extension is
Bernoulli, which translates as: If f is Hölder-continuous with margins then
T is Bernoulli. This was also shown in [Bose, 1989] under more general hypothesis. Recently [Rahe, 1993] has obtained a similar result which allows
$m\{x \mid f(x) = 0 \text{ or } f(x) = 1\} = 0$ but requires different continuity assumptions for the same conclusion. As another example, if f is continuous and
of bounded variation with margins, T is Bernoulli. This follows from an
analogous interval map result of [Wong, 1978].

In this article, we take a different approach, studying the effect of
monotonicity in f under no continuity assumptions.

Proposition 1.1. *Let f be monotone non-increasing with margins. Then
T is ergodic. Moreover T is isomorphic to the Bernoulli shift with entropy*

$$H = \int_0^1 f \log_2 f \, dm + \int_0^1 (1 - f) \log_2 (1 - f) \, dm.$$

The second part of this result follows from the weak-Bernoulli property for the partition P.

As background and motivation we consider, given f with margins and associated T and P, an isomorphic shift system on $X = \prod_{\mathbb{Z}}\{0,1\}$, the isomorphism $\Psi : I \times I \to X$ being provided by the $P - T$ name of $(x,y) \in I \times I$. Put on X the product of the discrete topology on $\{0,1\}$. Ψ carries Lebesgue measure to a shift invariant Borel probability measure $\mu = (m \times m) \circ \Psi^{-1}$ on X.

Setting $X^- = \prod_{\mathbb{Z}^-}\{0,1\}$ and for $x^- \in X^-$, $g(x^-) = f(\Psi^{-1}(x^-))$ one easily verifies that μ is a g-measure in the sense of [Keane, 1974], *i.e.* that it is a Borel probability measure on X satisfying

$$\mu\{x_0 = 0 \mid x_{-1}, x_{-2}, \ldots\} \overset{\text{a.e.}}{=} g(x_{-1}, x_{-2}, \ldots).$$

(Here we allow for a slight abuse of notation as $\Psi^{-1}(x^-)$ is a vertical fibre in the square rather than a point in I.) Conversely, given a measurable g on X^- with $0 < c \leq g \leq 1 - c$ (g with *margins*) and a g-measure μ_g there is a generalized baker's transformation isomorphic to the shift on (X, μ_g) so that

$$g(x^-) = f(\Psi^{-1}(x^-)).$$

If g has a unique g-measure then T is ergodic; on the other hand from a non-ergodic μ_g one may construct a non-ergodic T as above.

Given a continuous g with margins on X^- let $\mathcal{M}_g$ denote the set of g-measures. $\mathcal{M}_g$ is a non-empty, convex, compact subset of the shift invariant probability measures on X. [Walters, 1975] has provided a "topological" criteria on g which is sufficient for $\mathcal{M}_g$ be a singleton, and proves that the shift is Bernoulli with respect to the unique g-measure. Incidentally, Walters' condition may also be used to show Hölder-continuous f with margins generate Bernoulli T.

Equip X^- (and X) with the partial order $x \leq y$ iff $x_i \leq y_i$ $\forall i \in \mathbb{Z}^-$ ($\forall i \in \mathbb{Z}$). [Hulse, 1991] has shown that if g is continuous, monotone non-increasing in $\leq$ and has margins then $\mathcal{M}_g$ has a restricted structure: There exist two canonical g-measures μ_g^+ and μ_g^- so that $\mathcal{M}_g$ is a singleton if and only if $\mu_g^+ = \mu_g^-$. However, no examples where $\mu_g^+ \neq \mu_g^-$ were provided. Recently [Bramson, Kalikow, 1993] have produced a continuous, monotone non-increasing g with margins for which $\mu_g^+ \neq \mu_g^-$. Their method is related to the production of phase transition examples in statistical mechanics and the monotonicity of g plays a crucial role in the success of their example.

When carried over to the generalized baker's transformation, the example of Bramson and Kalikow implies that there exists a cut function f with margins so that T is not ergodic. This was also observed in [Bose, 1989] where a more explicit example was produced. Both of these non-ergodic examples have discontinuous (in interval topology) cut functions with unbounded variation. The production of a non-ergodic T with continuous f is an interesting and apparently difficult problem. Our present result may be viewed as a first step in understanding the restriction implied by bounded variation. We do not know at this time whether or not our results may generalize to bounded variation f. Curiously, our method does not work for *non-decreasing* f (see Remark 1.3).

We complete this section by establishing ergodicity of T. Mixing, and the weak-Bernoulli property for P are shown in Section 2.

The Perron-Frobenius operator P on $L_1(I)$ associated with ϕ is defined by

$$\int_{\phi^{-1}A} h \, dm = \int_A Ph \, dm \qquad \forall h \in L_1, \ \forall A \in \mathcal{B},$$

which in our setup may be written explicitly:

$$Ph(x) = f(x) \, h\left(\phi_0^{-1}(x)\right) + (1 - f(x)) \, h\left(\phi_1^{-1}(x)\right).$$

Lemma 1.2. *If $h \in L_1$ is monotone and $c > 0$ satisfies $c \leq f \leq 1-c$ then Ph is monotone and*

$$|Ph(1) - Ph(0)| \leq (1 - c) \, |h(1) - h(0)|.$$

Proof. We assume h is non-decreasing, the argument in the other case being identical. Let $x \leq y$. Then

$$Ph(y) - Ph(x) = f(y) \, h(\phi_0^{-1} y) + (1 - f(y)) \, h(\phi_1^{-1} y)$$

$$- f(x) \, h(\phi_0^{-1} x) - (1 - f(x)) \, h(\phi_1^{-1} x)$$

$$\geq f(y) \, h(\phi_0^{-1} x) + (1 - f(y)) \, h(\phi_1^{-1} x)$$

$$- f(x) \, h(\phi_0^{-1} x) - (1 - f(x)) \, h(\phi_1^{-1} x)$$

$$= (f(x) - f(y)) \left(h(\phi_1^{-1} x) - h(\phi_0^{-1} x) \right) \geq 0,$$

where the first inequality uses the fact that the ϕ_i^{-1} are non-decreasing. For the estimate,

$$Ph(1) = f(1) \, h(a) + (1 - f(1)) \, h(1) \leq c \, h(a) + (1 - c) \, h(1)$$

$$Ph(0) = f(0) \, h(0) + (1 - f(0)) \, h(a) \geq (1 - c) \, h(0) + c \, h(a)$$

and the result follows. ∎

Remark 1.3. The above lemma is false for f non-decreasing.

We denote by $BV(I)$ and $C(I)$ classes of bounded variation and continuous functions on $[0,1]$ respectively. Var_J denotes variation over an interval $J \subseteq I$.

Corollary 1.4. If $h \in BV(I)$, $\|P^n h - \int_I h\, dm\|_\infty \leq (1-c)^n \, \text{Var}_I \, h$. If $h \in C(I)$, $\|P^n h - \int_I h\, dm\|_\infty \longrightarrow 0$.

Proof. Since

$$\int_I Ph\, dm = \int_I h\, dm$$

and $\text{Var}_I \, Ph \leq (1-c) \, \text{Var}_I \, h$ one has

$$\left\| P^n h - \int_I h\, dm \right\|_\infty \leq \sup P^n h - \inf P^n h$$

$$\leq (1-c)^n \, \text{Var}_I \, h.$$

The result for $h \in C(I)$ now follows by approximation with bounded variation functions and the fact that

$$\|Ph\|_\infty \leq \|h\|_\infty \qquad \forall h \in C(I). \qquad \blacksquare$$

We may now prove the first claim of Proposition 1.1. Let μ be any regular Borel probability measure on I which satisfies

$$\mu(Ph) = \mu(h) \qquad \forall h \in C(I). \tag{1.5}$$

Using Corollary 1.4 one has $\mu(h) = \lim_{n\to\infty} \mu(P^n h) = m(h) \quad \forall h \in C(I)$, so Lebesgue measure m is the unique (regular, Borel probability) measure satisfying (1.5). It follows that m is ergodic for ϕ. Conclude that $m \times m$ is ergodic for T since T is the natural extension of ϕ.

2. T Is Bernoulli

As in Section 1, $\bar{P} = T^{-1} P = \{I_0, I_1\}$ is a generator for ϕ on I. If $0 \leq s \leq t < \infty$ are integers we denote by $\bar{P}_{-t}^{-s}$ the partition $\phi^{-s}\bar{P} \vee \phi^{-s-1}\bar{P} \vee \cdots \vee \phi^{-t}\bar{P}$. Observe that if $J \in \bar{P}_{-t}^0$, $c^t \leq m(J) \leq (1-c)^t$. If $s \geq 0$, $\bar{P}_{-\infty}^{-s}$ is the smallest σ-algebra containing all the $\bar{P}_{-t}^{-s}$, $t \geq s$.

Lemma 2.1. ϕ is mixing.

Proof. Let $A \in \bar{P}_{-k}^0$ and $B \in \bar{P}_{-\ell}^0$. Then

$$\left| m(\phi^{-t}A \cap B) - m(A)m(B) \right| = \left| \int_I \chi_A(\phi^t(t))\chi_B(x)\, dm(x) - m(A)mB \right|$$

$$= \left| \int_I \chi_A P^t \chi_B\, dm - m(A)m(B) \right|$$

$$\leq \int_I \chi_A \left| P^t \chi_B - m(B) \right| dm$$

$$\overset{(*)}{\leq} 2m(A)(1-c)^t,$$

where inequality $(*)$ follows from Corollary 1.3 and the fact that $\operatorname{Var}_I \chi_B = 2$. Mixing for arbitrary A and B in $\mathcal{B}$ now follows by standard arguments. ∎

Remark 2.2. One may easily modify the previous proof to show ϕ is exact, but this will follow from the next result.

Lemma 2.3. P is weak-Bernoulli for T.

Proof. We use the fact, due to [del Junco, Rahe, 1979] that if T is mixing and the finite partition P satisfies

$$\sum_n h\left(P \mid P_{-n}^{-1}\right) - h\left(P \mid P_{-\infty}^{-1}\right) < \infty$$

then P is weak-Bernoulli. Set $\Gamma(t) = -t\log_2 t$, and define

$$f_n = E\left(f \mid P_{-n}^{-1}\right) = \sum_{J \in P_{-n}^{-1}} \chi_J \frac{1}{m(J)} \int_J f\, dm.$$

We estimate

$$\sum_n h\left(P \mid P_{-n}^{-1}\right) - h\left(P \mid P_{-\infty}^{-1}\right).$$

Each term in this sum may be written as

$$\int_I \Gamma(f_n) - \Gamma(f)\, dm \quad + \quad \int_I \Gamma(1-f_n) - \Gamma(1-f)\, dm$$

$$= \qquad\qquad A_n \qquad\qquad + \qquad\qquad B_n.$$

For each A_n we may write

$$A_n = \sum_{J \in P_{-n}^{-1}} m(J) \left\{ \frac{1}{m(J)} \int_J \Gamma(f_n) - \Gamma(f)\, dm \right\}$$

$$= \sum_{J \in P_{-n}^{-1}} m(J)\beta(J) \tag{2.4}$$

$$= \sum_{\{J|\beta(J) > \frac{A_n}{2}\}} m(J)\,\beta(J) \quad + \quad \sum_{\{J|\beta(J) \leq \frac{A_n}{2}\}} m(J)\,\beta(J)$$

$$= \sum_{J \in \mathcal{J}_1} m(J)\,\beta(J) \qquad + \qquad \sum_{J \in \mathcal{J}_2} m(J)\,\beta(J).$$

We now observe the following points.

1. $\sum_{J \in \mathcal{J}_1} m(J) \geq A_n$.

2. $\#\mathcal{J}_1 \geq A_n(1-c)^{-n}$.

3. If $\beta(J) = \gamma > 0$ then there exists $x_0 \in J$ satisfying $|f_n(x_0) - f(x_0)| \geq \frac{\gamma}{M}$ where

$$M = \max_{c \leq t \leq 1-c} \Gamma'(t).$$

1. follows from (2.4) and the fact that $0 \leq \beta(J) \leq \frac{1}{2}$. Since $m(J) \leq (1-c)^n$, 2. follows from 1. 3. is elementary. From 3. it follows that if $J \in \mathcal{J}_1$, $\mathrm{Var}_J f \geq \frac{A_n}{2M}$, after which

$$\mathrm{Var}_I f \geq \sum_{J \in \mathcal{J}_1} \mathrm{Var}_J f \geq \sum_{J \in \mathcal{J}_1} \frac{A_n}{2M} \geq \frac{A_n^2}{2M(1-c)^n}$$

so $\sum A_n < \infty$. A similar argument shows $\sum B_n < \infty$ and the lemma is proved. ∎

REFERENCES

1. C. Bose, *Generalized baker's transformations*, Ergod. Th. & Dynam. Sys. **9** (1989), 1-17.

2. M. Bramson and S. Kalikow, *Non-uniqueness in g-functions*, to appear in Israel J. Math.

3. A. del Junco and M. Rahe, *Finitary codings and weak Bernoulli partitions*, Proc. Amer. Math. Soc. **75** (1979), 259-264.

4. P. Halmos, *Lectures on Ergodic Theory*, Chelsea Publishing Company, New York, 1956.

5. P. Hulse, *Uniqueness and ergodic properties of attractive g-measures*, Ergod. Th. & Dynam. Sys. **11** (1991), 65-77.

6. M. Keane, *Strongly mixing g-measures*, Invent. Math. **16** (1974), 309-324.

7. G. Keller, *Generalized bounded variation and applications to piecewise monotonic transformations*, Z. Wahrsch. Verw. Gebiete **69** (1985), 461-478.

8. M. Rahe, *On a class of generalized baker's transformations*, Can. J. Math. **45** (1993), 638-649.

9. P. Walters, *An Introduction to Ergodic Theory*. Springer-Verlag, G.T.M. #79, 1981.

10. 10. P. Walters, *Ruelle's operator theorem and g-measures*, Trans. Amer. Math. Soc. **214** (1975), 375-387.

11. 11. S. Wong, *Some metric properties of piecewise monotonic mappings of the unit interval*, Trans. Amer. Math. Soc. **246** (1978), 493-500.

DEPARTMENT OF MATHEMATICS AND STATISTICS, UNIVERSITY OF VICTORIA, P.O. BOX 3045, VICTORIA, B. C ., CANADA V8W 3P4

E-mail address: cbose@entropy.uvic.ca

ALMOST SURE CONVERGENCE OF PROJECTIONS TO SELFADJOINT OPERATORS IN $L_2(0,1)$

Lech Ciach, Ryszard Jajte, Adam Paszkiewicz

It is well known and rather easy to prove that every selfadjoint operator A in $L_2(0,1)$ with spectrum on the interval $(0,1)$ is a limit of orthogonal projections P_n in the weak operator topology. In this paper we are interested in the almost sure approximation of Af by $P_n f$ for every $f \in L_2(0,1)$. To elucidate a little bit a general situation, let us start with the following example.

Example. Let (e_k) be an orthonormal basis in $L_2(0,1)$ and let λ_k be positive numbers with $0 < \lambda_k < 1$ $(k = 1, 2, \dots)$. We put

$$A = \sum_{k=1}^{\infty} \lambda_k \hat{e}_k,$$

where $\hat{e}_k(f) = (f, e_k)e_k$, for $f \in L_2(0,1)$.

Now we construct a suitable sequence (P_n) of orthogonal projections. By a well-known theorem of Marcinkiewicz [Marcinkiewicz, 1933] there exists a sequence $k_1 < k_2 < \dots$ of positive integers such that the sequence

$$s_n(f) = \sum_{k=1}^{k_n} \hat{e}_k(f)$$

converges almost surely, for every $f \in L_2(0,1)$. Then the sequence

$$\sigma_n(f) = \sum_{k=1}^{k_n} \lambda_k \hat{e}_k(f)$$

also converges almost surely, for every $f \in L_2(0,1)$, i.e. $\sigma_n(f) \to Af$ a.s., for every $f \in L_2$. For every n, it is possible to find an orthonormal system $\{\psi_1^{(n)}, \dots, \psi_{k_n}^{(n)}\}$ which is orthogonal to $\{\varphi_1, \dots, \varphi_{k_n}\}$ and enjoys the following properties:

$$\operatorname{supp} \psi_i^{(n)} \subset \left(\frac{1}{n+1}, \frac{1}{n}\right), \qquad n = 1, 2, \dots, \quad i = 1, 2, \dots, k_n,$$

Research supported by KBN Grant 211529101.

and supp $\psi_i^{(n)} \cap$ supp $f_j^{(n)} = \emptyset$ for $i \neq j$. The existence of such a system $\{\psi_1^{(n)}, \ldots, \psi_{k_n}^{(n)}\}$ follows easily from the general relation between two projections [Stratila, Zsido, 1979, p. 94] $P - P \wedge Q \sim Q^\perp - Q^\perp \wedge P^\perp$ (where $\sim$ denotes unitary equivalence), if we notice that $P - P \wedge Q \neq 0$ for P infinite-dimensional and Q finite-dimensional.

Now it is enough to put, for $f \in L_2(0,1)$,

$$P_n(f) = \sum_{k=1}^{k_n} [\lambda_k \hat{e}_k(f) + (1-\lambda_k)\hat{\psi}_k(f) + \sqrt{\lambda_k(1-\lambda_k)}((f,\psi_k)e_k + (f,e_k)\psi_k)].$$

(P_n) is a sequence of orthogonal projections such that $P_n f \to Af$ a.s., for every $L_2(0,1)$.

The above example embraces to some extent the crucial points of our further considerations. Our main goal is the following

Theorem A. *Let A be a selfadjoint operator on $L_2(0,1)$ with spectrum in the interval $(0,1)$. Then there exists a sequence (P_n) of finite-dimensional orthogonal projections in $L_2(0,1)$ such that*

$$P_n f \to Af \qquad \text{almost surely, for every } f \in L_2(0,1). \tag{$*$}$$

Instead of $(*)$ we shall also write briefly $P_n \to A$ a.s.

The above theorem is an easy consequence of the following general result.

Theorem B. *Let (A_n) be a sequence of finite-dimensional selfadjoint operators on $H = L_2(0,1)$. Assume that (A_n) strongly converges to a selfadjoint operator A. Then there exists a sequence $k_1 < k_2 < k_3 < \ldots$ of positive integers such that $A_n f \to Af$ a.s., for all $f \in H$.*

Theorem B can be treated as an extension of the theorem of Marcinkiewicz mentioned in the example. Having Theorem B, it is easy to prove Theorem A. Indeed, let $0 \leq A \leq I$ and let A_n be of the form

$$A_n f = \sum_{k=1}^{k_n} \lambda_k^{(n)}(f, \varphi_k^{(n)})\varphi_k^{(n)},$$

where $(\varphi_1^{(n)}, \ldots, \varphi_{k_n}^{(n)})$ is an orthogonal basis in H and $A_n \to A$ strongly. By Theorem B we may assume that $A_n \to A$ a.s. Now to conclude the proof it is enough to imitate the procedure described in the example.

Sketch of the proof of Theorem B.

Step 1. Let F be a finite-dimensional projection on $H = L_2(0,1)$ and let (Q_n) be a sequence of projections on H strongly convergent to $\mathbf{1}$. Then, for every $\varepsilon > 0$ there exist a projection R and a positive integer s such that

$$F \leq R \qquad \text{and} \qquad \|R - Q_s\| < \varepsilon.$$

Indeed, we have

$$\|F - FQ_sF\| < \varepsilon^2 \qquad \text{for some } s.$$

We put $R = F + F^\perp \wedge Q_s$. Then $R \geq F$.

Moreover, by [Paszkiewicz, 1986, 2.4 (i) and Paszkiewicz, 1990, 3.2 (f)], we have

$$\|F - [Q_s - F^\perp Q_s]\|^2 = \|F - FQ_sF\| < \varepsilon^2.$$

Step 2. Let (Q_n) be a sequence of finite-dimensional orthogonal projections on $H = L_2(0,1)$ such that $Q_n \to 1$ strongly and $Q_n f \to f$ a.s., for all $f \in H$. Then one can define (by induction, using Step 1) the sequences

$$n(1) < n(2) < \dots$$
$$s(1) < s(2) < \dots$$

of positive integers and the sequences

$$P_1 \leq P_2 \leq \dots$$
$$R_1 \leq R_2 \leq \dots$$

of finite-dimensional orthogonal projections such that

$$R_{k-1} \leq P_k, \qquad k = 2, 3, \dots,$$
$$\|(A_{n(k)} - A)P_k\| < 2^{-k},$$
$$R_k A P_k = A P_k, \qquad R_k A_{n(k)} = A_{n(k)},$$
$$\|(R_k - Q_{s(k)})\| < 2^{-k},$$
$$R_k A P_{k+1}^\perp = 0, \qquad A_{n(k)} P_{k+1}^\perp = 0.$$

Step 3. Using Step 2, we verify that

$$(A - A_{n(k)})f = (A - A_{n(k)})P_k f + A P_k^\perp f - A_{n(k)} P_k^\perp f \to 0$$
$$\text{a.s., for every } f \in L_2(0,1).$$

This is done, among others, by the decomposition $f = f_1 + f_2$, where

$$f_1 = [P_1 f + (P_3 - P_2)f + (P_5 - P_4)f + \dots] + (\bigvee_j P_j)^\perp f$$
$$f_2 = (P_2 - P_1)f + (P_4 - P_3)f + \dots (P_{2k} - P_{2k-1})f + \dots .$$

In particular, we have

$$AP_{2k-1}f_1 = R_{2k-1}AP_{2k-1}f_1 = R_{2k-1}A(P_{2k-1}f_1 + P_{2k}^\perp f_1)$$
$$= R_{2k-1}Af_1 \to Af_1 \qquad \text{a.s.}$$

Analogously,

$$AP_{2k}f_1 \to Af_1.$$

Also, in a similar way, we obtain

$$AP_k f_2 \to Af_2 \qquad \text{a.s.}$$

Consequently,

$$AP_k^\perp f \to 0 \qquad \text{a.s.}$$

By the construction of (P_k) we have immediately $(A - A_{n(k)})f \to 0$ a.s. To show that $A_{n(k)}P_k^\perp \to 0$ a.s. we use the relations

$$A_{n(k)}P_k^\perp = R_{k-1}A_{n(k)}P_k^\perp + R_{k-1}^\perp A_{n(k)}P_k^\perp$$

and

$$R_{k-1}^\perp A_{n(k)}P_k^\perp = (R_k - R_{k-1})A_{n(k)}(P_{k+1} - P_k).$$

Final remarks. The main point in the theorem just proved is to find a single subsequence that works simultaneously for all $f \in L_2(0,1)$. The diagonal method obviously gives the result for f running over a countable subset of $L_2(0,1)$. Such routine arguments are not efficient for the whole L_2-space. In this case, the assumption that the A_n are finite-dimensional (or at least compact) is essential. Namely, as is shown in [Ciach, Jajte, Paszkiewicz, to appear], there exists an increasing sequence (P_n) of (infinite-dimensional) orthogonal projections in $L_2(0,1)$, such that, for any sequence of indices $n(k) \nearrow \infty$, one can find a vector $f \in L_2(0,1)$ such that $P_{n(k)}f$ does not converge a.s.

Marcinkiewicz's idea (developed in our proof of Theorem B) is fruitful and rather necessary. It depends heavily (and, in fact, only) on the existence of finite-dimensional projections P_k tending strongly and almost surely to the identity operator. The approximation $\|(A_n - A)P_k\| < \varepsilon_k$ for n large enough is then obvious. In the paper, we wanted to be as close as possible to the classical result of Marcinkiewicz. That is why Theorem A and Theorem B have been formulated for the Hilbert space $L_2(0,1)$. Clearly, both theorems hold in much more general situations. They are valid at least for $L_2(X, \mathcal{A}, \mu)$ with μ being separable, σ-finite and such that there exists a sequence $(Y_n) \subset \mathcal{A}$ with $0 < \mu(Y_n) \to 0$. For details, see [Jajte, Paszkiewicz, to appear].

1. L. Ciach, R. Jajte, A. Paszkiewicz, *On the almost sure approximation and convergence of linear operators in L_2-spaces* (to appear).
2. R. Jajte, A. Paszkiewicz, *Almost sure approximation of unbounded operators in $L_2(X, \mathcal{A}, \mu)$* (to appear).

3. J. Marcinkiewicz, *Sur la convergence de séries orthogonales*, Studia Math. **6** (1933), 39–45.
4. A. Paszkiewicz, *Measures on projectors in W^*-factors*, J. Funct. Anal. **69** (1986), 87–117.
5. _____, *Convergence in W^*-algebras*, ibid. **90** (1990), 143–154.
6. S. Stratila, L. Zsido, *Lectures on von Neumann algebras*, Abacus Press, 1979.

INSTITUTE OF MATHEMATICS, LÓDŹ UNIVERSITY, UL. BANACHA 22, 90-238 LÓDŹ, POLAND
E-mail address: rjajte@plunlo51.bitnet

QUASI-UNIFORM LIMITS OF
UNIFORMLY RECURRENT POINTS

TOMASZ DOWNAROWICZ

ABSTRACT. A quasi-uniformly convergent sequence which is uniformly recurrent as a point in the product system admits a uniformly recurrent limit. In general, a quasi-uniformly convergent sequence of uniformly recurrent points need not have a uniformly recurrent limit.

NOTATION

By a *dynamical system* we will mean a pair (X, T), where X is a compact metric space and T is a homeomorphism of X onto itself. The *enveloping semigroup* $E \subset X^X$ of the system (X, T) is the pointwise closure of the iterates $\{T^n : n \in \mathbf{Z}\}$. A subset $Y \neq \emptyset$ of X is called *invariant* if it is closed and $TY = Y$. An invariant set is called *minimal* if it contains no proper invariant subsets. The *orbit-closure* $Ex = \{\tau x : \tau \in E\}$ of any element x of a minimal set Y is equal to Y. The points contained in minimal sets will be called *uniformly recurrent*, which refers to a well-known characterization of minimal sets [Gottschalk–Hedlund, 1955]. Since the orbit-closure of any point x, as well as any other invariant set, contains a minimal subset, we conclude that there is a $\tau \in E$ with τx uniformly recurrent. Now, for x uniformly recurrent, each τx is uniformly recurrent and there is a $\tau' \in E$ such that $\tau' \tau x = x$.

For a given dynamical system (X, T) and $A = \mathbf{N}$ or $\mathbf{N}_0$ ($\mathbf{N}_0 = \mathbf{N} \cup \{0\}$, where $\mathbf{N}$ is the natural numbers) we will consider the product system (X^A, T^A), where X^A is endowed with the Tychonov topology, and T^A is given by

$$T^A(x_n) = (Tx_n)$$

(here (x_n) denotes a sequence over $n \in A$). Now, the enveloping semigroup of the system (X^A, T^A) is easily seen to be isomorphic to E, since each of its elements has the form τ^A:

$$\tau^A(x_n) = (\tau x_n),$$

for a certain $\tau \in E$. Since a continuous factor of a minimal set is also minimal, it is obvious that whenever τ^A sends a point (x_n) to a uniformly

Research supported by the KBN grant PB 666/2/91.

recurrent point in X^A, then the same holds for τ and each individual x_n in X.

Below we cite some basic definitions and facts on quasi-uniform convergence. More details on this topic may be found in [Downarowicz–Iwanik, 1988]. Quasi-uniform convergence in a dynamical system (X, T) is given by the *Weyl pseudometric*:

$$D_W(x, y) = \inf\{\delta : BD^*\{n \in \mathbf{Z} : d(T^n x, T^n y) > \delta\} < \delta\},$$

where BD^* denotes the *upper Banach density* defined for subsets B of $\mathbf{Z}$ by the formula

$$BD^*(B) = \varlimsup_{L > 0} \sup_{k \in \mathbf{Z}} L^{-1} |B \cap [k, k + L)|.$$

The quasi-uniform convergence does not depend on the choice of the metric d in X. In symbolic dynamics concerning subshifts over a finite alphabet Λ the quasi-uniform convergence is given by an equivalent pseudometric

$$D_W''(x, y) = BD^*\{n \in \mathbf{Z} : x(n) \neq y(n)\},$$

where $x, y \in \Lambda^{\mathbf{Z}}$.

Another important pseudometric (*Besicovitch*) is given by

$$D_B(x, y) = D^*\{n \in \mathbf{Z} : x(n) \neq y(n)\},$$

where D^* denotes the usual upper density. Clearly $D_B \leq D_W''$, and the convergence in D_B is essentially weaker.

For $x \in X$ let $P(x)$ denote the weakly compact convex set of T-invariant probability measures carried by Ex. Theorem 2 in [Downarowicz–Iwanik, 1988] says that $P(x)$ is a quasi-uniformly continuous function of x for the Hausdorff distance between compact sets of measures.

In [Jacobs–Keane, 1969], a paper on Toeplitz sequences, the authors raised a question whether a quasi-uniform limit of periodic points has to be uniformly recurrent. Recall that their notation of quasi-uniform convergence was slightly different, as it was given by the metric $D_W + d$. A negative answer to their question is given in [Downarowicz–Iwanik, 1988]; neverthe less, a question of this type remains interesting for the convergence in D_W: is a quasi-uniform limit of uniformly recurrent points D_W-equivalent to a uniformly recurrent point? In this note we show that the answer in an important case including convergence of periodic points is positive and we give an example for a negative answer in the general case.

QUASI-UNIFORM LIMITS

Notice that, by the definition of D_W, for any $x, y \in X$ and $n \in \mathbf{N}$, we have

$$D_W(T^n x, T^n y) = D_W(x, y).$$

The following is now easy to prove by a standard approximation argument:

Lemma 1. *For any $x, y \in X$ and $\tau \in E$*
$$D_W(\tau x, \tau y) \leq D_W(x, y).$$

$\square$

Theorem 1. *Let x_0 be a quasi-uniform limit of uniformly recurrent points. Then x_0 is also a quasi-uniform limit of elements of a single minimal subset of Ex_0.*

Proof. Let $(x_n)_{n \in \mathbf{N}}$ be a sequence of uniformly recurrent points quasi-uniformly convergent to x_0. Consider the point $(x_n)_{n \in \mathbf{N}_0}$ and let $\tau \in E$ be such that $\tau^{\mathbf{N}_0}(x_n)_{n \in \mathbf{N}_0}$ is uniformly recurrent in $(X^{\mathbf{N}_0}, T^{\mathbf{N}_0})$. For each $n \in \mathbf{N}$, by minimality of the orbit closure of x_n, we have

$$\tau_n \tau x_n = x_n,$$

for some $\tau_n \in E$. Let

$$y_n = \tau_n \tau x_0.$$

We have

$$\begin{aligned} D_W(x_0, y_n) &\leq D_W(x_0, x_n) + D_W(x_n, y_n) \\ &= D_W(x_0, x_n) + D_W(\tau_n \tau x_n, \tau_n \tau x_0) \leq 2 D_W(x_0, x_n) \to 0, \end{aligned}$$

the last inequality following from Lemma 1. Clearly, each y_n is contained in the minimal orbit-closure of τx_0, which ends the proof. $\square$

Theorem 2. *Let x_0 be a quasi-uniform limit of a sequence $(x_n)_{n \in \mathbf{N}}$ which is uniformly recurrent as an element of $(X^{\mathbf{N}}, T^{\mathbf{N}})$. Then there exists a uniformly recurrent point $x_0' \in Ex_0$, which is D_W-equivalent to x_0.*

Proof. As previously, let $\tau \in E$ be such that $\tau^{\mathbf{N}_0}(x_n)_{n \in \mathbf{N}_0}$ is uniformly recurrent in $(X^{\mathbf{N}_0}, T^{\mathbf{N}_0})$. By minimality of the orbit-closure of $(x_n)_{n \in \mathbf{N}}$, there exists a $\tau' \in E$ such that

$$\tau' \tau x_n = x_n,$$

for each $n \in \mathbf{N}$. Denote

$$x_0' = \tau' \tau x_0.$$

Since τx_0 is uniformly recurrent, so is x_0'. Now, by Lemma 1,

$$D_W(x_0', x_n) = D_W(\tau' \tau x_0, \tau' \tau x_n) \leq D_W(x_0, x_n),$$

hence we obtain that x_0' is another quasi-uniform limit of $(x_n)_{n \in \mathbf{N}}$. $\square$

Important examples of uniformly recurrent points are *almost periodic points*, defined as those points x for which (Ex, T) is a minimal equicontinuous dynamical system. The simplest case of almost periodic points are those with periodic orbits. By the well-known Halmos–von Neumann Theorem, the minimal equicontinuous system (Ex, T) may be viewed as a compact monothetic group rotation (G, g), where g denotes a fixed generator of G. A point x_0 which is a quasi-uniform limit of almost periodic points will be called *Weyl almost periodic*, which refers to a characterization of Weyl almost periodic points in [Iwanik, 1988].

Theorem 3. *Every Weyl almost periodic point x_0 is D_W-equivalent to a uniformly recurrent point x_0'.*

Proof. In view of Theorem 2 it suffices to show that any sequence $(x_n)_{n\in\mathbf{N}}$ of almost periodic points is uniformly recurrent in $(X^{\mathbf{N}}, T^{\mathbf{N}})$. We will show that $(x_n)_{n\in\mathbf{N}}$ is even almost periodic. For each $n \in \mathbf{N}$, let (G_n, g_n) be the monothetic group rotation isomorphic to (Ex_n, T). It is easily seen that the orbit-closure of $(x_n)_{n\in\mathbf{N}}$ in $(X^{\mathbf{N}}, T^{\mathbf{N}})$ is isomorphic to (G, g), where $G \subset \prod_{n\in\mathbf{N}} G_n$ is the compact monothetic subgroup generated by $g = (g_n)_{n\in\mathbf{N}}$. □

In the sequel we will show in Example 2 that the assertion of Theorem 2 need not hold under the weaker assumption of Theorem 1. That example will be based on the following construction:

Example 1. There exists a subshift $Y \subset \{0,1\}^{\mathbf{Z}}$ containing a quasi-uniformly convergent sequence (y_n) but none of its quasi-uniform limits.

Construction. Let $y_0 \in \{0,1\}^{\mathbf{Z}}$ be defined by $y_0(k) = 0$ for all $k \leq 0$ and

$$(y_0(1), y_0(2), y_0(3), \dots) = (1,0,0,1,1,1,1,0,0,0,0,0,0,0,0,\dots),$$

so that the blocks $y[2^{n-1}, 2^n)$ are constant-valued and both values 0 and 1 are assumed infinitely many times. For $n \in \mathbf{N}$, let $A_n \subset \mathbf{N}_0$ be the 2^{2n}-periodic set consisting of the interval $[0, 2^n)$ and all its translates by the multiples of 2^{2n}. Let y_n be given by

$$y_n(k) = \begin{cases} 0, & \text{for all } k \in A_n \\ y_0(k), & \text{elsewhere.} \end{cases}$$

We define Y to be the closed shift-invariant set generated by $\{y_n : n \in \mathbf{N}\}$. Clearly, by the construction of y_n, we have

$$D_W''(y_0, y_n) \leq \frac{2^n}{2^{2n}} \to 0.$$

Let $y \in Y$. We will show that

$$D_W''(y, y_0) > 0.$$

To this end observe that

$$y = \lim_k T^{m_k} y_{n_k},$$

for certain subsequences m_k and n_k. First suppose that $n_k \to \infty$. Since each y_{n_k} is built of constant-valued blocks of lengths at least 2^{n_k}, so is $T^{m_k} y_{n_k}$, and hence y is constant-valued starting at some place. This clearly implies

$$D_W''(y, y_0) = 1.$$

In the case of n_k bounded, y is in the orbit-closure of some y_n, hence the blocks of zeros of the length 2^n appear in y at least along a periodic set. Since y_0 contains arbitrarily long blocks of the symbol 1, we conclude that $D_W''(y, y_0) > 0$. $\square$

Remark. Note that $D_B(y, y_0) > 0$ in both cases considered in the above proof.

The following observation will be useful in the construction of Example 2:

Lemma 2. *Whenever a uniformly recurrent point x_0 occurs as a quasi-uniform limit of a sequence (x_n) contained in a single minimal set Y, then also $x_0 \in Y$.*

Proof. Otherwise the minimal orbit-closure of x_0 would be disjoint from Y, and the collection of invariant measures $P(x_0)$ would be different from $P(Y)$. On the other hand, by Theorem 2 in [Downarowicz–Iwanik, 1988], $P(x_n) \to P(x_0)$ in the Hausdorff distance. Since clearly $P(x_n) = P(Y)$ for each n, we obtain a contradiction. $\square$

We are in a position to give an example which shows, in terms of the topology given by D_W in the appropriate quotient space, that the (classes of) uniformly recurrent points do not form a closed set.

Example 2. There exists a sequence of uniformly recurrent points $(x_n)_{n \in \mathbf{N}}$ which is quasi-uniformly convergent but has no uniformly recurrent limit.

Construction. Let Y, y_n and y_0 be as in Example 1. Let η be a non-regular 0-1 valued Oxtoby sequence in which all the blocks appearing in the subshift Y occur along p-periodic parts (see [Williams, 1984, sec. 4]). More precisely, each block of Y (repeated periodically) is used to fill up the p-periodic part of η, for some period p. We can select a sequence $(x_n)_{n \in \mathbf{N}}$ of elements of the (minimal) orbit closure $E\eta$, all having the same periodic part $B \subset \mathbf{Z}$, such that

$$D^*(\mathbf{Z} \setminus B) > 0,$$

and all equal to one another on B, but differing on $\mathbf{Z} \setminus B$, namely, $\mathbf{Z} \setminus B$ in (x_n) is filled up with y_n (c.f. [Downarowicz, 1991, sec. 5]). It is clear, that $(x_n)_{n \in \mathbf{N}}$ converges quasi-uniformly to the element $x_0 \in \{0, 1\}^{\mathbf{Z}}$ with the same periodic part as in all the (x_n)s, and with $\mathbf{Z} \setminus B$ filled out with y_0. Suppose there exists a uniformly recurrent quasi-uniform limit x of $(x_n)_{n \in \mathbf{N}}$. Then, by Lemma 2, $x \in E\eta$. We will arrive to a contradiction by showing that

$$D_W''(x, x_0) > 0.$$

First suppose that

$$a = x_0(j) \neq x(j) = b,$$

for some $j \in B$, $a, b \in \{0, 1\}$. There is a period p for which

$$x_0(j + kp) = a,$$

for each $k \in \mathbf{Z}$. Clearly, if

$$x(j + kp) = b,$$

for all k, then x differs from x_0 on a periodic set and we are done. Otherwise, the numbers $j + kp$ are all contained in the complement to the p-periodic part of x. Now, from the construction of the Oxtoby sequence η in [Williams, 1984], it is not hard to see that the numbers $j + kp$ visit all the q-periodic parts of x with $q > p$ (then also q is a multiplicity of p). Since in Y there appear arbitrarily long constant-valued blocks, we can find a $q > p$ for which the q-periodic part of x is filled up entirely with the symbol b. Again x differs from x_0 on a periodic set. The only remaining case is when $x = x_0$ on B. Since $x \in E\eta$, the set $\mathbf{Z} \setminus B$ is filled up in x with an element $y \in Y$. Now, by the Remark made to Example 1 and the fact that $\mathbf{Z} \setminus B$ has positive upper density, we conclude that

$$D_B(x, x_0) > 0,$$

and the more $D''_W(x, x_0) > 0$, as desired. $\square$

All the results stated in this work remain valid for noninvertible mappings T on X. In this case E denotes the closure of positive iterates of T, and the upper Banach density is defined analogously for subsets of $\mathbf{N}$. Also, for this notation of E and BD^*, both the examples presented above maintain their desired properties.

REFERENCES

1. Downarowicz, T., *The Choquet simplex of invariant measures for minimal flows*, Isr. J. Math. **74** (1991), 241–256.
2. Downarowicz, T. and Iwanik, A., *Quasi-uniform convergence in compact dynamical systems*, Studia Math. **89** (1988), 11–25.
3. Gottschalk, W. H. and Hedlund, G. A., *Topological dynamics*, vol. 36, Amer. Math. Soc. Colloq. Publ., 1955.
4. Iwanik, A., *Weyl almost periodic points in Topological Dynamics*, Colloq. Math. **56** (1988), 107–119.
5. Jacobs, K. and Keane, M., *0-1 sequences of Toeplitz type*, Z. Wahr. **13** (1969), 123–131.
6. Williams, S., *Toeplitz minimal flows which are not uniquely ergodic*, Z. Wahr. **67** (1984), 95–107.

INSTITUTE OF MATHEMATICS, TECHNICAL UNIVERSITY, 50-370 WROCŁAW, POLAND
E-mail address: downar@math.im.pwr.wroc.pl

STRICTLY NONPOINTWISE MARKOV
OPERATORS AND WEAK MIXING

Tomasz Downarowicz

ABSTRACT. Strictly nonpointwise Markov operators are those which admit no nontrivial pointwise factors. The asymptotic set-system which is responsible for the pointwise factors is then weakly mixing. Examples of nontrivial behavior in this case are given.

INTRODUCTION

Many results proved for Markov operators, especially for their asymptotic behavior, are generalizations of analogous results obtained formerly for topological dynamical systems (see e.g. [Downarowicz–Iwanik, 1988], [Jamison–Sine, 1969], [Lotz, 1968]). A way of extending these results is by viewing the operator as a pointwise transformation on some compact space (the space of probability measures, the space of trajectories, etc.). Sometimes, however, this analogy fails to work. For example, the multiple recurrence theorem for n commuting operators has only been proved in the case of $n = 2$, since the passage to the space of trajectories used in the proof is not possible in higher dimensions [Downarowicz–Iwanik, 1988].

In order to better understand the nature of Markov operators in distinction from topological dynamical systems, this note is devoted to selecting and investigating operators whose behavior has as little in common with a pointwise transformation as possible. For most of the time, we will concentrate on the case of a minimal (irreducible) action of the operator.

In Section 1, we introduce the definition of strictly nonpointwise Markov operators and derive a basic spectral property. Section 2 contains preliminaries for further deductions. In Section 3, the notion of the asymptotic set-system associated with a Markov operator is introduced, the carrier of the pointwise type behavior that still remains. Then dynamical properties of the asymptotic set-system are investigated. For strictly nonpointwise Markov operators, the asymptotic set-system is proved to be mixing in a rather strong sense (stronger than weakly mixing). Section 4 contains examples illustrating that, even with trivial asymptotic set-system, the action of the operator may remain involved. Finally, in Section 5, we give an example of a strictly nonpointwise Markov operator with nontrivial asymptotic set-system.

Research supported by the KBN grant PB 666/2/91.

SECTION 1. TRANSITION SYSTEMS

Let X be a compact metric space and let 2^X denote the compact space of all nonempty compact subsets of X, endowed with the Hausdorff distance.

Definition 1. By a *transition map* on X we mean a lower semi-continuous map $\psi : X \mapsto 2^X$, i.e. such that if $U \in X$ is open, then $\{x : \psi(x) \cap U \neq \emptyset\}$ is also open in X.

The domain of ψ can be extended to 2^X by letting

$$\psi(F) = \overline{\bigcup_{x \in F} \psi(x)},$$

$F \in 2^X$. The pair (X, ψ) will be called a *transition system*. Examples of transition systems are provided by Markov operators: a *Markov operator* on $C(X)$, the space of all continuous functions on X, is a linear nonnegative operator $T : C(X) \mapsto C(X)$ leaving the constant functions invariant. It is a classical fact that each Markov operator T is described by a *Feller transition probability*, a continuous map $P_T : X \mapsto P(X)$ into the space of all Borel probability measures on X, endowed with the weak* topology. The relation is given by the formula

$$Tf(x) = \int f \, dP_T(x),$$

for each $f \in C(X)$ (see [Jamison–Sine, 1969]). Defining

$$\psi_T(x) = \operatorname{supp} P_T(x),$$

where supp denotes the topological support of a measure, we obtain a transition map ψ_T associated with the Markov operator T. Conversely, using a continuous selection from the multifunction sending each $x \in X$ to the set of all probability measures with support equal to $\psi(x)$ (see Theorem 3.1''' in [Michael, 1956]), it can be proved that every transition map ψ is associated with some Markov operator T. It is straightforward that, for each natural $n \geqslant 1$, the iterate $(\psi_T)^n$ is the same as the transition map ψ_{T^n} associated with the iterate T^n.

A *topological dynamical system* is a special case of a transition system which occurs when the transition map ψ sends points to points. The map ψ can then be viewed as a continuous transformation $\psi : X \mapsto X$ (see [Gottschalk–Hedlund, 1955] for a general reference on topological dynamical systems). The associated Markov operator is then called *deterministic* or *pointwise generated*, and it is given by the formula

$$T_\psi f(x) = f\psi(x)$$

(see also [Jamison–Sine, 1969]).

The transition system carries all topological properties of the associated Markov operator, hence, in the considerations of this work, we may concentrate almost exclusively on transition systems.

Let a transition system (X, ψ) be given. A nonempty closed subset F of X is called *invariant* if $\psi(F) \subset F$. An invariant subset F is called *minimal* if it contains no proper invariant subsets. The above notions of invariant and minimal sets agree with the classical notions of the same objects in topological dynamics and theory of Markov operators (in the later the names self-supporting and irreducible are also used, respectively).

Let Y be another compact metric space, and let a map $\tau : X \mapsto Y$ be continuous. By Π_τ we denote the upper semi-continuous partition of X by the level sets of τ. We say that τ is a *factor map* of the transition system (X, ψ) if the partition Π_τ is invariant, i.e., for every $F \in \Pi_\tau$ we have $\psi(F) \subset E$, for some $E \in \Pi_\tau$. Suppose τ is a factor map. Then a transformation $\phi : Y \mapsto Y$ can be defined by

$$\phi = \tau \psi \tau^{-1},$$

and, by an elementary argument, it can be verified that ϕ is continuous. So, (Y, ϕ) is a topological dynamical system.

Definition 2. (cf. the definition of a cycle in [Jamison–Sine, 1969]). The triplet (Y, ϕ, τ) is called a *pointwise factor* of (X, ψ).

We agree to identify factors which yield the same partition of X.

Definition 3. A pointwise factor (Y, ϕ, τ) of (X, ψ) is called *maximal* if there exists no other pointwise factor (Y', ϕ', τ') with the partition $\Pi_{\tau'}$ inscribed into Π_τ.

A proof of the following statement is a standard argument and it will be omitted:

Theorem 1. *For every transition system (X, ψ) there exists a unique maximal pointwise factor of (X, ψ).* $\square$

Definition 4. A transition system (X, ψ) will be called *strictly nonpointwise* if the maximal pointwise factor of (X, ψ) is trivial, i.e., the corresponding partition is $\{X\}$.

We agree to call a Markov operator T *strictly nonpointwise* if so is its associated transition system. Strictly nonpointwise Markov operators appear as an antitype to topological dynamical systems, and hence the study of their behavior may provide information complementary to that arising from topological dynamics.

Theorem 2. *If T is a strictly nonpointwise Markov operator on $C(X)$ then 1 is the only unimodular eigenvalue of T and the only eigenfunctions are the constant functions.*

Proof. The partition of X by the level sets of all the eigenfunctions pertaining to the unimodular eigenvalues provides a pointwise factor of (X, ψ_T) (see Theorem 4.2 in [Lotz, 1968]). By assumption, this factor must be trivial, hence all the eigenfunctions are constant. $\square$

The case of (X, ψ) minimal is of special interest. Like in the classical case of a topological dynamical system, the existence of a minimal subset in every transition system (X, ψ) is guaranteed by essentially the same argument. The same applies to the fact that (X, ψ) is minimal if and only if

$$\overline{\bigcup_{n \geqslant 1} \psi^n(x)} = X,$$

for every $x \in X$. It is now obvious that if (X, ψ) is minimal, so is its maximal pointwise factor.

Remark. In topological dynamics a *maximal uniformly continuous factor* has been proved to exist for every minimal topological dynamical system (see [Ellis–Gottschalk, 1960]). Now, for an irreducible Markov operator T, it can be proved that the factor defined by the eigenfunctions pertaining to the unimodular eigenvalues is equal to the maximal uniformly continuous factor of the maximal pointwise factor of (X, ψ_T).

SECTION 2. S-SEQUENCES

Some of the objects and properties discussed in this section appear in the literature in various contexts (see e.g. [Auslander, 1963], [Clay, 1963], [Glasner–Maon, 1989]). However, in order to establish a uniform notation for our deductions, we introduce the following definitions. The facts stated without a proof are to be derived by standard arguments.

Let $\mathbf{N}$ and $\mathbf{R}_+$ denote the sets of natural numbers (positive integers) and of positive real numbers, respectively.

Definition 5. A subset $M \subset \mathbf{N}$ is called an *S-set* if there exists a topological dynamical system (Y, ϕ) containing a unique minimal subset F, an open set U, $F \subset U$, and a point $y \in Y$ such that

$$M = \{n : \phi^n(y) \in U\}.$$

The S-sets are also characterized by the following property: There exists a function $k : \mathbf{N} \mapsto \mathbf{N}$ such that for every $n \in \mathbf{N}$ the intersection of M with an arbitrary interval of length $k(n)$ contains an interval of length n. The S-sets which satisfy the above with respect to an a priori preset function $k : \mathbf{N} \mapsto \mathbf{N}$ will be called $k(n)$-*sets*.

By an *S-sequence* (or $k(n)$-*sequence*), we shall mean the characteristic 0-1 valued function of an S-set (or $k(n)$-set, respectively) viewed as an element of the one-sided 0-1 shift system. Note that for a given $k : \mathbf{N} \mapsto \mathbf{N}$, the collection of all $k(n)$-sequences forms a compact shift-invariant set with the unique minimal subset $\{(1,1,1,\dots)\}$.

Recall that two points y_1 and y_2 of a topological dynamical system (Y,ϕ) are called *proximal* provided the orbit-closure of the pair (y_1,y_2) in the product system $(Y \times Y, \phi \times \phi)$ intersects the diagonal.

Definition 6. Two points y_1 and y_2 of a topological dynamical system (Y,ϕ) are called *S-proximal* if the orbit-closure of the pair (y_1,y_2) in the product system contains a unique minimal subset F, and F is contained in the diagonal.

Analogously, we have the following equivalent condition: there exists a function $k : \mathbf{R}_+ \mapsto \mathbf{N}$ such that for every $\epsilon > 0$, every interval of length $k(\epsilon)$ contains an n for which $\phi^n(y_1)$ and $\phi^n(y_2)$ are within ϵ-distance from each other. The pairs satisfying the above with respect to an a priori preset function $k : \mathbf{R}_+ \mapsto \mathbf{N}$ will be called $k(\epsilon)$-*proximal* . The relation of S-proximality is transitive but not in general closed, while $k(\epsilon)$-proximality is closed but not transitive.

Recall that one of equivalent conditions for a minimal topological dynamical system to be *weakly mixing* is that the smallest equivalence relation containing proximality is dense in the product. Analogously, we introduce the following notations:

Definition 7. A minimal topological dynamical system (X,ψ) is called *S-mixing* ($k(\epsilon)$-*mixing*) if S-proximality (or the smallest equivalence relation containing $k(\epsilon)$-proximality, respectively) is dense in the product $X \times X$.

Obviously, $k(\epsilon)$-mixing implies S-mixing implies weakly mixing, while the converse implications do not hold in general.

In symbolic dynamics, in particular in the case when (Z,s) is a one-sided 0-1 subshift, we may replace the notions of $k(\epsilon)$-proximality and $k(\epsilon)$-mixing by analogous notions of $k(n)$-proximality and $k(n)$-mixing:

Definition 8. Given a function $k : \mathbf{N} \mapsto \mathbf{N}$. Two elements of a symbolic system (Z,s) are said to be $k(n)$-*proximal* if they match along a $k(n)$-set. (Z,s) is said to be $k(n)$-*mixing* if the smallest equivalence relation containing $k(n)$-proximality is dense in the product.

Clearly, $k(n)$-proximality ($k(n)$-mixing) is equivalent to $k'(\epsilon)$-proximality ($k'(\epsilon)$-mixing) for an appropriately chosen function $k' : \mathbf{R}_+ \mapsto \mathbf{N}$. Two elements of a symbolic system are S-proximal if and only if they match along an S-set.

Section 3. Enveloping and asymptotic set-systems

We now return to transition systems. For the reminder of this paper we will assume that the transition map $\psi : X \mapsto 2^X$ is continuous. By extending ψ, $(2^X, \psi)$ becomes a topological dynamical system. In order to distinguish the systems whose elements are sets we shall call them *set-systems*.

Definition 9. Let (X, ψ) be a continuous transition system . By the *enveloping set-system* we will mean the action of ψ on the family

$$S = \overline{\{\psi^n(x) : x \in X, \ n \in \mathbf{N}\}},$$

where closure is taken in 2^X.

As a consequence of Zorn's Lemma, it is seen that each element v of S is contained in a maximal (in terms of inclusion) element v_0 of S. We shall call these elements *maximal sets*.

Throughout the reminder of this work we will also assume that the transition system (X, ψ) is minimal.

Lemma 1. *For every* $x \in X$ *and every maximal set* $v_0 \in S$ *there exists a sequence* n_k *such that*

$$\lim_k \psi^{n_k}(x) = v_0.$$

Proof. Since $v_0 \in S$, we have

$$v_0 = \lim_i \psi^{n_i}(x_i),$$

for some sequences $n_i \in \mathbf{N}$ and $x_i \in X$. By minimality, for each i

$$x_i = \lim_j \psi^{n_{i,j}}(x),$$

$n_{i,j} \in \mathbf{N}$. The sequence n_k can now be chosen diagonally from $n_{i,j}$. $\square$

Note also that maximal sets cover all of X, since otherwise their union (equal to the union of all elements of S) would provide a closed invariant subset of X.

The following observation is basic for the further deductions:

Theorem 3. *Let* (X, ψ) *be a continuous and minimal transition system. There exists a unique minimal subsystem* (Z, ψ) *of the enveloping set-system* (S, ψ), *and* Z *contains all maximal sets.*

Proof. Suppose we are given two minimal subsets Z_1 and Z_2 of S, and let $x_1 \in v_1 \in Z_1$ and $x_2 \in v_2 \in Z_2$. Choose an arbitrary maximal set v_0. By Lemma 1,

$$v_0 = \lim_k \psi^{n_{1,k}}(x_1) = \lim_k \psi^{n_{2,k}}(x_2),$$

for some sequences $n_{1,k}$ and $n_{2,k}$. If x_1 and x_2 are replaced by v_1 and v_2, in the above formula, respectively, then the two limits become supersets of v_0. By maximality of v_0, equalities are retained. Thus, $v_0 \in Z_1 \cap Z_2$. Since minimal sets are either disjoint or equal, we obtain that $Z_1 = Z_2$. $\square$

Definition 10. The unique minimal subsystem (Z, ψ) of (S, ψ) will be called the *asymptotic set-system* for (X, ψ).

The name "asymptotic" is motivated by the fact that for each $\epsilon > 0$ and $x \in X$, the images $\psi^n(x)$ are within ϵ-distance from the elements of Z along an S-set of n's. Using an argument like in the proof of Theorem 4, it can be derived that there exists a common function $k : \mathbf{R}_+ \mapsto \mathbf{N}$ which governs the distances of the consecutive $\psi^n(x)$'s from Z, for every $x \in X$. This fact, however, will not be used in the sequel.

Lemma 2. *Nondisjoint elements of the enveloping set-system are proximal.*

Proof. Let $x \in v_1 \cap v_2$, where $v_1, v_2 \in S$. By Lemma 1, we have

$$\lim_k \psi^{n_k}(x) = v_0,$$

with v_0 a maximal set and $n_k \in \mathbf{N}$. Substituting x in the above formula subsequently by v_1 and v_2, we obtain two supersets of v_0. By maximality, equality holds each time. Thus, the sequence of iterates ψ^{n_k} brings v_1 and v_2 to the same limit. $\square$

Theorem 4. *Let (X, ψ) be a transition system which is continuous and minimal. Then there exists a function $k : \mathbf{R}_+ \mapsto \mathbf{N}$ such that any two nondisjoint elements v_1 and v_2 of the enveloping set-system (S, ψ) are $k(\epsilon)$-proximal.*

Proof. The family of nondisjoint pairs of sets (v_1, v_2) in the product $S \times S$ is closed and invariant. By Lemma 2, all minimal subsets of this family are contained in the diagonal. But, by Theorem 3, there is a unique minimal subset of the diagonal. Thus, all nondisjoint pairs are S-proximal. In order to prove the existence of a common function $k : \mathbf{R}_+ \mapsto \mathbf{N}$ suppose the opposite. Then we could find an $\epsilon > 0$ and a sequence of nondisjoint pairs $(v_{1,n}, v_{2,n})$ which remain ϵ-apart for n consecutive iterates, $n \in \mathbf{N}$. A cluster point (v_1, v_2) of these pairs would provide a nondisjoint and nonproximal pair, a contradiction to Lemma 2. $\square$

In the context of studying pointwise factors it is the asymptotic set-system that plays the most important role:

Theorem 5. *The maximal pointwise factor of a continuous and minimal transition system (X, ψ) is identical with the factor (Y, ϕ), of the asymptotic set-system (Z, ψ), induced by the smallest closed (invariant) equivalence relation containing nondisjointness.*

Proof. Let $\pi : Z \mapsto Y$ be the corresponding factor map. Let $\tau : X \mapsto Y$ be given by

$$\tau(x) = \pi(v),$$

where $x \in v \in Z$. It is obvious that the above definition does not depend on the choice of v containing x. For continuity of τ let $x_n \to x_0$ in X. Pick $v_n \in Z$ such that $x_n \in v_n$ and choose a subsequence of v_n convergent to some v_0. Obviously $x_0 \in v_0$. By continuity of π, we obtain

$$\tau(x_n) = \pi(v_n) \to \pi(v_0) = \tau(x_0).$$

Clearly, τ preserves the system action, and hence it provides a pointwise factor of the transition system (X, ψ). Maximality follows, since no invariant partition can cut the elements of Z. $\square$

Combining the last two results, we deduce the following:

Theorem 6. *If (X, ψ) is a continuous, minimal, and strictly nonpointwise transition system then the asymptotic set-system (Z, ψ) is $k(\epsilon)$-mixing for some function $k : \mathbf{R}_+ \mapsto \mathbf{N}$ (the same as in Theorem 4).* $\square$

SECTION 4. TRIVIAL ASYMPTOTIC SET-SYSTEM.

Example 1. Let X be a connected compact metric space and let $\psi(x)$ be the ϵ-ball centered at x (ϵ is given). We have $\psi^n(x) \underset{n}{\to} X$, for each $x \in X$, hence $Z = \{X\}$.

Generally, whenever $X \in S$, we have $Z = \{X\}$, by uniqueness of the minimal subset of S, and, by Theorem 5, (X, ψ) is strictly nonpointwise. We shall call this case *trivial asymptotic set-system*. The following example of trivial asymptotic set-system will also be used in further construction in Section 5.

Example 2. Let a function $k : \mathbf{N} \mapsto \mathbf{N}$ be given. A subset M of natural numbers is called a $\widetilde{k(n)}$-*set* if for each n the intersection of M with every interval I of length $k(n)$ contains an interval of length n, provided the interval I starts from a number at least $n + 1$. Obviously, every $k(n)$-set is also a $\widetilde{k(n)}$-set and each $\widetilde{k(n)}$-set is a $(k(n) + n)$-set. Let $\widetilde{K}$ denote the compact shift-invariant set of all $\widetilde{k(n)}$-*sequences* (characteristic 0-1 functions of the $\widetilde{k(n)}$-sets). We define the transition map ψ on $\widetilde{K}$ by the following formula:

$$\psi(x) = \begin{cases} \{y \in \widetilde{K} : y(j) = x(j+1) \text{ for } j \geqslant \frac{m(x)}{2}\}, & m(x) < \infty \\ \widetilde{K}, & x = (1, 1, 1, \dots), \end{cases}$$

where $m(x)$ denotes the position of the earliest digit 0 in x.

It is not very hard to see, from the definition of the $\widetilde{k(n)}$-sequences, that if we replace in some $x \in \widetilde{K}$ an initial block of length at most $\frac{m(x)}{2} - 1$ by any block of the same length appearing anywhere else in the $\widetilde{k(n)}$-sequences, we still obtain a $\widetilde{k(n)}$-sequence. The map ψ is the composition of the above operation with the shift s. Continuity of ψ is straightforward. The transition system $(\widetilde{K}, \psi)$ is minimal. Indeed, every $x \in \widetilde{K}$ contains arbitrarily long blocks of ones, thus $\widetilde{K}$ can be obtained as a limit of a subsequence of $\psi^n(x)$. This also proves that $\widetilde{K}$ itself appears as an element of the enveloping set-system, which implies trivial asymptotic set-system. Nevertheless, the sequence $\psi^n(x)$ does not converge to X, except for those x's which are, starting at some place, constantly 1.

Example 3. We now present a construction of a transition system (X, ψ), with trivial asymptotic set-system, where, unlike in the previous two examples, $\psi^n(x)$ does not converge to X for any $x \in X$. The consecutive images of each point alter between being close to X, in Hausdorff distance, and being small. Consider the following 0-1 matrices:

$$\begin{bmatrix} 0 & 1 & 0 \\ 0 & 0 & 0 \end{bmatrix}, \begin{bmatrix} 0 & 1 & 1 & 0 \\ 1 & 1 & 1 & 1 \end{bmatrix}$$

and let us call them the 2-*blocks*. Suppose, we have already defined all n-*blocks* $(n \geqslant 2)$, and they are of dimensions either $n \times (L_n - 1)$ or $n \times L_n$. Then we let $L_{n+1} = 2L_n(L_n - 1) - 1$ and we list all possible matrices, of dimensions $n \times (L_{n+1} - 1)$ and $n \times L_{n+1}$, which are concatenations of the n-blocks. Finally, to each such matrix we attach the $n + 1^{st}$ bottom row which consists entirely either of zeros, for the shorter matrices, or of ones, for the longer matrices. In this manner all $(n+1)$-blocks have been defined. Below are examples of 3-blocks:

$$\begin{bmatrix} 0110 & 0110 & 010 & 0110 & 010 & 0110 \\ 1111 & 1111 & 000 & 1111 & 000 & 1111 \\ 0000 & 0000 & 000 & 0000 & 000 & 0000 \end{bmatrix} \quad (3 \times 22),$$

$$\begin{bmatrix} 010 & 0110 & 010 & 010 & 010 & 010 & 0110 \\ 000 & 1111 & 000 & 000 & 000 & 000 & 1111 \\ 111 & 1111 & 111 & 111 & 111 & 111 & 1111 \end{bmatrix} \quad (3 \times 23).$$

We now define X to be the space of all $(\mathbf{N} \times \mathbf{N})$-arrays:

$$x = \{x(i,j) : \ i \in \mathbf{N}, \ j \in \mathbf{N}\},$$

such that, for every $n \geqslant 2$, the $(n \times \mathbf{N})$-subarray (the upper n rows of x) is a concatenation of the n-blocks, except that at the beginning an incomplete n-block may appear (with few initial columns missing). We call this portion the *incomplete n-opening* of x. If x starts with a complete n-block, then we call it the *complete n-opening* of x. Since L_{n+1} and L_n are relatively prime, every row contains both zeros and ones, hence the length of the n-opening can be determined by looking only at the nth row. Endow X with any metric for the coordinatewise convergence. The transition map ψ is now defined as follows:

$$\psi(x) = \{y \in X : y(i, j+1) = x(i,j), \; i \in \mathbf{N}, \; j \in \mathbf{N}\}.$$

In other words, $\psi(x)$ is the preimage of x by the horizontal shift. For continuity of ψ we need to show that, if x and x' are close enough, then the arrays appearing in $\psi(x)$ admit the same patterns on the upper n positions of the first column as those in $\psi(x')$. Let x and x' be so close that they have the same $(n+1)$-opening. If it is incomplete, then the patterns admitted at the upper $n+1$ positions of the first column in $\psi(x)$ and in $\psi(x')$ are the same: they arise from all possible completions of the common $(n+1)$-opening to an $(n+1)$-block. If the common $(n+1)$-opening is complete, then the $(n+1)$-opening of every array both in $\psi(x)$ and in $\psi(x')$ is a single column identical with the last column of some $(n+1)$-block. The length of the $(n+1)$-block admitted here may be different for $\psi(x)$ and $\psi(x')$, as it depends on the $n + 2^{nd}$ row. Regardless to this possible difference, the upper n positions of the first column are admitted the same patterns in $\psi(x)$ and $\psi(x')$: namely all appearing as last columns of the n-blocks. This follows from the observation that all possible n-blocks are admitted at the end of both shorter and longer $(n+1)$-blocks. So, continuity of ψ is proved. Next, pick an arbitrary $x \in X$. Let k' be the index of the last column in the $(n+1)$-opening of x. Obviously, $k' \leqslant L_{n+1}$. We let

$$k = L_{n+1} - k' + L_n(L_n - 1) - 1.$$

Observe that the $(n+1)$-opening of each array in $\psi^k(x)$ is about a half of an $(n+1)$-block (its length equals $L_n(L_n - 1)$ or $L_n(L_n - 1) - 1$). The n-openings of such arrays may be admitted any portion of any n-block: it follows from the fact that any number $\leqslant L_n$ can be completed as well to $L_n(L_n - 1)$ as to $L_n(L_n - 1) - 1$ by adding multiplicities of the numbers L_n and $L_n - 1$. This implies that $\psi_k(x)$ is close to X in the Hausdorff distance. Thus, the trivial asymptotic set-system follows. On the other hand, we will show that, for each $n \in \mathbf{N}$, there exists $k \geqslant L_n$ such that all arrays in $\psi^k(x)$ have zero at the initial position of the first row. Namely, let k' be the index of the last column in the n-opening of x. Since the last row of this opening determines the length of the n-block from which it comes, a number k'' can be found, for which $\psi^{k''}(x)$ admits but complete n-openings ($k'' = L_n - k'$ or $k'' = L_n - 1 - k'$). Since k'' is too small, we let

$$k = k'' + L_n.$$

Now, the n-opening of an element of $\psi^k(x)$ is either complete or it is a single column, identical with the last column of an n-block. In either case, the initial entry is zero. Therefore, as desired, $\psi^n(x)$ does not converge to X.

Section 5. The nontrivial case

On the extremity opposite to the trivial asymptotic set-system occur the transition systems whose asymptotic set-system consists of pairwise disjoint sets. Then these are all maximal sets, and, by Theorem 5, the asymptotic set-system itself provides the maximal pointwise factor. Most interesting seems to be the intermediate case. For a long time, an example of a minimal and strictly nonpointwise transition system which admits a nontrivial asymptotic set-system was not known. The construction presented below was inspired by Theorem 6.

Example 4. Let (Z, s) be some minimal one-sided 0-1 subshift which is $k(n)$-mixing for some function $k : \mathbf{N} \mapsto \mathbf{N}$. An explicit example of such a subshift is given at the end of this paper. Next, let $(\widetilde{K}, \psi)$ be the transition system of Example 2. For each pair $z \in Z$, $x \in \widetilde{K}$, let $z \circ x$ denote the sequence on three symbols 0, 1, 2, given by the formula

$$z \circ x \, (i) = x(i)(z(i) + 1),$$

$i \in \mathbf{N}$. Note that $z_1 \circ x_1 = z_2 \circ x_2$ if and only if $x_1 = x_2$, and z_1 matches with z_2 along the set $\{i \in \mathbf{N} : x_1(i) = 1\}$, the $\widetilde{k(n)}$-set corresponding to x_1. We assign X to be the space of all sequences of the form $z \circ x$, $z \in Z$, $x \in \widetilde{K}$, endowed with the following transition map ϕ:

$$\phi(z \circ x) = s(z) \circ \psi(x) \quad (= \{s(z) \circ y : y \in \psi(x)\}).$$

It is easily seen that the map ϕ is well defined and continuous. Observe that $\phi^n(z \circ x) \subset s^n(z) \circ \widetilde{K}$, for each z, x, and n, whereas for $x = (1, 1, 1, \dots)$ we have equality. Consider two arbitrary elements, $z \circ x$ and $z_1 \circ x_1$, of X. Since $\psi^n(x)$ has a subsequence convergent to $\widetilde{K}$ (see Example 2), $\phi^n(z \circ x)$ has a subsequence convergent to some $z' \circ \widetilde{K}$, which then, by minimality of Z can be sent to $z_1 \circ \widetilde{K} \ni z_1 \circ x_1$, and minimality of the transition system (X, ϕ) follows. Moreover, it is now seen that the sets $z \circ \widetilde{K}$ are the maximal sets, and they constitute the asymptotic set-system isomorphic to (Z, s). Finally, notice that two sets $z_1 \circ \widetilde{K}$ and $z_2 \circ \widetilde{K}$, corresponding to some $k(n)$-proximal elements z_1 and z_2 of Z, are nondisjoint, since z_1 matches with z_2 along some $k(n)$-set, hence a $\widetilde{k(n)}$-set, induced by an element of $\widetilde{K}$. By the assumption of $k(n)$-mixing made on (Z, s), and by Theorem 5, the maximal pointwise factor of (X, ϕ) is trivial.

Example of a minimal $k(n)$-mixing 0-1 subshift. We start the construction with a recursive definition of two sequences of blocks:

$$a_0 = 1 \quad \text{and} \quad b_0 = 01,$$

then, for $n \geqslant 0$,

$$a_{n+1} = b_n a_n,$$

and

$$b_{n+1} = b_n b_n \dots b_n a_n,$$

where b_n is repeated p_n times. The quantities p_n will be specified later. For example, if $p_0 = 2$, then

$$a_1 = 011 \quad \text{and} \quad b_1 = 01011.$$

We let b be the one-sided sequence:

$$b = \lim_n b_n,$$

and we let Z be the shift orbit-closure of b. Note that for each $n \geqslant 0$, every $z \in Z$ is a concatenation of the blocks b_n and a_n, except perhaps for the initial incomplete block. By a *regular occurrence* of a_n in z we shall mean an occurrence of a_n in this concatenation, in distinction from occurrences within the blocks b_n. Minimality of Z follows, since, for each n, b_n occurs in b with bounded gaps. In order to prove the $k(n)$-mixing property, we first inductively check the following statement:

For each $z \in Z$, $n \geqslant 0$, and $m \leqslant n$, $s^{|a_n|}(z)$ matches with z at every regular occurrence of a_m in $s^{|a_n|}(z)$,

where $|\cdot|$ denotes the length of a block. The statement holds for $n = 0$, since every regular occurrence of $a_0 = 1$ is preceded by the symbol 1. Suppose the statement holds for some $n \geqslant 0$. Pick a regular occurrence of a_m in $s^{|a_{n+1}|}(z)$, so far for $m \leqslant n$. Possible are three cases:

(1) $m < n$ and our a_m is a part of some b_n preceded by another b_n. Then $s^{|a_{n+1}|}(z)$ matches with $s^{|a_n|}(z)$ at the entire block b_n containing our a_m, since the two sequences differ in shift exactly by $|b_n|$. At the same time our a_m meets a regular occurrence of a a_m in $s^{|a_n|}(z)$, and hence, by the inductive assumption, it matches with the corresponding part of z (see Fig. 1).

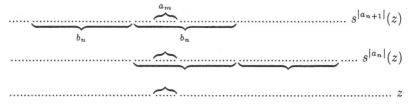

FIG. 1

(2) $m < n$ and our a_m is a part of a b_n preceded by a regular occurrence of a_n. Every such occurrence of a_n is then preceded by another b_n, thus matching of the corresponding blocks b_n in $s^{|a_{n+1}|}(z)$ and z follows directly, since $|a_{n+1}|$ is the sum of $|a_n|$ and $|b_n|$ (see Fig. 2).

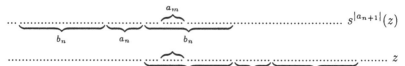

FIG. 2

(3) $m \leqslant n$ and our a_m is a part (or whole) of some regular a_n. Then the a_n is preceded by a b_n and next by either another a_n or another b_n. But the ending of b_n is identical with a_n, thus, in either case, the desired matching follows, as in the case (2) (see Fig. 3).

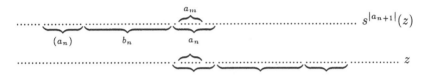

FIG. 3

Finally, for $m = n + 1$, pick a regular occurrence of a_{n+1} in $s^{|a_{n+1}|}(z)$. It is preceded by b_{n+1}, thus, our a_{n+1} meets the ending of a b_{n+1} in z, identical with a_{n+1} (see Fig. 4).

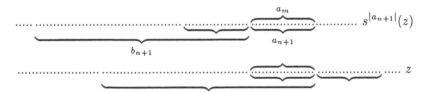

FIG. 4

Let now t be a limit transformation for a convergent subsequence of $s^{|a_n|}$. From the statement, proved above, it follows that $t(z)$ matches with z at every regular occurrence of a_n in $t(z)$, for each $n \geqslant 0$, which means that $t(z)$ and z are $k(n)$-proximal, with the function $k : \mathbf{N} \mapsto \mathbf{N}$ independent from the choice of $z \in Z$. We shall now point out a subsequence of $t^n(b)$ convergent to $s(b)$, which is sufficient for $k(\epsilon)$-mixing. Fix some $n \geqslant 0$. Recall that b starts with

$$b_n b_n \dots b_n a_n,$$

where b_n is repeated p_n times. Observe that $t(b)$ starts with

$$b_n^1 b_n b_n \dots b_n a_n,$$

where b_n is repeated $p_n - 2$ times, and b_n^1 is the part of b_n with the initial $|a_n|$ symbols missing. (This statement is obvious if t is replaced by $s^{|a_{n+1}|}$. But further shifting by $|b_{n+2}|$, $|b_{n+3}|$, etc., does not alter this beginning). Now, $t^2(b)$ starts with

$$b_n^2 b_n b_n \dots b_n a_n,$$

where b_n repeats at least $p_n - 4$ times, while b_n^2 is the part of b_n with the initial $(2|a_n|)_{mod|b_n|}$ symbols missing. We now specify the quantities p_n.

Suppose $p_0, p_1, \ldots, p_{n-1}$ are such that $|a_n|$ and $|b_n|$ are relatively prime. Choose p_n large enough ($p_n > 2|a_n||b_n|$ would suffice), so that, for some m, $t^m(b)$ starts with

$$b_n^m b_n b_n \ldots b_n a_n,$$

where b_n^m is the part of b_n with exactly one symbol missing, and b_n is still repeated at least once. At the same time p_n can be chosen be so that $|a_{n+1}|$ and $|b_{n+1}|$ are again relatively prime. In this manner $t^m(b)$ is obtained close to $s(b)$, as desired.

Acknowledgments. The author would like to thank H. Furstenberg for his inspiration and D. Rudolph for giving a clue of Example 3.

REFERENCES

1. Auslander, J., *On the proximal relation in topological dynamics*, Proc. Amer. Math. Soc. **11** (1960), 890–895.
2. Clay, J., *Proximality relation in transformation groups*, Trans. Amer. Math. Soc. **108** (1963), 88–96.
3. Downarowicz, T. and Iwanik, A., *Multiple recurrence for discrete time Markov processes, II*, Colloq. Math. **55** (1988), 311–316.
4. Ellis, R. and Gottschalk, W. H., *Homomorphisms of transformation groups*, Trans. Amer. Math. Soc. **94** (1960), 258–271.
5. Glasner, S. and Maon, D, *Rigidity in topological dynamics*, Ergod. Th. & Dynam. Sys. **9** (1989), 309–320.
6. Gottschalk, W. H. and Hedlund, G. A., *Topological Dynamics*, vol. **36**, Amer. Math. Soc. Colloq. Publ., 1955.
7. Jamison, B. and Sine, R., *Irreducible almost periodic Markov operators*, Math. and Mech. **18** (1969), 1043–1057.
8. Lotz, H. P., *Über das Spektrum positiver Operatoren*, Math. Z. **108** (1968), 15–32.
9. Michael, E., *Continuous selections*, Ann. Math. **63** (1956), 361–382.

INSTITUTE OF MATHEMATICS, TECHNICAL UNIVERSITY, 50-370 WROCŁAW, POLAND
E-mail address: downar@math.im.pwr.wroc.pl

TWO TECHNIQUES IN MULTIPLE RECURRENCE

A. H. FORREST

ABSTRACT. This paper describes a programme whose aim is to reprove the multi-dimensional Furstenberg Szemerédi theorem using one-dimensional techniques. Several steps along the way are completed, developing notions of uniform multiple recurrence and making a construction based on the odometer which may be of wider application.

1. INTRODUCTION

The proof of Szemerédi's Theorem on arithmetic progressions [Szemerédi, 1975] by means of dynamics [Furstenberg, 1977, 1981] marks an important development in ergodic theory. As a reminder,

Theorem 1 [Szemerédi, 1975]. *For all $\delta > 0$, $k \in \mathbb{N}$, there is an $N \in \mathbb{N}$ so that any subset, E, of $\{1, 2, ..., N\}$, with at least δN elements, contains an arithmetic progression of length at least k.*

Although Furstenberg's proof [1977] of this theorem is considerably shorter and has a clearer structure than that of Szemerédi, to date only Szemerédi's argument gives explicit bounds on the size of N as a function of δ. The possibility of quantifying the dynamical proof of Furstenberg remains open.

Other natural questions remain open likewise. For example, questions about the existence, for any measure-preserving system $(X, \mathbb{B}, \mu, T)$ and any $A \in \mathbb{B}$, of the limit of

$$\frac{1}{N} \sum_{n=0}^{N-1} \mu(A \cap T^{-n}A \cap .. \cap T^{-kn}A)$$

as $N \to \infty$, k fixed. (Here, and throughout this paper, a measure-preserving system is a probability space $(X, \mathbb{B}, \mu)$ with a bimeasurable bijection $T : X \to X$ such that $\mu(A) = \mu(TA) = \mu(T^{-1}A)$ for all measurable sets A). Thus, it is of some interest to follow the general line of Furstenberg's proof [1977 or 1981] with different and hopefully more quantitative arguments.

Also, Furstenberg's techniques gain over existing combinatorial methods by proving successive generalisations not proved yet by purely combinatorial means: a multi-dimensional version [Furstenberg, 1981] and an IP version [Furstenberg, Katznelson, 1985]. For example:

Theorem 2 [Furstenberg, 1981]. *Suppose that E is a subset of $\mathbb{Z}^k$ of positive upper density. Write e_i for the vector $(0, 0, .., 0, 1, 0, .., 0)$ with the 1 in the i'th place. Then there is an $n > 0$ and a $v \in \mathbb{Z}^k$ such that $v + ne_i \in E$ for all $1 \le i \le k$.*

The connection with dynamics is outlined below:

Definitions: Several commuting measure-preserving bijections can act on a probability space to form a measure-preserving system with $\mathbb{Z}^k$ action.

A measure-preserving system $(X, \mu, S_1, ..., S_k)$ with $\mathbb{Z}^k$ action is said to be *multi-dimensional Szemerédi* if for all A with $\mu(A) > 0$ there is an $n > 0$ such that

$$\mu(S_1^n A \cap S_2^n A \cap ... \cap S_k^n A) > 0.$$

Theorem 3. *Every probability measure-preserving system with $\mathbb{Z}^k$ action is multi-dimensional Szemerédi.*

A stronger version of Theorem 3 is proved in [Furstenberg, 1981], where it is shown to imply Theorem 2. This proof is subtle and inherits much complication from the fact that several different transformations have to be treated.

The purpose of this paper is to describe two techniques which aim to simplify the arguments involved in the proof of Theorem 3. The basic idea is to take the one-dimensional Furstenberg-Szemerédi theorem as a start and to try and derive the multi-dimensional generalisation with not too much extra effort.

The first technique mentioned in the title of this paper connects a density version of the definition of the Szemerédi property with a condition involving an IP Szemerédi property.

The second technique exploits a multiple recurrence generalisation of a construction developed by Schmidt [1976] for the study of recurrence of cocycles. Here cocycles are used to show that uniform multiple recurrence passes through the class of compact extensions.

These are applied to a construction based on the odometer by which it is shown that uniform one-dimensional recurrence results imply their multi-dimensional analogues. This construction is developed abstractly in the penultimate section of the paper.

The final section discusses the possibility of using the IP structure theory developed in [Furstenberg, Katznelson, 1985] and notes that a collection of results and constructions can be made in this case analogously to the arguments above.

Several sections of this paper appeal to unpublished results from the author's Ph.D. Thesis [Forrest, 1990], and the author wishes to thank his

advisor, Professor V. Bergelson, for his help and guidance in the preparation of that thesis. This paper was written while the author was a Fellow
of Corpus Christi College, Cambridge, U.K., and he wishes to thank the
College for its generous support during that time. He is also grateful to the
two referees whose constructive suggestions have been adopted with much
benefit to this paper.

2. IP RECURRENCE PROPERTIES

This section describes the first technique, which connects the IP Szemerédi property with the following well-known definition [Furstenberg,
1977]:

Definition: Suppose that $(X, \mathbb{B}, \mu, T)$ is a measure-preserving dynamical
system. It is said to be *SZ* (Szemerédi) if for all A with $\mu(A) > 0$ and all
$k > 0$

$$\liminf \frac{1}{N} \sum_{n=1}^{N} \mu(\bigcap_{i=0}^{k} T^{-in} A) > 0.$$

As in [Furstenberg, 1977], the dynamical theorem which proves Szemerédi's Theorem (Theorem 1) and a little bit more is simply stated:

Theorem 4. *All ergodic measure-preserving systems are SZ.*

A particularily convenient argument developed by Bergelson [1985]
makes a measure-zero adaption of the measure space X to deduce positive measure of intersection of sets from a given countable algebra merely
from their non-empty intersection. However, the SZ property requires a
uniformly large intersection $\mu(\bigcap_{i=0}^{k} T^{-in} A)$ for a positive lower density set of
n's, and there is no hope of using Bergelson's technique directly to achieve
such uniformity. Therefore, it would be useful to drop the requirement of
uniformity, and this may, indeed, be done, but at the expense of replacing
the property of positive lower density with the stronger IP* property. The
following definitions make this clear:

Definitions: Given a sequence of integers $n_1, n_2, ...$, define the *set of finite
sums*:

$$FS(n_1, n_2, ...) = \{\sum_{i \in C} n_i : C \text{ finite and } \subset \mathbb{N}\}.$$

By convention this set will include the empty sum, which is defined to be
0.

A subset E of the integers is said to be IP^* if $E \setminus \{0\}$ intersects every infinite finite-sums set (see [Furstenberg, 1981, Ch. 9]).

A measure-preserving system $(X, \mathbb{B}, \mu, T)$ is *IPSZ* if for all A with $\mu(A) > 0$ and $k \in \mathbb{N}$, the set

$$\{n : \mu(\bigcap_{i=0}^{k} T^{-in} A) > 0\}$$

is IP*.

The following lemma shows the usefulness of the IPSZ property in proving Theorem 4. The proof is given here in full since it is short and its detail amenable to generalisation. It is amplified in the next section.

Lemma 5. *If* $(X, \mathbb{B}, \mu, T)$ *is IPSZ, then it is also SZ.*

Proof. Suppose, for a contradiction, that there are a subset A of X of positive measure and a k so that, for all $\alpha > 0$, the set

$$\{n : \mu(\bigcap_{i=0}^{k} T^{-in} A) \le \alpha\}$$

has upper density 1. Then one can find a sequence of blocks of consecutive positive integers $\{a_m, a_m + 1, ..., b_m\}, m = 1, 2, ...,$ with $b_m - a_m \to \infty$ on which the value of $n \to \mu(\bigcap_{i=0}^{k} T^{-in} A)$ is particularly small, say less than α_m on the block $\{a_m, a_m + 1, ..., b_m\}$. In particular, the following conditions may be imposed without loss of generality:

$$b_m - a_m > b_{m-1}^2$$

and

$$\sum_{m=1}^{\infty} \alpha_m 2^{m-1} < \mu(A)/2.$$

The first condition ensures that the set $R = FS(a_1, a_2, ...)$ is contained in the union of the blocks, and the two conditions together show that the set

$$A^* = A \setminus \bigcup_{n \in R} \bigcap_{i=1}^{k} T^{-in} A$$

has measure at least $\mu(A)/2 > 0$. However,

$$\mu(\bigcap_{i=0}^{k} T^{-in} A^*) = 0.$$

for all $n \in R$ by construction, a contradiction to the IPSZ property in (X, μ, T), as required.

By removing the uniformity implicit in the lim inf of the SZ property, this argument is especially useful when dealing with the compact extension part (see structure theorem section below) of the original proof [Furstenberg, 1977] of Theorem 1. The details are omitted here since they will be presented in the more complicated context of uniform recurrence in the next few sections. The reader can reconstruct the simpler non-uniform argument from these without trouble and thereby a slight variation on the line of Furstenberg's proof.

3. Uniform Szemerédi Conditions

This section describes the uniform Szemerédi condition, which will allow the constructions of the next few sections to say something about the mutual recurrence properties of several commuting measure-preserving transformations $S_n, n = 1, .., k$.

Suppose that $D_o = \{d_1 < d_2 < ...\}$ is a sequence of positive integers with the property that if $d, d' \in D_o$ and $d < d'$ then $d | d'$. This set, to be chosen later, is assumed to be fixed for the purposes of this section.

Definitions: A measure-preserving system $(X, \mathbb{B}, \mu, T)$ is *uniformly SZ* if for all A with $\mu(A) > 0$, $k \in \mathbb{N}$, there is an $n \neq 0$ such that

$$\limsup_{d \in D_o} \mu(A \cap \bigcap_{i=1}^{k} T^{-d^i n} A) > 0.$$

As in the first section, the uniformly large size of intersection implicit in the lim sup is hard to treat conveniently and so it must be removed with some refinement. The following definitions lead up to this:

Given an infinite set E, write $E^{(\omega)}$ for the collection of infinite subsets of E, and $E^{(<\omega)}$ for the collection of finite subsets of E.

A measure-preserving system $(X, \mathbb{B}, \mu, T)$ has the uniform IPSZ property if the following is true:

For all $D_1 \in D_o^{(\omega)}$, all infinite finite-sums sets R, all A with $\mu(A) > 0$ and all k, there exist $D_2 \in D_1^{(\omega)}$ and $n \in R$ such that $\mu(A \cap \bigcap_{i=1}^{k} T^{-d^i n} A) > 0$ for all $d \in D_2$.

Both this condition and the uniform SZ condition dictate a degree of uniformity over d for the choice of n such that $\mu(A \cap \bigcap_{i=1}^{k} T^{d^i n} A) > 0$, and neither is a direct consequence of the one-dimensional Furstenberg-Szemerédi

Theorem. However, it is a consequence of a straightforward compactness argument, although not immediate (see [Forrest, 1990], for example), that if the multi-dimensional Szemerédi property holds for all measure-preserving $\mathbb{Z}^k$ actions, then every measure-preserving system satisfies the uniform IPSZ condition.

In the light of Lemma 5 the following is no surprise:

Lemma 6. *If a measure-preserving system satisfies the uniform IPSZ condition, then it satisfies the uniform SZ condition.*

Proof. Suppose that there is a measure-preserving system $(X, \mathbb{B}, \mu, T)$ and a set A of positive measure for which,

$$\limsup_{d \in D_o} \mu(A \cap \bigcap_{i=1}^{k} T^{-d^i n} A) = 0$$

for all $n > 0$. Form inductively a sequence of nested infinite subsets of D_o, $D_1 \supset D_2 \supset ...$, such that, for all $n > 0$,

$$\sum_{d \in D_n} \mu(A \cap \bigcap_{i=1}^{k} T^{-d^i n} A) < \mu(A)/2^{n+1}.$$

As in Lemma 5, form

$$A' = A \setminus \bigcup_{n>0} \bigcup_{d \in D_n} \bigcap_{i=1}^{k} T^{-d^i n} A,$$

which, by construction, is a set of measure at least $\mu(A)/2$. Let D be an infinite subset of D_o for which $D \setminus D_n$ is finite for each $n > 0$. Now note that this construction ensures that for each $n > 0$ there is only a finite collection of $d \in D$ for which

$$\mu(A' \cap \bigcap_{i=1}^{k} T^{-d^i n} A') > 0,$$

a contradiction to the uniform IPSZ condition.

4. A STRUCTURE THEOREM FOR $\mathbb{Z}$ ACTIONS

Furstenberg's proof [1977] of the one-dimensional Szemerédi Theorem exploits a technical structure theorem which conveniently separates the two extreme cases to be examined: compactness and weak mixing.

Theorem 7. *Suppose that $(X, \mathbb{B}, \mu, T)$ is an ergodic measure-preserving system. Then there is a set of factors, $X_\alpha \leftarrow X$, well-ordered by extension and indexed by a set of ordinals $\{\alpha : \alpha \leq \beta\}$ with β countable and with the further properties: (i) X_0 = single-point system and $X_\beta = X$. (ii) If $\alpha + 1 = \alpha' < \beta$, then $X_\alpha \leftarrow X_{\alpha'}$ is a compact extension. (iii) If $\beta = \alpha + 1$, then $X_\alpha \leftarrow X_\beta$ is either a compact extension or a weak-mixing extension. (iv) If α is a limit ordinal, then $X_\alpha = \lim_{\alpha' < \alpha} X_{\alpha'}$, the direct limit.*

The following result is implicit in [Furstenberg , 1977] and explicit in [Zimmer, 1976]:

Theorem 8. *Suppose that $X \to Y$ is a compact extension and that X is ergodic. Then there is a compact group extension of Y by G, which also is an extension of X, so that $Y \times G \to X \to Y$ is the natural first-coordinate projection.*

Thus the compact case is subsumed by the case of a compact group extension, and the next section exploits this to show that the uniform IPSZ condition passes through compact extensions.

Uniform IPSZ conditions pass to limits as well. So the one check remaining is that of the weak-mixing extension, and this becomes the subject of further investigation in the last section, although this is still incomplete at the moment.

5. THE UNIFORM SZEMERÉDI CONDITIONS AND COMPACT EXTENSIONS

Definition: Given an infinite subset $E = \{n_1 < n_2 < ...\}$ of $\mathbb{N}$, write $n_C = \sum_{i \in C} n_i$, where C is a finite subset of $\mathbb{N}$, the empty sum being 0 by definition. So $FS(E) = \{n_C : C \in \mathbb{N}^{(<\omega)}\}$, the finite-sums set generated by E.

Suppose that R is an infinite finite-sums set and that $M = \{r_B : B \in S\}$ is a finite subset of R. Then the notation $t >> M, t \in R$, refers to the condition that $t = r_C$ and $\min C > \max B$ for all $B \in S$. The notation $m_1 << m_2 << ..., m_k \in R$, refers to the condition that $m_k >> FS(m_1, m_2, ..., m_{k-1})$ for all k.

Remarks: The meaning of $>>$ depends on the set R, but this will be made clear from the context. The set $\{r \in R : r >> M\}$ is a finite-sums set provided that M is finite. In the case that $n >> m, n, m \in R, n+m \in R$ also.

As noted in the previous section, the following lemma is one of three checks needed to show that the uniform IPSZ property is found in every system.

Lemma 9. *If* $(X \times G, T) \to (X, T)$ *is an ergodic compact group extension, and* X *has the uniform IPSZ property, then* $X \times G$ *has the uniform IPSZ property.*

Proof. Recall the cocycle notation: $T^n(x, g) = (T^n x, a(x, n)g)$. It is sufficient to prove the following statement about cocycles on a compact group:

Suppose that X has the uniform IPSZ property. Then for all infinite finite-sums sets R, all $D \in D_o^{(\omega)}$, U open neighbourhood of $e \in G$, $\mu(A) > 0$, there are a $D' \in D^{(\omega)}$ and $n \in R$ such that for all $d \in D'$ there is an $x \in A$ with $T^{d^i n} x \in A$ and $a(x, d^i n) \in U$ for all $1 \le i \le k$.

Suppose that A and U have been fixed as above. Let $\mathbb{U}$ be a countable base of open sets in G which contains U. Use the argument of [Bergelson, 1985] to construct $\mathbb{B}_o$, a countable T-invariant algebra of measurable subsets of X which includes the sets $\{x : a(x, n) \in V\}(n \in \mathbb{Z}, V \in \mathbb{U})$ and all of whose non-empty elements are of strictly positive measure.

So, fix R and D, and consider the following set:

$$G_x = \{(g_1, g_2, .., g_k) \in G^k, \ \forall B \in \mathbb{B}_o, x \in B, \ \forall U_i \in \mathbb{U} : g_i \in U_i, \ \forall m \in R,$$

$$\exists D' \in D^{(\omega)}, \ \exists n >> m \in R : \ \forall d \in D' \ T^{d^i n} x \in B,$$

$$\forall 1 \le i \le k, \ a(x, d^i n) \in U_i\}.$$

The claim is that for some $x \in A$ this is a non-empty closed semi-group. Therefore, it contains $(e, e, .., e)$ and, appealing to the construction of $\mathbb{U}$ and $\mathbb{B}_o$, the statement above is proved.

In fact, G_x is a non-empty closed semi-group for almost every x as follows:

Non-emptiness and closure of G_x for almost every x follow by examining the sets

$$H_x(B) = \{(g_1, g_2, .., g_k) \in G^k : \ \forall U_i \in \mathbb{U} : g_i \in U_i, \ \forall m \in R, \ \exists D' \in D^{(\omega)},$$

$$\exists n >> m \in R : \ \forall d \in D' \ T^{d^i n} x \in B, \ \forall 1 \le i \le k, \ a(x, d^i n) \in U_i\}$$

for fixed $B \in \mathbb{B}_o$.

These are closed easily. The following argument demonstrates a finite intersection property for the $H_x(B)$, so that, by using the compactness of G^k, G_x, which is an intersection of $H_x(B)$ sets, is shown to be non-empty.

The uniform IPSZ property shows that for all R and D, almost every $x \in B$ has an $n \in R$ and $D' \in D^{(\omega)}$ such that for all $d \in D'$, $T^{d^i n} x \in B$ for all $1 \le i \le k$; for, if not, there is a set, B_1 of positive measure such that for all $x \in B_1$ and all $n \in R$, there are $d_n \in D$ such that for all $d \in D, d > d_n$,

$$T^{d^i n} x \notin B \text{ for some } 1 \le i \le k. \text{ This implies that } \mu(B_1 \cap \bigcap_{i=1}^{k} T^{-d^i n} B_1) = 0$$

for all $n \in R$ and all $d \in D, d > d_n$, a contradiction to the uniform IPSZ property.

Thus, if the U_i were all equal to G, the non-emptiness would follow immediately. However, it is not hard to refine the set D' obtained from the definition to an infinite subset D'' to ensure convergence of the cocycles $a(x, d^i n)$ to points in G as $d \in D''$ tends to infinity. So $H_x(B) \ne \emptyset$ for almost every $x \in B$.

Let $B' = \{x \in B : H_x(B) \ne \emptyset\}$. Let $B'' = B' \cup (X \setminus B)$, a set of measure 1, and let $C = \bigcap_{B \in \mathbb{B}_o} B''$, also of measure 1. Thus, noting that $H_x(\cap B_j) \subset \cap H_x(B_j)$ and using the compactness of G,

$$G_x = \bigcap_{B: x \in B \in \mathbb{B}_o} H_x(B)$$

is non-empty for all $x \in C$, as required.

The semi-group property for all x follows: Suppose that $(g_i), (h_i) \in G_x$ and that $W_i \in \mathbb{U}$: $g_i h_i \in W_i$ and $x \in B \in \mathbb{B}_o$. Let $U_i, V_i \in \mathbb{U}$: $g_i \in U_i$ and $h_i \in V_i$ such that $U_i V_i \subset W_i$. Then for all $m \in R$, there are a $D' \in D^{(\omega)}$ and $n >> m \in R$ such that, for all $d \in D'$,

$$x \in B' = \bigcap_{i=1}^{k} T^{-d^i n} B \cap \{y \in B : a(y, d^i n) \in U_i \ \forall 1 \le i \le k\}.$$

Now note that $B' \in \mathbb{B}_o$ by construction, and so there are an $n' >> n + m \in R$ and $D'' \in D'^{(\omega)}$ such that for all $d \in D''$

$$x \in \bigcap_{i=1}^{k} T^{-d^i n'} B' \cap \{y \in B' : a(y, d^i n') \in V_i \ \forall 1 \le i \le k\}.$$

Thus $T^{d^i (n+n')} x \in B$ for all $d \in D''$ and

$$a(x, d^i(n + n')) = a(T^{d^i n'} x, d^i n) a(x, d^i n') \in U_i V_i \subset W_i,$$

the membership occurring for all $d \in D''$ since $T^{d^i n'} x \in B'$. Further, $n + n' >> m \in R$, as required.

6. THE ODOMETER CONSTRUCTION

As noted in the introduction, the aim of the second part of this paper is to use the uniform one-dimensional Szemerédi property to prove the multi-dimensional Szemerédi Theorem. This section describes the construction to which the one-dimensional recurrence results will be applied to produce this result. The following proposition [Forrest, 1990] illustrates the basic idea.

Proposition 10. *Suppose that D_o is an infinite subset of the positive integers and that $k > 0$. Suppose that for all measure-preserving systems (X, μ, T) and all $\alpha > 0$ there are an $\epsilon > 0$ and N such that, for all $d \in D_o$ and all A with $\mu(A) > \alpha$ there is $0 < n \leq N$ such that*

$$\mu(A \cap \bigcap_{i=1}^{k} T^{-d^i n} A) \geq \epsilon.$$

Then all measure-preserving systems with $\mathbb{Z}^k$ action are multi-dimensional Szemerédi.

Proof. The following argument will show that this works for $k = 3$; the complications of general k are all accounted for here.

By a simple product argument, the fact that every measure-preserving system has the property described implies that the N and ϵ implicit in the definition can be made independent of the measure-preserving system in question.

Given (X, μ, S_1, S_2, S_3) and $\mu(A) = \alpha > 0$, find N and ϵ above for this α. Let $t \geq 5N/\epsilon, t \in D_o$. Write $[t] = \{0, 1, .., t-1\}$ and assign to this set the probability measure σ_t with even weights.

Construct the following measure-preserving system: $Y = X \times [t] \times [t]$ with measure $\nu = \mu \times \sigma_t \times \sigma_t$. Let U be defined by

$U(x, n, m) = (S_1 x, n+1, m)$ if $n < t - 1$
$U(x, t-1, m) = (S_2 S_1^{-t+1} x, 0, m+1)$ if $m < t - 1$
$U(x, t-1, t-1) = (S_3 S_2^{-t+1} S_1^{-t+1} x, 0, 0)$,

which is measure-preserving. Further, let $A' = A \times [t] \times [t]$, a set of measure α.

By the conditions given, there is $n \leq N$ such that

$$\nu(U^n A' \cap U^{tn} A' \cap U^{t^2 n} A') \geq \epsilon.$$

However, observe that for all $n > 0$, $\nu(U^n A' \Delta S_1^n A \times [t] \times [t]) < 2n/t$, $\nu(U^{tn} A' \Delta S_2^n A \times [t] \times [t]) < 2n/t$ and $\nu(U^{t^2 n} A' \Delta S_3^n A \times [t] \times [t]) = 0$.
Thus

$$\mu(S_1^n A \cap S_2^n A \cap S_3^n A) > \nu(U^n A' \cap U^{tn} A' \cap U^{t^2 n} A') - 4\epsilon/5 \geq \epsilon/5 > 0,$$

as required.

This proposition motivates the following general construction based on the odometer:

Construction. Recall D_o used to define the uniform Szemerédi conditions before; this will be written as $D_o = \{d_1', d_2', ...\}$ for the purposes of this definition. Choose D_o so that $d_i'^k | d_{i+1}'$ for all $i > 0$, and let $d_i'^k / d_{i+1}' \to 0$ as $i \to \infty$. Let $D = \{d_1' < d_1'^2 < ... < d_1'^k < d_2' < d_2'^2 < ... < d_2'^k < d_3' < ...\} = \{d_1 < d_2 < ..\}$ be fixed as the set from which the odometer constructions will be made from now on. Then the D-odometer is a dynamical system constructed as follows:

Let $\Omega = \prod_{i \geq 0} [d_{i+1}/d_i]$, which is a compact set in the product topology and a probability space with the natural product, λ, of normalised counting measures. Points $(w_n)_{n \geq 0}$ in Ω may be thought of as representing infinite D-ary expansions:

$$(w_n)_{n \geq 0} \leftrightarrow \sum_{n \geq 0} w_n d_n.$$

The map $K : \Omega \to \Omega$ acts on a point in Ω as if it were adding 1 with carry left to its D-ary expansion. In this way (Ω, λ, K) becomes a measure-preserving system called the *D-odometer*.

Now suppose that $S_0, S_1, S_2, ...$ is a sequence of commuting measure-preserving transformations on a probability space (X, μ). For $w = (w_n) \in \Omega$, let

$$h(w) = \min\{n : w_n < d_{n+1}/d_n - 1\},$$

which is finite almost everywhere.

The following measure-preserving transformation on $(\Omega \times X, \lambda \times \mu)$ is an extension of Ω: Let

$$T(w, x) = (Kw, S_h S_{h-1}^{-d_h/d_{h-1}+1} S_{h-2}^{-d_{h-1}/d_{h-2}+1} ... S_0^{-d_1+1} x),$$

where $h = h(w)$. Compare this with the construction of Proposition 10. The following useful properties are immediate from the construction:

Lemma 11. *If A is a subset of X and $A' = \Omega \times A$, then*

$$\lambda \times \mu(T^{-d_j n} A' \Delta (\Omega \times S_j^{-n} A)) < 2n d_j/d_{j+1}$$

for all j. Also,

$$T^m((0), x) = ((w_n), S_0^{w_0} S_1^{w_1} ... S_N^{w_N} x),$$

where $m = \sum_{n=0}^{N} w_n d_n$.

Beyond these elementary properties, however, the motion of individual points becomes hard to express in general. In particular, the spectral measure seems hard to find in the case that the S_n are of contrasting spectral type. However, if the transformations S_n tend to a measure-preserving transformation in the Rohlin topology, then the construction above has T^{d_n} tending to the same limit producted with the odometer. Other connections between the S_n and T will be made later.

The construction of Proposition 10 is contained in this construction (put $S_{k+1} = S_{k+2} = ...id$). However, the uniform recurrence properties which will be considered over the next few paragraphs will be applied more naturally to the system constructed above with a periodic sequence of S's: $S_1, S_2, .., S_k, S_1, S_2, ...$, these S's being generators of a given $\mathbb{Z}^k$ action $(X, S_1, ..., S_k)$. This will be called the *natural odometer construction* made from a given measure-preserving system with $\mathbb{Z}^k$ action, and it will be written $(\overline{X}, T)$.

The advantage of this odometer construction over that of Proposition 10 is that the error of approximating a multi-dimensional action by a one-dimensional action can be made arbitrarily close to zero (by the $d_j'^k/d_{j+1}' \to 0$ property) without changing the dynamical system, and that there is control over the rate of convergence.

Thus

Theorem 12. *Suppose that $(\overline{X}, T)$ (constructed from D) is uniformly SZ (with respect to D_o). Then $(X, S_1, .., S_k)$ is multi-dimensional Szemerédi .*

Proof. Suppose that $A \subset X$ with $\mu(A) > 0$. Let $A' = \Omega \times A$ so that $\lambda \times \mu(A') > 0$ as well. Then there are a $\delta > 0$ and an $n > 0$ so that for infinitely many $d \in D_o$

$$\lambda \times \mu(A' \cap \bigcap_{i=1}^{k} T^{-d^i n} A') \geq \delta.$$

Choose $d = d_t'$ sufficiently large that $d_t'^k/d_{t+1}', 1/d_t' < \delta/4kn$, so that, by Lemma 11,

$$\lambda \times \mu(T^{-d^i n} A' \Delta \Omega \times S_i^{-n} A) < \delta/2k$$

for all $1 \leq i \leq k$. Thus

$$\mu(A \cap \bigcap_{i=1}^{k} S_i^{-n} A) \geq \delta/2,$$

as required.

This argument together with an argument that all measure-preserving systems are uniformly SZ would therefore provide an alternative proof of the Furstenberg-Szemerédi Theorem.

7. FURTHER PROPERTIES OF THE ODOMETER CONSTRUCTION

This section seeks to give some reason for the difficulty in pushing the uniform Szemerédi property through the final weak-mixing extension in the system built over the odometer.

Given fixed D whose connection with D_o is now removed, the process of making the natural D-odometer extension explained before is in many ways a natural transformation of the category of measure-preserving systems with $\mathbb{Z}^k$ action to the category of measure-preserving systems with $\mathbb{Z}$ action.

The odometer construction will be written $\Omega \rtimes X$ to emphasise the skew nature of the transformation. In general, $\rtimes$ is used when expressing an extension as a skew-product of the base (on the left) with some probability space (on the right), as may always be done (see [Rohlin, 1966]).

Definitions: Suppose that D is a fixed sequence of natural numbers with the multiplicative property needed for the construction of the D-odometer described before.

Assume from now on that d_n/d_{n+1} tends to zero monotonically.

Given $(X, S_1, S_2, ..., S_k)$, a measure-preserving system with $\mathbb{Z}^k$ action, let $(\overline{X}, T)$ be the measure-preserving system constructed naturally from X on the D-odometer described before. (The mention of D may be suppressed since it is fixed in the discussion which follows).

The following lemma describes some of the 'categorical' properties of the correspondence $X \to \overline{X}$. The relative product involved in part (iii) is defined in [Furstenberg, 1981].

Lemma 13. *Suppose that $\pi : X \to Y$ is an extension of measure-preserving systems with $\mathbb{Z}^k$ action; then (i) $\overline{X} \to \overline{Y}$ is an extension with natural projection. (ii) If $\overline{X} \to Z \to \overline{Y}$, then $Z \equiv \overline{W}$ for some $(W, \mathbb{Z}^k)$. (iii) $\overline{X \times_Y X} \equiv \overline{X} \times_{\overline{Y}} \overline{X}$.*

Proof. (i) The natural projection $\overline{\pi} : \Omega \rtimes X \to \Omega \rtimes Y \colon \overline{\pi}(w, x) = (w, \pi(x))$ is an extension.

(ii) With no loss of generality, the spaces involved in the extensions can be represented as skew-products:

$$\overline{X} \equiv \Omega \rtimes Y \rtimes Z' \rtimes X' \to \Omega \rtimes Y \rtimes Z' \to \Omega \rtimes Y \equiv \overline{Y}$$

(see [Rohlin, 1966] for a rigorous assessment of this reduction). In all cases but the odometer rotation itself (K), the $\mathbb{Z}$ action will be represented by T without confusion.

Suppose that $w = (w_n) \in \Omega$ such that $w_0 < d_1 - 1$. Then, considering the T action on $\overline{X}$ as the skew action over Ω,

$$T(w, y, z', x') = (Kw, S_1(y, z', x')),$$

where the S_1 is the action on X. On the other hand, $\overline{X}$ is an extension of Z:

$$
\begin{aligned}
T(w,y,z',x') &= (T(w,y,z'),\sigma((w,y,z');x')) \\
&= (Kw,\tau(w;y,z'),\sigma((w,y,z');x')),
\end{aligned}
$$

where σ is a skewing function which describes $\overline{X}$ as an extension of Z and τ a skewing function for Z as an extension of Ω (*via* $\overline{Y}$). Equating coordinates in both these expressions shows that $\tau(w;y,z')$ (an element of $Y \times Z'$) i s independent of $w \in \{(w_n) \in \Omega : w_0 < d_1 - 1\}$, and this common value can be defined to be $S_1(y,z')$. Thus a transformation S_1 is defined on $Y \times Z'$ so that the natural projections $Y \times Z' \times X' \to Y \times Z' \to Y$ are extensions with respect to the S_1 action. And so $(Y \times Z', S_1)$ is a measure-preserving system which is intermediate to (X, S_1) and (Y, S_1).

Similarily, by considering the action of T^{d_j}, with w restricted to the space $\{(w_n) \in \Omega : w_i = d_{i+1}/d_i - 1 \ \forall i < j \ , w_j < d_{j+1}/d_j - 1\}$, for each $j \geq 1$ one obtains commuting actions $S_2, .., S_j, ..$ on $Y \times Z'$.

Recall the periodic sequence of S's involved in the natural odometer constructions of $\overline{X}$ and $\overline{Y}$. This shows that the S_{k+1} obtained to act on $Y \times Z'$ by putting $j = k + 1$ in the argument above is, in fact, the same as S_1, and so on periodically. Thus a $\mathbb{Z}^k$ action on $Y \times Z'$ is obtained which is intermediate to X and Y. By construction, $Z = \overline{Y \times Z'}$, and so $W = Y \times Z'$, as required.

(iii) In general, if $(X, S) \equiv (Y \rtimes X', S)$, $S(y, x') = (Sy, \tau(y; x'))$, then $(X \times_Y X, S) \equiv (Y \times X' \times X', S)$, with natural relative product action

$$
S : (y, x_1', x_2') \to (Sy, \tau(y; x_1'), \tau(y; x_2')),
$$

and this may be extended to $\mathbb{Z}^k$ actions naturally.

The natural D-odometer construction does not interfere unduly with this equivalence: $\overline{X} \equiv (\Omega \rtimes Y) \rtimes X'$, $\overline{Y} \equiv \Omega \rtimes Y$, and so the natural equivalences

$$
\overline{X \times_Y X} \equiv \Omega \rtimes (X \times_Y X) \equiv \Omega \rtimes Y \rtimes (X' \times X') \equiv \overline{X} \times_{\overline{Y}} \overline{X}
$$

are all naturally equivalences of measure-preserving systems, as required.

The next lemma examines the spectral properties of the correspondence:

Lemma 14. *Suppose that* $(X, \mathbb{Z}^k) \to (Y, \mathbb{Z}^k)$ *is an extension.* (i) *If there is some* $S \in \mathbb{Z}^k$ *for which* $(X, S) \to (Y, S)$ *is weak-mixing, then* $(\overline{X}, T) \to (\overline{Y}, T)$ *is weak-mixing.* (ii) *If, for all* $S \in \mathbb{Z}^k$, $(X, S) \to (Y, S)$ *is compact, then* $(\overline{X}, T) \to (\overline{Y}, T)$ *is compact.*

Proof. (i) A weak-mixing extension $X \to Y$ is one for which $X \times_Y X \to Y$ is an ergodic extension. Thus part (iii) of Lemma 14 above reduces the problem to one of showing that $(X, S) \to (Y, S)$ ergodic for some $S \in \mathbb{Z}^k$ implies that $(\overline{X}, T) \to (\overline{Y}, T)$ is ergodic.

A necessary and sufficient condition that $(X, \mu, S) \equiv (Y \times X', \nu \times \lambda, S) \to (Y, \nu, S)$ is ergodic is that, for $A = C \times A_o, B = C \times B_o$, subsets of X of positive measure, there is an $n > 0$ such that $\mu(S^{-n} A \cap B) > \mu(A)\mu(B)/2$.

The aim is to prove this quantitative bound for all A_o, B_o and C. However, since the bound is quantitative it may be inferred from the same result shown true for a collection of basic measurable sets which generate the whole sigma field. So there is no loss of generality in assuming that C, a subset of $\Omega \times Y$, has the form $\Omega' \times C'$, where Ω' is a clopen ball in Ω. Say $\Omega' = \{(w_n) \in \Omega : w_i = a_i \ \forall i \leq I\}$ for some fixed $0 \leq a_i < d_{i+1}/d_i$.

By the ergodicity of $(X, S) \to (Y, S)$, there is an n for which $\mu(S^{-n} A' \cap B') > \mu(A')\mu(B')/2$, where $A' = C' \times A_o$ and $B' = C' \times B_o$.

Now suppose that S can be written $S = \prod S_i^{h_i}$. Given $\epsilon > 0$, pick $j > I$ sufficiently large that $\sum_{i=0}^{k} n h_i (d_{jk+i}/d_{jk+i+1}) < \epsilon$. Let $m = \sum_{i=0}^{k} n h_i d_{jk+i}$.

Then $\lambda(K^{-m} \Omega' \Delta \Omega') < \epsilon$ and

$$\lambda \times \mu(T^{-m} A \cap B) > (\lambda(\Omega') - 2\epsilon)\mu(A')\mu(B')/2.$$

So the condition required above is satisfied for ϵ small enough, and we are done.

(ii) A $\mathbb{Z}^k$ extension which obeys the hypothesis of this part is a compact extension with respect to the $\mathbb{Z}^k$ action. This would be described in [Furstenberg, 1981, Ch. 6], as $\Gamma_c = \mathbb{Z}^k$. In full:

There are a subset Y' of Y of full measure and a dense subset $\mathbb{F}$, of $L_2(X)$ so that for all $f \in \mathbb{F}$ and $\epsilon > 0$ there is a finite collection of $g_i \in L_2(X), 0 < i \leq I$, so that for all $y \in Y'$ and all $S \in \mathbb{Z}^k$ there is $0 < i \leq I$ for which $\|Sf - g_i\|_y < \epsilon$.

However, this is enough to apply the result of Zimmer [1976] mentioned before to show that there is a compact group extension $Y \times G$ of Y which is also an extension of X, $\pi' : Y \times G \to X$. Further, the composition $(Y \times G, \mathbb{Z}^k) \to (X, \mathbb{Z}^k) \to (Y, \mathbb{Z}^k)$ is the natural projection.

Suppose that $a : Y \times \mathbb{Z}^k \to G$ is the $\mathbb{Z}^k$ cocycle implicit in this compact group extension. The idea is to construct a $\mathbb{Z}$ cocycle on $\overline{Y}$ which generates a G-extension of $\overline{Y}$ which is also an extension of $\overline{X}$. This would show directly that $\overline{X} \to \overline{Y}$ is a compact extension.

Recall the function $h(w) = \min\{n : w_n < d_{n+1}/d_n - 1\}$, which is finite almost everywhere on Ω. Let i be the uniquely defined integer-valued function on Ω such that $h(w) = jk + i(w)$, where $0 \leq i(w) < k, j \in \mathbb{Z}^k$.

Define $f((w_n), y) = a(y, v(w))$, where $v : \Omega \to \mathbb{Z}^k$ is defined via its coordinates:

$$v_i(w) = 1 + \sum_{j : jk+i < h(w)} (-d_{jk+i+1}/d_{jk+i} + 1) \quad \text{if } i = i(w),$$

$$= \sum_{j:jk+i<h(w)} (-d_{jk+i+1}/d_{jk+i} + 1) \quad \text{otherwise.}$$

The skew-product $\overline{Y} \times_f G$ induced by this f is an extension of $\overline{X}$ by means of $\overline{\pi'} : ((w,y),g) \to (w,\pi'(y,g))$. Further, $\overline{(Y \times G, \mathbb{Z}^k)} \equiv \overline{(Y, \mathbb{Z}^k)} \times_f G$, as required.

Remarks: The argument above shows that, in general, good spectral properties, i.e. those which can be deduced from consideration of a generating set of elements of a sigma-algebra, can be transferred from individual elements of $\mathbb{Z}^k$ acting on X and Y to certain powers of T in $\overline{X}$ and $\overline{Y}$.

These two lemmas explain the incompleteness of the argument of the second section of this paper:

Consider the final weak-mixing extension $\overline{X} \to Z$, say, (which by Lemma 13 is of the form $\overline{X} \to \overline{Y}$) in the structure decomposition (Theorem 7) of $\overline{X}$. The fact that this is weak-mixing may, by Lemma 14, be due to the fact merely that one of the elements S of $\mathbb{Z}^k$ makes $(X, S) \to (Y, S)$ weak-mixing.

Indeed, it is possible to construct an extension $(X, \mathbb{Z}^k) \to (Y, \mathbb{Z}^k)$ etc. so that $\overline{X} \to \overline{Y}$ is weak-mixing but $(X, S) \to (Y, S)$ is not weak-mixing for any element S of $\mathbb{Z}^k$.

Thus the correspondence $X \to \overline{X}$ hides all the structural complications in the weak-mixing extension, and a general argument for uniform IPSZ recurrence in the case of weak-mixing extensions would be close to an argument valid in the case of general extensions.

This does, however, give a relatively effortless way of producing a rigid weak-mixing extension of the odometer. One applies the construction above to the $\mathbb{Z}^2$ action generated by a weak-mixing transformation and the identity transformation.

Quite apart from its use in multiple recurrence, therefore, the odometer system above could provide a clean method of constructing subtle examples in ergodic theory.

8. The IP-Szemerédi Theorem

The constructions above can be made with reference to the general-isation of $\mathbb{Z}^k$ actions to IP-systems developed in [Furstenberg, 1981] and [Furstenberg, Katznelson, 1985]. The following is a rough outline of the procedure:

Given a commuting family of IP-systems of measure-preserving maps $S_\alpha^{(1)}, ..., S_\alpha^{(k)}$, α finite subsets of $\mathbb{N}$, on a probability space X, a natural odometer construction can still be made: $(\Omega \rtimes X, T)$. It will be based on the sequence of maps

$$S_{\{1\}}^{(1)}, S_{\{1\}}^{(2)}, ..., S_{\{1\}}^{(k)}, S_{\{2\}}^{(1)}, ..., S_{\{2\}}^{(k)}, S_{\{3\}}^{(1)}, ...,$$

and to imitate $S_\alpha^{(j)}$ it is sufficient to consider powers of T such as T^{α_j}, where $\alpha_j = \sum_{i \in \alpha} d_i'^j$, for example. Thus it is possible to formulate refinements of one-dimensional recurrence which prove established results about commuting IP systems.

A structure theory for IP-systems is developed in [Furstenberg, Katznelson, 1985] in complete analogy with Theorem 7, and the refinements mentioned above can be pushed through the corresponding compact extensions by similar arguments to the work of this paper (and a notion of IP-cocycle is developed in the process). But, like the weak-mixing extensions before, the mixing extensions appropriate to this case remain impenetrable to the uniform recurrence properties that are most naturally defined.

Perhaps not many more ideas are needed to push this programme through the mixing extensions, and, if so, a one-dimensional theory of multiple recurrence may grow in parallel to that developed already by Furstenberg and Katznelson.

REFERENCES

1. V. Bergelson, *Sets of recurrence of $\mathbb{Z}^m$ actions and properties of sets of differences in $\mathbb{Z}^m$*, J.London Math. Soc. (2) **31** (1985), 295-304.

2. A. H. Forrest, *Recurrence in Dynamical Systems: A Combinatorial Approach.*, Ph.D. Thesis, The Ohio State University, 1990.

3. H. Furstenberg, *Ergodic behavior of diagonal measures and a theorem of Szemerédi on arithmetic progressions*, J. d'Analyse Math. **31** (1977), 204-256.

4. H. Furstenberg, *Recurrence in Ergodic Theory and Combinatorial Number Theory*, Princeton University Press, Princeton, New Jersey, 1981.

5. H. Furstenberg and Y. Katznelson, *An ergodic Szemerédi theorem for IP-systems and combinatorial theory*, J. d'Analyse Math. **45** (1985), 117-168.

6. V. A. Rokhlin, *Selected topics from the metric theory of dynamical systems*, AMS Transl. (2) **49** (1966), 171-240.

7. K. Schmidt, *On Recurrence*, Lecture Notes, Mathematics Institute, University of Warwick, Coventry, U.K., 1976.

8. E. Szemerédi, *On sets of integers containing no k elements in arithmetic progression*, Acta. Arith. **27** (1975), 199-245.

9. R. J. Zimmer, *Extensions of ergodic group actions*, Ill. J. Math. **20** (1976), 373-409.

DEPARTMENT OF MATHEMATICS AND STATISTICS, THE UNIVERSITY OF EDINBURGH, THE KING'S BUILDINGS, EDINBURGH, EH9 3JZ, SCOTLAND, U. K.

E-mail address: forrest@abel.maths.ed.ac.uk

FOR BERNOULLI TRANSFORMATIONS THE SMALLEST NATURAL FAMILY OF FACTORS CONSISTS OF ALL FACTORS

ELI GLASNER

ABSTRACT. A direct proof is given of the theorem which appears as the title of this paper, concerning the notion of natural family of factors of an ergodic dynamical system introduced in [del Junco et al.].

The notion of natural family of factors of a measure-preserving ergodic system was introduced in [del Junco et al.]. Given such a system $\mathcal{X} = (X, \mathcal{B}, \mu, T)$ a collection of factors (interpreted here as T-invariant sub σ-algebras of $\mathcal{B}$), $\mathfrak{A} = \{A_\iota\}_{\iota \in I}$, is called a **family** if (i) $\mathcal{B} \in \mathfrak{A}$ and (ii) $\mathfrak{A}$ is closed under intersections.

Given a family $\mathfrak{A}$ of factors of $\mathcal{X}$, and any factor $\mathcal{F} \subset \mathcal{B}$ we define the $\mathfrak{A}$-cover of $\mathcal{F}$ to be the factor

$$\hat{\mathcal{F}} = \cap \{A : A \in \mathfrak{A}, A \supset \mathcal{F}\}.$$

A family $\mathfrak{A}$ is called **natural** if:

(iii) for every pair of factors $\mathcal{F}, \mathcal{G} \subset \mathcal{B}$, whenever λ is a 2-fold selfjoining of $\mathcal{X}$ (i.e. a $T \times T$-invariant probability measure on $X \times X$ with projections μ), which induces an isomorphism between $\mathcal{F}$ and $\mathcal{G}$, λ also induces an isomorphism between their $\mathfrak{A}$-covers.

It is not hard to see that the intersection of any number of natural families is again a natural family. Since the family of all factors is clearly a natural family, it follows that for every ergodic system $\mathcal{X}$ there is a unique minimal natural family of factors.

In his presentation of the paper [del Junco et al.] at the Alexandria conference, Professor Lemańczyk outlined a proof of the claim that for a Bernoulli system the minimal natural family of factors is the family of all factors. In the present note I present an alternative proof of this statement. I wish to thank Benjamin Weiss for his crucial help in pointing out the relevance of Rudolph's result [Rudolph, 1983].

We shall use the following notations: $\mathcal{X} = (X, \mu, T)$, $\mathcal{Y} = (Y, \nu, T)$, and $\mathcal{Z} = (Z, \eta, T)$ will denote ergodic systems,(we usually omit the σ-algebra). Let $\pi : \mathcal{X} \to \mathcal{Y}$ be a homomorphism of ergodic systems. Following [Furstenberg, 1977], we denote by $\mathcal{E}(X/Y)$ the closed subspace of $L_2(X)$

spanned by generalized eigenfunctions with respect to Y. Thus $\mathcal{E}(X/Y)$ is the L_2 subspace which corresponds to the largest isometric extension of $\mathcal{Y}$ in $\mathcal{X}$. The symbol $\mathcal{X} \underset{Y}{\times} \mathcal{X}$ will be used to denote the relatively independent self joining of $\mathcal{X}$ over $\mathcal{Y}$. If $S : Y \to Y$ is an automorphism of $\mathcal{Y}$ we write $\mathcal{X} \underset{Y,S}{\times} \mathcal{X}$ for the relatively independent self joining of $\mathcal{X}$ over the S-graph selfjoining of $\mathcal{Y}$. Thus if $\mu = \int \mu_y d\nu(y)$ is the disintegration of μ over Y, then the first joining is $\mu \underset{Y}{\times} \mu = \int \mu_y \times \mu_y d\nu(y)$ and the latter is $\mu \underset{Y,S}{\times} \mu = \int \mu_y \times \mu_{Sy} d\nu(y)$. Sometimes, when no confusion can arise, we use the spaces to denote the systems: X for $\mathcal{X}$, $X \underset{Y}{\times} X$ for $\mathcal{X} \underset{Y}{\times} \mathcal{X}$ etc. Again when no confusion can arise we use, as above, the letter T to denote the acting transformation on various systems. Given a compact group K, we let m denote its normalized Haar measure and we let R_k denote the map $R_k(l) = lk$.

Lemma 1. *Let $\mathcal{X} \to \mathcal{Z} \to \mathcal{Y}$ be a sequence of ergodic systems such that $\mathcal{Z} \to \mathcal{Y}$ is an isometric extension of $\mathcal{Y}$. Then this extension is the largest isometric extension of $\mathcal{Y}$ in $\mathcal{X}$ iff every invariant L_2 function of the system $X \underset{Y}{\times} X$ is $Z \underset{Y}{\times} Z$ measurable.*

Proof. Let $\mathcal{W} \to \mathcal{Y}$ be the largest isometric extension of $\mathcal{Y}$ in $\mathcal{X}$. By [Furstenberg, 1977] Theorem 7.1, every $T \times T$ invariant function on $X \underset{Y}{\times} X$ is in $\mathcal{E}(X/Y) \underset{Y}{\otimes} \mathcal{E}(X/Y)$; i.e $W \underset{Y}{\times} W$ measurable. If $\mathcal{W} \to \mathcal{Z}$ is a proper (necessarily isometric) extension, there will be $T \times T$ invariant functions which are $W \underset{Y}{\times} W$ but not $Z \underset{Y}{\times} Z$ measurable. □

Lemma 2. *Let $\mathcal{X} \to \mathcal{Z} \to \mathcal{Y}$ be a sequence of ergodic systems such that $\mathcal{Z} \to \mathcal{Y}$ is the largest isometric extension of $\mathcal{Y}$ in $\mathcal{X}$ and assume moreover that this extension is an Abelian compact group extension, (say K). Then every $T \times T$ invariant L_2 function f on $X \underset{Y}{\times} X$ is of the form $f(x, x') = h(kl^{-1})$ for some L_2 function h on K, where the space X is represented as a product space $X = Y \times K \times U$, $T(y, k, u) = (Ty, \phi(y)k, \alpha(y, k)(u))$. Here $\phi : Y \to K$ is a measurable cocycle, (U, θ) a measure space, $\alpha : Z = Y \times K \to Aut(U, \theta)$ a measurable cocycle, and $x = (y, k, u), x' = (y, l, v)$.*

Proof. By [Furstenberg, 1977], Theorem 7.1, $f \in \mathcal{E}(X/Y) \underset{Y}{\otimes} \mathcal{E}(X/Y)$. By assumption $\mathcal{E}(X/Y) = L_2(Z)$, hence we deduce that $f = f(y, k, l)$ does not depend on u and v. We now expand f as a Fourier series

$$f(y, k.l) = \sum a_{\chi,\psi}(y)\chi(k)\psi(l)$$

where we sum over all pairs χ, ψ of elements of Γ, the dual group of K. From the invariance of f we get

$$f(y, k, l) = \sum a_{\chi,\psi}(Ty)\chi(\phi(y))\chi(k)\psi(\phi(y))\psi(l).$$

Hence for every pair χ, ψ

$$a_{\chi,\psi}(Ty)(\chi\psi)(\phi(y)) = a_{\chi,\psi}(y). \tag{*}$$

By ergodicity of $\mathcal{Y}$, $a_{\chi,\psi}$ has constant modulus and for $\psi \neq \chi^{-1}$, (*) represents a nontrivial coboundary equation satisfied by ϕ which, unless $a_{\chi,\psi} = 0$, contradicts the ergodicity of $\mathcal{Z}$. If $\psi = \chi^{-1}$, (*) yields $a_{\chi,\psi}(Ty) = a_{\chi,\psi}(y)$, and the ergodicity of $\mathcal{Y}$ implies that $a_{\chi,\psi}$ is a constant c_χ. Thus

$$f(y, k.l) = \sum c_\chi \chi(kl^{-1}) = h(kl^{-1}). \qquad \square$$

Proposition 3. *Let $\mathcal{X} \to \mathcal{Z} \to \mathcal{Y}$ be a sequence of ergodic systems such that $\mathcal{Z} \to \mathcal{Y}$ is the largest isometric extension of $\mathcal{Y}$ in $\mathcal{X}$ and assume moreover that this extension is an Abelian compact group extension, (say K). Then there exists $k_0 \in K$ such that the relative independent product over the graph of k_0, $X \underset{Z,R_{k_0}}{\times} X$ is ergodic.*

Proof. Notations as in Lemma 2, let $a(y), b(k), c(u)$ and $d(u)$ be bounded measurable functions on Y, K, U and U respectively; let $F(y, k, u, l, v) = a(y)b(k)c(u)d(v)$ be defined on $X \underset{Y}{\times} X$. By the ergodic theorem the limit

$$\frac{1}{N} \sum_{n=1}^{N} (T \times T)^n F = f(y, k, u, l, v)$$

exists $\mu \underset{Y}{\times} \mu = \nu \times m \times \theta \times m \times \theta$ a.e. and f is a $T \times T$-invariant function on $X \underset{Y}{\times} X$. By Lemma 2, $f(y, k, u, l, v) = h(kl^{-1})$.

Suppose now that $\int b \, dm = 0$, then

$$< F, f > = \int a(y)b(k)c(u)d(v)h(kl^{-1})d(\mu \underset{Y}{\times} \mu) = 0$$

since

$$\int \int b(k)h(kl^{-1})dm(k)dm(l) = 0.$$

Since f is $T \times T$-invariant we also have $< (T \times T)^n F, f > = 0$ for every n, and therefore also $< f, f > = 0$; i.e. $f = 0$.

We have shown that the ergodic sums $\frac{1}{N} \sum (T \times T)^n F$ tend to zero for every F as above with $\int b \, dm = 0$. If $\bar{b} = \int b \, dm \neq 0$ we write $F = a(y)(b - \bar{b})(k)c(u)d(v) + \bar{b}a(y)c(u)d(v)$. Thus we can assume that $F = a(y)c(u)d(v)$ and again we get $< F, f > = 0$ if $\int a \, d\nu = 0$, and so on. Finally we see that for every bounded function F on $X \underset{Y}{\times} X$ which does not depend on l,

$$\lim \frac{1}{N} \sum F(T^n y, \phi_n(y)k, \alpha_n(y, k)(u), \phi_n(y)l, \alpha_n(y, l)(v))$$

$$= \int F d\nu(y)dm(k)d\theta(u)d\theta(v)$$

for $\mu \underset{Y}{\times} \mu$ almost every (y, k, u, l, v). (Here $\phi_n(y) = \phi(T^{n-1}y) \cdots \phi(Ty)\phi(y)$ and α_n is defined by the equation $T^n(y, k, u) = (T^n y, \phi_n(y)k, \alpha_n(y.k)u)$.). In particular if we write $l = kk_0$ we can conclude that there exists k_0 such that for $\nu \times m \times \theta \times \theta$ almost every (y, k, u, v)

$$\lim \frac{1}{N} \sum F(T^n y, \phi_n(y)k, \alpha_n(y, k)(u), \alpha_n(y, kk_0)(v)) = \int F.$$

In other words the map $T(y, k, u, v) = (Ty, \phi(y)k, \alpha(y, k)(u), \alpha(y, kk_0)(v))$ on

$$X \underset{Z, R_{k_0}}{\times} X = Y \times K \times U \times U$$

is $\mu \underset{Z, R_{k_0}}{\times} \mu = \nu \times m \times \theta \times \theta$ ergodic. $\square$

Lemma 4. *Let $\mathcal{Y} = (Y, \nu, T)$ be a weakly mixing system, K a compact Abelian group, and $\phi : Y \to K$ a measurable cocycle so that the system $\mathcal{Z} = (Z, \eta, T)$ defined on $Z = Y \times K$ by $T(y, k) = (Ty, \phi(y)k)$ with respect to the measure $\eta = \nu \times m$ is also weakly mixing. Define a system $\mathcal{X} = (X, \mu, T)$ by $X = Y \times K \times K$, $T(y, k, l) = (Ty, \phi(y)k, kl)$ and $\mu = \nu \times m \times m$. Then*

(1) *$\mathcal{X}$ is weakly mixing.*
(2) *$\mathcal{Z}$ is the largest isometric extension of $\mathcal{Y}$ in $\mathcal{X}$.*

Proof. (1) Suppose $f = f(y, k, l)$ is an eigenfunction of T in $L_2(X)$ with eigenvalue λ. Write $f(y, k, l) = \sum a_{\chi, \psi}(y)\chi(k)\psi(l)$, where the sum is taken over all pairs of characters in Γ, the dual group of K. Then the eigenvalue equation

$$f(Ty, \phi(y)k, kl) = \sum a_{\chi, \psi}(Ty)\chi(\phi(y))\chi(k)\psi(k)\psi(l)$$
$$= \lambda f(y, k, l)$$

yields

$$a_{\chi, \psi}(Ty)\chi(\phi(y)) = \lambda a_{\chi\psi, \psi}(y) \qquad (*)$$

for all $\chi, \psi \in \Gamma$. These equations imply that $\|a_{\chi, \psi}\| = \|a_{\chi\psi^n, \psi}\|$ for every $n \in \mathbb{Z}$. If ψ has infinite order this implies that $a_{\chi, \psi} = 0$. If $\psi \neq 1$ is of finite order, say $\psi^n = 1$, then we get

$$a_{\chi, \psi}(T^n y)\chi(\phi_n(y)) = \lambda a_{\chi\psi, \psi}(T^{n-1}y)\chi(\phi_{n-1}(y))$$
$$= \cdots = \lambda^{n-1} a_{\chi\psi^{n-1}, \psi}\chi(\psi(y))$$
$$= \lambda^n a_{\chi, \psi}(y). \qquad (**)$$

It follows that the function $h(y, k) = a_{\chi, \psi}(y)\chi(k)$ is a T^n eigenfunction with eigenvalue λ^n. If $\lambda^n \neq 1$ this contradicts the weak mixing of T^n on Z unless $a_{\chi, \psi} = 0$. If $\lambda^n = 1$, $h(y, k)$ is a T^n-invariant function on Z, hence a constant. If $\chi \neq 1$ this implies that $a_{\chi,\psi} = 0$. If $\chi = 1$, $a_{1,\psi}(y)$ is a constant, but from (*) we get $a_{1,\psi}(Ty) = \lambda a_{\psi,\psi}$ and we have already seen that $a_{\psi,\psi} = 0$, hence also $a_{1,\psi} = 0$. Finally for $\chi = \psi = 1$ we get from (*), $a_{1,1} = \lambda a_{1,1}(y)$ and weak mixing of Y implies $a_{1,1} = 0$, or $\lambda = 1$ and $f(y,k,l) = a_{1,1}$ is constant. This completes the proof of (1).

(2) Let $W \to Y$ be the maximal isometric extension of Y in X. Then clearly $W \to Z$. Let f be a $T \times T$-invariant function on $X \times X$. The action on $X \underset{Y}{\times} X$ is given by: $T(y,k,p,l,q) = (Ty, \phi(y)k, kp, \phi(y)l, lq)$. Expand f as

$$f = \sum a_{\chi,\psi,\alpha,\beta}(y)\chi(k)\psi(l)\alpha(p)\beta(q),$$

then the invariance of f implies that for every $\chi, \psi, \alpha, \beta \in \Gamma$

$$a_{\chi,\psi,\alpha,\beta}(Ty)(\chi\psi)(\phi(y)) = a_{\chi\alpha,\psi\beta,\alpha,\beta}(y).$$

If α or β have infinite order this implies $a_{\chi,\psi,\alpha,\beta}(y) = 0$. Otherwise we can find an n such that $\alpha^n = \beta^n = 1$ and then, as

$$T^n(y,k,p,l,q) = (T^n y, \phi_n(y)k, k^n \phi_{n-1}(y) \cdots \phi(y)p,$$
$$\phi_n(y)l, l^n \phi_{n-1}(y) \cdots \phi(y)q),$$

we conclude by the uniqueness of the expansion that

$$a_{\chi,\psi,\alpha,\beta}(T^n y)(\chi\psi)(\phi_n(y))(\alpha\beta)(\phi_{n-1} \cdots \phi(y)) = a_{\chi,\psi,\alpha,\beta}(y).$$

In that case the function $h(y,k,p) = a(y)(\chi\psi)(k)(\alpha\beta)(p)$, where $a = a_{\chi,\psi,\alpha,\beta}$, satisfies

$$h(T^n(y,k,p)) = h(T^n y, \phi_n(y)k, k^n \phi_{n-1}(y) \cdots \phi(y)p)$$
$$= a(T^n y)(\chi\psi)(\phi_n(y))(\alpha\beta)(\phi_{n-1}(y) \cdots \phi(y))(\chi\psi)(k)(\alpha\beta)(p)$$
$$= a(y)(\chi\psi)(k)(\alpha\beta)(p) = h(y,k,p).$$

so that h is a constant function. This implies that $a(y)$ is a constant and if moreover $\chi\psi \neq 1$ or $\alpha\beta \neq 1$ then this constant is necessarily 0. If $\psi = \chi^{-1}$ and $\beta = \alpha^{-1}$ then $c_{\chi,\alpha} = a_{\chi,\chi^{-1},\alpha,\alpha^{-1}}(Ty) = a_{\chi\alpha,\chi^{-1}\alpha^{-1},\alpha,\alpha^{-1}}(y) = c_{\chi\alpha,\alpha}$.

Let F be the torsion subgroup of Γ and for $1 \neq \alpha \in F$ of order n let χ^* be some choice of representatives in Γ for elements of $\Gamma/\{1, \ldots, \alpha^{n-1}\}$. Then

$$f(y,k,p,l,q) = \sum_{\alpha \in F} \sum_{\chi \in \Gamma} c_{\chi,\alpha}\chi(kl^{-1})\alpha(pq^{-1})$$
$$= \sum_{\alpha} \sum_{\chi^*} (c_{\chi^*,\alpha}\chi^*(kl^{-1}) + c_{\chi^*\alpha,\alpha}(\chi^*\alpha)(kl^{-1}) + \cdots$$
$$+ c_{\chi^*\alpha^{n-1},\alpha}(\chi^*\alpha^{n-1})(kl^{-1}))\alpha(pq^{-1})$$
$$= \sum_{\alpha} \sum_{\chi^*} c_{\chi^*,\alpha}(1 + \alpha(kl^{-1}) + \cdots + \alpha^n(kl^{-1}))\chi^*(kl^{-1})\alpha(pq^{-1})$$
$$= \sum_{\chi \in \Gamma} c_{\chi,1}\chi(kl^{-1}).$$

Thus we conclude that our invariant function f has the form

$$f = h(kl^{-1}) = \sum_\chi c_{\chi,1}\chi(kl^{-1}),$$

and is therefore $Z \underset{Y}{\times} Z$ measurable. By Lemma 1 we conclude that $Z = W$. This concludes the proof of (2). $\square$

Corollary 5. *For $\mathcal{X}$ as in Lemma 4 with nontrivial K, the automorphism R_{k_0} of $\mathcal{Z}$ (from proposition 3), can not be lifted to an automorphism of $\mathcal{X}$.*

Proof. By Lemma 4 $\mathcal{Z}$ is the largest isometric extension of $\mathcal{Y}$ in $\mathcal{X}$, and by proposition 3 the measure $\mu \underset{Z,R_{k_0}}{\times} \mu$ is ergodic. Now suppose S is a lift of R_{k_0}, i.e. the diagram

$$
\begin{array}{ccc}
X & \xrightarrow{\ S\ } & X \\
\downarrow & & \downarrow \\
Z & \xrightarrow{R_{k_0}} & Z
\end{array}
$$

commutes. Let us write Δ for the diagonal measure on $X \times X$ and let $L_q : X \to X$ denote the map $L_q(y,k,p) = (y,k,qp)$. Then each measure $(I \times SL_q)(\Delta)$ is an ergodic selfjoining of $\mathcal{X}$ and

$$\int (I \times SL_q)(\Delta)dm(q) = (I \times S)\int (I \times L_q)(\Delta)dm(q)$$

$$= (I \times S)\int \delta_{(y,k,p)} \times (\delta_{(y,k)} \times m)d\mu(y,k,p)$$

$$= (I \times S)\int (\delta_{(y,k)} \times m) \times (\delta_{(y,k)} \times m)d\nu(y)dm(k)$$

$$= \int \mu_{(y,k)} \times \mu_{(y,kk_0)}d\nu(y)dm(k)$$

$$= \mu \underset{Z,R_{k_0}}{\times} \mu.$$

This contradicts the ergodicity of $\mu \underset{Z,R_{k_0}}{\times} \mu$ and the proof is complete. $\square$

Theorem 6. *For a Bernoulli system $\mathcal{W}$, the smallest natural family of factors coincides with the family of all factors.*

Proof. Assume on the contrary that $\mathcal{Y}$ is a factor of $\mathcal{W}$ which does not belong to the smallest natural family and let $\mathcal{Z}$ be its natural cover. Then by [del Junco et al.] Corollary 4.1, the extension $\mathcal{Z} \to \mathcal{Y}$ is a compact group extension, say with compact group K. If we choose any $k_0 \neq 1$ in K and let K_0 be the closed subgroup of K generated by k_0, then the factor map $\mathcal{Z} \to \mathcal{Z}/K_0$ is an Abelian compact group extension and we still have $\mathcal{Z}$ as the natural cover of $\mathcal{Z}/K_0$. Thus we may assume that K itself is Abelian.

We now write $Z = Y \times K$ and $T(y,k) = (Ty, \phi(y)k)$. Next we define the system $\mathcal{X} = (X, \mu, T)$ by $X = Y \times K \times K$ and $T(y, k, l) = (Ty, \phi(y)k, kl)$. Of course $\mathcal{X}$ as defined is no longer a factor of $\mathcal{W}$. However by Lemma 5, $\mathcal{X}$ is weakly mixing, hence by [Rudolph, 1978], Bernoulli of the same entropy as $\mathcal{Z}$. Also, by Lemma 5, $\mathcal{Z}$ is the largest isometric extension of $\mathcal{Y}$ in $\mathcal{X}$.

Next we consider the following two actions of the Abelian topological group $G = \mathbb{Z} \times K$. The first is defined on X by:

$$(T^n, p)(y, k, l) = (T^n y, \phi_n(y)k, k^n \phi_{n-1}(y) \cdots \phi(y)lp), \quad n \in \mathbb{Z}, \ p \in K$$

and the second is defined on Z by:

$$(T^n, p)(y, k) = (T^n y, \phi_n(y)kp), \quad n \in \mathbb{Z}, \ p \in K.$$

In both cases G acts freely and the G-action is Bernoulli of the same entropy as that of the $\mathbb{Z}$-action on $\mathcal{Z}$. By a theorem of Rudolph, [1983], the two systems (X, G) and (Z, G) are isomorphic. In particular there exists a factor map $(Y, T) \to (V, T)$ so that the diagrams $(X, T) \to (Z, T) \to (Y, T)$ and $(Z, T) \to (Y, T) \to (V, T)$ are isomorphic. Thus the extension $(Y, T) \to (V, T)$ is the largest isometric extension of (V, T) in (Z, T) and by Corollary 4, there exists an element $k_0 \in K$ such that the automorphism R_{k_0} of $\mathcal{Y}$, can not be lifted to an automorphism of $\mathcal{Z}$. This contradicts the definition of a natural cover and our proof is complete. $\square$

Problem. *Does the same result hold for K-automorphisms?*

REFERENCES

1. H. Furstenberg, *Ergodic behavior of diagonal measures and a theorem of Szemerédi on arithmetic progressions*, Journal d'Analyse Math. **31** (1977), 204-256.
2. A. del Junco, M. Lemańczyk and M. K. Mentzen, *Semisimplicity, joinings and group extensions*, preprint.
3. D. Rudolph, *Classifying the compact extensions of a Bernoulli shift*, J. d'Analyse Math. **34** (1978), 36-59.
4. D. Rudolph, *An isomorphism theory for Bernoulli free z-skew-compact group extensions*, Adv. Math. **47** (1983), 241-257.

SCHOOL OF MATHEMATICS, TEL AVIV UNIVERSITY, TEL AVIV 69978, ISRAEL
E-mail address: glasner@math.tau.ac.il

TOPOLOGICAL ENTROPY OF EXTENSIONS

ELI GLASNER AND BENJAMIN WEISS

ABSTRACT. The notions of u.p.e. and c.p.e. extensions as well as null-entropy and entropy-generated extensions are introduced and their inter-relations are studied. In particular we show that even for minimal flows a c.p.e. extension need not be a u.p.e. extension, and that, as a result of this, there exists a minimal flow for which the set $E \cup \Delta$ of entropy pairs (with the diagonal added) is not an equivalence relation. Also, the converse implication, which does hold in the absolute case, is no longer true for extensions; i.e., there exists a u.p.e. extension (of minimal flows) which is not a c.p.e. extension.

1. INTRODUCTION

The class of uniform positive entropy (u.p.e.) flows (or dynamical systems) was introduced in [Blanchard, 1992], as an analogue in topological dynamics of the notion of a K-process in ergodic theory. In particular every non-trivial factor of a u.p.e. flow has positive topological entropy. The precise definition of u.p.e. is as follows.

Let (X, T) be a flow; an open cover $\mathcal{U} = \{U, V\}$ of X is called a *standard cover* if both U and V are non-dense in X. (X, T) has *uniform positive entropy (u.p.e.)* if for every standard cover $\mathcal{U}$ of X, the topological entropy $h(\mathcal{U}, T) > 0$. A pair $(x, x') \in X \times X$ is an *entropy pair* if for every standard cover $\mathcal{U}$ with $x \in int(U^c)$ and $x' \in int(V^c)$, $h(\mathcal{U}, T) > 0$. Thus (X, T) is u.p.e. iff every nondiagonal pair in $X \times X$ is an entropy pair. A flow every nontrivial factor of which has positive entropy is called a *completely positive entropy (c.p.e.) flow*.

Denote the set of entropy pairs by $E = E_X$, let $\Delta = \{(x, x) : x \in X\}$. The following facts were shown in [Blanchard, 1992 and 1993].

(1) $h(X, T) > 0$ implies $E \neq \emptyset$.
(2) $\overline{E} \subset E \cup \Delta$.
(3) E (hence also $\overline{E}$) is $T \times T$ invariant.
(4) If $(X, T) \xrightarrow{\pi} (Y, T)$ is a homomorphism, then $\pi \times \pi \overline{[E_X]} = \overline{E_Y}$.
(5) u.p.e. implies c.p.e., but the converse does not hold in general.

Further developments of the theory of entropy pairs were obtained in [Blanchard, 1993], [Blanchard and Lacroix, 1993], [Glasner and Weiss, BSMF] and [Blanchard et al.]. The question whether there exists a minimal flow which is c.p.e. but not u.p.e., however, remains open. In the present note we would like to examine some natural relative versions of the above notions and their interrelations.

Given an extension of flows $(X,T) \xrightarrow{\pi} (Y,T)$, let

$$R_\pi = \{(x,x') : \pi(x) = \pi(x')\}.$$

Put $E_\pi = E \cap R_\pi$, and let F_π be the smallest closed invariant equivalence relation containing E_π. The extension π is

(1) a *null-entropy extension* if $E_\pi = \emptyset$;
(2) a *u.p.e. extension* if $E_\pi \cup \Delta = R_\pi$;
(3) an *entropy-generated extension* if $F_\pi = R_\pi$;
(4) a *completely positive entropy extension*, or a *c.p.e. extension*, if whenever $(X,T) \xrightarrow{\sigma} (Z,T) \xrightarrow{\rho} (Y,T)$, where $\rho \circ \sigma = \pi$ and ρ is not an isomorphism, then $h(Z,T) > h(Y,T)$.

Obviously, every u.p.e. extension is an entropy-generated extension. Taking in these definitions (Y,T) to be the trivial, one-point flow, we see that π is a null, a u.p.e., or a c.p.e. extension iff (X,T) has zero entropy, is a u.p.e. flow, or is a c.p.e. flow, respectively. However—as we will see—unlike the absolute case, where it is easy to check that a flow is c.p.e. iff it is an entropy-generated extension of the trivial flow, in the relative case, even when one assumes minimality, an entropy-generated extension and a c.p.e. extension are, in general, not one and the same thing. Also unlike the absolute case, even with minimality, a u.p.e. extension need not be a c.p.e. extension and vice versa. In particular, it follows from the example we construct to prove the last assertion that the relation $E \cup \Delta$ need not be an equivalence relation. The situation is summarized by the following diagram:

where no other implication is true.

The technique we use for the construction of the minimal flows which demonstrate the latter statement is reminiscent of the one we used in [Glasner and Weiss, JAMS] to construct a minimal flow of zero entropy having a minimal quasifactor of positive entropy; however, the complex device we had to develop there in order to deal with the space of closed subsets is not needed here, and consequently the constructions here are considerably less complicated. We end up with some remarks on the relative disjointness theorem for u.p.e. extensions and about entropy n-tuples.

2. The relative Pinsker factor of X over Y

We start with a relative version of claim (1) in the previous section.

Lemma 2.1. *Suppose that in the extension* $(X,T) \overset{\pi}{\to} (Y,T)$, $h(X,T) > h(Y,T)$. *Then* $E_\pi \neq \emptyset$.

Proof. Assume the contrary is true, and let $\delta > 0$ be given. We will prove the lemma by constructing an open cover $\mathcal{V}$ of X with $\max\{\text{diam}(V) : V \in \mathcal{V}\} < \delta$ and $h(\mathcal{V}, T) \leq h(Y, T)$.

For an $\eta > 0$ let

$$K_{\delta,\eta} = \{(x, x') \in X \times X : d(\pi(x), \pi(x')) < \eta, d(x, x') \geq \delta\}.$$

If for every $\eta > 0, K_{\delta,\eta} \cap E \neq \emptyset$, then there exist sequences η_n and $(x_n, x'_n) \in K_{\delta,\eta_n} \cap E$, with $\lim(x_n, x'_n) = (x, x')$ for some (x, x'). Then clearly $d(x, x') \geq \delta, \pi(x) = \pi(x')$, and it follows that $(x, x') \in E_\pi$, and we are done. Thus we may assume that an $\eta > 0$ exists for which $K_{\delta,\eta} \cap E = \emptyset$.

Under this assumption we proceed as in the proof of Proposition 1.1 in [Glasner and Weiss, BSMF] to get an open cover $\mathcal{V}_1$ of X with the following properties:

(1) $h(\mathcal{V}_1, T) = 0$;
(2) for every $V \in \mathcal{V}_1$ and $x, x' \in V$, if $d(\pi(x), \pi(x')) < \eta$ then $d(x, x') < \delta$.

Next let $\mathcal{V}'_2$ be any open cover of Y satisfying $\max\{\text{diam}(V) : V \in \mathcal{V}'_2\} < \eta$, and put $\mathcal{V}_2 = \{\pi^{-1}(V) : V \in \mathcal{V}'_2\}$ and $\mathcal{V} = \mathcal{V}_1 \vee \mathcal{V}_2$. Then $\max\{\text{diam}(V) : V \in \mathcal{V}\} < \delta$, and

$$h(\mathcal{V}, T) \leq h(\mathcal{V}_1, T) + h(\mathcal{V}_2, T) = h(\mathcal{V}_2, T).$$

This completes the proof. $\square$

Proposition 2.2.

(1) *Every extension* $(X,T) \overset{\pi}{\to} (Y,T)$ *can be decomposed uniquely as follows:*

$$(X,T) \overset{\sigma}{\to} (Z,T) \overset{\rho}{\to} (Y,T),$$

where $\rho \circ \sigma = \pi$, ρ *is a null-entropy extension, and* σ *is an entropy-generated extension.*

(2) *Every c.p.e. extension is an entropy-generated extension.*

(3) $h(X,T) = h(Y,T)$ *when* $(X,T) \overset{\pi}{\to} (Y,T)$ *is a null-entropy extension.*

Proof. We have $E_\pi \subset F_\pi \subset R_\pi$, and we let $Z = X/F_\pi$ and let ρ and σ be the natural maps. Then it follows directly that $E_\rho = \emptyset$; i.e. that ρ is a null-entropy extension. The fact that σ is an entropy-generated extension follows from the definitions. The uniqueness is also clear.

If π is a null-entropy extension, then Lemma 2.1 implies that $h(X,T) = h(Y,T)$. Finally, if π is a c.p.e extension then in the above decomposition of π, ρ is an isomorphism; i.e., $F_\pi = R_\pi$, and π is an entropy-generated extension. $\square$

In this decomposition we call $(X,T) \overset{\sigma}{\to} (Z,T)$ the *relative Pinsker factor of X over Y*. Thus π is an entropy-generated extension iff the corresponding relative Pinsker factor is trivial, i.e. equals (X,T).

Proposition 2.3. *Let* $(X,T) \overset{\pi}{\to} (Y,T)$ *be a homomorphism with* $h(Y,T) = 0$; *then* π *is a c.p.e. extension iff* π *is an entropy-generated extension.*

Proof. Our assumption implies that $E_\pi = E$. Suppose π is an entropy-generated extension and let

$$(X,T) \overset{\theta}{\to} (W,T) \overset{\lambda}{\to} (Y,T)$$

be an intermediate factor with $h(W,T) = h(Y,T) = 0$. Then necessarily $E_\pi = E \subset R_\theta$, and it follows that also $F_\pi = R_\pi \subset R_\theta$, i.e. λ is an isomorphism. This shows that π is c.p.e.. The converse follows from Proposition 2.2.(2). $\square$

The question—in the absolute case—whether every c.p.e. flow is a u.p.e. flow (or equivalently, whether $F = X \times X$ implies that $E \cup \Delta$ is an equivalence relation) was answered negatively in [Blanchard, 1992]; namely there is a c.p.e. flow which is not u.p.e.. However, the question whether this can happen for a minimal flow (X,T) is still open. The analogous relative question is whether for minimal flows every c.p.e. extension is necessarily a u.p.e. extension. We will show in the next section that the answer to that question is negative. We shall construct an example of a minimal flow (X,T) and a homomorphism $(X,T) \overset{\pi}{\to} (Z,T)$, with (Z,T) of zero entropy, for which $F_\pi = X \underset{Z}{\times} X = R_\pi$, while $E_\pi \cup \Delta$ is a proper subset of $X \underset{Z}{\times} X$. In particular, since in that case also $E_\pi = E$, we conclude that $E \cup \Delta$ is not an equivalence relation.

3. A MINIMAL C.P.E. EXTENSION WHICH IS NOT A U.P.E. EXTENSION.

Proposition 3.1. *There exists a minimal flow* (X,T) *and an almost 1-1 homomorphism* $(X,T) \overset{\pi}{\to} (Z,T)$ *such that* (Z,T) *is the Kronecker factor of* (X,T) *and the relative Pinsker factor of* π, *so that* π *is a c.p.e. extension, yet* π *is not a u.p.e. extension. In particular,* $E \cup \Delta$ *is not an equivalence relation.*

Proof. The minimal flow (X,T) will be a subset of $\{0,1,2\}^{\mathbb{Z}}$ with the shift. The construction will be inductive. At stage n, we will have a set of allowable words $\mathcal{W}_n$, each a concatenation of words from $\mathcal{W}_{n-1}$. The length of the words in $\mathcal{W}_n$ is fixed and equals l_n. Thus, if X_n denotes all infinite concatenations of words from $\mathcal{W}_n$, we have $X_0 \supset X_1 \supset X_2 \supset \ldots$ and $X = \cap_0^\infty X_n$.

The minimality of (X, T) is guaranteed by incorporating in each word $w \in \mathcal{W}_n$ an initial segment which contains all pairs of words from $\mathcal{W}_{n-1}$. Thus in X any block of length l_{n-1} that occurs at all will be seen in every block of length $2l_n$; this implies that (X, T) is minimal.

This procedure will establish a hierarchical block structure in elements of X which will, most probably, be determined uniquely by the initial segment in each of the basic blocks. If we want to ensure that this block structure is indeed uniquely determined by encoding this information into the content of the blocks, our construction will become more complicated and may hinder the reader from concentrating on the more essential points of the construction. It is, however, always possible to couple our system with a Kronecker system which consists of an inverse limit of cyclic groups of orders l_n, to mark the block structure of elements of X. (To be more explicit, the coupled system will consist of the orbit closure of the point $(x_0, \underline{0}) \in X \times Z$ where Z is the compact group $\lim_{\leftarrow} \mathbb{Z}_{l_n}$, $\underline{0}$ its zero element, and x_0 is some point in X for which the blocks $x[0, \ldots, l_n - 1]$ are basic n-blocks for every n.) For that reason we shall, from now on, ignore the problem of recognition of the block frameworks and assume that the block structure can be uniquely recognized in elements of X. Thus there is now a natural Kronecker factor $(X, T) \xrightarrow{\pi} (Z, T)$, where $Z \subset \prod_{n=1}^\infty \{0, 1, \ldots, l_n - 1\}$ is defined by $Z = \{z : z(n+1) \equiv z(n)(mod \ l_n), n = 1, 2, \ldots\}$.

The words of $\mathcal{W}_0$ come in two types: $\mathcal{W}_0(1)$ consists of ten words of length $l_0 = 10$:

$$w_0 = 100 \ldots 0 \ ; \ w_1(1) = 110 \ldots 0 \ ; \ w_2(1) = 101 \ldots 0 \ ; \ \ldots w_9(1) = 100 \ldots 1;$$

and $\mathcal{W}_0(2)$ is the set

$$w_0 = 100 \ldots 0 \ ; \ w_1(2) = 120 \ldots 0 \ ; \ w_2(2) = 102 \ldots 0 \ ; \ \ldots w_9(2) = 100 \ldots 2.$$

Note that $\mathcal{W}_0(1) \cap \mathcal{W}_0(2) = \{w_0\}$, the *zero word of stage zero*.

The words of $\mathcal{W}_n$ will have two parts, the *initial segment* of length r_n and the *main part* of length s_n, so that $l_n = r_n + s_n$. Assume that $\mathcal{W}_n$ is already constructed as $\mathcal{W}_n(1) \cup \mathcal{W}_n(2)$ with $\mathcal{W}_n(1) \cap \mathcal{W}_n(2) = \{w_0^{(n)}\}$, the zero word at stage n. We now form one word α by concatenating all pairs of words from $\mathcal{W}_n$. This word α, of length r_{n+1}, will be the first segment of each of the words in $\mathcal{W}_{n+1}$; this, of course, is in order to take care of the minimality of (X, T).

Now for the main parts we will have two types of these, so that

$$\mathcal{W}_{n+1} = \mathcal{W}_{n+1}(1) \cup \mathcal{W}_{n+1}(2).$$

Choose s_{n+1} so that $1 - s_{n+1}/l_{n+1} < 1/(10)^{n+1}$ (where $l_{n+1} = r_{n+1} + s_{n+1}$), and let $k_{n+1} = s_{n+1}/l_n$. Then the main parts of words in $\mathcal{W}_{n+1}(j), j = 1, 2$, will consist of all possible concatenations of k_{n+1} words from $\mathcal{W}_n(j)$. The zero word of stage $n + 1$ is the word $w_0^{(n+1)} = \alpha(w_0^{(n)})^{k_{n+1}}$. Again $w_0^{(n+1)}$ is the only word common to $\mathcal{W}_{n+1}(1)$ and $\mathcal{W}_{n+1}(2)$.

The construction of X is now complete, and we proceed to prove the assertions of the proposition.

Lemma 3.2. *The homomorphism π is almost one-to-one; i.e. there exists a dense G_δ subset, Z_0 of Z, such that $|\pi^{-1}(z)| = 1$ for $z \in Z_0$.*

Proof. Let

$$Z_0 = \{z \in Z : \text{ for some } n_i \to \infty, \lim z(n_i) = \lim r_{n_i} - z(n_i) = \infty\},$$

where $z \in \prod_{n=1}^{\infty}\{0, 1, \ldots, l_n - 1\}$ and r_n is the length of the initial segment in words of $\mathcal{W}_n$.

Since there is a unique initial segment at each stage, it is clear that for each $z \in Z_0$ there is one and only one $x \in X$ for which $\pi(x) = z$. It is easily seen that, by minimality, the set of points $z \in Z$ for which $|\pi^{-1}(z)| = 1$—since it is not empty—is a dense G_δ subset of Z. □

Next let us recall the definition of "E-pairs" from [Glasner and Weiss, JAMS]. In an arbitrary flow (Y, T), we call a pair of points $(y_0, y_1) \in Y \times Y$ an E-pair if $y_0 \neq y_1$ and for every pair of disjoint neighborhoods U_0, U_1 of y_0, y_1, respectively, there exist $\delta > 0$ and k_0 such that for every $k \geq k_0$ there exists a sequence $0 \leq n_1 < n_2 \cdots < n_k < k/\delta$ such that for every $s \in \{0, 1\}^k$ there exists $y \in Y$ with

$$T^{n_1}y \in U_{s(1)}, T^{n_2}y \in U_{s(2)}, \ldots, T^{n_k}y \in U_{s(k)}.$$

It follows directly from the definition that for every pair U_0, U_1 as above the entropy $h(\mathcal{U}, T) \geq \delta$, where $\mathcal{U}$ is the open cover $\{(\overline{U_0})^c, (\overline{U_1})^c\}$ of Y. Hence every E-pair is an entropy-pair.

Suppose now $z \notin Z_0$; then in every $x \in \pi^{-1}(z)$ at either one direction or both going towards ∞ we eventually see a nesting of main parts. These should be either all of type 1 or all of type 2. We associate with z the unique element $x_0(z) \in \pi^{-1}(z)$ where in all of these main parts we substitute the zero word of the corresponding order. Now one easily shows that with the condition

$$\prod_{n=1}^{\infty} s_n/l_n = \gamma > 0,$$

each pair of the form $(x, x_0(z))$, where $z = \pi(x)$ and $x \neq x_0(z)$, is an E-pair and hence an entropy pair, i.e. an element of E. We can now conclude that F_π, the smallest closed invariant equivalence relation containing E, is all of R_π; i.e., π is an entropy-generated extension. By Proposition 2.3—as $h(Z, T) = 0$—π is also a c.p.e. extension.

In order to complete the proof of Proposition 3.1 it remains to show that not every pair in $R_\pi \backslash \Delta$ is an entropy pair. For this it is enough to produce a standard open cover $\mathcal{V} = \{V_1, V_2\}$ such that there exist $x_1, x_2 \in X$ with $\pi(x_1) = \pi(x_2), x_1 \notin \overline{V_2}$ and $x_2 \notin \overline{V_1}$, for which $h(\mathcal{V}, T) = 0$.

Let U_j be the set of all $x \in X$ for which $x(0)x(1)\dots x(s_1 - 1)$ is the main part of a word of $\mathcal{W}_1(j)$ other than the zero word, $j = 1, 2$, and U_0 if this word is the zero word. We let $V_j = (U_j)^c$ and $\mathcal{V} = \{V_1, V_2\}$. Note that $U_0 \cup U_1 \cup U_2 := B$ is the inverse image of a point in $\mathbb{Z}_{l_1}$ under the natural map of X onto that space; thus $\{B, TB, T^2B, \dots, T^{l_1-1}B\}$, is a clopen partition of X.

In order to compute $h(\mathcal{V}, T)$ it is enough to consider the number of elements in a minimal subcover of $(\mathcal{V})_0^{l_n}, n = 1, 2, \dots$. Now for an arbitrary point $x \in X$, consider the sequence $x, Tx, T^2x, \dots, T^{l_n}x$. We choose an element $D = D_0 \cap T^{-1}D_1 \cap \dots \cap T^{l_n}D_{l_n}$ where $D_m \in \{V_1, V_2\}, 0 \le m \le l_n$, containing x as follows.

If $T^mx \notin B$, choose $D_m = V_1$ for x. If $T^mx \in B$, we choose $D_m = V_1$ whenever we see the zero word of level 1 in our B "window". Our choice $D_m = V_1$ or $D_m = V_2$ is determined in the remaining cases by the type (1 or 2) of the l_1-word we see in the B "window". Now, depending on the position of the zero coordinate of x in the $\mathcal{W}_n$ word, and the type (1 or 2) of this word, and that which follows it, it is easy to see that we end up this way with a subcover of $(\mathcal{V})_0^{l_n}$ with at most $4l_n$ elements.

Finally, since here $E_\pi = E$, we conclude that $E \cup \Delta$ is not an equivalence relation. This completes the proof of Proposition 3.1. □

4. A MINIMAL U.P.E. EXTENSION WHICH IS NOT A C.P.E. EXTENSION.

Proposition 4.1. *There exists a minimal flow (X, T) and a homomorphism $(X, T) \xrightarrow{\sigma} (Y, T)$ such that π is a u.p.e. extension but not a c.p.e. extension.*

Proof. To begin with, let (X_0, T) be the disjoint union of two Bernoulli shifts, $X_0 = \{0, 1, 2\}^{\mathbb{Z}} \cup \{0, 3\}^{\mathbb{Z}}$, and let (Y_0, T) be the flow obtained from X_0 by identifying the symbol 3 with 0. Thus (Y_0, T) is isomorphic to the shift on $\{0, 1, 2\}^{\mathbb{Z}}$, and one easily checks that the extension $(X_0, T) \xrightarrow{\sigma_0} (Y_0, T)$ is a u.p.e. extension (see Example 8 in [Blanchard, 1992]). Since obviously $h(X_0, T) = h(Y_0, T) = \log 3$, we conclude that σ_0 is not a c.p.e. extension. Of course (X_0, T) is not even topologically transitive.

Next we construct a minimal flow $(X, T) \subset \{0, 1, 2, 3\}^{\mathbb{Z}}$ and a factor $(X, T) \xrightarrow{\sigma} (Y, T)$ such that σ is a u.p.e. but not a c.p.e. extension. The construction of X is very similar to the one given in the previous section. The words in $\mathcal{W}_{n+1}$ (for X) will be of two types, $\mathcal{W}_{n+1}(0, 1, 2)$ and $\mathcal{W}_{n+1}(0, 3)$, where in building the main parts of words in $\mathcal{W}_{n+1}(0, 1, 2)$ we substitute all possible concatenations of k_{n+1} words from $\mathcal{W}_n(0, 1, 2)$, etc.. It is now easy to see that the subset

$$R = \{(x, x') : x \text{ and } x' \text{ have the same block structure}$$
$$\text{and, when distinct, are both of "type" } 0, 3\},$$

is a closed invariant equivalence relation. We let $Y = X/R$ and let $\sigma : X \to Y$ be the quotient map. We leave it to the reader to check that π indeed has the desired properties. $\square$

In [Blanchard, 1993] (Proposition 6), it is shown that every u.p.e. flow is disjoint from every minimal flow with zero topological entropy. Our example in [Glasner and Weiss, JAMS] (Corollary 5.3) provided also an example for two minimal flows (X, T) and (Ω, T) and a common factor $(X, T) \overset{\pi}{\to} (Z, T)$ and $(\Omega, T) \overset{\sigma}{\to} (Z, T)$, where (X, T) has zero entropy (hence π is a null-entropy extension), σ is a u.p.e. extension and yet (X, T) and (Ω, T) are not disjoint over (Z, T).

We will show next that a relative disjointness theorem can be obtained if we assume that our extension satisfies a stronger condition than u.p.e.. Generalizing the notion of an entropy pair, given a flow (Y, T), let us call an n-tuple $(y_1, y_2, \ldots, y_n) \in Y \times Y \cdots \times Y = Y^{(n)}$ an *entropy n-tuple* if at least two of the points $\{y_j\}_{j=1}^n$ are different and if whenever U_j are closed mutually disjoint neighborhoods of the distinct points y_j, the open cover $\mathcal{U} = \{U_j^c : 0 \leq j \leq n\}$ satisfies $h(\mathcal{U}, T) > 0$. Let us call an extension $(Y, T) \overset{\sigma}{\to} (Z, T)$ a *u.p.e. extension of order n* if every point $(y_1, y_2, \ldots, y_n) \in Y^{(n)}$, not on the diagonal $\Delta_n = \{(y, y, \ldots, y) : y \in Y\}$, with $\sigma(y_i) = \sigma(y_j)$ for $0 \leq i, j \leq n$, is an entropy n-tuple. We will call σ *u.p.e. of all orders* if it is u.p.e. of order n, for every $n \geq 2$.

Proposition 4.2. *If an extension $(Y, T) \overset{\sigma}{\to} (Z, T)$ is u.p.e. of all orders and open, and if $(X, T) \overset{\pi}{\to} (Z, T)$ is a null-entropy extension with (X, T) minimal, then (X, T) and (Y, T) are disjoint over (Z, T).*

Proof. Let J be any closed invariant subset of $X \underset{Z}{\times} Y := \{(x, y) \in X \times Y : \pi(x) = \pi(y)\}$. We have the natural projections π_X and π_Y of J on X and Y, respectively; we also have the induced maps from $J^{(n)}$ to $X^{(n)}$ and $Y^{(n)}$. For each n let

$$J_n = \{(x, y_1, y_2, \ldots, y_n) : \pi(x) = \sigma(y_j), j = 1, 2, \ldots, n\}$$

be a subset of $X \times Y^{(n)}$. As in [Blanchard, 1993], one can show that over every entropy n-tuple in $Y^{(n)}$ lies at least one entropy n-tuple in $J_n^{(n)}$, and that an entropy n-tuple in $J_n^{(n)}$ is projected onto either a diagonal n-tuple or an entropy n-tuple in $X^{(n)}$. Since all the points in this n-tuple have the same π image, and since, by assumption, there are no entropy n-tuples with this property, it must be a diagonal n-tuple. Now, by assumption, every

n-tuple $(y_1, y_2, \ldots, y_n) \in Y^{(n)} \backslash \Delta_n$ with $\sigma(y_i) = \sigma(y_j) = z$ for all i, j is an entropy n-tuple. Whence, for every such n-tuple, there exists an $x \in X$ with $(x, y_1, y_2, \ldots, y_n) \in J_n$. Since this is true for every n, we conclude that for every $z \in Z$ there exists an $x \in X$ with $\{x\} \times \sigma^{-1}(z) \subset J$. Now use the minimality of (X, T) and the openness of σ to conclude that $J = X \underset{Z}{\times} Y$, proving the relative disjointness of X and Y over Z. $\square$

Remark. Several natural questions arise in connection with the notion of entropy n-tuples. In particular the following is a question of B. Host. Is it true that an n-tuple $(y_1, y_2, \ldots, y_n)$, such that (y_i, y_j) is an entropy pair whenever $i \neq j$, is necessarily an entropy n-tuple? A positive answer to this question will yield the assertion that a u.p.e. flow (extension) is a u.p.e. flow (extension) of all orders. This in turn will yield a stronger version of Proposition 4.2, in which one assumes that σ is u.p.e. rather than u.p.e. of all orders, and then, taking Z to be the trivial one-point flow, a new proof of the disjointness result in [Blanchard, 1993].

References

1. F. Blanchard, *Fully positive topological entropy and topological mixing*, Symbolic Dynamics and Applications; AMS Contemporary Mathematics, vol. 135, 1992, pp. 95-105.
2. F. Blanchard, *A disjointness theorem involving topological entropy*, Bull. Soc. Math. France. **121** (1993), 465-478.
3. F.Blanchard, B.Host, A.Maass, S. Martínez, D.J.Rudolph, *Entropy pairs for a measure*, to appear in Erg. Th. Dyn. Sys..
4. F. Blanchard and Y. Lacroix, *Zero-entropy factors of topological flows*, Proc. Amer. Math. Soc. **119** (1993), 985-992.
5. M. Denker, C. Grillenberger and K. Sigmund, *Ergodic Theory on Compact Spaces*, vol. 527, Springer-Verlag, Lecture Notes in Math., 1976.
6. E. Glasner and B. Weiss, *Strictly ergodic uniform positive entropy models*, to appear in Bull. Soc. Math. France..
7. E. Glasner and B. Weiss, *Dynamics and entropy of the space of measures*, C. R. Acad. Sci. Paris **317** (1993), 239-243.
8. E. Glasner and B. Weiss, *Quasi-factors of zero entropy systems*, to appear in J. Amer. Math. Soc..

SCHOOL OF MATHEMATICS, TEL AVIV UNIVERSITY, RAMAT AVIV, ISRAEL
E-mail address: glasner@math.tau.ac.il

INSTITUTE OF MATHEMATICS, HEBREW UNIVERSITY, JERUSALEM, ISRAEL
E-mail address: weiss@math.huji.ac.il

FUNCTIONAL EQUATIONS ASSOCIATED
WITH THE SPECTRAL PROPERTIES
OF COMPACT GROUP EXTENSIONS

GEOFFREY GOODSON

ABSTRACT. New criteria for compact abelian group extensions of discrete spectrum transformations to have non-simple spectrum are given. The relation between the maximal spectral multiplicity of such automorphisms and the existence of solutions to certain functional equations is studied. These results are applied to Morse automorphisms arising from Morse sequences involving constant blocks and also to give a new and elementary proof that the Mathew-Nadkarni transformation has a Lebesgue component of multiplicity 2 in its spectrum.

0. INTRODUCTION

In [Goodson et al., 1992] it was noted that if $T_\phi : X \times G \to X \times G$,

$$T_\phi(x, g) = (Tx, \phi(x) + g),$$

is a compact abelian group extension of a measure-preserving transformation T for which there exists $S \in C(T)$, the centralizer of T, $v : G \to G$ a continuous group automorphism, and a measurable solution $u : X \to G$ to the equation

$$\phi(Sx) - v(\phi(x)) = u(Tx) - u(x),$$

then the maximal spectral multiplicity of T_ϕ is non-simple if there exists $\chi \in \widehat{G}$, the character group of G, for which $\chi \neq \chi \circ v$. Using this idea in [Goodson et al., 1992], examples of ergodic automorphisms with arbitrary maximal spectral multiplicity were constructed. In these examples, the multiplicity comes from a non-abelian centralizer, which always results from a non-trivial solution to the above equation [Newton, 1978]. It has been noted that other types of examples have non-simple spectrum of finite multiplicity, even though their centralizer is abelian (these include certain Morse automorphisms, substitutions and more general extensions of discrete-spectrum transformations such as the Mathew-Nadkarni example [Mathew-Nadkarni, 1984]).

This paper came about as an attempt to understand such examples and to show that in many cases the non-simplicity of the spectrum arises at the measure-theoretic level from a unitary equivalence. All the group extensions

T_ϕ we shall consider have in common the property of being isomorphic to their inverses. Their spectral multiplicity arises from the existence of a group automorphism $v : G \to G$ and a measurable function $u : X \to G$ satisfying

$$v(\phi(x)) - \phi(kTx) = u(Tx) - u(x),$$

where k is a conjugation between T and T^{-1}.

We start Section 2 by generalizing some results from [Queffelec, 1987], [Newton, 1978] and [Goodson et al., 1992], which are to be used in the proofs of our main theorems. In particular, we give the general form of the automorphism which conjugates two isomorphic compact abelian group extensions T_ϕ and S_ψ in the case that T and S are isomorphic with discrete spectrum. The functional equation arising in this situation is shown to have implications concerning the unitary equivalence of certain unitary operators, which are the building blocks of the unitary operator U_{T_ϕ} induced by the group extension T_ϕ. New general results (Theorems 4 and 5) giving sufficient conditions for a compact abelian group extension to have a non-simple spectrum are given. Note that it is shown in [Robinson, 1988b] that non-abelian extensions always have non-simple spectrum.

In Section 3 we show that the above functional equation can always be solved for certain Morse automorphisms arising from constant blocks. Using Theorem 4 we give a new general condition for such Morse automorphisms to have maximal spectral multiplicity equal to 2 (Theorem 9). The question of the multiplicity is shown to be related to the existence of group automorphisms $v : G \to G$ of order 2 for which the above-mentioned functional equation holds. It is still an open question whether or not for Morse automorphisms arising from constant blocks (i.e bijective and commutative substitutions of constant length) the multiplicity can be greater than 2. We feel that the results of this paper are in the right direction to answering this question.

In Section 4 we give a number of examples to illustrate our methods. In particular, Theorem 5 is used to give a new proof that the Mathew-Nadkarni transformation has a Lebesgue component of multiplicity 2 in its spectrum.

I wish to acknowledge that this paper benefitted from conversations with E. A. Robinson.

1. PRELIMINARIES

Throughout this paper $S, T : (X, \mathcal{B}, \mu) \to (X, \mathcal{B}, \mu)$ denote ergodic automorphisms (invertible measure-preserving transformations) on a Lebesgue probability space. Both the identity automorphism and the identity group automorphism will be denoted by I. The context should make it clear which is being used.

Let G be a compact metric abelian group with Haar measure m, and denote by $\widehat{G}$ the character group of G. By a *cocycle* we mean a measurable function $\phi : X \to G$. A cocycle ϕ defines an automorphism T_ϕ on $(X \times G, \tilde{\mu})$ by $T_\phi(x, g) = (Tx, \phi(x) + g)$, $x \in X$, $g \in G$, where $\tilde{\mu} = \mu \times m$. Such an automorphism is called a *G–extension* of T. It need not be ergodic. In fact T_ϕ is ergodic iff *for no non-trivial character* $\chi \in \widehat{G}$ *is there a measurable solution* $f : X \to S^1$ *of the functional equation* $\chi(\phi(x)) = f(Tx)/f(x)$.

The cocycle ϕ is said to be *ergodic* if T_ϕ is an ergodic automorphism, and it is said to be *continuous* if T_ϕ has no eigenvalues other than those for T. The eigenvalues of T_ϕ are those of the induced unitary operator $U_{T_\phi} : L^2(X \times G, \tilde{\mu}) \to L^2(X \times G, \tilde{\mu})$ defined by $U_{T_\phi}(\tilde{f}) = \tilde{f} \circ T_\phi$. The space $L^2(X \times G, \tilde{\mu})$ can be decomposed as

$$L^2(X \times G, \tilde{\mu}) = \oplus_{\chi \in \widehat{G}} L_\chi,$$

where each of the subspaces $L_\chi = \{f \otimes \chi : f \in L^2(X, \mu)\}$ is U_{T_ϕ}-invariant. Notice that $U_{T_\phi} : L_\chi \to L_\chi$ is unitarily equivalent to the unitary operator $V_{\phi, T, \chi} : L^2(X, \mu) \to L^2(X, \mu)$, where $V_{\phi, T, \chi}(f)(x) = \chi(\phi(x))f(Tx)$, $x \in X$.

By the spectral theorem, U_{T_ϕ} is determined up to unitary equivalence by a spectral measure class ρ and a $\{+\infty, 1, 2, \dots\}$-valued multiplicity function M on the circle. The set $\mathcal{M}_T$ of essential spectral multiplicities of T_ϕ is the set of all ρ-essential values of M. The maximal spectral multiplicity (or just multiplicity) of T_ϕ is $\mathrm{msm}(T_\phi) = \sup \mathcal{M}_T$. T_ϕ has simple spectrum if $\mathrm{msm}(T_\phi) = 1$, otherwise T_ϕ has non-simple spectrum.

Given a finite abelian group G, we can define a Morse sequence over G as follows. Let $Y = G^{\mathbb{Z}}$ with its usual topology. Each finite sequence $b = b[0]b[1] \dots b[n-1]$, $b[i] \in G$, $i = 0, 1, \dots, n-1$, $n \geq 1$, is called a *block* over G of length $|b| = n$. Given $g \in G$, $\sigma_g(b)$ is the block defined by

$$\sigma_g(b)[j] = b[j] + g, \quad j = 0, 1, \dots, n-1.$$

$\sigma_g(y)$ is defined similarly for any $y \in Y$.

If $c = c[0]c[1] \dots c[m-1]$ is another block over G, then we define

$$b \times c = \sigma_{c[0]}(b) \sigma_{c[1]}(b) \dots \sigma_{c[m-1]}(b).$$

Assume $b^0, b^1, \dots$, are finite blocks with $|b^t| \geq 2$, $t \geq 0$, each starting with e (the identity of G); then we can define a one sided infinite sequence x as

$$\mathrm{x} = b^0 \times b^1 \times \dots.$$

Conditions (1) and (2) from [Martin, 1976], which we assume throughout, guarantee that x is aperiodic and almost periodic and has an almost periodic extension, which we also denote by x $\in G^{\mathbb{Z}}$. Let $O(\mathrm{x})$ denote the orbit closure of x in $G^{\mathbb{Z}}$ under the shift map $\tilde{T}$. Then it is well known that the restriction of $\tilde{T}$ to the invariant compact metric space $O(\mathrm{x})$ is minimal and uniquely ergodic, with unique invariant measure denoted by $\tilde{\mu}$. In fact, $\tilde{T} : O(x) \to O(x)$ can be represented as a G-extension $T_\phi : X \times G \to X \times G$, of a discrete-spectrum transformation $T : X \to X$ (see for example [Robinson, 1988a]). T is actually the rotation $Tx = x + \hat{1}$ on the group of $(r|b^0|, |b^1|, |b^2|, \dots)$-adic integers, (for some integer r dividing $|G|$, the order of G), and $\phi : X \to G$ is a cocycle which is constant on all but the top level of each of the towers used in the stacking construction of T. T_ϕ is called a *Morse automorphism* and ϕ is called a *Morse cocycle*. In particular, if X is the group of q-adic integers and ϕ is a Morse cocycle over G with a certain kind of self-similarity property, then the corresponding Morse automorphism T_ϕ arises from a Morse sequence of the form $x = b \times b \times \dots$ over G, where $|b| = q$. In this case T_ϕ can also be represented as a bijective and commutative substitution of constant length q over the finite alphabet G.

2. GENERAL THEOREMS

In this section we generalize some results from [Goodson et al., 1992] and [Newton, 1978], and we give some new results giving conditions for the non-simplicity of the spectrum of group extensions.

We start by giving a new general result which says that for fixed ϕ and T having discrete spectrum, the maximal spectral types ρ_χ of $V_{\phi,T,\chi}$ are either equivalent ($\rho_\chi \sim \rho_\gamma$) or mutually singular ($\rho_\chi \perp \rho_\gamma$, $\chi, \gamma \in \widehat{G}$). This result was previously known only in the case of Morse automorphisms and substitutions (see Theorem 2 of [Kwiatkowski-Sikorski, 1987] and also [Queffelec, 1987]). It would be interesting to know if the condition that T has discrete spectrum can be replaced by requiring that T be rank one. Part (ii) is due to [Helson, 1975] (see also [Queffelec, 1987]). Part (iii) is well known, see for example [Goodson, 1985], [Kwiatkowski-Sikorski, 1987] or [Queffelec, 1987].

Theorem 1. *Suppose that T_ϕ is ergodic and that T has discrete spectrum. Let ρ_χ denote the maximal spectral type of $V_{\phi,T,\chi}$.*

(i)   For each $\chi, \gamma \in \widehat{G}$, either $\rho_\chi \sim \rho_\gamma$ or $\rho_\chi \perp \rho_\gamma$.

(ii)   For each $\chi \in \widehat{G}$, ρ_χ is either purely discrete, purely singular continuous or purely Lebesgue.

(iii)   If T_ϕ is a Morse automorphism, then for each $\chi \in \widehat{G}$, $V_{\phi,T,\chi}$ has simple singular spectrum.

Proof. (i)   If ρ is a Borel probabilty measure on the unit circle S^1 and D is a countable subgroup of S^1, then ρ is said to be D-*quasi-invariant* if $\delta_\alpha * \rho \ll \rho$ for every $\alpha \in D$. (δ_α denotes the Borel probability measure with support α). ρ is said to be D-*ergodic* if every Borel set A in S^1 with $\alpha A = A$ for all $\alpha \in D$ satisfies $\rho(A) = 0$ or 1. Theorem 1 now follows from the following two lemmas (Lemma 1 can be found in [Queffelec, 1987] and Lemma 2 generalizes Proposition 3.23 of [Queffelec, 1987]). $\square$

Lemma 1. *If ρ_1 and ρ_2 are two D-quasi-invariant and D-ergodic Borel probability measures on the circle S^1, then $\rho_1 \sim \rho_2$ or $\rho_1 \perp \rho_2$.*

Lemma 2. *Suppose that $T : (X, \mathcal{B}, \mu) \to (X, \mathcal{B}, \mu)$ is ergodic with discrete spectrum and that $\omega : X \to S^1$ is a measurable function. If $V : L^2(X, \mu) \to L^2(X, \mu)$ is the unitary operator defined by $Vf(x) = \omega(x)f(Tx)$, then ρ_m, the maximal spectral type of V, is D-ergodic and D-quasi-invariant, where D is the eigenvalue group of T.*

Now let $\psi : X \to G$ be another cocycle, but this time defining a different G-extension S_ψ, $S_\psi(x, g) = (Sx, \psi(x) + g)$.

Theorem 2. *Let $\chi, \gamma \in \widehat{G}$. Suppose that there exists a continuous group epimorphism $v : G \to G$ and $k : X \to X$ an automorphism satisfying*

(i) $\gamma = \chi \circ v$; $Tk = kS$,

(ii)   there exists a measurable solution $u : X \to G$ to the functional equation

$$\phi(kx) - v(\psi(x)) = u(Sx) - u(x). \tag{1}$$

Then the unitary operator $W = V_{u,k,\chi}$ (i.e. $Wf(x) = \chi(u(x))f(kx))$ satisfies $WV_{\phi,T,\chi} = V_{\psi,S,\gamma}W$. Consequently, $V_{\phi,T,\chi}$ and $V_{\psi,S,\gamma}$ are unitarily equivalent.

Proof. It is easy to check that

$$WV_{\phi,T,\chi}f = V_{\psi,S,\gamma}Wf$$

for all $f \in L^2(X, \mu)$. $\square$

Notice that if T_ϕ is ergodic and equation (1) holds for some continuous group homomorphism $v : G \to G$, then v is necessarily an epimorphism. It follows that if G is a finite abelian group then v is a group automorphism. Also notice that the transformation $\tilde{K}$ acting on $X \times G$ and defined by the formula

$$\tilde{K}(x, g) = (kx, u(x) + v(g)) \tag{2}$$

preseves $\tilde{\mu}$ and satisfies $T_\phi \tilde{K} = \tilde{K} S_\psi$. Consequently S_ψ and T_ϕ are isomorphic when v is an automorphism. Actually, we shall see in the next theorem that every isomorphism between T_ϕ and S_ψ is of the form given in equation (2), if both S_ψ and T_ϕ are ergodic, and S and T are isomorphic with discrete spectrum.

This theorem is a straightforward generalization of Theorem 2.1 of [Newton, 1978] concerning the lifting of members of the centralizer $C(T)$ to members of the centralizer $C(T_\phi)$, so we omit the proof. See also [Anzai, 1951].

Theorem 3. *Suppose that S and T are isomorphic ergodic transformations having discrete spectrum. If also S_ψ and T_ϕ are ergodic, then an onto map $\tilde{K} : X \times G \to X \times G$ satisfies $T_\phi \tilde{K} = \tilde{K} S_\psi$ if and only if*

$$\tilde{K}(x, g) = (kx, u(x) + v(g))$$

for some automorphism $k : X \to X$, measurable function $u : X \to G$ and continuous group epimorphism $v : G \to G$ satisfying

(i) $Tk = kS$

(ii) $\phi(kx) - v(\psi(x)) = u(Sx) - u(x)$.

Furthermore, $\tilde{K}$ is an isomorphism if and only if v is a group automorphism.

If S and T as above are ergodic with discrete spectrum, they can be represented as rotations S_{g_0} and T_{g_1} on a compact abelian group G, i.e. $S_{g_0}(g) = g_0 + g$ and $T_{g_1}(g) = g_1 + g$ for all $g \in G$. The ergodicity of S_{g_0} and T_{g_1} implies that the sets $\{ng_0 : n \in \mathbb{Z}\}$ and $\{ng_1 : n \in \mathbb{Z}\}$ are dense in G.

We use Theorem 3 to determine the form of all conjugations between two isomorphic and ergodic discrete-spectrum transformations S and T. This result is classical, but does not seem to be widely known.

Corollary 1. *Suppose S and T are ergodic with discrete spectrum. If in addition, they are isomorphic with $Tk = kS$, then $k : G \to G$ is of the form*

$$k(g) = h_0 + v(g) \tag{3}$$

for some $h_0 \in G$ and some group automorphism $v : G \to G$ satisfying $v(g_0) = g_1$.

Corollary 2. *(i)   If $T = S^n$ for some $n \in \mathbb{Z}$, then $v(g) = ng$ for all $g \in G$.*

(ii)   If $n = -1$, then k is an involution, i.e. $k^2(g) = g$ for all $g \in G$.

Proof. We apply Theorem 3 with $X = \{a\}$, a 1-point space. Suppose that $\tilde{K}(a, g) = (a, k(g))$, $S_\psi(a, g) = (a, S_{g_0}(g))$ and $T_\phi(a, g) = (a, T_{g_1}(g))$, where $T_{g_1} k = k S_{g_0}$ and k is a measure-theoretic automorphism of G.

Then $\tilde{K} S_\psi = T_\phi \tilde{K}$, and Theorem 3 implies that $\tilde{K}(a, g) = (a, h_0 + v(g))$ for some $h_0 \in G$ and continuous group automorphism $v : G \to G$ satisfying $v(g_0) = g_1$.

Corollary 2 now follows, since $v(ng_0) = ng_1$, $g_1 = ng_0$ and the ergodicity of S_{g_0} and T_{g_1} imply that $v(g) = ng$ for all $g \in G$. If $n = -1$, $v(g) = -g$, $k(g) = h_0 - g$, and so $k^2(g) = g$ for all $g \in G$.   □

In the case of $n = 1$ we obtain the well-known result that $v(g) = g$, or that every member of the centralizer $C(T)$ is a rotation on G. Note that a discrete-spectrum transformation is always isomorphic to its inverse. Also, if G is the group of p-adic integers and $T : G \to G$ is $T(x) = x + \hat{1}$, then T is isomorphic to T^n if and only if $(n, p) = 1$.

From now on we are primarily concerned with functional equations of the form $\phi(kx) - v(\psi(x)) = u(Sx) - u(x)$ when $S = T^{-1}$ and $\psi = \phi T^{-1}$. This is because solutions to these equations give rise to multiplicity for group extensions and also give a conjugation between the group extension and its inverse. Such an equation can be written as

$$v(\phi(x)) - \phi(kTx) = u(Tx) - u(x). \tag{4}$$

Theorem 4. *(a) Suppose that T_ϕ is ergodic and that k is an automorphism satisfying*

$$Tk = kT^{-1} \quad v(\phi(x)) - \phi(kTx) = u(Tx) - u(x), \tag{5}$$

for some continuous group epimorphism $v : G \to G$ and measurable function $u : X \to G$. If $v \neq I$, then T_ϕ has non simple spectrum.

(b) If in addition T has discrete spectrum, then $V_{\phi,T,\chi}$ and $V_{\phi,T,\chi \circ v}$ have equivalent maximal spectral types.

Proof. By Theorem 2, the operators $V_{\phi,T,\chi}$ and $V_{\phi T^{-1},T^{-1},\chi \circ v}$ are unitarily equivalent.

Let $f \in L^2(X, \mu)$; then

$$(V_{\phi T^{-1},T^{-1},\chi \circ v}^n f, f) = \int_X \chi \circ v(\phi T^{-1}x) \dots \chi \circ v(\phi T^{-n}x) f(T^{-n}) \overline{f(x)} \, d\mu$$

$$= \int_X \chi \circ v(\phi T^{n-1}x) \dots \chi \circ v(\phi(x)) f(x) \overline{f(T^n x)} \, d\mu$$

$$= (V_{\phi,T,\chi \circ v}^n \bar{f}, \bar{f}) \quad \forall\, n \in \mathbb{Z}. \tag{6}$$

If we let $f = 1$, the characteristic function of X, we see that the spectral measures of 1 for each of the operators $V_{\phi T^{-1},T^{-1},\chi \circ v}$ and $V_{\phi,T,\chi \circ v}$ are equivalent, and hence the maximal spectral types of these operators are not mutually singular. It follows that $V_{\phi,T,\chi}$ and $V_{\phi,T,\chi \circ v}$ have maximal spectral types which are not mutually singular. This proves (a).

(b) If in addition T has discrete spectrum, then Theorem 1(i) implies that the maximal spectral types are equivalent. $\square$

The following theorem applies whenever T is isomorphic to its inverse via a conjugating automorphism k satisfying $k^2 = I$. We have seen this is true for any discrete-spectrum transformation, but a recent unpublished result shows this to be true for any rank one transformation isomorphic to its inverse.

Proposition 1. *If T_ϕ is ergodic and satisfies equation 5 of Theorem 4 with $k^2 = I$, then*

   (i) $v^2 = I$;

   (ii) $u(kx) - v(u(x)) = h$, *a.e. a constant, for some fixed $h \in G$.*

Proof. (i)   Suppose that $v(\phi(x)) - \phi(kTx) = u(Tx) - u(x)$; then

$$v(\phi(kTx)) - \phi(kTkTx) = u(TkTx) - u(kTx)$$

$$v(\phi(kTx)) - \phi(k^2T^{-1}Tx) = u(kT^{-1}Tx) - u(kTx)$$

$$v(\phi(kTx)) - \phi(x) = u(kx) - u(kTx). \tag{7}$$

Operating on both sides of equation (5) by v, we obtain

$$v^2(\phi(x)) - v(\phi(kTx)) = v(u(Tx)) - v(u(x)). \tag{8}$$

From equations (7) and (8) we get

$$v^2(\phi(x)) - \phi(x) = g(Tx) - g(x), \tag{9}$$

where $g = v \circ u - u \circ k$. Now let $\chi \in \widehat{G}$; then by equation (9)

$$\chi \circ v^2(\phi(x))/\chi(\phi(x)) = \chi(g(Tx))/\chi(g(x)), \tag{10}$$

and T_ϕ ergodic implies $\chi \circ v^2 = \chi$ for all $\chi \in \widehat{G}$, and so $v^2 = I$.

   (ii) If we now add equations (7) and (8) with $v^2 = I$, we obtain

$$u(kTx) - v(u(Tx)) = u(kx) - v(u(x)) \tag{11}$$

Since T is ergodic, the result follows. $\square$

   In view of Theorem 4 it seems likely that we cannot conclude anything about the non-simplicity of the spectrum of T_ϕ in the case that $v = I$. However, this is not the case. The next theorem shows that a non-simple spectrum results when the measurable function $u : X \to G$ is non-trivial in the sense that $u(x) - u(kx) \neq 0$. Its proof shows that this construction is possible only for groups G containing a copy of $\mathbb{Z}_2$.

Theorem 5. *Suppose that T_ϕ is ergodic and that there exists an automorphism $k : X \to X$ and a non-trivial measurable function $u : X \to G$ satisfying*

   (i) $Tk = kT^{-1}$ and $k^2 = I$,

   (ii) $\phi(x) - \phi(kTx) = u(Tx) - u(x)$.

   Then T_ϕ has non-simple spectrum.

Proof.

   The theorem will follow from the following lemma which is of independent interest.

Lemma 3. *If $h = u(x) - u(kx) \neq 0$ is the constant from Proposition 1, and $\chi \in \widehat{G}$ satisfies $\chi(h) \neq 1$, then $V_{\phi,T,\chi}$ has non simple-spectrum. In fact the functions 1 and $\overline{\chi}(u(x))$ generate orthogonal cyclic subspaces having the same spectral types.*

Proof. Theorem 2 implies that $V_{\phi,T,\chi}$ and $V_{\phi T^{-1}, T^{-1},\chi} = V_{\phi,T,\overline{\chi}}^{-1}$ are unitarily equivalent via $Wf(x) = \chi(u(x))f(kx)$.

Let $f_1(x) = W1 = \chi(u(x))$ and $f_2(x) = \overline{f_1}(x)$; then

$$W f_2(x) = \chi(u(x))\overline{\chi}(u(kx)) = \chi(u(x) - u(kx)) = \chi(h).$$

We see as follows that 1 and f_2 have the same spectral type:

$$(V_{\phi,T,\chi}^n 1, 1) = (W V_{\phi,T,\chi}^n 1, W1) = (V_{\phi,T,\overline{\chi}}^{-n} W1, W1)$$

$$= (W1, V_{\phi,T,\overline{\chi}}^n W1) = (f_1, V_{\phi,T,\overline{\chi}}^n f_1) = \overline{(V_{\phi,T,\overline{\chi}}^n f_1, f_1)}$$

$$= (V_{\phi,T,\chi}^n \overline{f_1}, \overline{f_1}) = (V_{\phi,T,\chi}^n f_2, f_2) \quad \forall n \in \mathbb{Z}.$$

Furthermore,

$$(V_{\phi,T,\chi}^n f_2, 1) = (W V_{\phi,T,\chi}^n f_2, W1)$$

$$= (V_{\phi,T,\overline{\chi}}^{-n} W f_2, W1) = (W f_2, V_{\phi,T,\overline{\chi}}^n W1)$$

$$= (\chi(h), V_{\phi,T,\overline{\chi}}^n f_1) = \chi(h)(1, V_{\phi,T,\overline{\chi}}^n f_1)$$

$$= \chi(h)\overline{(V_{\phi,T,\overline{\chi}}^n f_1, 1)} = \chi(h)(V_{\phi,T,\chi}^n \overline{f_1}, 1)$$

$$= \chi(h)(V_{\phi,T,\chi}^n f_2, 1).$$

Now choose $\chi \in \widehat{G}$ satisfying $\chi(h) \neq 1$. This can be done since the characters separate the points of G.

It follows that $(V_{\phi,T,\chi}^n f_2, 1) = 0$ for all $n \in \mathbb{Z}$. This implies that for the operator $V_{\phi,T,\chi} : L^2(X,\mu) \to L^2(X,\mu)$, $Z(1)$ and $Z(\overline{\chi}(u(x))$ are two orthogonal cyclic subspaces with 1 and $\overline{\chi}(u(x))$ having the same spectral type. It follows that $V_{\phi,T,\chi}$ has non-simple spectrum. The lemma and hence the theorem now follow. $\square$

Note that since $k^2 = I$, then $h = u(x) - u(kx) = u(kx) - u(k^2x) = u(kx) - u(x) = -h$, so $2h = 0$ and G must contain a copy of $\mathbb{Z}_2$.

The following corollary is often useful in practice. In particular, we will use it in Example 5.

Corollary 3. *If condition* (ii) *of Theorem 5 is replaced by*

$$\phi(x) - \phi(kTx) = g \quad a.e.,$$

a constant in G, then $(V^m_{\phi,T,\chi}1,1) = 0$ *for all* $m \in \mathbb{Z}$ *and* $\chi \in \hat{G}$ *for which*
$\chi^m(g) \neq 1.$

Proof.

$$(V^m_{\phi,T,\chi}1,1) = \int_X \chi(\phi(x) + \ldots + \phi(T^{m-1}x))\, d\mu$$

$$= \int_X \chi(\phi(kTx) + \ldots + \phi(kT^mx) + mg)\, d\mu$$

$$= \chi^m(g) \int_X \chi(\phi(kTx) + \ldots + \phi(kT^mx))\, d\mu$$

$$= \chi^m(g)(V^m_{\phi,T,\chi}1,1),$$

from which the result follows. □

Note that if $u(Tx) - u(x) = g$ a.e., then for each $\chi \in \hat{G}$, $\chi(g)$ is an
eigenvalue for T with corresponding eigenfunction $\chi(u(x))$.

The following theorem shows that the group extensions for which equation (5) holds, are isomorphic to their inverses. The conjugating automorphism K' satisfies $K'^2 = \sigma_h$ for some $h \in G$, where $\sigma_h(x,g) = (x, h+g)$.

Theorem 6. *(i) Suppose that T_ϕ is ergodic. Then T_ϕ is isomorphic to
T_ϕ^{-1}, with K' the conjugating automorphism, if and only if $K'(x,g) =
(kx, u(x) - v(g))$ for k, u and v satisfying equation (5), where $v : G \to G$
is a continuous group automorphism.*

(ii) If in addition $k^2 = I$ then $K'^2 = \sigma_h$ for some $h \in G$.

Proof. (i) If $T_\phi K' = K'T_\phi^{-1}$ then we apply Theorem 3 with $S = T^{-1}$
and $\psi = \phi T^{-1}$ to obtain the result.

Conversely if $K'(x,g) = (kx, u(x) - v(g))$, then

$$T_\phi K'(x,g) = T_\phi(kx, u(x) - v(g)) = (Tkx, \phi(kx) + u(x) - v(g)).$$

But

$$K'T_\phi^{-1}(x,g) = K'(T^{-1}x, -\phi(T^{-1}x) + g)$$

$$= (kT^{-1}x, u(T^{-1}x) - v(-\phi(T^{-1}x) + g))$$

$$= (kT^{-1}x, u(T^{-1}x) + v(\phi(T^{-1}x) - v(g)),$$

and the result follows from equation (5).

(ii) $K'^2(x,g) = (k^2x, u(kx) - v(u(x)) + v^2(g)) = (x, h+g) = \sigma_h(x,g)$ for
some $h \in G$, by Proposition 1. □

Finally in this section we mention a partial converse of Theorem 2, generalizing Proposition 2 of [Goodson et al., 1992] . The proof is omitted, as
it is similar to the proof of that result. It is assumed that T_ϕ and S_ψ are
ergodic.

Theorem 7. *Let* $\chi, \gamma \in \widehat{G}$, *where* $G = \mathbb{Z}_n$, $n \geq 2$. *Suppose that* $V_{\phi,T,\chi}$ *and* $V_{\psi,S,\gamma}$ *are unitarily equivalent via* $W : L^2(X,\mu) \to L^2(X,\mu)$, *a unitary operator of the form*

$$(Wf)(x) = h(x)f(kx) \tag{12}$$

for some measurable $h : X \to \mathbb{C}$ *and* $k : X \to X$. *Then*

(i) $Tk = kS$, $|h(x)| = 1$ (S *and* T *are isomorphic*).

(ii) *There exists a continuous group automorphism* $v : G \to G$ *such that* $\gamma = \chi \circ v$.

(iii) *If* χ *is a generator for* $\widehat{G}$, *then*

$$\phi(kx) - v(\psi(x)) = u(Sx) - u(x)$$

for some measurable $u : X \to G$. *Moreover,* $W = cV_{u,k,\chi}$ *for some* $|c| = 1$.

3. MORSE AUTOMORPHISMS ARISING FROM CONSTANT BLOCKS

Our aim now is to show that there are ergodic Morse automorphisms satisfying the conditions of Theorem 4. In particular, the functional equation (5) can be solved using Morse automorphisms arising from constant blocks. Note that it follows from Theorem 4 of [Goodson et al., 1992] that $C(T_\phi) = \{(T_\phi)^n \sigma_g : n \in \mathbb{Z}, \ g \in G\}$, where $\sigma_g(x,h) = (x, g+h)$, for Morse automorphisms T_ϕ arising from constant blocks.

Theorem 8. *Let* $H = \{g_1, g_2, \ldots, g_{q-1}\}$ *be an ordered set of members of* G, *and let* $v : G \to G$ *be an automorphism satisfying* $v(g_i) = g_{q-i}$, $1 \leq i \leq q - 1$. *If* T_ϕ *is the Morse automorphism defined by*

$$x = b \times b \times \ldots, \quad over \quad G, \quad where$$

$$b = e, g_1, g_1 + g_2, g_1 + g_2 + g_3, \ldots, g_1 + g_2 + \ldots + g_{q-1} \tag{13}$$

and e *is the identity of* G, *then* T_ϕ *can be represented as a group extension* $T_\phi(x,g) = (Tx, \phi(x) + g)$ *satisfying equations (5) of Theorem 4.*

Proof. First suppose that T has rational discrete spectrum with respect to the sequence of towers $D^t = \{D_i^t : 0 \leq i \leq n_t - 1\}$, $t \in \mathbb{N}$.

Define a set transformation Φ by $\Phi(D_i^t) = D_{n_t-i-1}^t$, $0 \leq i \leq n_t - 1$, $t \in \mathbb{N}$, and extend Φ to all of $\mathcal{B}$. It can be seen that Φ is a σ-isomorphism and hence defines an automorphism $k : X \to X$ satisfying $k^{-1}(A) = \Phi(A)$ for all $A \in \mathcal{B}$. It is now easy to see that $Tk(D_i^t) = kT^{-1}(D_i^t)$ and $k^2(D_i^t) = D_i^t$ for all $0 \leq i \leq n_t - 1$, $t \in \mathbb{N}$. It follows that $Tk(x) = kT^{-1}(x)$ and $k^2(x) = x$ for a.e. $x \in X$.

We will show that under the conditions of this theorem, equation (5) has a solution with $u =$constant, i.e. that $v(\phi(x)) = \phi(kTx)$ for all $x \in X$. To do this we use induction to show that

$$v(\phi(D_i^t) = \phi(kT(D_i^t)) \quad \text{for all} \quad 0 < i < q^t - 1, \ t \in \mathbb{N}.$$

This is clearly true when $t = 1$, since $v(\phi D_i^1) = v(g_{i+1}) = g_{q-i-1}$, $0 < i < q - 1$, and $\phi(kTD_i^1) = \phi(kD_{i+1}^1) = \phi(D_{q-i-2}^1) = g_{q-i-1}$.

Suppose it is true that $v(\phi D_i^t) = \phi(kTD_i^t)$, $0 < i < q^t - 1$; then we show that $v(\phi D_i^{t+1}) = \phi(kTD_i^{t+1})$, $0 < i < q^{t+1} - 1$.

Denote the sequence of values of ϕ on D^t as θ^t; then θ^{t+1} is formed by concatenation as follows:

$$\theta^{t+1} = \theta^t g_1 \theta^t g_2 \theta^t g_3 \ldots \theta^t g_{q-1} \theta^t$$

(where $\theta^1 = g_1 g_2 \ldots g_{q-1}$, so that $\theta^t[0] = g_1$, $\theta^t[1] = g_2$, etc. and g_1 occurs in the $\theta^{t+1}[q^t - 1]$ position; it may be assumed that $g_1 + g_2 + \ldots + g_{q-1} = e$, for if not, replace b by $b \times b \times \ldots \times b$, for which this holds).

Then for $0 \le k \le q - 1$,

$$\theta^{t+1}[i] = \begin{cases} \theta^t[i - kq^t] & kq^t \le i < (k+1)q^t - 1, \\ g_k & i = kq^t - 1, \ k \ne 0. \end{cases}$$

Now

$$v(\phi D_i^{t+1}) = v(\theta^{t+1}[i]) = \begin{cases} v(\theta^t[i - kq^t]) \\ v(g_k) \end{cases}$$

$$= \begin{cases} \theta^t[q^t - i + kq^t - 2] & kq^t \le i < (k+1)q^t - 1 \\ g_{q-k} & i = kq^t - 1, \ k \ne 0 \end{cases}$$

(since from above $v(\theta^t[i]) = \theta^t[q^t - i - 2]$).

Now also $\phi(kTD_i^{t+1}) = \phi(kD_{i+1}^{t+1}) = \phi(D_{q^{t+1}-i-2}^{t+1}) = \theta^{t+1}[q^{t+1} - i - 2]$

$$= \begin{cases} \theta^t[q^{t+1} - i - 2 - kq^t] & kq^t \le q^{t+1} - i - 2 < (k+1)q^t - 1 \\ g_k & q^{t+1} - i - 2 = kq^t - 1, \ k \ne 0. \end{cases}$$

$$= \begin{cases} \theta^t[q^t(q - k) - i - 2] & q^t(q - k) - q^t - 1 < i \le q^t(q - k) - 2 \\ g_k & i = q^t(q - k) - 1, \ k \ne 0. \end{cases}$$

Now it can be seen that the above two expressions are the same, that is $v(\phi D_i^{t+1}) = \phi(kTD_i^{t+1})$, $t \in \mathbb{N}$, by induction.

The result now follows, for if $x \in X$, there exists $t \in \mathbb{N}$ for which $x \in D_i^t$, $0 < i < q^t - 1$, and then

$$v(\phi x) = v(\phi D_i^t) = \phi(kTD_i^t) = \phi(kTx). \qquad \square$$

By putting minimal restrictions on H in Theorem 8 it is possible to ensure that T_ϕ is ergodic (see [Goodson-Lemanczyk, 1990] or [Kwiatkowski-Sikorski, 1987]). Recall that the maximal spectral multiplicity of T_ϕ is denoted by $\mathrm{msm}(T_\phi)$. Using Theorem 9 we can calculate $\mathrm{msm}(T_\phi)$ for certain Morse automorphisms T_ϕ.

Theorem 9. *Suppose that an ergodic Morse automorphism T_ϕ satisfies the conditions of Theorem 8. Then*

(i)   if $v(g) = g$ for all $g \in G$ then $msm(T_\phi) = 1$;

(ii) if $v \neq I$ then T_ϕ has even multiplicity; if in addition g_1 is a generator for G and the order of G is odd, then $msm(T_\phi) = 2$;

(iii) if g_1 is a generator for G and $v(g) = -g$ for all $g \in G$, then $msm(T_\phi) = 2$.

Proof. We first state a lemma from [Kwiatkowski-Sikorski, 1987]. In this lemma, the underlying group extension is an ergodic Morse automorphism arising from a Morse sequence of the form $x = b \times b \times \ldots$ over a finite abelian group G of order n. It is assumed that every member of G appears in the block b.

Lemma 4. *If $n \geq 3$ and $\chi \neq \gamma$ in $\widehat{G}$, then the operators $V_{\phi,T,\chi}$ and $V_{\phi,T,\gamma}$ are unitarily equivalent if and only if*

$$\sum_{r \in G} s_a(r)(\chi(r) - \gamma(r)) = 0 \quad \forall \, a \in \{1, 2, \ldots, q-1\},$$

where $s_a(r) = \mathrm{card}\,\{i : 0 \leq i \leq q - a - 1 \quad and \quad b[i+a] - b[i] = r\}$.

We now prove Theorem 9. Let $\chi, \gamma \in \widehat{G}$ and suppose that

$$\sum_{r \in G} s_a(r)(\chi(r) - \gamma(r)) = 0, \quad 1 \leq a \leq q - 1.$$

We will show that $\chi = \gamma$ (in case (i)) and $\chi = \gamma$ or $\chi = \gamma \circ v$ (in cases (ii) and (iii)), and the result will follow from Theorems 1 and 8. Since T_ϕ is ergodic, we may assume that every member of G appears in b, and thus it suffices to show these equalities on H. Note that

$$\sum_{r \in G} s_a(r)(\chi(r) - \gamma(r)) = \sum_{i=0}^{q-a-1} \chi(b[a+i] - b[i]) - \gamma(b[a+i] - b[i])$$

$$= \sum_{i=0}^{q-a-1} \chi(g_{i+1} + g_{i+2} + \ldots + g_{a+i}) - \gamma(g_{i+1} + g_{i+2} + \ldots + g_{a+i}) = 0.$$

On substituting $a = 1, 2, \ldots, q - 1$, we obtain $q - 1$ equations, the first two of which are:

$$\chi(g_1 + g_2 + \ldots + g_{q-1}) = \gamma(g_1 + g_2 + \ldots + g_{q-1}) \tag{14}$$

$$\chi(g_1 + \ldots + g_{q-2}) + \chi(g_2 + \ldots + g_{q-1}) = \gamma(g_1 + \ldots + g_{q-1}) + \gamma(g_2 + \ldots + g_{q-1}) \tag{15}$$

To prove (i), we suppose that $v(g) = g$ for all $g \in G$, so that $g_{q-1} = g_1$, $g_{q-2} = g_2$, etc.

Equations (14) and (15) imply that

$$1/\chi(g_{q-1}) + 1/\chi(g_1) = 1/\gamma(g_{q-1}) + 1/\gamma(g_1), \qquad (16)$$

so that $\chi(g_1) = \gamma(g_1)$. We now proceed using a finite induction and the other $q - 3$ equations to deduce that $\chi(g_j) = \gamma(g_j)$ for $1 \leq j \leq q - 1$, and hence $\chi = \gamma$. In particular, $\mathrm{msm}(T_\phi) = 1$.

To prove (ii) we first make the observation that if $z_1, z_2, w_1, w_2 \in S^1$, the unit circle in the complex plane, and $0 \neq z_1 + z_2 = w_1 + w_2$, then $z_1 = w_1$, $z_2 = w_2$ or $z_1 = w_2$, $z_2 = w_1$.

Now again from equations (14) and (15)

$$1/\chi \circ v(g_1) + 1/\chi(g_1) = 1/\gamma \circ v(g_1) + 1/\gamma(g_1) \qquad (17)$$

and hence $\chi(g_1) = \gamma(g_1)$ or $\chi(g_1) = \gamma \circ v(g_1)$ (the third possibility, namely that $\chi(g_1) + \chi \circ v(g_1) = 0$, cannot occur, as the order of G is odd). Since g_1 is a generator, $\chi = \gamma$ or $\chi = \gamma \circ v$, so that $\mathrm{msm}(T_\phi) = 2$, since there exists $\chi \in \widehat{G}$ with $\chi \circ v \neq \chi$, as $v \neq I$. Part (iii) is done in the same way, except that additional arguments show that $\chi(g_1) + \chi \circ v(g_1) = 0 = \gamma(g_1) + \gamma \circ v(g_1)$ implies $\chi = \gamma$ or $\chi = \gamma \circ v$. $\square$

If $G = \mathbb{Z}_n$ where n is prime (so that $v(g) = -g$), we recover part of a theorem from [Kwiatkowski-Sikorski, 1987], that $\mathrm{msm}(T_\phi) = 2$ if and only if b is a symmetric block, otherwise $\mathrm{msm}(T_\phi) = 1$.

Remarks.

(1) If T_ϕ is an ergodic Morse automorphism which arises from constant blocks over G and $v : G \to G$ is a group automorphism for which $V_{\phi,T,\chi}$ is unitarily equivalent to $V_{\phi,T,\chi \circ v}$ for all $\chi \in \widehat{G}$, then it can be shown that $v^2(g) = g$ for all $g \in G$.

(2) The converse to Theorem 8 is also true, namely that if T_ϕ is an ergodic Morse automorphism arising from a Morse sequence of the form

$$\mathbf{x} = b \times b \times \ldots \quad over \quad G,$$

where $b = e, g_1, g_1 + g_2, \ldots, g_1 + g_2 + \ldots + g_{q-1}$, and if there exists a measurable solution $u : X \to G$ to the functional equation

$$v(\phi x) - \phi(kTx) = u(Tx) - u(x),$$

where $k : X \to X$ is an automorphism satisfying $Tk = kT^{-1}$ and $v : G \to G$ is a group automorphism, then $v(g_i) = g_{q-i}$ for $1 \leq i \leq q - 1$.

This result can be used to construct a continuum of Morse automorphism with simple spectrum, non-isomorphic to their inverses.

4. EXAMPLES

It is easy to construct a wide variety of examples of ergodic group extensions isomorphic to their inverses and satisfying Theorem 4 or Theorem 5. Theorem 8 gives a special class of such examples, two of which are studied in more detail below (Examples 2 and 3). In Example 5 we use these ideas to give a new proof that the Mathew-Nadkarni transformation has a finite Lebesgue component of multiplicity two in its spectrum. Example 4 is the simplest example we know of an ergodic automorphism having non-simple spectrum of finite multiplicity.

Example 1. *A $\mathbb{Z}_2$ extension of an irrational rotation.*

Let $G = Z_2$ and define $T_\phi : [0,1) \times \mathbb{Z}_2 \to [0,1) \times \mathbb{Z}_2$ by

$$T_\phi(x,g) = (T_\alpha x, \phi(x) + g),$$

where $T_\alpha x = x + \alpha \pmod 1$, $x \in [0,1)$, is an irrational rotation, and $\phi : [0,1) \to G$ is defined by

$$\phi(x) = \left\{ \begin{array}{ll} 0 & 0 \leq x < \frac{1}{4};\ \frac{3}{4} \leq x < 1, \\ 1 & \frac{1}{4} \leq x < \frac{3}{4}. \end{array} \right.$$

If $K'(x,g) = (kx, u(x) + g)$, where $u : [0,1) \to G$, $u(x) = $ constant, $x \in [0,1)$, and $k(x) = 1 - x + \alpha \pmod 1$, $x \in [0,1)$, then $T_\alpha k = kT_\alpha^{-1}$, and $\phi(x) - \phi(kT_\alpha x) = 0$, so that T_ϕ is isomorphic to T_ϕ^{-1}. In this case $K'^2 = I$. A nice proof of the ergodicity of T_ϕ and the continuity of the spectrum of $V_{\phi,T,\chi}$ for all α irrational and $\chi \neq 1$, can be given using the methods of [Cornfeld et al., 1982, p. 348]. The simplicity of spectrum of $V_{\phi,T,\chi}$ is known for well-approximable α.

Example 2. *The Thue-Morse Sequence.*

Let $x = 01 \times 01 \times 01 \times \ldots$ over $\mathbb{Z}_2$, the Thue-Morse sequence; then it is well known that the corresponding Morse automorphism can be written as a $\mathbb{Z}_2$-extension $T_\phi : [0,1) \times \mathbb{Z}_2 \to [0,1) \times \mathbb{Z}_2$,

$$T_\phi(x,g) = (Tx, \phi(x) + g),$$

where $T : [0,1) \to [0,1)$ is the von Neumann–Kakutani adding machine and $\phi : [0,1] \to \mathbb{Z}_2$ arises from the Toeplitz sequence :

```
1 . 1 . 1 . 1 . 1 . 1 . 1 . 1 . 1 . . . .
  0   .   0   .   0   .   0   .   0  . .
    1           .           1          .          1 .
1 0 1 1 1 0 1 0 1 0 1 1 1 0 1 1 1 0 1 .
```

i.e., define $\phi(x) = 1$, $0 \leq x < 1/2$, and leave it undefined on the top level of the stack at the 1st stage in the construction of T. At the 2nd stage, ϕ

is already defined to be 1 on the 1st and 3rd levels. Define ϕ to be 0 on the 2nd level and continue the construction, alternately inserting the values 0 and 1 for ϕ, resulting in the Toeplitz sequence.

It is not difficult to see that the equation

$$\phi(x) - \phi(kTx) = 0 \quad \text{in} \quad \mathbb{Z}_2$$

holds, where $k : [0,1) \to [0,1)$ satisfies $Tk = kT^{-1}$ and can be defined by $kD_i^n = D_{2^n - i - 1}$, $i = 1, 2, \ldots$, and where $D^n = \{D_i^n : 0 \le i < 2^n\}$ are the levels in the nth stage of the construction of T.

It now follows that T_ϕ is isomorphic to T_ϕ^{-1} via $K'(x,g) = (kx, h+g)$ for some $h \in \mathbb{Z}_2$ (and all such K' are of this form). Again we have $K'^2 = I$; however, Theorem 9(i) implies the (known) result that T_ϕ has simple spectrum.

Example 3. *The Morse automorphism* $x = 010 \times 010 \times \ldots$ *over* $\mathbb{Z}_3$.

As before, this automorphism is a group extension $T_\phi : [0,1) \times \mathbb{Z}_3 \to [0,1) \times \mathbb{Z}_3$, where $T : [0,1) \to [0,1)$ is the 3-adic adding machine and $\phi : [0,1) \to \mathbb{Z}_3$ is defined by the Toeplitz sequence:

$$
\begin{array}{llllllll}
1\,2\,. & 1\,2\,. & 1\,2\,. & 1\,2\,. & 1\,2\,. & 1\,2\, \ldots \\
\quad 1 & \quad 2 & \quad . & \quad 1 & \quad 2 & \quad \ldots \\
& & \quad 1 & & & \quad 2\,.\,. \\
\end{array}
$$

$$1\,2\,1\,1\,2\,2\,1\,2\,1\ 1\,2\,1\ 1\,2\,2\,1\,2\,2\,.\,.$$

This time we have $v(\phi(x)) - \phi(kTx) = 0$, where $v : \mathbb{Z}_3 \to \mathbb{Z}_3$, $v(g) = -g$, and it follows that $K'(x,g) = (kx, u(x) + g)$, where k is defined in a similar way to Example 2. Also we see that $u(kx) - v(u(x)) = h$ a constant in $\mathbb{Z}_3$.

It follows that T_ϕ and T_ϕ^{-1} are isomorphic and $K'^2(x,g) = (x, h+g)$, since $h \in \mathbb{Z}_3$. Theorem 9(iii) (taking $H = \{1,2\}$) implies that T_ϕ has maximal spectral multiplicity equal to 2.

Example 4. *A* $\mathbb{Z}_3$*-extension of an irrational rotation.*

Let us now define $T_\phi : [0,1) \times \mathbb{Z}_3 \to [0,1) \times \mathbb{Z}_3$ in a similar manner to Example 1, where T_α is again an irrational rotation and k is defined in the same way, but now require that

$$
\phi(x) = \begin{cases} 1 & 0 \le x < \frac{1}{2}, \\ 2 & \frac{1}{2} \le x < 1. \end{cases}
$$

As above, T_ϕ is isomorphic to its inverse, since $v(\phi(x)) - \phi(kT_\alpha x) = 0$ and $v : \mathbb{Z}_3 \to \mathbb{Z}_3$, $v(g) = -g$. Also, $K'^2 = \sigma_1$. In fact, using Theorem 4 we see that for suitable α, T_ϕ has maximal spectral multiplicity equal to 2. (A result of [Cornfeld et al., 1982, p. 350], shows that for certain well-approximated α's, T_ϕ is ergodic and $V_{\phi, T, \chi}$ has continuous spectrum ($\chi \ne 1$). Again $V_{\phi, T, \chi}$ will have simple spectrum for these α's).

Example 5. *The Mathew-Nadkarni Transformation.*

Again $T_\phi : [0,1) \times \mathbb{Z}_2 \to [0,1) \times \mathbb{Z}_2$ is defined by $T_\phi(x,g) = (Tx, \phi(x) + g)$, where T is the von Neumann-Kakutani adding machine, but now ϕ is defined using a Toeplitz sequence with "2 holes", i.e. at the nth stage in the construction of the 2^n levels of T, ϕ is left undefined on the 2^{n-1}th and top levels of the stack. We define k in the same way as in Example 2.

The Toeplitz sequence is :

$$
\begin{array}{l}
0\,.\,1\,.\,0\,.\,1\,.\,0\,.\,1\,.\,0\,.\,1\,.\,0\,.\,.\,.\,. \\
\quad 0\ \ .\ \ 1\ \ .\ \ 0\ \ .\ \ 1\ \ .\ \ 0\,.\,. \\
\quad\ \ 0\qquad.\qquad 1\qquad.\qquad 0\,. \\
0\,0\,1\,0\,0\,1\,1\,0\,0\,0\,1\,1\,0\,1\,1\,0\,0\,0\,0\,.
\end{array}
$$

This time we see that ϕ satisfies the equation

$$\phi(x) - \phi(kTx) = 1 \quad \text{in} \quad \mathbb{Z}_2.$$

If we define $u : [0,1) \to \mathbb{Z}_2$ alternately on the levels of T as 0 on $[0,1/2)$ and 1 on $[1/2,1)$, then we see that $u(Tx) - u(x) = 1$ for all $x \in [0,1)$.

So $\phi(x) - \phi(kTx) = u(Tx) - u(x)$, and hence T_ϕ is isomorphic to T_ϕ^{-1} via $K'(x,g) = (kx, u(x) - g)$, where $K'^2(x,g) = (x, u(kx) - u(x) + g) = (x, 1 + g)$.

We now give a new and elementary proof of the result of Mathew and Nadkarni that T_ϕ has a Lebesgue component of finite multiplicity in its spectrum. Actually, they gave a continuum of such examples, the following methods being applicable to each of them.

Theorem 10. [Mathew-Nadkarni, 1984] *The Mathew-Nadkarni transformation has spectrum consisting of a Lebesgue component with multiplicity 2, together with a discrete component.*

Proof. $U_{T_\phi} : L^2(X \times \mathbb{Z}_2, \tilde{\mu}) \to L^2(X \times \mathbb{Z}_2, \tilde{\mu})$ has a direct sum decomposition

$$L^2(X \times \mathbb{Z}_2, \tilde{\mu}) = L_0 \oplus L_1,$$

where $L_0 = \{f \in L^2(X \times \mathbb{Z}_2) : f(x,g) = f(x)\}$, $L_1 = \{f \in L^2(X \times \mathbb{Z}_2) : f(x,g) = -f(x)\}$, and $U_{T_\phi}|L_0$ is unitarily equivalent to U_T, so gives rise to the discrete component, and $U_{T_\phi}|L_1$ is unitarily equivalent to

$$V_{\phi,T,\chi} : L^2(X,\mu) \to L^2(X,\mu)$$

defined by $V_{\phi,T,\chi}f(x) = \chi(\phi(x))f(Tx)$, where $\chi : \mathbb{Z}_2 \to S^1$, $\chi(g) = (-1)^g$. Denote by 1 the characteristic function of $X = [0,1)$ and write $f_1 = \chi_{[0,1/2)} - \chi_{[1/2,1)}$ (where χ_A denotes the characteristic function of A).

Lemma 5. *The functions* 1 *and* f_1 *generate orthogonal cyclic subspaces* $Z(1)$ *and* $Z(f_1)$ *of* $L^2(X, \mu)$ *and have identical spectral types.*

Proof. Since the functional equation

$$\phi(x) - \phi(kTx) = u(Tx) - u(x)$$

holds, Lemma 3 is applicable. The result then follows from the observation that $f_1 = \overline{\chi}(u(x)) = \chi(u(x))$.

Lemma 6. *For all* $n \in \mathbb{Z}$,

$$(V_{\phi,T,\chi}^{2n} 1, 1) = \begin{cases} (-1)^{\frac{n}{2}} (V_{\phi,T,\chi}^{n} 1, 1) & n \text{ even,} \\ 0 & n \text{ odd.} \end{cases}$$

Proof. This is a direct consequence of the self-replication used in the construction of T and ϕ.

First suppose that n is even; then we have

$$(V_{\phi,T,\chi}^{2n} 1, 1) = \int_X \chi(\phi(x) + \phi(Tx) + \cdots + \phi(T^{2n-1}x))\, d\mu.$$

Now we write this integral as a sum over the levels of the stack D^2 and use the fact that ϕ is constant on every level except for the top and 2nd levels. The above integral is then equal to

$$\sum_{i=0}^{3} \int_{D_i^2} \chi(\phi(x) + \phi(Tx) + \ldots + \phi(T^{2n-1}x))\, d\mu$$

$$= \chi^{\frac{n}{2}}(1)\Big\{ \int_{D_0^2 \cup D_2^2} \chi(\phi(Tx) + \phi(T^3x) + \ldots + \phi(T^{2n-1}x))\, d\mu$$

$$+ \int_{D_1^2 \cup D_3^2} \chi(\phi(x) + \phi(T^2x) + \ldots + \phi(T^{2n-2}x))\, d\mu \Big\}$$

$$= 2(-1)^{\frac{n}{2}} \int_{D_1^2 \cup D_3^2} \chi(\phi(x) + \phi(T^2x) + \ldots + \phi(T^{2n-2}x))\, d\mu$$

$$= (-1)^{\frac{n}{2}} \int_X \chi(\phi(x) + \phi(Tx) + \ldots + \phi(T^{n-1}))\, d\mu$$

$$= (-1)^{\frac{n}{2}} (V_{\phi,T,\chi}^{n} 1, 1).$$

Here we have used the obvious isomorphism between $T : X \to X$ and $T^2 : D_1^2 \cup D_3^2 \to D_1^2 \cup D_3^2$ (modulo a factor of 2).

A similar argument shows that for n odd $(V_{\phi,T,\chi}^{2n} 1, 1) = 0$. $\square$

Lemma 7. *The spectral types of 1 and f_1 are those of Lebesgue measure.*

Proof. Corollary 3 implies that $(V^n_{\phi,T,\chi}1,1) = 0$ for all odd n, since $\phi(x) - \phi(kTx) = 1$ and $\chi^n(1) \neq 1$ when n is odd.

Lemma 6 now gives $(V^n_{\phi,T,\chi}1,1) = 0$ for all $n \neq 0$, and also $(V^n_{\phi,T,\chi}1,1) = 1$ for $n = 0$, so that 1 has the spectral type of Lebesgue measure. It follows from above that the same is true of f_1. $\square$

We have shown that T_ϕ has a discrete component and also a Lebesgue component of multiplicity 2. Theorem 5 now follows from a general result from [Lemańczyk, 1988] that a Toeplitz extension with 2 holes can have maximal spectral multiplicity at most 2.

Remarks.

The above example is just the 2-Morse sequence (see [Lemańczyk, 1988])

$$x = 0001 \times_2 0010 \times_2 0010 \times_2 \ldots \quad \text{over} \quad \mathbb{Z}_2.$$

The same methods can be applied to other examples of this type, e.g. the Rudin-Shapiro sequence, studied by [Queffelec, 1987], defined by

$$x = 0001 \times_2 0001 \times_2 0001 \times_2 \ldots \quad \text{over} \quad \mathbb{Z}_2.$$

REFERENCES

1. H. Anzai, *Ergodic skew product tranformations on the torus*, Osaka J. Math. **3** (1951), 83-99.
2. I.P. Cornfeld, S.V. Fomin, Ya.G. Sinai, *Ergodic Theory*, Springer-Verlag,, New York, 1982.
3. G.R. Goodson, *On the spectral multiplicity of a class of finite rank transformations*, Proc. Amer. Math. Soc. **93** (1985), 303-306.
4. G.R. Goodson, M. Lemańczyk, *On the rank of a class of bijective substitutions*, Studia Math. **96** (1990), 219-230.
5. G.R. Goodson, J. Kwiatkowski, M. Lemańczyk, P. Liardet, *On the multiplicity function of ergodic group extensions of rotations*, Studia Math. **102** (1992), 157-174.
6. H. Helson, *Analiticity in compact abelian groups*, Algebra in Analysis, Academic Press (1975), 1-62.
7. J. Kwiatkowski, A. Sikorski, *Spectral properties of G-symbolic Morse shifts*, Bull. Soc. Math. France **115** (1987), 19-33.
8. M. Lemańczyk, *Toeplitz-$\mathbb{Z}_2$ extensions*, Ann. Inst. Henri Poincaŕe **24** (1988), 1-43.
9. J.C. Martin, *Generalized Morse sequences on n-symbols*, Proc. Amer. Math. Soc. **54** (1976), 379-383.
10. J. Mathew, M.G. Nadkarni, *A measure preserving transformation whose spectrum has a Lebesgue component of multiplicity two*, Bull. Lond. Math. Soc. **16** (1984), 402-406.
11. D. Newton, *On canonical factors of ergodic dynamical systems*, J. London Math. Soc. **19** (1978), 129-136.
12. M. Queffelec, *Substitution Dynamical Systems, Spectral Analysis*. Lecture Notes in Math., vol. 1294, 1987.
13. E.A. Robinson, *Spectral multiplicity for nonabelian Morse sequences*, Lecture Notes in Math. **1342** (1988), 645-652.

14. E.A. Robinson, *Nonabelian extensions have non-simple spectra*, Compositio Math. **63** (1988), 155-170.

MATHEMATICS DEPARTMENT, TOWSON STATE UNIVERSITY, TOWSON, MARYLAND 21204 USA

E-mail address: e7m2grg@toe.towson.edu

MULTIPLE RECURRENCE FOR NILPOTENT GROUPS OF AFFINE TRANSFORMATIONS OF THE 2-TORUS

DANIEL A. HENDRICK

ABSTRACT. A topological multiple recurrence theorem is proved for finite collections of affine transformations of the 2-torus that generate a nilpotent group.

1. INTRODUCTION

In [Furstenberg and Weiss, 1978], H. Furstenberg and B. Weiss prove the following *multiple Birkhoff recurrence theorem*.

Theorem [Furstenberg and Weiss, 1978, Th. 1.4]. *If X is a compact metric space and $T_1, T_2, \ldots, T_\ell$ are commuting continuous transformations of X to itself, then there exists a point $x \in X$ and a sequence of positive integers $n_k \to \infty$ such that*

$$T_1^{n_k} x \to x, T_2^{n_k} x \to x, \ldots, T_\ell^{n_k} x \to x,$$

i.e., there exists a multiply recurrent point for $T_1, T_2, \ldots, T_\ell$.

In [Furstenberg, 1981, p. 40], Furstenberg provides an example of two continuous transformations of the compact metric space $\{-1, 1\}^{\mathbb{Z}}$ to itself with no multiply recurrent point. A reformulation of this example as two affine transformations of the compact abelian group $(\mathbb{Z}/2\mathbb{Z})^{\mathbb{Z}}$ is given below.

For an affine transformation T of a compact abelian group X with group operation $+$, the notation $T = \alpha \oplus A$ will be used to indicate that T is the composition of an automorphism $A \in \mathrm{Aut}(X)$ followed by a rotation by an element $\alpha \in X$. With this notation, for $x \in X$, $Tx = (\alpha \oplus A)x = \alpha + Ax$.

Example 1.1. Let θ denote the left-shift transformation of the compact abelian group $(\mathbb{Z}/2\mathbb{Z})^{\mathbb{Z}}$ (with the usual product topology and the discrete topology on $\mathbb{Z}/2\mathbb{Z}$), i.e., for $\omega \in (\mathbb{Z}/2\mathbb{Z})^{\mathbb{Z}}$, $\theta\omega(n) = \theta\omega(n+1)$. Define two transformations of $(\mathbb{Z}/2\mathbb{Z})^{\mathbb{Z}}$ by

$$T_1 = \theta \quad \text{and} \quad T_2 = \alpha \oplus \theta,$$

where $\alpha = (\ldots, 0, 0, \hat{1}, 0, 0, \ldots) \in (\mathbb{Z}/2\mathbb{Z})^{\mathbb{Z}}$ (the notation $\hat{1}$ indicates that $\alpha(0) = 1$). Then, for every $\omega \in (\mathbb{Z}/2\mathbb{Z})^{\mathbb{Z}}$ and every integer n, we have $T_1^n\omega(0) = \omega(n)$ while $T_2^n\omega(0) = 1 + \omega(n)$. Thus the sequence $T_1^n\omega$ is bounded away from the sequence $T_2^n \omega$, for every $\omega \in (\mathbb{Z}/2\mathbb{Z})^{\mathbb{Z}}$, and so the two transformations T_1 and T_2 have no multiply recurrent point.

The transformations T_1 and T_2 in the above example are members of the group generated by the shift and rotation transformations of $(\mathbb{Z}/2\mathbb{Z})^{\mathbb{Z}}$, a solvable group whose structure is given by $(\mathbb{Z}/2\mathbb{Z})^{\mathbb{Z}} \rtimes \mathbb{Z}$. The group generated by the transformations T_1 and T_2 is not, however, nilpotent. Our first theorem says, in fact, that a finite collection of transformations of the type occurring in this example that generate a nilpotent group have a multiply recurrent point. It is Professor Sergey Yuzvinsky's conjecture that the multiple Birkhoff recurrence theorem is true for any finite collection of transformations that generate a nilpotent group.

Theorem A_0. *Let $T_1 = \alpha_1 \oplus \theta^{n_1}, T_2 = \alpha_2 \oplus \theta^{n_2}, \dots, T_\ell = \alpha_\ell \oplus \theta^{n_\ell}$ be transformations of the compact abelian group $(\mathbb{Z}/k\mathbb{Z})^{\mathbb{Z}}$, for some integer k, where θ is the left-shift transformation and $\alpha_1, \alpha_2, \dots, \alpha_\ell \in (\mathbb{Z}/k\mathbb{Z})^{\mathbb{Z}}$. If the group generated by the transformations $T_1, T_2, \dots, T_\ell$ is nilpotent, then there exists a multiply recurrent point for these transformations.*

Theorem A_0 is an immediate consequence of the following theorem that is proved in Section 2.

Theorem A. *Let $T_1 = \alpha_1 \oplus A_1, T_2 = \alpha_2 \oplus A_2, \dots, T_\ell = \alpha_\ell \oplus A_\ell$ be invertible affine transformations of a compact abelian torsion group X to itself. If the group generated by the transformations $T_1, T_2, \dots, T_\ell$ is nilpotent and if the automorphisms $A_1, A_2, \dots, A_\ell$ commute, then there exists a multiply recurrent point for $T_1, T_2, \dots, T_\ell$.*

To prove Theorem A, we will show the hypotheses imply that, for some integer s, the transformations $T_1^s, T_2^s, \dots, T_\ell^s$ commute. This mostly involves counting commutators that arise in a commutator collection process. The theorem is then an immediate consequence of the multiple Birkhoff recurrence theorem.

In Section 3, the same approach is applied to affine transformations of the 2-torus with better results. The main result is the following theorem.

Theorem B. *Let $T_1, T_2, \dots, T_\ell$ be invertible affine transformations of the 2-torus $\mathbb{T}^2 = \mathbb{R}^2/\mathbb{Z}^2$ to itself. If the group generated by the transformations $T_1, T_2, \dots, T_\ell$ is nilpotent, then there exists a multiply recurrent point for these transformations.*

The following example, an analogue of Furstenberg's example for affine transformations of the 2-torus, shows that Theorem B cannot be extended to solvable groups of affine transformations.

Example 1.2. Let A be an ergodic automorphism of the 2-torus $\mathbb{T}^2 = \mathbb{R}^2/\mathbb{Z}^2$, let α be a non-recurrent point for A, and then define two transformations of $\mathbb{T}^2$ by

$$T_1 = A \qquad \text{and} \qquad T_2 = \alpha \oplus A.$$

For example, we could use

$$A = \begin{pmatrix} 2 & 1 \\ 1 & 1 \end{pmatrix} \quad \text{and} \quad \alpha = \begin{pmatrix} \frac{1}{2\sqrt{5}} \\ \frac{2+\sqrt{5}}{2\sqrt{5}} \end{pmatrix} + \mathbb{Z}^2.$$

(The point $\begin{pmatrix} \frac{1}{2\sqrt{5}} \\ \frac{2+\sqrt{5}}{2\sqrt{5}} \end{pmatrix} + \mathbb{Z}^2$ is homoclinic to the periodic point $\begin{pmatrix} \frac{1}{2} \\ \frac{1}{2} \end{pmatrix} + \mathbb{Z}^2$
and hence is non-recurrent for A. See [Devaney, 1989] for more details.)
Then, for every $w \in \mathbb{T}^2$ and every integer n, we have $T_1^n w = A^n w$ while
$T_2^n w = \sum_{i=0}^{n-1} A^i \alpha + A^n w$. Note that we can write $\sum_{i=0}^{n-1} A^i \alpha = (A - I)^{-1}(A^n - I)\alpha$. Now, since α is not a recurrent point for A, the sequence
$(A^n - I)\alpha$ is bounded away from 0, and, further, since $(A - I)^{-1}$ is a
continuous automorphism, the sequence $(A-I)^{-1}(A^n-I)\alpha$ is also bounded
away from 0. Thus the sequence $T_1^n w$ is bounded away from the sequence
$T_2^n w$, for every $w \in \mathbb{T}^2$, and so the two transformations T_1 and T_2 have no
multiply recurrent point.

I am grateful to Professors Kenneth Ross and Sergey Yuzvinsky for their
help and advice on an earlier version of this paper. This paper has also benefitted from several discussions I have had with Professor Vitaly Bergelson.
In particular, Bergelson suggested some of the examples given here. I would
also like to thank Professors Thomas Ward and David Yuen for several useful conversations.

2. Proof of Theorem A

Define the *commutator* of two group elements x and y to be $[x,y] = xyx^{-1}y^{-1}$. The commutator of two affine transformations $T_1 = \alpha_1 \oplus A_1$
and $T_2 = \alpha_2 \oplus A_2$ is then

$$[T_1, T_2] = T_1 T_2 T_1^{-1} T_2^{-2} = (A_1 - [A_1, A_2])\alpha_2 - (A_1 A_2 A_1^{-1} - I)\alpha_1 \oplus [A_1 A_2].$$

When the automorphisms A_1 and A_2 commute, this reduces to

$$[T_1, T_2] = (A_1 - I)\alpha_2 - (A_2 - I)\alpha_1 \oplus I. \tag{2.1}$$

For higher weight commutators, we will use the notation

$$[_i y, _j x, y] = \overbrace{[y, [y, \ldots, [y,}^{i} \overbrace{[x, [x, \ldots, [x, y]}^{j} \cdots].$$

The weight of a commutator of the form $[_i y, _j x, y]$ is $i + j + 1$.

A group G is *metabelian* if its first commutator subgroup, $\gamma_2(G) = [G, G]$,
is abelian. A group G is *nilpotent of class q* if the subgroup in position
$q + 1$ in the lower central series of G, $\gamma_{q+1}(G) = [G, \gamma_q(G)]$, contains only
the identity. The following two lemmas on the structure of commutators in
metabelian nilpotent groups form the basis for the proof of Theorem A.

Lemma 2.1. *Let G be a metabelian nilpotent group of class q. For any two elements A and B of the group G and any positive integer k,*

$$[A^k, B^k] = \prod_{n=1}^{q-1} \prod_{m=1}^{n} [_{n-m}B, {}_mA, B]^{\binom{k}{m}\binom{k}{n-m+1}} \tag{2.2}$$

Proof. Since $A^k B^k = [A^k, B^k] B^k A^k$, we can verify equation (2.2) by considering the following commutator collection process that transforms the sequence $A^k B^k$ into a sequence of the form $C A^k B^k$, where C is a product of commutators. First, enumerate $A^k B^k$ as $A_1 A_2 \cdots A_k B_1 B_2 \cdots B_k$. An element B_j can be *moved past* an element A_i by replacing $A_i B_j$ with $[A_i, B_j] B_j A_i$. Next, consider moving each B_j past all A_i's to the left and, at the same time, moving all the commutators produced past any additional A_i's to the left. This will result in a sequence of the form $C_1 B_1 C_2 B_2 \cdots C_k B_k A_1 A_2 \cdots A_k$, where each C_j is a product of commutators of the form $[A_{i_m}, \ldots, A_{i_1}, B_j]$, where $1 \leq i_m < i_{m-1} < \cdots < i_2 < i_1 \leq k$. Finally, consider moving each collection of commutators C_j past all B_ℓ's to the left and, at the same time, moving all the commutators produced past any additional B_ℓ's to the left. This will result in a sequence of the form $C B_1 B_2 \cdots B_k A_1 A_2 \cdots A_k$ where C is a product of commutators of the form $[B_{j_{n-m+1}}, \ldots, B_{j_2}, A_{i_m}, \ldots, A_{i_1}, B_{j_1}]$, $1 \leq m \leq n$, where $1 \leq i_m < i_{m-1} < \cdots < i_2 < i_1 \leq k$ and $1 \leq j_{n-m+1} < j_{n-m} < \cdots < j_2 < j_1 \leq k$. The number of commutators of the form $[B_{j_{n-m+1}}, \ldots, B_{j_2}, A_{i_m}, \ldots, A_{i_1}, B_{j_1}]$ in C is then $\binom{k}{m}\binom{k}{n-m+1}$, for $1 \leq m \leq n$. Since the group G is nilpotent of class q, it follows that the number of commutators of the form $[_{n-m}B, {}_mA, B]$ in C is given by $\binom{k}{m}\binom{k}{n-m+1}$, for $1 \leq m \leq n \leq q-1$. Since G is metabelian, all commutators of the same type may be collected together. $\square$

Lemma 2.2. *Let G be a metabelian nilpotent group of class q. For any two elements A and B in G and any integer r,*

$$[A^{r(q-1)!}, B^{r(q-1)!}] = \prod_{n=1}^{q-1} \prod_{m=1}^{n} [_{n-m}B, {}_mA, B]^{r^2 a_{m,n}} \tag{2.3}$$

for some sequence of integers $a_{m,n}$.

Proof. Let $k = r(q-1)!$ in equation (2.2). Since $r^2 \left| \binom{r(q-1)!}{m}\binom{r(q-1)!}{n-m+1} \right.$, for $1 \leq m \leq n \leq q-1$, the exponents appearing on the right side of equation (2.2) may be written in the form $r^2 a_{m,n}$, for some sequence of integers $a_{m,n}$. $\square$

Proof of Theorem A. Let $T_1 = \alpha_1 \oplus A_1, T_2 = \alpha_2 \oplus A_2, \ldots, T_\ell = \alpha_\ell \oplus A_\ell$ be invertible affine transformations of a compact abelian torsion group X to itself such that the automorphisms $A_1, A_2, \ldots, A_\ell$ commute and such that the group generated by the transformations $[T_1, T_2, \ldots, T_\ell]$ is nilpotent.

Let G denote the group generated by $T_1, T_2, \ldots, T_\ell$ and let q denote the nilpotence class of G. Since the automorphisms $A_1, A_2, \ldots, A_\ell$ commute, by equation (2.1), the commutator of any two transformations $S = \alpha \oplus A$ and $T = \beta \oplus B$ in G has the form $[S, T] = \xi \oplus I$, for some element $\xi \in X$. Thus G is a metabelian group.

We will use Lemma 2.2 to show that there exists an integer s such that each pair of transformations T_i^s and T_j^s commute, for $1 \leq i < j \leq \ell$. The multiple Birkhoff recurrence theorem then implies the transformations $T_1^s, T_2^s, \ldots, T_\ell^s$ have a multiply recurrent point and this, in turn, implies the transformations $T_1, T_2, \ldots, T_\ell$ have a multiply recurrent point.

First, for any two transformations T_i and T_j, $1 \leq i < j \leq \ell$, consider the commutators of the form $[_{n-m}T_j, {}_mT_i, T_j]$, for $1 \leq m \leq n \leq q - 1$. There are a finite number of these commutators and each has the form

$$[_{n-m}T_j, {}_mT_i, T_j] = \xi_{n,m} \oplus I,$$

for some element $\xi_{n,m} \in X$. Since X is a torsion abelian group, there exists an integer $r_{i,j}$ such that $r_{i,j}\xi_{n,m} = 0$, for all $1 \leq m \leq n \leq q - 1$ (where 0 represents the identity in the group X). Thus,

$$[_{n-m}T_j, {}_mT_i, T_j]^{r_{i,j}} = r_{i,j}\xi_{n,m} \oplus I = I, \tag{2.4}$$

for all $1 \leq m \leq n \leq q - 1$.

Next, let $r = \prod_{1 \leq i < j \leq \ell} r_{i,j}$ and consider any two transformations of the form $T_i^{r(q-1)!}$ and $T_j^{r(q-1)!}$, $1 \leq i < j \leq \ell$. By Lemma 2.2 and equation (2.4),

$$[T_i^{r(q-1)!}, T_j^{r(q-1)!}] = \prod_{n=1}^{q-1} \prod_{m=1}^{n} [_{n-m}T_j, {}_mT_i, T_j]^{r^2 a_{n,m}} = I,$$

so the transformations $T_i^{r(q-1)!}$ and $T_j^{r(q-1)!}$ commute, for all $1 \leq i < j \leq \ell$. $\square$

With only slight modifications to the proof, Theorem A can also be proved under the somewhat weaker hypothesis that the automorphisms $A_1, A_2, \ldots, A_\ell$ generate a metabelian group whose first commutator subgroup has finite exponent.

The following example presents two affine transformations $T_1 = \alpha_1 \oplus \theta^{n_1}$ and $T_2 = \alpha_2 \oplus \theta^{n_2}$ of $(\mathbb{Z}/2\mathbb{Z})^{\mathbb{Z}}$ the generate a nilpotent group. Note that if α_1 and α_2 are periodic elements with respect to the shift transformation with periods p_1 and p_2, respectively, then the transformations $T_1^{2p_1p_2}$ and $T_2^{2p_1p_2}$ commute, regardless of the structure of the group generated by T_1 and T_2. Non-periodic elements α_1 and α_2 are used in the example.

Example 2.3. Let $T_1 = \alpha_1 \oplus \theta^{n_1}$ and $T_2 = \alpha_2 \oplus \theta^{n_2}$ be two transformations of $(\mathbb{Z}/2\mathbb{Z})^{\mathbb{Z}}$ with $n_1 = n_2 = 1$, and

$$\alpha_1 = (\ldots, \overbrace{0,0,0,0}^{4}, \overbrace{1,1,1}^{3}, \hat{0}, 0, \overbrace{1,1,1}^{3}, \overbrace{0,0,0,0}^{4}, \overbrace{1,1,1,1,1}^{5}, \overbrace{0,0,0,0,0,0}^{6}, \ldots).$$

To define α_2, let

$$a(n) = \overbrace{(0,0,1,0,1,0,\ldots,1,0,1,0)}^{n+1} \quad \text{and} \quad b(n) = \overbrace{(1,1,0,1,0,1,\ldots,0,1,0,1)}^{n+1},$$

so that $a(n)$ and $b(n)$ are both sequences of length $n+1$ each consisting of alternating zeros and ones except for the first term. Now let

$$\alpha_2 = (\ldots, \overbrace{b(4), b(3), a(2), 0}^{bbaa}, \hat{0}, \overbrace{b(2), b(3), a(4), a(5)}^{bbaa}, \overbrace{b(6), b(7), a(8), a(9)}^{bbaa}, \ldots).$$

The commutator of the transformations T_1 and T_2 is given by

$$[T_1, T_2] = (\theta - I)\alpha_2 - (\theta - I)\alpha_1 \oplus I,$$

where

$$(\theta - I)\alpha_1 = (\ldots, \overbrace{0,0}^{2}, 1, \hat{0}, 1, \overbrace{0,0}^{2}, 1, \overbrace{0,0,0}^{3}, 1, \overbrace{0,0,0,0}^{4}, 1, \overbrace{0,0,0,0,0}^{5}, 1, \ldots)$$

and

$$(\theta - I)\alpha_2 = (\ldots, \overbrace{1,1}^{2}, 0, \hat{1}, 0, \overbrace{1,1}^{2}, 0, \overbrace{1,1,1}^{3}, 0, \overbrace{1,1,1,1}^{4}, 0, \overbrace{1,1,1,1,1}^{5}, 0, \ldots).$$

Subtracting $(\theta - I)\alpha_1$ from $(\theta - I)\alpha_2$ gives

$$[T_1, T_2] = (\ldots, 1, 1, 1, \ldots) \oplus I.$$

Since the commutator $[T_1, T_2]$ commutes with T_1 and T_2, we see that the group generated by the transformations T_1 and T_2 is nilpotent of class 2.

3. Proof of Theorem B

Theorem B follows easily from the three lemmas proved in this section. Its proof comes after Lemma 3.7. Several examples are also presented here that demonstrate the limitations of the various results.

It is well-known that an automorphism A of the n-torus $\mathbb{T}^n = \mathbb{R}^n/\mathbb{Z}^n$ is ergodic if and only if none its eigenvalues are roots of unity. On the other hand, if all of A's eigenvalues are roots of unity, it can be shown that A is a distal transformation (a transformation A of a metric space X is *distal* if, for distinct points $x, y \in X$, there exists $\epsilon > 0$ with $d(T^n x, T^n y) \geq \epsilon$ for all integers n). We will use the following two facts about distal systems in the proof of our first lemma (see [Furstenberg, 1981] for details): first, every point in a compact distal system is a recurrent point (in fact, uniformly recurrent), and, second, products of distal systems are distal.

Lemma 3.1. *Let $T_1 = \alpha_1 \oplus A_1, T_2 = \alpha_2 \oplus A_2, \ldots, T_\ell = \alpha_\ell \oplus A_\ell$ be invertible affine transformations of a compact abelian group X to itself. If each of the automorphisms $A_1, A_2, \ldots, A_\ell$ is distal, then every point $x \in X$ is a multiply recurrent point for the maps $T_1, T_2, \ldots, T_\ell$.*

Proof. If each of the automorphisms $A_1, A_2, \ldots, A_\ell$ is distal, it follows that the transformations $T_1, T_2, \ldots, T_\ell$ are also distal. Since products of distal systems are distal, the system $(X^\ell, T_1 \times T_2 \times \cdots \times T_\ell)$ is distal. Every point in a compact distal system is recurrent, and so, in particular, every point on the diagonal of the group X^ℓ is a recurrent point for the map $T_1 \times T_2 \times \cdots \times T_\ell$. Thus, every point of X is a multiply recurrent point for the transformations $T_1, T_2, \ldots, T_\ell$.   □

Lemma 3.2. *Let $T_1 = \alpha_1 \oplus A_1, T_2 = \alpha_2 \oplus A_2, \ldots, T_\ell = \alpha_\ell \oplus A_\ell$ be invertible affine transformations of the n-torus $\mathbb{T}^n = \mathbb{R}^n/\mathbb{Z}^n$ to itself such that the automorphisms $A_1, A_2, \ldots, A_\ell$ commute. If the group generated by the transformations $T_1, T_2, \ldots, T_\ell$ is nilpotent and if at least one of the automorphisms $A_1, A_2, \ldots, A_\ell$ is ergodic, then, for some integer s, the maps $T_1^s, T_2^s, \ldots, T_\ell^s$ commute.*

Proof. Let G denote the nilpotent group generated by the transformations $T_1 = \alpha_1 \oplus A_1, T_2 = \alpha_2 \oplus A_2, \ldots, T_\ell = \alpha_\ell \oplus A_\ell$ and let q denote the nilpotence class of G. Since the automorphisms $A_1, A_2, \ldots, A_\ell$ commute, it follows that G is a metabelian group. We will use Lemma 2.2 to show that there exists an integer s such that each pair of transformations T_i^s and T_j^s commute, for $1 \leq i < j \leq \ell$.

First, note that for any two transformations T_i and T_j, $1 \leq i < j \leq \ell$, by equation (2.1), the commutator $[T_i, T_j]$ has the form

$$[T_i, T_j] = \xi_{i,j} \oplus I,$$

for some element $\xi_{i,j} \in \mathbb{T}^n$ (since $[A_i, A_j] = I$).

Next, suppose the automorphism A_k is ergodic, for some $1 \leq k \leq \ell$, and consider the commutators $[_{q-1}T_k, T_i, T_j]$, $1 \leq i < j \leq \ell$, which, by equation (2.1), have the form

$$[_{q-1}T_k, T_i, T_j] = (A_k - I)^{q-1}\xi_{i,j} \oplus I.$$

Since the group G is nilpotent of class q and each of these commutators has weight $q + 1$, each must equal the identity, and so $(A_k - I)^{q-1}\xi_{i,j} = 0$, for all $1 \leq i < j \leq \ell$. The transformation $(A_k - I)$ is invertible and the inverse matrix has rational entries, so it follows that each element $\xi_{i,j}$ is rational. So there exists an integer r such that $r\xi_{i,j} = 0$, for all $1 \leq i < j \leq \ell$. Now each commutator of the form $[_{n-m}T_j, _mT_i, T_j]$, $1 \leq m \leq n \leq q - 1$, can be expressed in terms of $\xi_{i,j}$ as

$$[_{n-m}T_j, _mT_i, T_j] = (A_j - I)^{n-m}(A_i - I)^{m-1}\xi_{i,j} \oplus I$$

and so

$$[_{n-m}T_j, {}_mT_i, T_j]^r = r(A_j - I)^{n-m}(A_i - I)^{m-1}\xi_{i,j} \oplus I$$
$$= (A_j - I)^{n-m}(A_i - I)^{m-1}r\xi_{i,j} \oplus I \qquad (3.1)$$
$$= I.$$

Finally, consider any pair of transformations of the form $T_i^{r(q-1)!}$ and $T_j^{r(q-1)!}$, $1 \leq i < j \leq \ell$. By Lemma 2.2 and equation (3.1),

$$[T_i^{r(q-1)!}, T_j^{r(q-1)!}] = \prod_{n=1}^{q-1} \prod_{m=1}^{n} [_{n-m}T_j, {}_mT_i, T_j]^{r^2 a_{n,m}} = I,$$

so the transformations $T_i^{r(q-1)!}$ and $T_j^{r(q-1)!}$ commute, for all $1 \leq i < j \leq \ell$. □

The following example demonstrates the necessity of the hypothesis in Lemma 3.2 that at least one of the automorphisms is ergodic.

Example 3.3. Let

$$T_1 = \begin{pmatrix} 1 & 1 \\ 0 & 1 \end{pmatrix} \qquad \text{and} \qquad T_2 = \left(\begin{pmatrix} 0 \\ \sqrt{2} \end{pmatrix} + \mathbb{Z}^2 \right) \oplus I.$$

By equation (2.1), the commutator of T_1 and T_2 is given by

$$[T_1, T_2] = \left(\begin{pmatrix} 1 & 1 \\ 0 & 1 \end{pmatrix} - I \right) \left(\begin{pmatrix} 0 \\ \sqrt{2} \end{pmatrix} + \mathbb{Z}^2 \right) \oplus I = \left(\begin{pmatrix} \sqrt{2} \\ 0 \end{pmatrix} + \mathbb{Z}^2 \right) \oplus I.$$

Since the commutator $[T_1, T_2]$ commutes with T_1 and T_2, the group generated by the transformations T_1 and T_2 is nilpotent of class 2. But, since $[T_1^s, T_2^s] = [T_1, T_2]^{s^2} \neq I$, for all integers s, the transformations T_1^s and T_2^s do not commute for any integer s.

The next example demonstrates that Lemma 3.2 cannot be generalized to arbitrary compact abelian groups.

Example 3.4. We will construct two affine transformations T and S of the infinite torus $\mathbb{T}^\infty$ that satisfy all the hypotheses of Lemma 3.2 but such that T^s and S^s do not commute for any integer s. First, let U_n, $n \geq 0$, denote the *Fibonacci sequence* (defined by the recurrence relation $U_0 = 0$, $U_1 = 1$, and $U_n = U_{n-1} + U_{n-2}$). If we let $A = \begin{pmatrix} 1 & 1 \\ 1 & 0 \end{pmatrix}$, then A has the property that $A^n = \begin{pmatrix} U_{n+1} & U_n \\ U_n & U_{n-1} \end{pmatrix}$, for $n \geq 1$. Next, let $B = A^6$, so that

$B^n = \begin{pmatrix} U_{6n+1} & U_{6n} \\ U_{6n} & U_{6n-1} \end{pmatrix}$, let $\alpha_n = \begin{pmatrix} \frac{1}{U_{3n}^2} \\ 0 \end{pmatrix} + \mathbb{Z}^2$, and define two sequences of affine transformations of the 2-torus $\mathbb{T}^2$ by

$$T_n = B^n \qquad \text{and} \qquad S_n = \alpha_n \oplus B^n,$$

for $n \geq 1$. Finally, define two affine transformations T and S of $\mathbb{T}^\infty$ by

$$T = (T_1, T_2, \dots) \qquad \text{and} \qquad S = (S_1, S_2, \dots),$$

viewing $\mathbb{T}^\infty$ as $\prod_{i=1}^\infty \mathbb{T}^2$. To show that T and S have the properties we desire, we will use the following proposition.

Proposition 3.5. *For the sequence of transformations T_n and S_n defined above,*

(1) *the order of the commutator $[T_n, S_n]$ is the Fibonacci number U_{3n}, and*

(2) *the group generated by T_n and S_n is nilpotent of class 2.*

It follows from this proposition that the transformations T and S generate a nilpotent group of class 2. Also, since A is an ergodic automorphism of $\mathbb{T}^2$, each transformation T_i is an ergodic automorphism, and hence T is an ergodic automorphism of $\mathbb{T}^\infty$. Thus T and S satisfy all the hypotheses of Lemma 3.2. But, for any given integer s, there exists an integer n such that $U_{3n} > s$ and, for this n, $[T_n, S_n]^s \neq I$, which implies that T^s and S^s do not commute.

The proof of Proposition 3.5 uses several facts about the Fibonacci sequence U_n as well as the *Lucas sequence* V_n (defined by the recurrence relation $V_0 = 2$, $V_1 = 1$, and $V_n = V_{n-1} + V_{n-2}$) that can be found in [Ribenboim, 1988].

Proof of Proposition 3.5. For each positive integer n, by equation (2.1), the commutator of the transformations $T_n = B^n$ and $S_n = \alpha_n \oplus B^n$ is given by

$$
\begin{aligned}
[T_n, S_n] &= (B^n - I)\alpha_n \oplus I \\
&= \begin{pmatrix} U_{6n+1} - 1 & U_{6n} \\ U_{6n} & U_{6n-1} - 1 \end{pmatrix} \left(\begin{pmatrix} \frac{1}{U_{3n}^2} \\ 0 \end{pmatrix} + \mathbb{Z}^2 \right) \oplus I \\
&= \left(\begin{pmatrix} \frac{U_{6n+1}-1}{U_{3n}^2} \\ \frac{U_{6n}}{U_{3n}^2} \end{pmatrix} + \mathbb{Z}^2 \right) \oplus I.
\end{aligned}
$$

We can further reduce the terms in this commutator by using the following relations (equations (IV.3) and (IV.7), respectively, in [Ribenboim, 1988]):

$$U_{m+n} = U_m V_n - (-1)^n U_{m-n} \quad \text{(for } m \geq n\text{)}, \tag{3.2}$$

$$U_m V_n - U_n V_m = 2(-1)^n U_{m-n} \quad \text{(for } m \geq n\text{)}. \tag{3.3}$$

Combining these two equations, we can write

$$U_{m+n} = U_n V_m + (-1)^n U_{m-n} \quad \text{(for } m \geq n\text{)}. \tag{3.4}$$

Equation (3.2) gives us that $U_{6n} = U_{3n} V_{3n}$, while equation (3.4) allows us to write $U_{6n+1} = U_{3n} V_{3n+1} + 1$, or $U_{6n+1} - 1 = U_{3n} V_{3n+1}$. With these relations, we have

$$[T_n, S_n] = (B^n - I)\alpha_n \oplus I = \left(\begin{pmatrix} \frac{V_{3n+1}}{U_{3n}} \\ \frac{V_{3n}}{U_{3n}} \end{pmatrix} + \mathbb{Z}^2 \right) \oplus I. \tag{3.5}$$

To see that the order of the commutator $[T_n, S_n]$ is U_{3n}, we will first show that V_{3n+1} and U_{3n} are relatively prime. The key fact needed to show this is the following (equation (IV.5) in [Ribenboim, 1988]):

$$V_n = 2U_{n+1} - U_n. \tag{3.6}$$

Equation (3.6) and the recurrence relation defining U_n give us that

$$V_{3n+1} = 2U_{3n+2} - U_{3n+1} = U_{3n+2} + U_{3n},$$

and this, combined with the fact that $\gcd(U_{3n}, U_{3n+2}) = U_{\gcd(3n,3n+2)} = U_1 = 1$, provides the result:

$$\gcd(U_{3n}, V_{3n+1}) = \gcd(U_{3n}, U_{3n+2} + U_{3n}) = \gcd(U_{3n}, U_{3n+2}) = 1.$$

Now, since

$$[T_n, S_n]^{U_{3n}} = U_{3n} \left(\begin{pmatrix} \frac{V_{3n+1}}{U_{3n}} \\ \frac{V_{3n}}{U_{3n}} \end{pmatrix} + \mathbb{Z}^2 \right) \oplus I = I,$$

and since $[T_n, S_n]^s \neq I$ for all positive integers $s < U_{3n}$ (because V_{3n+1} and U_{3n} are relatively prime), the order of the commutator $[T_n, S_n]$ is U_{3n}.

To show that the group generated by the transformations T_n and S_n is nilpotent of class 2, it will suffice to show that the two commutators $[T_n, [T_n, S_n]]$ and $[S_n, [T_n, S_n]]$ both equal the identity transformation. But, by equation (2.1), we have

$$[T_n, [T_n, S_n]] = [S_n, [T_n, S_n]] = (B^n - I)^2 \alpha_n \oplus I,$$

and, by equation (3.5),

$$(B^n - I)^2 \alpha_n = (B^n - I) \left(\begin{pmatrix} \frac{V_{3n+1}}{U_{3n}} \\ \frac{V_{3n}}{U_{3n}} \end{pmatrix} + \mathbb{Z}^2 \right)$$

$$= \begin{pmatrix} U_{6n+1} - 1 & U_{6n} \\ U_{6n} & U_{6n-1} - 1 \end{pmatrix} \left(\begin{pmatrix} \frac{V_{3n+1}}{U_{3n}} \\ \frac{V_{3n}}{U_{3n}} \end{pmatrix} + \mathbb{Z}^2 \right)$$

$$= \begin{pmatrix} \frac{(U_{6n+1}-1)V_{3n+1}}{U_{3n}} \\ \frac{U_{6n}V_{3n}}{U_{3n}} \end{pmatrix} + \mathbb{Z}^2$$

$$= \begin{pmatrix} \frac{U_{3n}V_{3n+1}^2}{U_{3n}} \\ \frac{U_{3n}V_{3n}^2}{U_{3n}} \end{pmatrix} + \mathbb{Z}^2$$

$$= \mathbb{Z}^2.$$

Thus $[T_n, [T_n, S_n]] = [S_n, [T_n, S_n]] = I$. $\square$

We will need the following theorem to prove our next lemma. Let F denote an arbitrary field, let $F[x]$ denote the ring of all polynomials $f(x)$ over the field F, and let $\mathbb{M}_n(F)$ denote the algebra of all $n \times n$ matrices over the field F.

Theorem 3.6 [Suprunenko and Tyshkevich, 1968, Ch. 1, Th. 5]. *Let A denote a member of $\mathbb{M}_n(F)$. The centralizer of A in $\mathbb{M}_n(F)$ coincides with the ring of matrices $F[A]$ if and only if the minimum polynomial of A coincides with the characteristic polynomial of A.*

Lemma 3.7. *For any finite collection of automorphisms $A_1, A_2, \ldots, A_\ell$ of the 2-torus $\mathbb{T}^2$ that generate a nilpotent group, there exists an integer n such that the automorphisms $A_1^n, A_2^n, \ldots, A_\ell^n$ commute.*

Proof. Our proof will use induction on the nilpotent class of the group G generated by the automorphisms $A_1, A_2, \ldots, A_\ell$. The result is trivially true when the nilpotence class of G is 1. So assume the result is true for the case of nilpotence class $q - 1$ and suppose the group G is nilpotent of class q.

Let C be any element of the subgroup $\gamma_q(G)$, the qth subgroup in the lower central series of G, and view C's matrix representation as an element of the matrix algebra $\mathbb{M}_2(\mathbb{Q})$. Since G is nilpotent of class q, C commutes with every element in G. Now, if C's minimum polynomial coincided with its characteristic polynomial, then, by Theorem 3.6, there would exist polynomials $f_1, f_2, \ldots, f_\ell \in \mathbb{Q}[x]$ such that $A_i = f_i(C)$ for $i = 1, 2, \ldots, \ell$. But this would imply that G is abelian and hence that C is the identity, contradicting the equivalence of C's minimum and characteristic polynomials. So the minimum polynomial of C cannot equal its characteristic polynomial and, hence, the minimum polynomial of C must be of degree one. Since the determinant of C's matrix must equal 1 (because C is the commutator of matrices whose determinants are ± 1), it follows that the minimum polynomial of C must have the form $x \pm 1$. Thus $C = \pm I$ and so $C^2 = I$, for all $C \in \gamma_q(G)$.

Now, denote the group generated by the automorphisms $A_1^2, A_2^2, \ldots, A_\ell^2$ by G_2 and let D be any element of the subgroup $\gamma_{q-1}(G_2)$. For any generator A_i^2 we have

$$[A_i^2, D] = [A_i, D]^2 = I,$$

since $[A_i, D]$ is an element of $\gamma_q(G)$. Thus, every element in $\gamma_{q-1}(G_2)$ commutes with every element of G_2, so G_2 is a nilpotent group of class $q - 1$. Then, by the induction hypothesis, there exists an integer m such that the automorphisms $A_1^{2m}, A_2^{2m}, \ldots, A_\ell^{2m}$ commute and this proves the lemma. $\square$

The next example shows that Lemma 3.7 cannot be extended to higher-dimensional tori.

Example 3.8. Let A and B be two automorphisms of the 3-torus $\mathbb{T}^3$ whose matrix representations are given by

$$A = \begin{pmatrix} 1 & 1 & 1 \\ 0 & 1 & 0 \\ 0 & 0 & 1 \end{pmatrix} \quad \text{and} \quad B = \begin{pmatrix} 1 & 2 & 0 \\ 0 & 1 & 0 \\ 0 & 1 & 1 \end{pmatrix}.$$

The commutator of A and B is then

$$[A, B] = ABA^{-1}B^{-1}$$

$$= \begin{pmatrix} 1 & 1 & 1 \\ 0 & 1 & 0 \\ 0 & 0 & 1 \end{pmatrix} \begin{pmatrix} 1 & 2 & 0 \\ 0 & 1 & 0 \\ 0 & 1 & 1 \end{pmatrix} \begin{pmatrix} 1 & -1 & -1 \\ 0 & 1 & 0 \\ 0 & 0 & 1 \end{pmatrix} \begin{pmatrix} 1 & -2 & 0 \\ 0 & 1 & 0 \\ 0 & -1 & 1 \end{pmatrix}$$

$$= \begin{pmatrix} 1 & 1 & 0 \\ 0 & 1 & 0 \\ 0 & 0 & 1 \end{pmatrix}.$$

Thus, although the group generated by A and B is nilpotent of class 2, $[A^s, B^s] = [A, B]^{s^2} \neq I$, for all integers s, so the automorphisms A^s and B^s do not commute for any integer s.

Proof of Theorem B. Let $T_1 = \alpha_1 \oplus A_1, T_2 = \alpha_2 \oplus A_2, \ldots, T_\ell = \alpha_\ell \oplus A_\ell$ be invertible affine transformations of the 2-torus $\mathbb{T}^2$ to itself such that the group generated by the transformations $T_1, T_2, \ldots, T_\ell$ is nilpotent.

If each of the automorphisms $A_1, A_2, \ldots, A_\ell$ is distal, then the transformations $T_1, T_2, \ldots, T_\ell$ have a multiply recurrent point by Lemma 3.1. So assume that at least one of the automorphisms $A_1, A_2, \ldots, A_\ell$ is ergodic. In this case, by Lemma 3.7, there exists an integer n such that the automorphisms $A_1^n, A_2^n, \ldots, A_\ell^n$ commute, and then, by Lemma 3.2, there exists an integer s such that the transformations $T_1^{ns}, T_2^{ns}, \ldots, T_\ell^{ns}$ commute. Thus, by the multiple Birkhoff recurrence theorem, the transformations $T_1^{ns}, T_2^{ns}, \ldots, T_\ell^{ns}$ have a multiply recurrent point, and, hence, so do the transformations $T_1, T_2, \ldots, T_\ell$. □

The next example presents two affine transformations $T_1 = \alpha_1 \oplus A_1$ and $T_2 = \alpha_2 \oplus A_2$ of the 2-torus $\mathbb{T}^2$ that generate a nilpotent group. Note that if the elements α_1 and α_2 are rational, then there exists an integer r such that T_1^r and T_2^r commute, regardless of the structure of the group generated by T_1 and T_2. Irrational elements α_1 and α_2 are used in the example.

Example 3.9. Let $T_1 = \alpha_1 \oplus A_1$ and $T_2 = \alpha_2 \oplus A_2$ be two affine transformations of the 2-torus $\mathbb{T}^2$ with

$$T_1 = \alpha_1 \oplus A_1 = \left(\begin{pmatrix} \sqrt{2} \\ \sqrt{2} \end{pmatrix} + \mathbb{Z}^2 \right) \oplus \begin{pmatrix} 1 & 2 \\ 2 & 5 \end{pmatrix}$$

and

$$T_2 = \alpha_2 \oplus A_2 = \left(\begin{pmatrix} 2\sqrt{2} + \frac{1}{4} \\ 3\sqrt{2} + \frac{1}{4} \end{pmatrix} + \mathbb{Z}^2 \right) \oplus \begin{pmatrix} 2 & 5 \\ 5 & 12 \end{pmatrix}.$$

By equation (2.1), the commutator of T_1 and T_2 is given by $[T_1, T_2] = ((A_1 - I)\alpha_2 - (A_2 - I)\alpha_1) \oplus I$, where

$$(A_1 - I)\alpha_2 - (A_2 - I)\alpha_1$$
$$= \begin{pmatrix} 0 & 2 \\ 2 & 4 \end{pmatrix} \left(\begin{pmatrix} 2\sqrt{2} + \frac{1}{4} \\ 3\sqrt{2} + \frac{1}{4} \end{pmatrix} + \mathbb{Z}^2 \right) - \begin{pmatrix} 1 & 5 \\ 5 & 11 \end{pmatrix} \left(\begin{pmatrix} \sqrt{2} \\ \sqrt{2} \end{pmatrix} + \mathbb{Z}^2 \right)$$
$$= \begin{pmatrix} \frac{1}{2} \\ \frac{1}{2} \end{pmatrix} + \mathbb{Z}^2.$$

Now

$$[T_1, [T_1, T_2]] = (A_1 - I) \left(\begin{pmatrix} \frac{1}{2} \\ \frac{1}{2} \end{pmatrix} + \mathbb{Z}^2 \right) \oplus I$$
$$= \begin{pmatrix} 0 & 2 \\ 2 & 4 \end{pmatrix} \left(\begin{pmatrix} \frac{1}{2} \\ \frac{1}{2} \end{pmatrix} + \mathbb{Z}^2 \right) \oplus I = I,$$

and

$$[T_2, [T_1, T_2]] = (A_2 - I) \left(\begin{pmatrix} \frac{1}{2} \\ \frac{1}{2} \end{pmatrix} + \mathbb{Z}^2 \right) \oplus I$$
$$= \begin{pmatrix} 1 & 5 \\ 5 & 11 \end{pmatrix} \left(\begin{pmatrix} \frac{1}{2} \\ \frac{1}{2} \end{pmatrix} + \mathbb{Z}^2 \right) \oplus I = I,$$

so the group generated by the transformations T_1 and T_2 is nilpotent of class 2.

References

1. R. Devaney, *An Introduction to Chaotic Dynamical Systems*, Addison-Wesley, New York, 1989.
2. H. Furstenberg, *Recurrence in Ergodic Theory and Combinatorial Number Theory*, Princeton University Press, Princeton NJ, 1981.
3. H. Furstenberg and B. Weiss, *Topological dynamics and combinatorial number theory*, J. d'Analyse Math. **34** (1978), 61–85.
4. P. Ribenboim, *The Book of Prime Number Records*, Springer-Verlag, New York, 1988.
5. D. Suprunenko and R. Tyshkevich, *Commutative Matrices*, Academic Press, New York, 1968.

DEPARTMENT OF MATHEMATICS, COLGATE UNIVERSITY, 13 OAK DRIVE, HAMILTON, NY 13346
Current address: Department of Mathematics, Ohio State University, Columbus, OH 43210
E-mail address: dhendrick@center.colgate.edu

A REMARK ON ISOMETRIC EXTENSIONS IN RELATIVELY INDEPENDENT JOININGS

EMMANUEL LESIGNE

ABSTRACT. Answering a question asked by D. Rudolph in his article in the present volume, we show that in an ergodic relatively independent joining of two measure-preserving systems over a common factor Z, the maximal group extension (resp. maximal abelian group extension) of Z does not necessarily coincide with the relatively independent joining of the maximal group extensions (resp. maximal abelian group extensions) of Z in each of the two measure-preserving systems.

1. NOTATIONS AND RECALLS

Let $(Z, \mathcal{H}, \nu)$ be a probability measure space and U a measure-preserving transformation of this space. We denote by $\underline{Z}$ the measure-preserving system $(Z, \mathcal{H}, \nu, U)$.

Extensions

A system $\underline{X} = (X, \mathcal{F}, \lambda, S)$ is an *extension* of $\underline{Z}$ if there is a sub-σ-algebra $\mathcal{H}_1$ of $\mathcal{F}$, invariant under the transformation S, such that the systems $\underline{Z}$ and $(X, \mathcal{H}_1, \lambda, S)$ are isomorphic. In this case we consider $\mathcal{H}$ as a sub-σ-algebra of $\mathcal{F}$. We say also that $\underline{Z}$ is a *factor* of $\underline{X}$.

Relatively Independent Joinings

Let $\underline{X} = (X, \mathcal{F}, \lambda, S)$ and $\underline{Y} = (Y, \mathcal{G}, \mu, T)$ be two extensions of $\underline{Z}$. The *relatively independent joining* of $\underline{X}$ and $\underline{Y}$ over $\underline{Z}$ is the system

$$\underline{X} \times_Z \underline{Y} := (X \times Y, \mathcal{F} \times \mathcal{G}, \lambda \times_Z \mu, S \times T)$$

where the measure $\lambda \times_Z \mu$ is defined by

$$(\lambda \times_Z \mu)(A \times B) := \int_Z P(A|\mathcal{H}) \, P(B|\mathcal{H}) \, \mathrm{d}\nu \qquad \text{if } A \in \mathcal{F} \text{ and } B \in \mathcal{G}.$$

This relatively independent joining (or fibered product) is itself an extension of $\underline{Z}$.

Isometric Extensions

Let G be a compact metrizable group and Γ a closed subgroup of G; we designate respectively by $\mathcal{B}$ and m the Borel algebra and the G-invariant probability on G/Γ. If φ is a measurable map (called a cocycle) from Z into G, let U_φ be the transformation of $Z \times G/\Gamma$ defined by

$$U_\varphi(z, g\Gamma) := (U(z), \varphi(z)g\Gamma).$$

An *isometric extension* of $\underline{Z}$ is (modulo isomorphisms) a measure-preserving system of the type

$$(Z \times G/\Gamma, \mathcal{H} \times \mathcal{B}, \nu \times m, U_\varphi).$$

In the sequel this system is simply denoted by $(Z \times G/\Gamma, U_\varphi)$.

In the case where Γ is a normal subgroup of G, this extension is called a *group extension* of $\underline{Z}$. In the case where the group G is abelian, the extension is called an *abelian extension* of $\underline{Z}$.

Maximal Isometric Extensions

Let $\underline{X}$ be an extension of $\underline{Z}$. Among the factors of $\underline{X}$, there is:

- a maximal isometric extension of $\underline{Z}$, denoted by $\mathcal{K}(X, Z)$,
- a maximal group extension of $\underline{Z}$, denoted by $\mathcal{G}(X, Z)$,
- a maximal abelian extension of $\underline{Z}$, denoted by $\mathcal{A}(X, Z)$.

Existence of these maximal extensions and some of their characterizations are dicussed in [Furstenberg, 1978], [Zimmer, 1976 (a) and (b)] and [Rudolph, 1994], and other references given in this last paper.

2. D. RUDOLPH'S QUESTIONS

Let $\underline{X}$ and $\underline{Y}$ be two extensions of $\underline{Z}$. We suppose that the relatively independent joining $\underline{X} \times_Z \underline{Y}$ is ergodic.

In [Rudolph, 1994] (Section I), the author proves that the maximal isometric extension of $\underline{Z}$ in this joining coincides with the joining of the maximal isometric extensions in each of the systems, that is to say:

$$\mathcal{K}(X \times_Z Y, Z) = \mathcal{K}(X, Z) \times_Z \mathcal{K}(Y, Z). \qquad (1)$$

In his paper, D. Rudolph uses the fact that, in a particular situation, this property is also satisfied by abelian extensions (Proposition 1.8). And he asks whether this property is always satisfied by group extensions, and (or) by abelian extensions.

As he thought, the answers are no. In the sequel of this note, we give a short analysis of these questions, then we make an explicit counterexample.

3. Problem Analysis

With the help of (1), the preceding questions can be formulated in the following way. Let

$$X_i = (Z \times G_i/\Gamma_i, U_{\varphi_i}), \quad i = 1, 2,$$

be two isometric extensions of $\underline{Z}$. We have

$$X_1 \times_Z X_2 = (Z \times G_1/\Gamma_1 \times G_2/\Gamma_2, U_{(\varphi_1, \varphi_2)}).$$

We wonder if, under the assumption that this joining is ergodic,

- $\mathcal{G}(X_1 \times_Z X_2, Z) = \mathcal{G}(X_1, Z) \times_Z \mathcal{G}(X_2, Z)$?
- $\mathcal{A}(X_1 \times_Z X_2, Z) = \mathcal{A}(X_1, Z) \times_Z \mathcal{A}(X_2, Z)$?

As is explained in [Rudolph, 1994], it is possible to choose a representation of the system X_i such that the extension $(Z \times G_i, U_{\varphi_i})$ is ergodic, and we have then

$$\mathcal{G}(X_i, Z) = (Z \times G_i/\Gamma_i', U_{\varphi_i}),$$

$$\mathcal{A}(X_i, Z) = (Z \times G_i/(\Gamma_i G_i^*), U_{\varphi_i}),$$

where Γ_i' is the smallest normal subgroup of G_i containing Γ_i, and where G_i^* is the subgroup of G_i generated by commutators.

But it is not possible to assume ergodicity of the extension $(Z \times G_1 \times G_2, U_{(\varphi_1, \varphi_2)})$. In fact there is a closed subgroup H of $G_1 \times G_2$ and there is, for $i = 1, 2$, a cocycle ψ_i cohomologous to φ_i such that (ψ_1, ψ_2) takes its values in H and such that the extension $(Z \times H, U_{(\psi_1, \psi_2)})$ is ergodic. Under the preceding conditions we have then

$$G_1 \times G_2 = H(\Gamma_1 \times \Gamma_2) \text{ and } p_i(H) = G_i \text{ for } i = 1, 2, \tag{2}$$

where p_i denotes the natural projection from $G_1 \times G_2$ onto G_i.

We have

$$X_1 \times_Z X_2 = (Z \times H/(H \cap (\Gamma_1 \times \Gamma_2)), U_{(\psi_1, \psi_2)}),$$

$$\mathcal{G}(X_1 \times_Z X_2, Z) = (Z \times H/(H \cap (\Gamma_1 \times \Gamma_2))', U_{(\psi_1, \psi_2)}),$$

$$\mathcal{A}(X_1 \times_Z X_2, Z) = (Z \times H/H^*(H \cap (\Gamma_1 \times \Gamma_2)), U_{(\psi_1, \psi_2)}).$$

Now we have a new formulation of the questions: under conditions (2), do we have

- $H \cap (\Gamma_1 \times \Gamma_2)' \subset (H \cap (\Gamma_1 \times \Gamma_2))'$?
- $H \cap (G_1 \times G_2)^*(\Gamma_1 \times \Gamma_2) \subset H^*(H \cap (\Gamma_1 \times \Gamma_2))$?

This last question is equivalent to

$$(G_1 \times G_2)^* \subset H^*(\Gamma_1 \times \Gamma_2) \quad ?$$

We shall see from an example that, in general, the answer is no. In this example, we shall have $G_1 = G_2 = G$, and H will be the diagonal subgroup of $G \times G$.

4. AN EXAMPLE

Let n be an even integer ≥ 6. Consider the symetric group $G = S_n$, which is the permutation group of $\{1, 2, ..., n\}$, the subgroup $\Gamma_1 = S_{n-1}$ of permutations which fix the point n, and the subgroup Γ_2 generated by the circular permutation $(1, 2, ..., n)$. We have

$$\Gamma_1' = \Gamma_2' = S_n,$$

for the unique nontrivial normal subgroup of S_n is the alternating group A_n, and, because n is even, $\Gamma_2 \not\subset A_n$. Moreover, the natural map from S_n into $S_n/\Gamma_1 \times S_n/\Gamma_2$ is a bijection, for $S_n = \Gamma_1\Gamma_2$ and $\Gamma_1 \cap \Gamma_2 = \{e\}$.

Let $\underline{Z}$ be a measure-preserving system and φ a measurable map from Z into S_n such that the extension $(Z \times S_n, U_\varphi)$ is ergodic. For example, if $\underline{Z}$ is an irrational rotation on the circle S^1, there is always a step function φ from S^1 into S_n which satisfies this condition; such cocycles φ are constructed in [Derrien, 1993].

We set

$$X_i := (Z \times S_n/\Gamma_i, U_\varphi).$$

We have

$$\mathcal{G}(X_i, Z) = (Z \times S_n/\Gamma_i', U_\varphi) = (Z, U),$$

so

$$\mathcal{G}(X_1, Z) \times_Z \mathcal{G}(X_2, Z) = (Z, U).$$

On the other hand,

$$X_1 \times_Z X_2 = (Z \times S_n/\Gamma_1 \times S_n/\Gamma_2, U_{(\varphi,\varphi)}) = (Z \times S_n, U_\varphi)$$

implies

$$\mathcal{G}(X_1 \times_Z X_2, Z) = (Z \times S_n, U_\varphi).$$

For the abelian extensions, we have

$$\mathcal{A}(X_1, Z) \times_Z \mathcal{A}(X_2, Z) = (Z, U),$$

but

$$\mathcal{A}(X_1 \times_Z X_2, Z) = (Z \times S_n/A_n, U_\varphi)$$

is a two-point extension of (Z, U).

P.S. I thank Michèle Raynaud, who helped me to find this simple example.

REFERENCES

1. J. M. Derrien, *Critères d'ergodicité de cocycles en escalier. Exemples.*, C. R. Acad. Sci. Paris **316, Série I** (1993), 73–76.
2. H. Furstenberg, *Ergodic behavior of diagonal measures and a theorem of Szemerédi on arithmetic progressions*, J. Analyse Math. **34** (1978), 275–291.
3. D. J. Rudolph, *Eigenfunctions of $T \times S$ and the Conze-Lesigne algebra*, these Proceedings (1994).
4. R. Zimmer, *Extensions of ergodic group actions*, Ill. J. Math. **20** (1976(a)), 373–409.
5. R. Zimmer, *Ergodic actions with generalized discrete spectrum*, Ill. J. Math. **20** (1976(b)), 555–588.

DÉPARTEMENT DE MATHÉMATIQUES, UNIVERSITÉ FRANÇOIS RABELAIS, PARC DE GRANDMONT, 37200 TOURS, FRANCE

E-mail address: lesigne@univ-tours.fr

THREE RESULTS IN RECURRENCE

RANDALL MCCUTCHEON

ABSTRACT. Proofs of some recent results of Kriz and Forrest are presented which distinguish between four different kinds of "sets of recurrence" arising naturally in connection with topological and measure-preserving dynamical systems.

0. Suppose that (X, Ψ, μ, T) is an invertible measure-preserving system, that is, (X, Ψ, μ) is a measure space with $\mu(X) = 1$, and $T : X \to X$ is a bijection mod 0, with $\mu(A) = \mu(TA) = \mu(T^{-1}A)$ for all measurable sets A. (By bijection mod 0 we mean that there exists $X' \in \Psi$, $\mu(X') = 1$, such that the restriction of T to X' is a bijection.) Then

Theorem 0.1 (Poincaré). *If $A \in \Psi$ with $\mu(A) > 0$, then for some natural number n, $\mu(A \cap T^n A) > 0$.*

This is the classical Poincaré recurrence theorem. A refinement is

Theorem 0.2. *If E is an infinite subset of the natural numbers and $\mu(A) > 0$, then for some k in the set*

$$E - E = \{n - m : n > m, n, m \in E\}$$

we have $\mu(A \cap T^k A) > 0$.

Proof. The sets $T^n A$, $n \in E$, being of equal positive measure, cannot be pairwise disjoint in the finite measure space X, hence for some $n, m \in E$, $n > m$, $\mu(T^n A \cap T^m A) > 0$. Since T is measure-preserving we have $\mu(A \cap T^{n-m} A) > 0$.

Theorem 0.2 motivates the following definition.

Definition 0.3. Suppose $E \subset \mathbb{N} =$ the set of natural numbers. E will be called a *set of measure-theoretic recurrence* if for any invertible measure-preserving (X, Ψ, μ, T) and A, $\mu(A) > 0$, there exists $n \in E$ such that $\mu(A \cap T^n A) > 0$.

Hence the set $E - E$ of Theorem 0.2 is a set of measure-theoretic recurrence. Our concern here is not really invertible measure-preserving systems but subsets of $\mathbb{N}$.

Definition 0.4. Suppose that $B \subset \mathbb{N}$. The *upper density* of B is the number

$$\overline{d}(B) = \limsup_{N \to \infty} \frac{\#(\{1, 2, \cdots, N\} \cap B)}{N}.$$

A subset $E \subset \mathbb{N}$ is called *density intersective* if whenever $\overline{d}(B) > 0$ we have $\overline{d}(B \cap (B + n)) > 0$ for some $n \in E$.

Proposition 0.5. *E is a set of measure-theoretic recurrence if and only if E is density intersective.*

The proof of Proposition 0.5 is implicit in [Furstenberg, 1981] and explicit in [Bergelson, 1987]. Suppose now that instead of an invertible measure-preserving system, we have a minimal topological dynamical system (X, T). Here X is a compact metric space and $T : X \to X$ is a homeomorphism such that for every non-empty proper closed subset $C \subset X$, $C \neq T(C)$.

Definition 0.6. A subset E of $\mathbb{N}$ is said to be a *set of topological recurrence* if whenever (X, T) is a minimal topological dynamical system, and U is an open subset of X, there exists $n \in E$ such that $U \cap T^n U \neq \phi$.

Proposition 0.7. *E is a set of topological recurrence if and only if for any minimal system (X, T), the set of points $x \in X$ with the property that whenever U is open and contains x, there exists $r \in E$ such that $T^r x \in U$, is residual in X.*

Proof. Set

$$A_n = \{x \in X : \text{ there exists } r \in E \text{ such that } d(x, T^r x) < \frac{1}{n}\}.$$

Then each A_n is open and dense, so that $\bigcap_{n=1}^{\infty} A_n$ is the required residual set. The converse is trivial, since residual sets are necessarily dense.

Definition 0.8. By an *r-coloring* of $\mathbb{N}$ we mean a partition $\mathbb{N} = C_1 \cup C_2 \cdots \cup C_r$. The indices i, $1 \leq i \leq r$, are the *colors*. A subset $E \subset \mathbb{N}$ is said to be *r-intersective* if for every *r*-coloring of $\mathbb{N}$ there exists a color i such that $(C_i - C_i) \cap E$ is non-empty. E is said to be *chromatically intersective* provided it is *r*-intersective for all $r \in \mathbb{N}$.

Definition 0.9. $E \subset \mathbb{N}$ is said to be *syndetic* if there exists $k \in \mathbb{N}$ such that for all $n \in \mathbb{N}$,

$$E \cap \{n, n + 1, \cdots, n + k\} \neq \phi.$$

If T is a homeomorphism of a compact metric space X, a point $x \in X$ is called a *uniformly recurrent point* if for any open neighborhood U of x, the set $\{n : T^n x \in U\}$ is syndetic.

Proposition 0.10 (Birkhoff). *If T is a homeomorphism of a compact metric space X, then there exists a uniformly recurrent point $x \in X$.*

Proposition 0.11. *Suppose X is a compact metric space and $T : X \to X$ a homeomorphism such that $x \in X$ is uniformly recurrent. Then the orbit closure $\overline{\{T^n x : n \in \mathbb{Z}\}}$ is a minimal T-invariant closed subset of X.*

For proofs see, for example, [Furstenberg, 1981, p.29]. We now have the following characterization of sets of topological recurrence, which is the topological analog of Proposition 0.5.

Proposition 0.12. *Suppose $E \subset \mathbb{N}$. Then E is a set of topological recurrence if and only if E is chromatically intersective.*

Proof. Suppose E is a set of topological recurrence. Let $X_k = \{1, \cdots, k\}^{\mathbb{Z}}$ with the product topology, a compact metrizable space. Define $T : X_k \to X_k$ by $(T\phi)_n = \phi_{n+1}$. Suppose we are given a k-coloring $\mathbb{N} = C_1 \cup \cdots \cup C_k$. Let $x \in X_k$ be the point with $x_n = j$ precisely when $n \in C_j$. Let $X_x = \overline{\{T^n x : n \in \mathbb{Z}\}}$. X_x is closed and T-invariant, so there exists a uniformly recurrent point $\psi \in X_x$. By the foregoing proposition, T acts minimally on X_ψ. If $U = \{\gamma \in X_\psi : \gamma_0 = \psi_0\}$, then U is an open neighborhood of ψ, and there exists $r \in E$ such that $U \cap T^r U \neq \phi$. Therefore $\gamma_0 = \gamma_r$ for some $\gamma \in X_x$. Since $\gamma \in X_x$, this in turn implies that $x_n = x_{n+r}$ for some n, so $n, n + r \in C_{x_n}$, and E is a set of chromatic recurrence.

The converse is from [Forrest, 1990]. Suppose E is chromatically intersective. If now (X, T) is minimal with $U \subset X$ open, we have (by minimality) $\bigcup_{i=1}^{\infty} T^i U = X$. By compactness, there exists k such that $\bigcup_{i=1}^{k} T^i U = X$. Let $x \in X$ and consider a k-coloring of $\mathbb{N}$ such that $n \in C_i$ implies $T^n x \in T^i U$. For some $r \in E$, and some i, and some $n \in C_i$, $n + r \in C_i$, therefore $T^n x$ and $T^{n+r} x$ lie in the same shift of U. It follows that $U \cap T^r U \neq \phi$, and E is a set of topological recurrence.

The first of three results we will prove is Theorem 1.2 below, which was first proved by Kriz [1987]. Because of Propositions 0.5 and 0.12, it implies that sets of topological recurrence need not be sets of measure-theoretic recurrence. His treatment of the problem was in graph-theoretic terms and resolved a question asked by V. Bergelson in 1985 (see [Bergelson, 1987] for the details of Bergelson's conjecture.) Forrest gives another proof [1990], where he also observes that the result also resolves negatively the following question of Furstenberg [1981, p.76].

Question 0.13. *If S is a subset of the natural numbers $\mathbb{N}$ with positive density, does there necessarily exist a syndetic set W such that $W - W \subset S - S$?*

Answer. No. By Theorem 1.2 we have a set R which is a set of chromatic intersectivity and a set $S \subset \mathbb{N}$ of positive density such that $R \cap (S - S) = \phi$. For any syndetic W, $R \cap (W - W) \neq \phi$, since finitely many shifts of W color N, and each of these shifts has the same difference set as W. It follows that $W - W$ cannot be contained in $S - S$.

1. All of the proofs presented here depend heavily on an elegant idea of Ruzsa's (indeed the proof of Theorem 1.2 follows [Ruzsa, 1987] more or less exactly.) *All* proofs of this theorem seem to make use of the following result of Lovász, which had been conjectured by Kneser.

Proposition 1.1 [Lovász, 1978]. *Let E be the family of r-element subsets of $\{1, 2, \cdots 2r + k\}$. Given any k-coloring of E, there exists a disjoint pair of elements from E which are of the same color.*

Theorem 1.2 [Kriz, 1987]. *For every $\epsilon > 0$ there exists $A \subset$ N, such that $\overline{d}(A) > \frac{1}{2} - \epsilon$, and $C \subset$ N, which is chromatically intersective, that satisfy $(A - A) \cap C = \phi$. (Therefore C is not density intersective.)*

Proof. We claim that for every $\epsilon > 0, k \in$ N there exist natural numbers $M, N, A' \subset \{0, 1, 2, ..., M(N-1)\}$, $B' \subset \{0, 1, 2, ..., M(N-1)\}$, and $C' \subset \{0, 1, 2, \cdots, M - 1\}$ such that C' is k-intersective, $(A' + C') \cap A' = \phi$, $(B' + C') \cap B' = \phi$, $\nu(A') > \frac{1}{2} - \epsilon$, and $\nu(B') > \frac{1}{2} - \epsilon$. (Here ν is normalized counting measure on $\{0, \cdots, MN - 1\}$.)

We will prove the claim. Let r and $N \in$ N be large, and let $p_1, \cdots, p_{2r+k}$ be large odd primes (how large will be specified shortly.) Putting $M = p_1 \cdots p_{2r+k}$ and taking 0 to be neither odd nor even, let

$$A' = \{a : M \le a < (N-1)M \text{ and } a \bmod p_i \text{ is even for } < r \text{ indices and odd for all other } i\}$$

$$B' = \{b : M \le b < (N-1)M \text{ and } b \bmod p_i \text{ is odd for } < r \text{ indices and even for all other } i\}.$$

We require N, r, and the p_i's to be so large that $\nu(A') > \frac{1}{2} - \epsilon$ and $\nu(B') > \frac{1}{2} - \epsilon$. Note that if $a \in (A' \cup B')$, then $a \ne 0 \bmod p_i, 1 \le i \le 2r + k$. We now let

$$C' = \{c : 0 \le c < M \text{ and } c \bmod p_i \text{ is 1 or } p_i - 1 \text{ for } \ge 2r \text{ indices } i\}.$$

If $a \in A'$ and $c \in C'$, then $(a+c) \bmod p_i$ is even or zero for at least r indices $i, 1 \le i \le 2r + k$, hence $(A' + C') \cap A' = \phi$. Similarly $(B' + C') \cap B' = \phi$. We need to show that C' is k-intersective. Let
$$D = \{d : 0 \le d < M - 1, d = 2 \bmod p_i \text{ for exactly } r \text{ indices } i \text{ and 1 for all other } i\}.$$
D is in obvious $1-1$ correspondence with the family E of r-element subsets of $\{1, 2, \cdots, 2r + k\}$. Therefore, by Proposition 1.1, given any k-coloring of D, there exists a monochromatic pair $d_1 > d_2$ such that $d_1 \bmod p_i = 2 = d_2 \bmod p_i$ never occurs, $1 \le i \le 2r + k$. It follows that $(d_1 - d_2) \bmod p_i$ is 1 or $p_i - 1$ for $2r$ indices, that is, $d_1 - d_2 \in C'$ and C' is k-intersective. This establishes our claim. We note that $A' + C', B' + C' \subset \{0, 1, \cdots, MN - 1\}$.

Let $\epsilon > 0$. Let $(\epsilon_k)_1^\infty$ converge to zero very quickly (how quickly will be specified shortly). For every k we have numbers M_k, N_k, and sets A_k, B_k, and C_k having the properties stated in the previous paragraph (with ϵ_k instead of ϵ). Notice each $n \in \mathbb{N}$ is uniquely expressible as a sum

$$n = \alpha_1 + M_1 N_1 \alpha_2 + M_1 N_1 M_2 N_2 \alpha_3 + \cdots + M_1 N_1 \cdots M_l N_l \alpha_{l+1},$$

where $\alpha_i \in \{0, 1, 2, \cdots, M_i N_i - 1\}$, $1 \leq i \leq l + 1$.

The (ϵ_k) are chosen so that one of the sets

$A_1 = \{\alpha_1 + M_1 N_1 \alpha_2 + M_1 N_1 M_2 N_2 \alpha_3 + \cdots + M_1 N_1 \cdots M_l N_l \alpha_{l+1} : \alpha_i \in A_i$
for an even number of i and $\alpha_i \in B_i$ or $\alpha_i = 0$ for all other $i\}$

$A_2 = \{\alpha_1 + M_1 N_1 \alpha_2 + M_1 N_1 M_2 N_2 \alpha_3 + \cdots + M_1 N_1 \cdots M_l N_l \alpha_{l+1} : \alpha_i \in A_i$
for an odd number of i and $\alpha_i \in B_i$ or $\alpha_i = 0$ for all other $i\}$

has upper density $> \frac{1}{2} - \epsilon$. Let A be that set. Let

$$C = C_1 \cup M_1 N_1 C_2 \cup M_1 N_1 M_2 N_2 C_3 \cup \cdots.$$

C is k-intersective for all k. (It is routine to show that if J is k-intersective then zJ is for any integer z.) Also, $(A + C) \cap A = \phi$. To see this, suppose that

$$n = \alpha_1 + M_1 N_1 \alpha_2 + \cdots + M_1 N_1 \cdots M_l N_l \alpha_{l+1} \in A$$

and

$$c = M_1 N_1 \cdots M_j N_j c_{j+1} \in C;$$

now if $\alpha_{j+1} = 0$, then $\alpha_{j+1} + c_{j+1}$ is non-zero and in neither A_{j+1} nor B_{j+1}, so $n + c \notin A$. If, on the other hand, $\alpha_{j+1} \in A_{j+1}$ (B_{j+1}), then $\alpha_{j+1} + c_{j+1} \notin A_{j+1}$ (B_{j+1}), and again $n + c \notin A$. This completes the proof of Theorem 1.2.

2. We prove in this section two results of Forrest, which again provided answers to questions of Bergelson.

Definition 2.1. A subset $C \subset \mathbb{N}$ is called a *set of strong recurrence* if, whenever $B \subset \mathbb{N}$ with $\overline{d}(B) > 0$,

$$\limsup_{c \in C, c \to \infty} \overline{d}\big((c + B) \cap B\big) > 0.$$

If in fact for all such B

$$\limsup_{c \in C, c \to \infty} \overline{d}\big((c + B) \cap B\big) \geq \big(\overline{d}(B)\big)^2,$$

then C is called *set of nice recurrence*.

By Theorem 2.5 below, these two properties are not equivalent. Bergelson proved ([Bergelson, 1985]) that if D is a set of strong recurrence and $A \subset \mathbb{N}^2$ has positive density, then there exists an infinite $B \subset D$ such that $B \times B \subset A - A$. He then asked whether there were sets of (measure-theoretic) recurrence which are not sets of strong recurrence. If not, then the hypotheses of the aforementioned result could be weakened. Forrest showed, however, that there are such sets.

Our proof of Theorem 2.4 differs from Forrest's in that it uses products of cyclic prime-order groups similar to that used in the proof of Theorem 1.2. We do follow Forrest in the use of the following lemmas.

Lemma 2.2 [Kleitman, 1966]. *If $C \subset \{0,1\}^{2M}$ and for each $a, b \in C$, $a_i = b_i$ for at least $2J$ indices i, then*

$$|C| \leq \sum_{i \leq M-J} \binom{2M}{i}.$$

Lemma 2.3 (Central Limit Theorem). *If (X, μ) is a probability space, $A_i \subset X$, $i \in \mathbb{N}$ are independent subsets of measure α, $0 < \alpha < 1$, then for every real number a*

$$\lim_{k \to \infty} \mu\{x : \sum_{i=1}^{k} \chi_{A_i}(x) - k\alpha \leq a\sqrt{k\alpha(1-\alpha)}\} = F(a),$$

where $F(a) = \frac{1}{\sqrt{2\pi}} \int_{-\infty}^{a} e^{\frac{-t^2}{2}} dt$ is the distribution function determined by the standard normal distribution. F is continuous and increasing, with $F(0) = \frac{1}{2}$, $F(a) \to 0$ as $a \to -\infty$, and $F(a) \to 1$ as $a \to \infty$. Another form for $F(a)$ is

$$\lim_{k \to \infty} \mu^k\{(x_1, x_2, \cdots, x_k) \in X^k : \sum_{i=1}^{k} \chi_{A_i}(x_i) - k\alpha \leq a\sqrt{k\alpha(1-\alpha)}\}.$$

Theorem 2.4 [Forrest, 1991]. *There exists a set of recurrence $C \subset \mathbb{N}$ which is not a set of strong recurrence.*

Proof. Let $\mathbb{T}$ be the unit circle in the complex plane. Let μ be the usual Lebesgue probability measure on $\mathbb{T}$. Then μ^k is the usual measure on the k-torus $\mathbb{T}^k$. We claim that

(1) For all $M > 0$ and $\epsilon > 0$, there exist arbitrarily large k, N, and J, with $J \approx M\sqrt{Nk}$, and $W \subset \mathbb{T}^k$, such that $\mu^k(W) = \frac{1}{2}$ and for all $r \in R$, where

$$R = \{(x_1, x_2, \cdots, x_k) \in \mathbb{T}^k : |\arg x_i - \pi| \leq \pi/2^N \tag{2.1}$$
$$\text{for at least } k - 2J \text{ indices } i\},$$

we have $\mu^k\left((Wr)\bigcap W\right) < 3\epsilon$. R has the property that for any measurable subset $B \subset \mathbb{T}^k$ with $BB^{-1}\bigcap R = \phi$, we have

$$\mu^k(B) \leq \frac{1}{2^{2Nk}}\sum_{i \leq Nk-J}\binom{2Nk}{i}. \tag{2.2}$$

By Lemma 2.3 (with $\alpha = \frac{1}{2}$) the right-hand side is asymptotically $F(-J/\sqrt{Nk/2})$, which will go to zero as $J/\sqrt{Nk} \to \infty$, or as $M \to \infty$.

We now make some observations before proving (1). For all N there exists a measure-preserving map $g_N : \mathbb{T} \to \{0,1\}^{2N}$ such that if $\left(g_N(x)\right)_i \neq \left(g_N(y)\right)_i$ for every $i, 1 \leq i \leq 2N$, then $|\arg xy^{-1} - \pi| \leq \pi/2^N$. It follows that if $|\arg xy^{-1} - \pi| > \pi/2^N$ then $\left(g_N(x)\right)_i = \left(g_N(y)\right)_i$ for at least one i. (The required maps are pie partitions with opposite slices going to reversed bit patterns. For example, if $N = 1$ send the first quadrant to $(1,1)$, the third to $(0,0)$, the second to $(1,0)$, and the fourth to $(0,1)$.) Let

$$h_{N,k} : \mathbb{T}^k \to \{0,1\}^{2Nk} = \left(\{0,1\}^{2N}\right)^k$$

be the map induced by g_N. Now, for $x, y \in \mathbb{T}^k$, if $\left(h_{N,k}(x)\right)_i \neq \left(h_{N,k}(y)\right)_i$ for at least $2Nk - 2J$ indices i, then $|\arg x_j y_j^{-1} - \pi| \leq \pi/2^N$ for at least $k - 2J$ indices j, hence $xy^{-1} \in R$.

We verify (2.2). Let $B \subset \mathbb{T}^k$. If $BB^{-1}\bigcap R = \phi$ then for all $x = (x_i)$, $y = (y_i) \in B$, $|\arg x_i y_i^{-1} - \pi| > \pi/2^N$ for at least $2J$ indices i, $1 \leq i \leq k$; hence $\left(h_{N,k}(x)\right)_i = \left(h_{N,k}(y)\right)_i$ for at least $2J$ indices i, $1 \leq i \leq 2Nk$, and by Kleitman's lemma

$$|h_{N,k}(B)| \leq \sum_{i \leq Nk-J}\binom{2Nk}{i}.$$

But $h_{N,k}$ is measure-preserving, so (2.2) is satisfied.

We now complete the proof of (1). By Lemma 2.3, we may pick $t > 0$ small enough that if

$$W_k = \{x \in \mathbb{T}^k : \text{Re } x_i \geq 0 \text{ for at least } \frac{k}{2} + t\sqrt{k} \text{ indices } i\}, \tag{2.3}$$

then $\lim_{k\to\infty}\mu^k(W_k) > \frac{1}{2} - \epsilon$. Fix a large N to be determined, and let $k, J \to \infty$ according to $J \approx M\sqrt{Nk}$. In particular, we assume that $J < k/2$. For each choice of k, J, consider the set R of (2.1). Notice that $\sup_{r \in R}\mu^k\left(W_k \cap (W_k r)\right)$ is attained for such an r with $r_i = 1$ for $2J$ indices i, and $|\arg r_i - \pi| = \pi/2^N$ for the other $k - 2J$ indices. (Actually, $\mu^k\left(W_k \cap (W_k x)\right)$ is increasing in Re x_i, $1 \leq i \leq k$.) For such an r, and $x \in W_k \cap (W_k r)$, we have $x \in W_k$ and $xr^{-1} \in W_k$. A look at (2.3) shows us that one of the following must occur:

(a) Re $x_i > 0$ for at least $J + \frac{1}{2}t\sqrt{k}$ indices i, $1 \leq i \leq 2J$, or

(b) Re $x_i r_i^{-1} > 0$ for at least $J + \frac{1}{2}t\sqrt{k}$ indices i, $1 \leq i \leq 2J$, or

(c) Re $x_i > 0$ for at least $\frac{k}{2} - J + \frac{1}{2}t\sqrt{k}$ indices i, $2j + 1 \leq i \leq k$, and the same is true of Re $x_i r_i^{-1}$.

Since J increases as $\sqrt{k}$ and $t > 0$, the probability of (a), which is to say the measure of the set of points $x = (x_i)_{i=1}^{k} \in \mathbb{T}^k$ satisfying (a), goes to zero, by Lemma 2.3. The same is true for (b). As for (c), when $2J + 1 \leq i \leq k$, the probability that Re $x_i > 0$ and Re $x_i r_i^{-1} > 0$ is $2^{-(N+1)}$, as is the probability that these are both negative. (c) implies that the former happens for at least $t\sqrt{k}$ more indices i than the latter. By Lemma 2.3, the probability of that happening for $x \in \mathbb{T}^k$ can be made arbitrarily small by choosing N large enough. In other words, N can be chosen so that for large enough k and $J \approx M\sqrt{Nk}$, $\mu^k(W_k \cap (W_k r))$ is small for all $r \in R$. The proof of claim (1) is completed by picking k odd and letting

$$W = \{x \in \mathbb{T}^k : \text{Re } x_1 > 0 \text{ for at least } \frac{k}{2} \text{ indices } i\}.$$

Thus far it has been convenient to work in the torus $\mathbb{T}^k$ because of the naturalness of applying Lemma 2.3. We will denote by C_q the cyclic group $\{0, \cdots, q - 1\}$ under addition mod q. Let μ_q be normalized counting measure on C_q. What we really will need is the following:

(2) For every $\epsilon > 0$, there exists $M = p_1 p_2 \cdots p_k$, where p_i are distinct primes, $1 \leq i \leq k$, and $R \subset C_M$, such that if $A \subset C_M$, $\mu_M(A) > \epsilon$, then there exists $r \in R$ such that $\mu_M(A \cap (A + r)) > 0$, and $W \subset C_M$, $\mu_M(W) \approx \frac{1}{2}$, such that for every $r \in R$, $\mu_M(W \cap (W + r)) < \epsilon$ and $\mu_M(W^c \cap (W^c + r)) < \epsilon$.

(2) is true for the same reasons as (1). Instead of working in $\mathbb{T}^k$, however, one works in the natural subgroup $C_{p_1} \times \cdots \times C_{p_k} \subset \mathbb{T}^k$, where $p_1, \cdots, p_k$ are large enough. This subgroup is of course isomorphic to C_M.

We now complete the proof of the theorem. For a sequence $\{\epsilon_k\}$ converging to zero (at a rate to be determined shortly) let M_k, W_k, and R_k be as in (2), and let N_k be so large that $\mu_{M_k N_k}(A_k) > \frac{1}{2} - \epsilon_k$ and $\mu_{M_k N_k}(B_k) > \frac{1}{2} - \epsilon_k$, where

$$A_k = \{a : M_k \leq a < (N_k - 1)M_k \text{ and } a \mod M_k \in W_k\}$$

$$B_k = \{b : M_k \leq b < (N_k - 1)M_k \text{ and } b \mod M_k \in W_k^c\}.$$

Let $C_k = R_k$, considered now as a subset of $\mathbb{N}$. For each $c \in C_k$, we have $\mu_{M_k N_k}(A_k \cap (A_k + c)) < \epsilon_k$ and $\mu_{M_k N_k}(B_k \cap (B_k + c)) < \epsilon_k$. The $\{\epsilon_k\}$ are chosen so that the following set has upper density greater than $1 - \epsilon$:

$$\{\alpha_1 + M_1 N_1 \alpha_2 + \cdots + M_1 N_1 \cdots M_k N_k \alpha_{k+1} : \alpha_i \in A_i \cup B_i \cup \{0\}\}.$$

Let F be the subset of the above set consisting of those numbers that have $\alpha_i \in A_i$ for an even number of indices i. Then $\bar{d}(F) > \frac{1}{2} - \epsilon$. Now let

$$C = C_1 \cup M_1 N_1 C_2 \cup M_1 N_1 M_2 N_2 C_3 \cup \cdots.$$

We claim that C_i has the property that if $D \subset \mathbb{N}$, $\bar{d}(D) > \epsilon_i$, then there exists $r \in C_i$ with $\bar{d}((D + r) \cap D) > \epsilon_i / |C_i| M_i$. Suppose then that $\bar{d}(D) > \epsilon_i$. Then there exists a sequence $(a_t) \subset \mathbb{N}$ satisfying $a_{t+1} - a_t > M_i$, with $\bar{d}((a_t)) > \epsilon_i / M_i$, such that for each t,

$$\left| D \cap \{a_t, a_t + 1, \cdots, a_t + M_i - 1\} \right| > \epsilon_i M_i.$$

Recall that $C_i = R_i$, viewed as a subset of $\mathbb{N}$. A glance at the definition of R_i shows us that $r \in C_i$ if and only if $M_i - r \in C_i$. This fact, coupled with the intersectivity property of R_i, shows there exist $a, b \in \{0, 1, \cdots, M_i - 1\}$, with $a < b$, such that $a_t + a \in D$, $a_t + b \in D$, and $r = b - a \in C_i$. It follows from the first observation that $a_t + b \in D \cap (D + r)$. Therefore we have

$$\bar{d}\left(\bigcup_{r \in C_i} (D \cap D + r) \right) > \bar{d}((a_t)) > \frac{\epsilon_i}{M_i}.$$

Hence, for some $r \in C_i$ we have $\bar{d}((D + r) \cap D) > \epsilon_i / |C_i| M_i$, establishing our claim. One routinely verifies that the claim is true for C_i replaced by any constant multiple of C_i. From this fact it follows that C is a set of recurrence.

However, by construction

$$\lim_{c \in C, c \to \infty} \bar{d}(F \cap (F + c)) = 0,$$

because, if $c = M_1 N_1 \cdots M_{n-1} N_{n-1} c_n$, with $c_n \in C_n$, then the only elements of $F \cap (F + c)$ are those numbers

$$\alpha_1 + M_1 N_1 \alpha_2 + \cdots + M_1 N_1 \cdots M_k N_k a_{k+1}$$

with $\alpha_n \in A_n$ and $\alpha_n - c_n \in A_n$ (or else both these in B_n). The set of natural numbers with this property is less than two times the density of $A_n \cap (A_n + c_n)$ in $\{0, 1, \cdots, M_n N_n - 1\}$, that is, less than $2\epsilon_n$. Hence C is not a set of strong recurrence. This completes the proof of Theorem 2.4.

Finally we have another result of Forrest.

Theorem 2.5 [Forrest, 1990]. *There exists a set $\tilde{C} \subset \mathbb{N}$ which is a set of strong recurrence but not a set of nice recurrence.*

Proof. Let us return to the notation of the proof of Theorem 2.4 and let

$$\tilde{C}_1 = C_1, \tilde{C}_2 = C_1, \tilde{C}_3 = C_2, \tilde{C}_4 = C_1, \tilde{C}_5 = C_2, \tilde{C}_6 = C_3, \tilde{C}_7 = C_1,, \text{ etc.}$$

Define similarly $\tilde{N}_k, \tilde{M}_k, \tilde{A}_k$, and $\tilde{B}_k$. Following the construction of the previous proof with these new sets, we obtain $\tilde{C}$ and $\tilde{F}$. According to the claim at the end of the proof of Theorem 2.4, $\tilde{C}$ is a set of strong recurrence. $\tilde{C}$ is not, however, a set of nice recurrence, since $\bar{d}(\tilde{F} \cap (\tilde{F} + c)) < 2\epsilon_1$ for all $c \in \tilde{C}$, just as at the end of the proof of Theorem 2.4.

REFERENCES

1. V. Bergelson, *Ergodic Ramsey theory*, Contemp. Math. **65** (1987), 63–87.
2. V. Bergelson, *Sets of recurrence of Z^m- actions and properties of sets of differences in Z^m*, J. London Math. Soc. (2) **31** (1985), 295–304.
3. A.H. Forrest, *Phd. Dissertation*, The Ohio State University (1990).
4. A.H. Forrest, *The construction of a set of recurrence which is not a set of strong recurrence*, Israel J. Math. **76** (1991), 215–228.
5. H. Furstenberg, *Recurrence in Ergodic Theory and Combinatorial Number Theory*, Princeton University Press, 1981.
6. D. J. Kleitman, *On a combinatorial conjecture of Erdös*, J. Comb. Theory **1** (1966), 209–214.
7. I. Kriz, *Large independent sets in shift-invariant graphs. Solution of Bergelson's problem*, Graphs and Combinatorics **3** (1987), 145–158.
8. L. Lovász, *Kneser's conjecture, chromatic number and homotopy*, J. Comb. Theory (A) **25** (1978), 319–324.
9. I. Ruzsa, *Difference sets and the Bohr topology*, preprint, 1987.

DEPARTMENT OF MATHEMATICS, THE OHIO STATE UNIVERSITY, COLUMBUS, OHIO 43210

CALCULATION OF THE LIMIT IN THE RETURN TIMES THEOREM FOR DUNFORD-SCHWARTZ OPERATORS

JAMES H. OLSEN

ABSTRACT. In this paper, we examine the Return Times Theorem (see [Bourgain, Furstenberg, Katznelson and Ornstein, 1989]) and show how to compute the limit of weighted ergodic averages for unitary operators on L_2 when a return times sequence is used for weights. We then use this result to obtain the almost everywhere limit in the Return Times Theorem. Much of the work presented is already contained in [Bourgain, Furstenberg, Katznelson and Ornstein, 1989], but we show how to calculate the limit directly once the a.e. convergence of the weighted averages is known. We then show how to apply this calculation to Dunford-Schwartz operators acting on L_2 and give some multiparameter results.

1. PRELIMINARIES

In what follows, a *dynamical system* $(\Omega, \Sigma, \nu, \tau)$ will be a probability space (Ω, Σ, ν) together with a measure-preserving automorphism $\tau : \Omega \to \Omega$. An *ergodic dynamical system* will be a dynamical system $(\Omega, \Sigma, \nu, \tau)$ with τ ergodic.

We begin by stating the Return Times Theorem [Bourgain, Furstenberg, Katznelson and Ornstein, 1989].

Theorem 1.1. *Let* $(\Omega, \Sigma, \nu, \tau)$ *be an ergodic dynamical system,* $A \in \Sigma$. *Then there exists a set* $\Omega' \in \Sigma$ *with* $\nu(\Omega') = 1$ *such that for every* $y \in \Omega'$ *the sequence* $\{n_k(y)\} = \{n_k\}$ *defined by* $n_k(y) = \inf_{\ell > n_{k-1}} \tau^\ell(y) \in A$ *has the following property: Let* $f \in L_\infty(X, \mathcal{F}, \mu)$, *where* $(X, \mathcal{F}, \mu, \sigma)$ *is a dynamical system. Then*

$$\lim_{N \to \infty} \frac{1}{N} \sum_{k=0}^{N-1} f(\sigma^{n_k}(x)) \tag{1}$$

exists a.e. μ.

Since by the Birkhoff ergodic theorem we have that

$$\lim_{N \to \infty} \frac{|\{k : 0 \le n_k \le N-1\}|}{N} = \lim_{N \to \infty} \frac{1}{N} \sum_{k=0}^{N-1} \chi_A(\tau^k(y)) = \nu(A)$$

for a.e. y, Theorem 1.1 is equivalent to

Theorem 1.2. *Let* $(\Omega, \Sigma, \nu, \tau)$ *and* A *be as in Theorem 1.1. Then there exists a set* $\Omega' \in \Sigma$, *with* $\nu(\Omega') = 1$ *such that for every* $x \in \Omega'$ *the sequence* $a_n = a_n(x) = \chi_A(\tau^n(x))$ *is a good sequence of weights for the pointwise ergodic theorem, i.e., for every* $f \in L_\infty(X, \mathcal{F}, \mu)$, *where* $(X, \mathcal{F}, \mu, \sigma)$ *is a dynamical system, we have*

$$\lim_{n \to \infty} \frac{1}{N} \sum_{k=0}^{N-1} a_k f(\sigma^k(x)) \qquad (2)$$

exists a.e. μ.

We will call any of the sequences $\{a_n\}$ in Theorem 1.2 a *return times sequence*. In the notation of Theorem 1.2, we define the linear operator, defined for all measurable functions on X, by $Tf = f \circ \tau$. With this notation, the limit (2) becomes

$$\lim_{n \to \infty} \frac{1}{N} \sum_{k=0}^{N-1} a_n T^n f((x). \qquad (3)$$

We note that by the Banach Principle, the convergence in Theorem 1.2 can be taken for all $f \in L_1(X)$.

If T is a linear operator on $L_p(X, \mathcal{F}, \mu) = L_p(X)$ for all p, $1 \le p \le \infty$, such that T is a contraction of $L_1(X)$ and $L_\infty(X)$(and hence of $L_p(X)$ for all p, $1 \le p\infty$) we say that T is *Dunford - Schwartz* .

The following theorem [Baxter and Olsen, 1983] allows us to conclude that the return times theorem implies that the limit (3) exists for Dunford-Schwartz operators and $f \in L_p$ for $1 \le p \le \infty$.

Theorem 1.3. *Let* $\{a_n\}$ *be a sequence of complex numbers such that for all dynamical systems* $(X, \mathcal{F}, \mu, \sigma)$ *the limit (1) exists a.e. for all* $f \in L_1(X, \mathcal{F}, \mu)$. *Then for all probability spaces* $(X, \mathcal{F}, \mu)$, *we have for each Dunford-Schwartz operator* T, *the limit (3) exists a.e. for all* $f \in L_p, 1 \le p \le \infty$.

We note that in particular, $\{a_n\}$ can be a return times sequence, that is, as in Theorem 1.2.

Our calculations in the next section will make heavy use of the properties of Besicovitch sequences, which we define next.

Let $\{a_n\}$ be a sequence of complex numbers, $1 \le r < \infty, r$ fixed. Put

$$S_r(\{a_n\}) = \limsup_{N \to \infty} \frac{1}{N} \sum_{k=0}^{N-1} |a_k|^r.$$

Then S_r is a semi-norm on the set F_r of all sequences $\{a_n\}$ such that $S_r(\{a_n\})$ is finite. We define B_r, the set of r-*Besicovitch* sequences, to be the S_r closure of the trigonometric polynomials, that is, the S_r closure of the set of sequences $\{w(n)\}$ where $w(x)$ is a trigonometric polynomial. The set of sequences in $B_r, r \geq 1$ that are also bounded will be called the *bounded Besicovitch sequences*. This terminology is consistent, since it has been shown in several different settings that $B_r \cap \ell_\infty = B_1 \cap \ell_\infty$ for all $r > 1$ [Bellow and Losert, 1985; Jones and Olsen, 1994; Lin and Olsen, 1994].

The following theorem, a paraphrasing of theorems 5.2 and 6.3 in [Baxter and Olsen, 1983] gives a connection between the sequence $\{a_n\}$ in the return times theorem and bounded Besicovitch sequences.

Theorem 1.4. *Let* $(\Omega, \Sigma, \nu, \tau)$ *be a dynamical system,* $g \in L_\infty(\Omega)$. *Then there exists* $h, \ell \in L_\infty(\Omega)$ *such that* $g = h + \ell$ *and for* ν *a.e.* x, *the sequence* $h(\tau^n x) = b_n$ *is bounded Besicovitch while the sequence* $\ell(\tau^n x)$ *has the following property:*

If T *is a linear operator on* $L_\infty(X)$ *which is power bounded with respect* $L_\infty(X)$ *norm and to one other* L_p *norm,* $1 \leq < \infty$ *then for this* p *the norm limit (3) exists and is 0 for all* $f \in L_p(X)$.

In the next section, we will actually calculate the limit (3) when $\{a_n\}$ is Besicovitch and T is a unitary operator. The calculation will depend on the following property of 2-Besicovitch sequences [see Besicovitch, 1954].

Theorem 1.5. *Let* $\{a_n\} \in B_2$. *Then there exists a sequence of complex numbers* λ_m, *called the frequencies for* $\{a_n\}$, *and complex numbers* $\{b_m\}$ *called the Fourier coefficients for* $\{a_n\}$ *such that the sequence of partial sums*

$$\sum_{k=1}^{m} b_k \lambda_k^n = L_m(n),$$

defined for all n, *converges to* $\{a_n\}$ *in the* S_2 *semi-norm.*

The relation between the frequencies λ_m and the Fourier coefficients is given by the familiar formula

$$b_m = \lim_{N \to \infty} \frac{1}{N} \sum_{k=1}^{N} a_k \lambda_m^k.$$

The formal series $\sum_{m=1}^{\infty} b_m \lambda_m^n$ will be called the *Fourier Series* of the sequence $\{a_k\}$. We remark that just as in the case for periodic functions, the formal Fourier Series exists for all sequences in all the Besicovitch classes, but it is only in B_2 that we are assured of convergence to $\{a_n\}$.

2. CALCULATION OF THE LIMIT IN THE RETURN TIMES THEOREM

We now show how to calculate the limit (2) in Section 1. The calculation in Theorem 2.2 below for 1-Besicovitch sequences can be found in a different setting in [Tempelman, 1974]. We are grateful to Professor Tempelman for showing us this calculation.

Lemma 2.1. *Let λ be a complex number, $|\lambda| = 1$, U a unitary operator on L_2 of some measure space. Then*

$$\lim_{n \to \infty} \frac{1}{n} \sum_{k=0}^{n-1} \lambda^k U^k f = P_U\{\bar{\lambda}\},$$

for all $f \in L_2$.

Here, the limit is taken in the L_2 sense, and P_U is the projection valued measure associated with the spectral measure of U.

Proof. Let $f \in L_2$, E_f be the spectral measure defined on Sp_U, the spectrum of U, such that $(Uf, f) = \int_{|z|=1} \lambda\, dE_f(\lambda)$. Then

$$\left(\frac{1}{n} \sum_{k=0}^{n-1} \lambda^k U^k f, f\right) = \frac{1}{n} \sum_{k=0}^{n-1} \lambda^k (U^k f, f) = \frac{1}{n} \sum_{k=0}^{n-1} \lambda^k \int_{|z|=1} x^k\, dE_f(x)$$

$$= \int_{|z|=1} \frac{1}{n} \sum_{k=0}^{n-1} (\lambda x)^k\, dE_f(x).$$

Now, if $\lambda x = 1$, the integrand is 1. The integrand is a sequence of functions that converges to zero when n goes to ∞, 1 otherwise. Hence, by the bounded convergence theorem we have

$$\lim_{n \to \infty} \left(\frac{1}{n} \sum_{k=0}^{n-1} \lambda^k U^k f, f\right) = E_f\{\lambda\}.$$

Therefore, $\displaystyle\lim_{n \to \infty} \frac{1}{n} \sum_{k=0}^{n-1} \lambda^k U^k$ is the bounded linear operator on L_2 $P_U\{\lambda\}$.

Theorem 2.2. *Let $\{a_n\} \in B_2$, and let $\sum_k b_k \lambda_k^n$ be its Fourier series. Then*

$$\lim_{n \to \infty} \frac{1}{n} \sum_{k=0}^{n-1} a_k U^k f = \sum b_k P_U\{\bar{\lambda_k}\}$$

for all $f \in L_2(X)$ and unitary operators U on L_2.

Proof. Let

$$L_M(n) = \sum_{k=0}^{m-1} b_k \lambda_k^n.$$

We have, since U is unitary, that

$$\lim_{n \to \infty} \frac{1}{n} \sum_{k=0}^{n-1} L_M(k) U^k f = \sum_{k \le M} P_U\{\overline{\lambda_k}\}.$$

Now we have

$$\left\| \frac{1}{n} \sum_{k=0}^{n-1} a_k U^k f - \frac{1}{n} \sum_{k=0}^{n-1} L_M(k) U^k f \right\| \le \frac{1}{n} \sum_{k=0}^{n-1} \| a_k - L_M(k) U^k f \|$$

$$\le \frac{1}{n} \sum_{k=0}^{n-1} \sqrt{\int |a_k - L_M(k)|^2 |U^k f|^2}$$

$$\le \frac{1}{n} \sum_{k=0}^{n-1} |a_k - L_M(k)| \| f \|.$$

Now, since given $\epsilon > 0$, we can find M large enough so that $\limsup_n \frac{1}{n} \sum_{k=0}^{n-1} |a_k - L_M(k)| < \epsilon$, convergence is now an easy consequence of the fact that $\lim_{k \to \infty} |b_k| = 0$ (see [Besicovitch, 1954]).

Corollary 2.3. *Let g, h, and ℓ be as in Theorem 3.2, $(X, \mathcal{F}, \mu, \sigma)$ a dynamical system, $f \in L_2(X, \mathcal{F}, \mu)$. Then*

$$\lim_{N \to \infty} \frac{1}{N} \sum_{k=0}^{N-1} g(\tau^k y) f(\sigma^k x) = \lim_{N \to \infty} \frac{1}{N} \sum_{k=0}^{N-1} h((\tau^k y) f(\sigma^k x)$$

a.e.

Proof.

$$\frac{1}{N} \sum_{k=0}^{N-1} g(\tau^k y) f(\sigma^k x) = \frac{1}{N} \sum_{k=0}^{N-1} h(\tau^k y) f(\sigma^k x) + \frac{1}{N} \sum_{k=0}^{N-1} \ell(\tau^k y) f(\sigma^k x)$$

for all N.

$$\lim_{N \to \infty} \frac{1}{N} \sum_{k=0}^{N-1} g(\tau^k y) f(\sigma^k x)$$

exists for a.e. x by assumption and

$$\lim_{N \to \infty} \frac{1}{N} \sum_{k=0}^{N-1} h(\tau^k y) f(\sigma^k x)$$

exists for a.e. x since bounded Besicovitch sequences are universally good (see [Ryll-Nardzewski, 1975; Olsen, 1982; Baxter and Olsen, 1983; Bellow and Losert, 1985], for example). Therefore,

$$\lim_{N \to \infty} \frac{1}{N} \sum_{k=0}^{N-1} \ell(\tau^k y) f(\sigma^k x)$$

exists a.e. x, and must equal 0 by Theorem 1.4.

The calculation of the limit will be complete once we have the frequencies and Fourier coefficients for the sequence $\{h \circ \tau^k)(y)\}$. However, the Fourier coefficient b corresponding to the frequency λ of the sequence $\{a_k\}$ is found by $b = \lim \frac{1}{N} \sum \lambda^k a_k$. Since we have

$$\lim_{N \to \infty} \frac{1}{N} \sum_{k=0}^{N-1} \ell(\sigma^k y) \lambda^k = 0$$

by the same argument as in Corollary 2.3 with X the unit circle with normalized Lebesgue measure and τ defined by $\tau x = \lambda x$ and f the constant function 1. Therefore, the limit is the same as in Theorem 2.2, with the coefficient b of the frequency λ computed in exactly the same manner: i.e., $b = \lim \frac{1}{N} \sum \lambda^k a_k$, where now $\{a_k\}$ is the return time sequence.

3. DUNFORD-SCHWARTZ OPERATORS ON L_2

In the last section, we were able to calculate the limit (2) in the return times theorem. To do this, we used the calculation in Theorem 2.2 in the case the operator T in the limit (2) is unitary and the sequence $\{a_n\}$ bounded Besicovitch and then related Besicovitch sequences to return times sequences to obtain the limit in the return times theorem. In this section, we consider Dunford-Schwartz operators on L_2, which are then contractions of L_2.

Let A be an operator on a Hilbert Space H. If there is another Hilbert Space K such that H is a subspace of K and a unitary operator U on K such that for every non-negative integer n we have $PU^n P = A^n P$, where P is the projection of K onto H, we say that U is a *unitary power dilation* of A. We have the following

Theorem 3.1 [Sz. - Nagy, 1953]. *Every contraction on a Hilbert space has a unitary power dilation.*

We are indebted to Professor Kornfeld for calling this to our attention.

Since if A is a unitary power dilation of U, a contraction of a Hilbert Space H and $f \in H$ we have $Pf = f$, we have the following easy extension of Lemma 2.1.

Theorem 3.2. *Let* $\{a_k\} \in B_2$, $\sum_n b_n \lambda_n^k$ *it's Fourier series, A a contraction on a Hilbert space. Let U be a unitary power dilation of A with $PU^n P = A^n P$ for all non-negative integers n, where P is the projection onto H. Then*

$$\lim_{n\to\infty} \frac{1}{n} \sum_{k=0}^{n-1} a_k A^k f = P(\sum b_k P_U\{\overline{\lambda_k}\}f)$$

for all $f \in H$.

Here, the notation is as in Theorem 2.2, and the convergence is in the Hilbert space sense. In [Schäffer, 1955] such an operator U is actually exhibited, so this is theorem allows the identification of the limit if the operator U and the sequence $\{a_k\}$ are known.

In order to apply the decomposition in [Baxter and Olsen, 1983], we will also need to require that the operator be at least power bounded in L_∞. This restriction will still allow us to consider some important classes of operators, since, for example Dunford-Schwartz operators are contractions of L_2 and L_∞.

Theorem 3.3. *Let $(\Omega, \Sigma, \nu, \tau)$ be a dynamical system, $A \in \mathcal{F}$. Then there exists a set $\Omega' \subset \Omega$ of full measure such that for all $\omega \in \Omega'$ the sequence $\{a_n\} = \chi_A(\tau^n x)$ has the property that the limit (3) exists μ a.e. for all $f \in L_2(X)$ and all Dunford-Schwartz operators T on $L_2(X)$, where $(X, \mathcal{F}, \mu)$ is an arbitrary probability space and further, this limit is*

$$P(\sum b_k P_U\{\overline{\lambda_k}\}f),$$

where U is a unitary power dilation of T with $PU^n P = T^n P$ for all non-negative integers n and P is a projection, $P_U(\cdot)$ the projection-valued spectral measure associated with U, and the coefficients b_m, and frequencies λ_m found by

$$b_m = \lim_{N\to\infty} \frac{1}{N} \sum_{k=1}^{N} a_k \lambda_m^k.$$

Proof. Let X'' be the subset of X of full measure guaranteed by Theorems 1.1 such that for the sequences $\{a(n)\} = \chi_A(\tau^n x)$ we have the limit (3) exists a.e. ν for all $x \in X''$. Let X''' be the subset of X of full measure such that for all $x \in X'''$ we have $\chi_A(\tau^n x) = h(\tau^n x) + \ell(\tau^n x)$ where h and ℓ are as in Theorem 1.4. Put $X' = X'' \cap X'''$. Then for $x \in X'$ we also have the limit (3) exists a.e. ν for all $f \in L_2$ with $a(n) = h(\tau^n x$ since $\{a(n)\}$ is then bounded Besicovitch so the limit (3) exists with $a(n) = \chi_A(\tau^n x)$ in fact for all $f \in L_p$, $1 \le p \le \infty$[Olsen, 1982]. The limit (3) now exists a.e. ν with $a(n) = \ell(\tau^n x)$ and in fact the limit is zero by Theorem 1.4. The theorem now follows since $PT^n f = PU^n Pf = PU^n f$ for all $f \in L_2(Y)$ and P is continuous.

4. MULTIPARAMETER RESULTS

In this section we consider multiparameter extensions of the results in the previous sections. In fact, standard applications of known results [Olsen, 1989; Frangos and Sucheston, 1986; Cogswell, 1993] give us that if $T_1, \ldots, T_n$ are Dunford-Schwartz operators considered as operators on $L_2(X, \mathcal{F}, \mu) = L_2(X)$, where $(X, \mathcal{F}, \mu)$ is a probability space, and $\{a_k^1\}, \ldots, \{a_k^d\}$ are return times sequences, we have the a.e. μ convergence of

$$\lim_{N_1, \ldots N_d \to \infty} \frac{1}{N_1 \cdots N_d} \sum_{k_1=0}^{N_1-1} \cdots \sum_{k_d=0}^{N_d-1} a_{k_1}^1 \cdots a_{k_d}^d T_1^{k_1} \cdots T_d^{k_d} f$$

for all $f \in L_2(X)$. Here, $N_1, \cdots N_d$ tend to infinity independently of each other.

An easy calculation now gives us the following theorem which we state in two dimensions to keep the notation from obscuring the idea.

Theorem 4.1. Let T_1 and T_2 be Dunford-Schwartz operators acting on the space $L_2(X, \mathcal{F}, \mu) = L_2(X)$ where $(X, \mathcal{F}, \mu)$ is a probability space. Let U_1 and U_2 be unitary power dilations of T_1 and T_2, with $P_i U_i P_i f = T_i P_i f = T_i f$ for all $f \in L_2(Y)$, $i = 1, 2$. Let $P_{U_i}(\cdot)$ be the projection-valued spectral measure of U_i. Further, let $\{a_k^i\}$ be return time sequences with frequencies $\lambda_{m,i}$ associated with Fourier coefficients $b_{m,i}$, $i = 1, 2$ as in Theorem 3.3. Then

$$\lim_{N_1, N_2 \to \infty} \frac{1}{N_1 N_2} \sum_{k_1=0}^{N_1-1} \sum_{k_2=0}^{N_2-1} a_{k_1}^1 a_{k_2}^2 T_1^{k_1} T_2^{k_2} f$$

$$= P_1 \left(\sum_m b_{1,m} P_{U_1}(\overline{\lambda_{1,m}}) \right) P_2 \left(\sum_m b_{2,m} P_{U_2}(\overline{\lambda_{2,m}}) \right)$$

μ a.e. for all $f \in L_2(x)$. The limit is taken in the sense that N_1 and N_2 tend to infinity independently of each other.

References

1. Baxter, J.R. and Olsen, J.H., *Weighted and subsequential ergodic theorems*, Can. J. Math **35** (1983), 145-166.
2. Bellow, A. and Losert, V., *The weighted pointwise ergodic theorem and the individual ergodic theorem along subsequences*, Trans. Amer. Math. Soc. **288** (1985), 307-345.
3. Besicovitch, A.S., *Almost Periodic Functions*, Dover, 1954.
4. Bourgain, J., Furstenberg, H., Katznelson, Y., and Ornstein, D., *Return times of dynamical systems*; an appendix toJ. Bourgain, *Pointwise ergodic theorems for arithmetic sets*, IHES Publs. **68** (1989), 5-45.
5. Cogswell, K., *Multiparamater subsequence ergodic theorems along zero density subsequences*, Can. Math. Bull. **36** (1993), 33-37.
6. Frangos, N.E., and Sucheston, L., *On multiparameter martingale theorems in infinite measure spaces*, Prob. and Related Fields **71** (1986), 477-490.
7. Jones, R., and Olsen, J., *Multiparameter weighted ergodic theorems*, Can. J. Math. (to appear).
8. Lin, M., and Olsen, J., *Besicovitch functions and weighted ergodic theorems for LCA group actions* (to appear).
9. Sz. - Nagy, B., *Sur les contractions de l'espace Hilbert*, Acta Szeged **15** (1953), 87-92.
10. Olsen, J. H., *The individual weighted ergodic theorem for bounded Besicovitch sequences*, Canad. Math. Bull. **25** (1982), 468-471.
11. Olsen, J., *Multi-parameter weighted ergodic theorems from their single parameter versions*, Almost Everywhere Convergence, Proc. Conf. on A. E. Conv. in Prob. and Erg. Th. (G.A. Edgar and L. Sucheston, eds.), Academic Press, 1989.
12. Ryll-Nardzewski, C., *Topics in ergodic theory*, Proc. Winter School in Prob., Karpacz, Poland Lecture Notes in Math., vol. 472, Springer-Verlag, Berlin, 1975, pp. 131-156.
13. Schäffer, J.J., *On unitary dilations of contractions*, Proc. Amer. Math. Soc. **6** (1955), 322.
14. Tempelman, A., *Ergodic Theorems for amplitude modulated homogeneous random fields*, Lithuanian Math. J. **14** (1974), 221-229; English transl. in, in Lith. Math. Trans. **14**, 698-704.

DEPARTMENT OF MATHEMATICS, NORTH DAKOTA STATE UNIVERSITY, FARGO, ND, 58105

E-mail address: jolsen@plains.nodak.edu

EIGENFUNCTIONS OF $T \times S$
AND THE CONZE-LESIGNE ALGEBRA

DANIEL J. RUDOLPH

ABSTRACT. Suppose $(X, \mathcal{F}, \mu, T)$ and $(Y, \mathcal{G}, \nu, S)$ are ergodic probability measure-preserving systems. What we investigate here is the form taken by the eigenfunctions of the ergodic components of $(X \times Y, \mathcal{F} \times \mathcal{G}, \mu \times \nu, T \times S)$. In particular we show they arise from the product of the Conze-Lesigne algebras of the two systems. We describe this algebra as an increasing limit of order-2 nil rotations. We investigate this limit in some detail, showing that the limit itself need not be an order-2 nil rotation. From this structure we associate algebraic invariants to the Conze-Lesigne functions from which one can read off precisely what the eigenfunctions of the ergodic components of $T \times S$ are.

0. INTRODUCTION

The Conze-Lesigne equation and Conze-Lesigne algebra first appear in the work of Conze and Lesigne concerning the mean convergence of Fursten-berg's multiple recurrence ergodic averages (see [Conze-Lesigne, 1984 and 1988], [Lesigne, 1993].) For ergodic averages of the form

$$\frac{1}{N} \sum_{j=0}^{N-1} f(T^j(x)) g(T^{2j}(x))$$

the limit lies in the Kronecker (maximal isometric) factor algebra of T, but for three-term averages

$$\frac{1}{N} \sum_{j=0}^{N-1} f(T^j(x)) g(T^{2j}(x)) h(T^{3j}(x))$$

the limiting behavior arises from a larger sub-σ-algebra, the Conze-Lesigne algebra.

The action of T on this algebra takes the form of a compact Abelian group extension of the Kronecker algebra of T; that is, it has the form $(R_{g_0, \phi}, G \times G_1)$, where $\phi : G \to G_1$ and

$$R_{g_0, \phi}(g, g_1) = (g_0 g, \phi(g) g_1).$$

Moreover, for any character $\chi \in \widehat{G_1}$ there must be measurable solutions $\lambda : G \to S^1$ and $f : G \times G \to G_1$ to the Conze-Lesigne equation

$$\chi \circ \phi(cg)^{-1} \chi \circ \phi(g) = \lambda(c) f(c, g_0 g)^{-1} f(c, g).$$

Conze and Lesigne demonstrate that whenever G_1 is a finite-dimensional torus, such a Conze-Lesigne action must actually be a rotation on an order-two nil-manifold.

Our interest in this area arises from a desire to generalize Bourgain's return time theorem [Bourgain, 1989] to Wiener-Wintner averages of the sort

$$\frac{1}{n} \sum_{j=0}^{N-1} f(T^j(x)) g(S^j(y)) e^{-2\pi i j \alpha}.$$

Here one finds the natural convergence to depend on the eigenvalues of the ergodic components of $T \times S$. Hence our desire to characterize these eigenfunctions.

We will first see that the eigenfunctions of the ergodic components can be organized as measurable functions on the Cartesian product. It follows then that they arise from the product of certain 2-step Abelian actions associated to T and S, i.e. from actions of the form $R_{g_0,\phi}$. The eigen-equation now leads directly to the Conze-Lesigne equation. In particular, if $T = (R_{g_0,\phi}, G \times G_1)$ and $S = (R_{g_0,\phi'}, G \times G_2)$, assuming (R_{g_0}, G) represents the common point spectrum of T and S, any eigenfunction of the ergodic components of $T \times S$ will have the form

$$e(g, g', g_1, g_2) = C_1(g_1) C_2(g_2) f(g, g'),$$

where C_1 and C_2 are characters of G_1 and G_2 satisfying the C-L equation.

From here we proceed to redevelop and extend much of the work of Conze and Lesigne relating the C-L equation to nil-rotations. We approach this by showing G_1 must be an inverse limit of finite-dimensional tori, and that when it is a torus, the extension is (as Conze and Lesigne have shown) a nil-rotation. We investigate the structure of this inverse limit of order-two nilpotent groups through a 'δ-homotopy' theory. From this work we will see that these order-two nilpotent groups all arise as a product of a 'quasi-Abelian' extension and a generalized 'Heisenberg' extension. This will attach to the inverse limit a natural invariant in the form of a two-dimensional array of matrices encoding the 'Heisenberg' structure of the groups. This study will show, in particular, that in general the action of T on its C-L algebra, although an inverse limit of nil-rotations, need not itself be a nil-rotation.

Finally, we will attach to each character of G_1 a pair of algebraic invariants in the form of a vector of matrices, from the 'Heisenberg' component, and a character of G, from the 'quasi-Abelian' component of $R_{g_0,\phi}$. In order for characters C_1 and C_2 of T and S to arise as terms in an eigenfunction of the ergodic components of $T \times S$, these invariants must 'cancel' in the direct product. This gives us a precise picture of these eigenfunctions.

I am greatly indebted to E. Lesigne, A. del Junco, and H. Furstenberg for showing me much of what I present here and for correcting my many missteps in exploring these ideas. I have left open a number of questions, especially at the end of Section 5.

1. EIGENFUNCTIONS OF $T \times S$ AND 2-STEP DISTAL FACTORS

Definition 1.1. Suppose $(X, \mathcal{F}, \mu, T)$ is a probability measure preserving system, not necessarily ergodic. We say e is an *eigenfunction* of T if

$$f \circ T(x) = \lambda(x) f(x)$$

where f and λ are defined on some T-invariant set A of positive measure, and λ is T-invariant.

This is the standard definition when T is ergodic. What will interest us here is when the dynamical system is the direct product of two ergodic systems in which case we will in fact characterize the origin of the eigenfunctions as arising from a natural extension of the Kronecker algebra we will refer to as the Conze-Lesigne algebra (see [Lesigne, 1993] for its origins in their work).

We now state a general structure theorem for the eigenfunctions of the ergodic components of a measure-preserving system.

Theorem 1.1. *Suppose $(X, \mathcal{F}, \mu, T)$ is a measure-preserving system and $\mu = \int \mu_c \, dm(c)$ is its ergodic decomposition. There is an at most countable collection of measurable T-invariant sets A_i and measurable eigenfunctions $e_i : A_i \to S^1$, where for a.e. c the collection of functions $\{e_i \mid c \in A_i\}$ forms an orthonormal basis for the span of the eigenfunctions in $L^2(\mu_c)$.*

Proof. The idea here is just to repeat the spectral theorem for T conditionally over its invariant algebra $\mathcal{I}$. We sketch this argument. For $f \in L^2(\mu)$ define $\mathcal{I}$-measurable functions

$$c_n(f)(c) = E(f \times f \circ T^n | \mathcal{I})(c).$$

These are of course the Fourier coefficients of $\mathrm{sp}(f, \mu_c)$, the spectral measure of f with respect to the measure μ_c. It follows that we can define the measure

$$\mathrm{sp}(f) = \int \mathrm{sp}(f, \mu_c) \, dc$$

on $X \times S^1$ as the integral of this measurable family of measures.

The eigenfunctions we seek correspond to point masses of the measures $\mathrm{sp}(f, \mu_c)$. It is possible to find $\mathcal{I}$-measurable subsets A_i and measurable functions $\lambda_i : A_i \to S^1$ so that for a.e. c, $\{\lambda_i(c) \mid c \in A_i\}$ are the atoms of $\mathrm{sp}(f, \mu_c)$. Following the spectral theorem, there is an L^2-isometry between the action of T on the subspace spanned by the $f \circ T^n$ and $L^2(\mathrm{sp}(f))$ under the map $g(c, z) \to zg(c, z)$. For each i, let $f_i(c, z) = 1_{\lambda_i(c)}(z)$, for $c \in A_i$. This function is identified with an $L^2(\mu)$ function $e_i(x)$ defined on A_i which, for a.e. c, is an eigenfunction for μ_c, and as i varies gives all of them in the span of the $f \circ T^n$. Working sequentially through a dense family of $L^2(\mu)$ functions one can build up all eigenfunctions of a.e. μ_c. We can normalize $|e_i| = 1$.

We will be working a great deal with compact, separable, metric groups. The following basic facts about them arise regularly in our work.

Lemma 1.2. *Let G be a compact, separable metric group.*

1) If H_1 and H_2 are closed subgroups, then $H_1 H_2$ is a closed subset and, if either is normal, a subgroup.

2) Letting G^* be the closed commutator subgroup of G (minimal closed subgroup containing all $g_1 g_2 g_1^{-1} g_2^{-1}$), for any closed subgroup Γ, $G/G^*\Gamma$ is the maximal Abelian factor group of the homogeneous space G/Γ.

3) Letting $\widehat{G}$ be the group of characters of G, $\widehat{G}$ is countable and discrete and $\widehat{G/H} = \{\chi \in \widehat{G} \mid \chi(H) = 1\}$ and $\widehat{H} = \widehat{G}/(\widehat{G/H})$.

4) A character in $\widehat{G}$ maps G to a closed subgroup of S^1. G is Abelian and all characters in fact map to all of S^1 iff G is an inverse limit of finite-dimensional tori. This is equivalent to G being Abelian and connected. We call such an Abelian group toral.

To begin, we want to remind the reader of how to identify the Kronecker and relative Kronecker algebras of an ergodic system. The material in this section and much of the material in later sections is not original with the author. Some of it is quite classical, and most of it can be found in [Zimmer, 1976 (a) and (b)], [Furstenberg, 1963], [Conze-Lesigne, 1984], or [Conze-Lesigne, 1988]. The author apologizes for not giving precise attribution for much of the work here. In attempting to present these ideas in as coherent a manner as possible, exact attribution has become difficult.

Suppose $(X, \mathcal{F}, \mu, T)$ is an ergodic system and $\mathcal{H} \subseteq \mathcal{F}$ is a T-invariant factor algebra. Using the Rohlin decomposition, we can write this measure space as a direct product $(Z_1 \times Z_2, \mathcal{F}_1 \times \mathcal{F}_2, \mu_1 \times \mu_2)$, where $\mathcal{H}$ is $\mathcal{F}_1$. The action of T on this product space takes the form

$$T(z_1, z_2) = (T_0(z_1), \phi_{z_1}(z_2)),$$

where T_0 is a μ_1-preserving map and ϕ_{z_1} is a z_1-measurable family of μ_2-preserving maps of Z_2.

Furstenberg [1963] and Zimmer [1976 (a) and (b)] have shown that for any such factor $\mathcal{H}$ there is a unique maximal factor action $\mathcal{KH} \subseteq \mathcal{F}$ containing $\mathcal{H}$ so that in the Rohlin representation of $(T, \mathcal{KH})$, the space Z_2 can be taken to be a compact metric space and the maps ϕ_{z_1} to be isometries of Z_2. The group of all isometries of Z_2 acts transitively, and hence Z_2 can be regarded as a homogeneous space of this group, and μ_2 the projection of Haar measure.

This factor algebra $\mathcal{KH}$ arises from the invariant algebras of the relatively independent joinings of $(X, \mathcal{F}, \mu, T)$ over this factor. To be precise, let $(Z, \mathcal{H}_0, \eta, U)$ be a version of the factor action of T on $\mathcal{H}$, ϕ_1 the projection to Z. Suppose $(Y, \mathcal{G}, \nu, S)$ is another ergodic action, with $\phi_2 : Y \to Z$ a factor map. Set $\mathcal{H}' = \phi_2^{-1}(\mathcal{H}_0)$ and define the *relatively independent joining* $\hat{\mu}$ of T and S over the common factor U (ϕ_1 and ϕ_2 must also be fixed here) by

$$\int f \otimes \bar{g} \, d\hat{\mu} = \int E(f|\mathcal{H})(\phi_1^{-1}(z)) E(\bar{g}|\mathcal{H}')(\phi_2^{-1}(z)) \, d\eta(z).$$

This measure is supported on triples of points (z_1, z_2, z_2') where (z_1, z_2) is a point in the representation of X fibered over $\mathcal{H}$ and (z_1, z_2') is a point in the representation of Y fibered over $\mathcal{H}'$, and $z_1 = \phi_1((z_1, z_2)) = \phi_2((z_1, z_2'))$. Represented this way,

$$\hat{\mu} = \mu_1 \times \mu_2 \times \nu_2.$$

When $T = S$ and $\phi_1 = \phi_2$ we call $\hat{\mu}$ the *relatively independent self-joining* of T over the factor $\mathcal{H}$.

The fundamental fact we will work with concerning the $\mathcal{H}$-relative Kronecker algebra is that in any relatively independent joining of T to an ergodic system S over the factor $\mathcal{H}$, if $\mathcal{I}$ is the $T \times S$ invariant algebra, then

$$\mathcal{KH} \times \mathcal{KH}' \supseteq \mathcal{I} \qquad \hat{\mu}\text{-a.s.},$$

and $\mathcal{KH}$ is minimal for this property in that there is a choice for S (in fact a group extension of T acting on $\mathcal{H}$) with the property that

$$\mathcal{KH} \times Y \subseteq \mathcal{I} \vee (\epsilon_X \times \mathcal{KH}'),$$

where ϵ_X is the trivial algebra of X.

That is to say, $\mathcal{I}$ is always contained in the span of the two relative Kronecker algebras, and if you pick the second action S properly, $\mathcal{KH}$ in turn is contained in the span of the invariant algebra and the relative Kronecker algebra of the second action.

Here is a second way to identify $\mathcal{KH}$, connecting it to ergodic averages. Consider

$$\mathcal{Z} = \{ f \in L^2(\mu) \mid E(f \otimes \bar{f}|\mathcal{I}) = 0$$
$$\text{for the relatively independent self-joining of } T \text{ over } \mathcal{H} \}.$$

The Birkhoff theorem tells us that for any measure-preserving map T and f in L^1 the averages

$$\frac{1}{n} \sum_{j=1}^{n-1} f(T^j(x)) \xrightarrow{n} E(f|\mathcal{I}).$$

Thus, if we set $\mathcal{I}_f$ to be the invariant algebra within the factor generated by f, $E(f|\mathcal{I}) = E(f|\mathcal{I}_f)$.

For f to belong to $\mathcal{Z}$ is the same as to say

$$\frac{1}{n} \sum_{j=0}^{n-1} f(T^j(x_1))\overline{f}(T^j(x_2)) \to 0 \text{ in } L^2 \text{ of the relatively independent joining.}$$

One can show rather easily that for any S (not necessarily ergodic) joined relatively independently over $\mathcal{H}$ to T by $\hat{\mu}$,

1) for each f, $\mathcal{Z}_{\hat{\mu}}(f) = \{g \in L^2(\nu) \mid E(f \otimes \overline{g}|\mathcal{I}) = 0\}$ is a closed subspace of $L^2(\nu)$ and

2) f is in $\mathcal{Z}$ iff for all relatively independent joinings $\hat{\mu}$ of T to any S, $\mathcal{Z}_{\hat{\mu}}(f) = L^2(\nu)$.

These then tell us that $f \in \mathcal{Z}$ iff for all $g \in L^2(\mu)$,

$$\frac{1}{n} \sum_{j=0}^{n-1} f(T^j(x_1))\overline{g}(T^j(x_2)) \to 0$$

almost surely in the relatively independent self-joining. As $\overline{g}$ here can be replaced by $\overline{f}h$ for any $\mathcal{H}$ measurable function h, this tells us $f \in \mathcal{Z}$ iff

$$E(f \otimes \overline{f}|\mathcal{I} \vee \mathcal{H}) = 0.$$

In particular, $\mathcal{Z}$ is a closed subspace of $L^2(\mu)$.

Lemma 1.3. *$\mathcal{Z}$ and $L^2(\mathcal{KH})$ are dual subspaces of $L^2(\mu)$.*

Proof. To begin, notice that $\langle f, g \rangle = 0$ for all $\mathcal{KH}$-measurable g says $\langle f, gh \rangle = 0$ for all $\mathcal{KH}$-measurable g and $\mathcal{H}$-measurable h. Thus $E(f\overline{g}|\mathcal{H}) = 0$ for all $\mathcal{KH}$-measurable g. Hence

$$L^2(\mathcal{KH})^\perp = \{f \mid E(f\overline{g}|\mathcal{H}) = 0 \text{ for all } \mathcal{KH}\text{-measurable } g\}.$$

Analogously, as $\mathcal{Z}$ is also invariant under multiplication by $\mathcal{H}$-measurable functions,

$$\mathcal{Z}^\perp = \{g \mid E(f\overline{g}|\mathcal{H}) = 0 \text{ for all } \mathcal{Z}\text{-measurable } f\}.$$

In any measure-preserving system,

$$\{f \in L^2 \mid \frac{1}{n} \sum_{j=0}^{n-1} f(T^j(x)) \to 0\} \text{ and } \{g \in L^2 \mid g \circ T = g\}$$

are dual. Let $\hat{\mu}$ be a relatively independent joining of T with S chosen so that $\mathcal{I} \subseteq \mathcal{KH} \times \mathcal{KH}'$ and $\mathcal{KH} \times Y \subseteq \mathcal{I} \vee (\epsilon_X \times \mathcal{KH}')$, both $\hat{\mu}$-a.s.

Hence, for $f \in \mathcal{Z}$, any $f' \in L^2(\nu)$, and any $\mathcal{I}$-measurable h we will have

$$E((f \otimes \overline{f}')\overline{h}|\mathcal{H}) = 0 \qquad \hat{\mu}\text{-a.s.}.$$

So for $f' \in L^\infty(\nu)$ we have

$$E((f \otimes 1)(1 \otimes \overline{f}')\overline{h}) = 0.$$

Functions $(1 \otimes \overline{f}')\overline{h}$ have in their span all functions $\overline{g} \otimes 1$ for all $\overline{g}$ that are $\mathcal{KH}$-measurable. Hence, for any such g,

$$E((f \otimes 1)(\overline{g} \otimes 1)|\mathcal{H}) = 0.$$

For the other direction, suppose $E(f\overline{g}|\mathcal{H}) = 0$ for all $g \in L^2(\mu, \mathcal{KH})$. Now take $\hat{\mu}$ to be the relatively independent self-joining over $\mathcal{H}$, and for any f_1 that is $\mathcal{KH} \times \mathcal{KH}$-measurable,

$$E((f \otimes \overline{f})\overline{f}_1|\mathcal{H}) = 0.$$

Hence this holds for all f_1 that are $\mathcal{I}$-measurable and we must have $f \in \mathcal{Z}$.

We let $\mathcal{K}(T)$ represent $\mathcal{KH}$, where $\mathcal{H}$ is the trivial algebra. This is, of course, the Kronecker algebra, and restricted to it, T is an ergodic Abelian group rotation. We will write this in the form (R_{g_0}, G), where the measure is necessarily normalized Haar measure on G. L^2 of $\mathcal{K}(T)$ is spanned by eigenfunctions which are just the characters of G.

As a special case of the above discussion, we can identify the ergodic decomposition of a direct product of two ergodic actions T and S. Write the action of T on $\mathcal{K}(T)$ as (R_{g_0}, G) and of S on $\mathcal{K}(S)$ as $(R_{g_0'}, G')$. We know the invariant algebra $\mathcal{I}$ of $T \times S$ is the invariant algebra of $R_{g_0} \times R_{g_0'}$. Let (R_{h_0}, H) represent the maximal common factor action of (R_{g_0}, G) and $(R_{g_0'}, G')$, i.e.

$$(R_{h_0}, H) \simeq (R_{g_0}, G/\Gamma) \simeq (R_{g_0'}, G'/\Gamma').$$

Let $\psi_1 : G \to H$, $\psi_2 : G' \to H$, $\phi_1 : X \to H$, and $\phi_2 : Y \to H$ be corresponding factor maps to this common action.

As $R_{g_0 \times g_0'}$ acts isometrically, this direct product is the union of disjoint minimal isometries acting on the family of fiber products

$$S_c = \{(g, g') : \psi_1^{-1}(g)\psi_2(g') = c\}.$$

Any invariant set of $R_{g_0 \times g_0'}$ must consist of a union of such S_c. Hence the ergodic components of $T \times S$, as point sets supporting the ergodic measures, are sets of the form

$$\hat{S}_c = \{(x, y) \mid \phi_1^{-1}(x)\phi_2(y) = c\}.$$

For each c, the relatively independent joining of X and Y over H via the projections $x \to \phi_1(x)$ and $y \to c\phi_2(y)$ is supported on $\hat{S}_c$.

Call this joining $\mu \times_c \nu$. It is a calculation that

$$\mu \times \nu = \int_H \mu \times_c \nu \, dh(c).$$

Hence this is the ergodic decomposition of $\mu \times \nu$ both as point sets and as measures.

$\mathcal{KK}(T) = \mathcal{K}^2(T)$ is the maximal 2-step distal factor of T (see Definition 1.2). Our identification of the source of eigenfunctions of $T \times S$ will take place in three reductions. First we will show they are $\mathcal{K}^2(T) \times \mathcal{K}^2(S)$-measurable. The action of $\mathcal{K}^2(T)$ can be written as a twisted product $(R_{g_0,\phi}, G \times G_1/\Gamma)$ where $(R_{g_0,\phi}, G \times G_1)$ can be chosen canonically, and is ergodic.

Let G_1^* be the commutator subgroup of G_1. We will see that the maximal 2-step Abelian factor of T is $(R_{g_0,\phi}, G \times G_1/G_1^*\Gamma)$. Our second reduction is to show that all eigenfunctions of $T \times S$ are measurable with respect to the product of their maximal 2-step Abelian factors. Our third and final step will be to identify the characters of $G_1/G_1^*\Gamma$ which generate the algebra.

Now for a sketch of the structure of $\mathcal{KH}$ and relatively independent joinings from a compact-group perspective. Suppose $\mathbb{Y} = (Y, \mathcal{G}, \nu, S)$ is ergodic and $\mathbb{Y} \xrightarrow{\alpha} \mathbb{X} = (X, \mathcal{F}, \mu, T)$ is a factor, where $\mathbb{Y}$ is an isometric extension of $\mathbb{X}$. The system $\mathbb{Y}$ then can be represented in the form

$$(X \times G/\Gamma, \mathcal{F} \times \mathcal{B}, \mu \times m, T_\phi),$$

where G is a compact group, Γ a closed subgroup, $\phi : X \to G$, and

$$T_\phi(x, g\Gamma) = (T(x), \phi(x)g\Gamma)$$

Such are commonly called *cocycle extensions*, or *skew products*, over $\mathbb{X}$ by the metric space G/Γ. We will refer to ϕ as the *generating function* of the extension.

We will generically use $\mathcal{B}$ and m for the Borel algebra and Haar measure on a compact group or homogeneous space of this form. If the measure or measure algebra on a compact group or homogeneous space are not mentioned, Haar measure and the Borel algebra are to be assumed.

In such a representation of $\mathbb{Y}$ the choices for G, Γ, and ϕ are of course not unique, and an action written in the form of such a skew product is not necessarily ergodic. In particular, if we lift T_ϕ to act on all of $X \times G$ (we call this the *associated full group extension*) it is not necessarily ergodic. For a group extension, though, one can identify the ergodic components explicitly as supported on certain invariant point sets. These sets are of the form

$$Y_g = \{(x, c(x)G_0 g) \mid x \in X\},$$

where G_0 is a closed subgroup, and the measurable function c satisfies

$$\phi(x)c(x)G_0 = c(T(x))G_0.$$

The subgroup G_0, or any of its conjugates, is referred to as the *Mackey group* of the extension or the *Mackey range* of the cocycle generated by ϕ. The corresponding ergodic component as a measure is the integral over X of the lift to this coset of $g^{-1}G_0 g$ of Haar measure. As indicated above, whenever such a coset-valued graph is given, the measure on the first coordinate will be obvious, and for the the measure on the graph, we will integrate translated Haar measure on the corresponding cosets.

Now each of these measures must project a.s. to ν on Y, and each is a group extension of X by $g^{-1}G_0 g$, so one can always choose a representation for $\mathbb{Y}$ as an isometric extension over X so that T_ϕ acts ergodically on $X \times G$. We call such a skew extension *fully ergodic*. Recently del Junco, Lemanczyk, and Mentzen have shown that there is an essentially unique minimal choice for G in such a representation [del Junco et al.].

If $\mathbb{Y}$ is an isometric extension of $\mathbb{X}$, then any factor $\mathbb{Z}$ lying between $\mathbb{Y}$ and $\mathbb{X}$, when viewed on a fully ergodic representation $(T_\phi, X \times G/\Gamma)$ is a projection on the second coordinate

$$(x, g\Gamma) \to (x, g\Gamma'),$$

where $\Gamma \subseteq \Gamma'$ is the closed subgroup of G which acts on the right as the identity on the factor algebra.

Suppose such a factor action $\mathbb{Z}$ lying between $\mathbb{X}$ and $\mathbb{Y}$ can be represented as a group extension of $\mathbb{X}$ by some group G_0. Then in fact in any fully ergodic representation $(T_\phi, X \times G/\Gamma)$ of $\mathbb{Y}$, $\mathbb{Z}$ sits as $(T_\phi, X \times G/\Gamma')$, where Γ' is normal in G. The action of G_0 on the right in the representation of $\mathbb{Z}$ as a G_0-extension is the action of G/Γ' on the right. This is a direct consequence of Veech's theorem [Veech, 1982] that an ergodic isometric extension is a group extension iff all ergodic self-joinings that are diagonal measure on $X \times X$ are supported on graphs. These graphs are precisely the graphs of elements of the centralizer of the extension which act as the identity on the factor $\mathbb{X}$.

Lemma 1.4. *Any ergodic isometric extension $\mathbb{Y}$ of $\mathbb{X}$ contains:*

1) *a unique maximal factor* $\mathbb{Z}_{\text{group}}$ *between* $\mathbb{X}$ *and* $\mathbb{Y}$ *that is a group extension of* $\mathbb{X}$.

In any fully ergodic representation $(T_\phi, X \times G/\Gamma)$ *of* $\mathbb{Y}$, $\mathbb{Z}_{\text{group}}$ *is*

$$(T_\phi, X \times G/\Gamma'),$$

where Γ' *is the minimal normal subgroup containing* Γ.

2) *a unique maximal factor* $\mathbb{Z}_{\text{Abelian}}$ *between* $\mathbb{X}$ *and* $\mathbb{Y}$ *that is an Abelian group extension of* $\mathbb{X}$.

In any fully ergodic representation $(T_\phi, X \times G/\Gamma)$ *of* $\mathbb{Y}$, $\mathbb{Z}_{\text{Abelian}}$ *is*

$$(T_\phi, X \times G/(\Gamma G^*)).$$

Proof. 1) follows from the existence of the unique minimal normal subgroup Γ' containing Γ and the discussion preceding the Lemma.

For 2) just note that ΓG^* is the minimal normal subgroup of G containing Γ for which the quotient is Abelian. $(g(\gamma g^*)g^{-1} = \gamma(\gamma^{-1}g\gamma g^{-1})(gg^*g^{-1}) \in \Gamma G^*)$.

Definition 1.2. If T acting on $(X, \mathcal{F}, \mu)$ is ergodic and $\mathcal{H} \subseteq \mathcal{F}$ is a factor algebra, Lemma 1.2 tells us that within $\mathcal{K}\mathcal{H}$ are maximal factors we call $\mathcal{G}\mathcal{H}$ and $\mathcal{A}\mathcal{H}$, which are, respectively, the *maximal group extension of* T *over* $\mathcal{H}$ and the *maximal Abelian group extension of* T *over* $\mathcal{H}$.

The algebra $\mathcal{G}\mathcal{H}$ will not interest us here, and its functorial properties seem much less well behaved than $\mathcal{A}\mathcal{H}$, which will interest us very much.

Lemma 1.5. *Suppose* $(X, \mathcal{F}, \mu, T)$ *is ergodic and* $\mathcal{H} \subseteq \mathcal{F}$ *is a factor algebra. Any* $f : X \to S^1$ *satisfying*

$$f(T(x)) = \lambda(x)f(x),$$

where λ *is* $\mathcal{H}$-*measurable, is itself* $\mathcal{A}\mathcal{H}$-*measurable. Moreover the set of* f's *satisfying this functional equation spans* $L^2(\mathcal{A}\mathcal{H})$.

Proof. Let U acting on Y be a point version of T restricted to $\mathcal{H}$ with $\psi : X \to Y$ the factor map. Consider the S^1 extension of U given by the generating function $\lambda \circ \psi^{-1} = \psi'$. The maps $\phi_z(x) = (\psi(x), f(x)z)$ satisfy $\phi_z \circ T = U_{\psi'} \circ \phi_z$, so $\mu_z = \phi_{z*}\mu$ are $U_{\phi'}$-invariant measures, and each ϕ_z factors T onto $U_{\psi'}$ with invariant measure μ_z.

As

$$\int \mu_z \, d(m(z)) = \nu \times m,$$

$\int \mu_z \, d(m(z))$ is the ergodic decomposition of $\nu \times m$ with respect to $U_{\psi'}$.

The ergodic components of $U_{\psi'}$ we know must themselves be Abelian group extensions of U by some subgroup of S^1. Thus the σ-algebra generated by f must lie in an Abelian group extension of $\mathcal{H}$ and so is contained in $\mathcal{AH}$.

For the other direction just note that in an Abelian group extension any character of the group will satisfy the functional equation.

A similar result of course holds for $\mathcal{GH}$ involving maps from X to finite-dimensional orthogonal groups and representations of the group, but we will not need this here.

We next examine how $\mathcal{KH}$ and $\mathcal{AH}$ lift to relatively independent joinings. Suppose $(X, \mathcal{F}, \mu, T)$ and $(Y, \mathcal{G}, \nu, S)$ are ergodic systems and

$$\psi_1 : X \to Z \text{ and } \psi_2 : Y \to Z$$

factor them onto a third action $(Z, \mathcal{H}, \eta, U)$.

Let $\mathcal{H}_1 = \psi_1^{-1}(\mathcal{H})$ and $\mathcal{H}_2 = \psi_2^{-1}(\mathcal{H})$, and let $\hat{\mu}$ be the relatively independent joining of X and Y over Z.

Lemma 1.6. *If $\hat{\mu}$ is an ergodic $T \times S$-invariant measure, then*

$$\mathcal{K}\hat{\mathcal{H}} = \mathcal{KH}_1 \times \mathcal{KH}_2 \quad \hat{\mu}\text{-a.s.;}$$

i.e., if a relatively independent joining is ergodic then the relative Kronecker algebra of the joining is the relative product of the two relative Kronecker algebras.

Proof. From Lemma 1.3 this reduces to showing that for any f in $\mathcal{Z}(T)$ and g in $L^2(\nu)$,

$$E((f \otimes g) \otimes (\overline{f} \otimes \overline{g})|\mathcal{H} \vee \mathcal{I}) = 0$$

for the relatively independent self-joining over the factor $\mathcal{H}$.

But this is just a four-fold relatively independent joining and in particular can be viewed as a relatively independent joining of the first copy of X with the other three spaces (two copies of Y and one of X). If the three-fold joining were ergodic we would be done. If it isn't, decompose it into its ergodic components. Each will still possess a copy of $\mathcal{H}$, as this factor is ergodic. The four-fold relatively independent joining is the integral over these ergodic components of the relatively independent joining of each to X. Now the result follows.

In the case where Z is a trivial space, this simply says the Kronecker algebra of an ergodic direct product is the product of the two Kronecker algebras. In this case one knows more detail. The direct product is ergodic iff the point spectra of the two actions are disjoint, and one gets all eigenfunctions of $T \times S$ as products of eigenfunctions from the two components. In this context our work can be viewed as trying to understand how this result changes when the direct product is no longer ergodic.

Suppose $\hat{\mu}$ is the relatively independent joining as above of X and Y over Z. Let $(U_{\phi_1}, Z \times G_1/\Gamma_1)$ and $(U_{\phi_2}, Z \times G_2/\Gamma_2)$ be fully ergodic representations of T acting on $\mathcal{KH}_1$ and S acting on $\mathcal{KH}_2$. Assuming $\hat{\mu}$ is ergodic, $(U_{\phi_1 \times \phi_2}, Z \times (G_1 \times G_2)/(\Gamma_1 \times \Gamma_2))$ will be ergodic but not necessarily fully ergodic. As we have seen an ergodic component of the full extension (as a point set), though, will have the form

$$Y_{(g_1, g_2)} = \{(z, c(z)H(g_1, g_2)) \mid z \in Z\},$$

where the conjugates of H form the Mackey range of $\phi_1 \times \phi_2$.

Letting $\pi_i(g_1, g_2) = g_i$, we must have $\pi_i(H) = G_i$, as the coordinate actions are assumed ergodic.

Lemma 1.7. *The group $H \subseteq G_1 \times G_2$ arises as follows. There is a compact metric group $Q = Q(\phi_1, \phi_2)$ and group homomorphisms $h_i : G_i \to Q$ so that $(g_1, g_2) \in H$ iff $h_1(g_1) = h_2(g_2)$.*

Proof. Set $N_1 = \{g_1 \mid (g_1, id) \in H\}$ and $N_2 = \{g_2 \mid (id, g_2) \in H\}$. As each π_i is onto, one checks that these are normal subgroups of G_1 and G_2, each

$$N_1^{g_2} = \{g_1 \mid (g_1, g_2) \in H\}$$

is a coset $c_1(g_2)N_1$, and

$$N_2^{g_1} = \{g_2 \mid (g_1, g_2) \in H\} = c_2(g_1)N_2.$$

One checks

$$c_1(g_2'g_2)N_1 = c_1(g_2')c_1(g_2)N_1 \text{ and}$$
$$c_2(g_1'g_1)N_2 = c_2(g_1')c_2(g_1)N_2,$$

and further that

$$c_1(g_2) = c_1(g_2N_2) \text{ and}$$
$$c_2(g_1) = c_2(g_1N_1),$$

and lastly that

$$c_1(c_2(g_1N_1)N_2) = g_1N_1 \text{ and}$$
$$c_2(c_1(g_2N_2)N_1) = g_2N_2.$$

Hence c_1 conjugates G_2/N_2 to G_1/N_1. Letting $h_1 : G_1 \to G_1/N_1 = Q$ and $h_2 : G_2 \to Q$ be $h_2(g_2) = c_1(g_2N_2)N_1$,

$$(g_1, g_2) \in H \text{ iff } g_1 \in C_1(g_2)N_1 \text{ iff } h_1(g_1) = h_2(g_2).$$

We now describe a particular circumstance where in a relatively independent joining, $\mathcal{A}\hat{\mathcal{H}} = \mathcal{A}\mathcal{H}_1 \times \mathcal{A}\mathcal{H}_2$. The conditions we place on the joining are quite strong, making the proof relatively easy. Luckily they are precisely what we will have in our application. The more general question of precisely when this identity on algebras holds, or when it holds for the corresponding maximal group extensions, we leave as open questions.

Proposition 1.8. *Suppose $\hat{\mu}$ is the ergodic relatively independent joining of T and S over $(Z, \mathcal{H}, \eta, U)$ as above. Suppose one can select fully ergodic representations U_{ϕ_1} and U_{ϕ_2} for the actions $(T, \mathcal{K}\mathcal{H}_1)$ and $(S, \mathcal{K}\mathcal{H}_2)$, respectively, so that the group $Q(\phi_1, \phi_2)$ of Lemma 1.7 is Abelian, and further $h_1(\Gamma_1) = h_2(\Gamma_2) = Q$. We conclude*

$$\mathcal{A}\hat{\mathcal{H}} = \mathcal{A}\mathcal{H}_1 \times \mathcal{A}\mathcal{H}_2.$$

Proof. The discussion reduces to examining the ergodic components of the full extension $(U_{\phi_1 \times \phi_2}, Z \times G_1 \times G_2)$, which we write as before as

$$Y_{(g_1, g_2)} = \{(z, c(z)H(g_1, g_2)) \mid z \in Z\}$$

each of which we see is a copy of the same ergodic group extension of U by H. The generating function of this common version is

$$c^{-1}(T(z))(\phi_1(z), \phi_2(z))c(z) = \overline{\phi}(z).$$

As $U_{\overline{\phi}}$ is fully ergodic, to finish we simply show that under the assumptions of the Proposition,

$$H^*(\Gamma_1 \times \Gamma_2 \cap H) = (G_1^* \times G_2^*)(\Gamma_1 \times \Gamma_2 \cap H).$$

If we can show for any commutators $g_i^* \in G_i^*$ that both (g_1^*, id) and $(id, g_2^*) \in H^*(\Gamma_1 \times \Gamma_2 \cap H)$, we are done.

We show the first, the second being completely symmetric. Let $g_1^* = g\overline{g}g^{-1}\overline{g}^{-1}$. As $h_2(\Gamma_2) = Q$, select $\gamma, \overline{\gamma} \in \Gamma_2$ with

$$h_2(\gamma) = h_1(g), \, h_2(\overline{\gamma}) = h_1(\overline{g}).$$

Now

$$(g_1^*, id) = (g, \gamma)(\overline{g}, \overline{\gamma})(g, \gamma)^{-1}(\overline{g}, \overline{\gamma})^{-1}(id, \overline{\gamma}\gamma\overline{\gamma}^{-1}\gamma^{-1}) = h^*(id, \gamma_1),$$

where h^* is a commutator of H and, as Q is Abelian, $h_2(\gamma_1) = 0$, $(id, \gamma_1) \in \Gamma_1 \times \Gamma_2 \cap H$.

Return now to the direct product $T \times S$ and the common point spectrum (R_{h_0}, H) of T and S. Assume for the moment that both T and S are 2-step distal, i.e. are of the form

$$(R_{h_0, \phi_1}, H \times G_1/\Gamma_1) \text{ and } (R_{h_0, \phi_2}, H \times G_2/\Gamma_2),$$

where we assume both representations are fully ergodic.

We know a.e. ergodic component of $T \times S$ is supported on a point set of the form

$$Z_c = \{(h, hc, g_1\Gamma_1, g_2\Gamma_2) \mid h \in H\},$$

with the action given as

$$(h, hc, g_1\Gamma_1, g_2\Gamma_2) \longrightarrow (h_0 h, h_0 hc, \phi_1(h)g_1\Gamma_1, \phi_2(hc)g_2\Gamma_2).$$

If we define $\phi_2^c(h) = \phi_2(hc)$, this is seen to be conjugate to

$$(R_{h_0, \phi_1 \times \phi_2^c}, H \times (G_1 \times G_2)/(\Gamma_1 \times \Gamma_2)),$$

the relatively independent joining of R_{h_0, ϕ_1} and R_{h_0, ϕ_1^c} over the common first-coordinate algebras.

We want to examine how these supports sit within the product of the full extensions and hence how the lifted measure further decomposes into ergodic components of the full extension. To do this, examine the product of the two full extensions,

$$(R_{h_0, \phi_1}, H \times G_1) \text{ and } (R_{h_0, \phi_2}, H \times G_2).$$

These two will have some common point spectrum we can write as $(R_{\overline{h}_0}, \overline{H})$, with $P_1 : H \times G_1 \to \overline{H}$ and $P_2 : H \times G_2 \to \overline{H}$ the two factor maps. Now P_1 and P_2 can be post-multiplied by any element of $\overline{H}$ and still be a factoring. Hence we can assume that $P_i(id, id) = id$. Right rotation by elements of the G_i must project to multiplication by some element in $\overline{H}$, and so $P_1(id, G_1)$ and $P_2(id, G_2)$ are closed subgroups of $\overline{H}$. As (R_{h_0}, H) is a part of the common point spectrum, there is a group homomorphism

$$\pi : \overline{H} \to H.$$

As the action of $R_{\overline{h}_0}$ will project to R_{h_0}, $\pi \circ P_i(h, g_i) = h$, and so

$$P_1(id, G_1) = P_2(id, G_2) = \ker(\pi).$$

The ergodic components of the direct product of these full extensions will be supported a.s. on sets of the form

$$\overline{Z}_{\overline{c}} = \{((h_1, g_1), (h_2, g_2)) \mid P_1^{-1}(h_1, g_1)P_2(h_2, g_2) = \overline{c}\}.$$

Hence the ergodic components of a lift $(R_{h_0, \phi_1 \times \phi_2^c}, H \times (G_1 \times G_2))$, for m-a.e. $c \in H$, will be supported on point sets

$$\tilde{Z}_{\overline{c}} = \{(h, g_1, g_2) \mid P_1^{-1}(h, g_1)P_2(hc, g_2) = \overline{c}\}.$$

Here $\overline{c}$ is restricted to lie in $\pi^{-1}(c)$, a coset of $\ker(\pi)$. Thus the Mackey range $\mathbb{H}$ of elements (g_1, g_2) which acting on the right fix such an ergodic component is

$$\mathbb{H} = \{(g_1, g_2) \mid P_1(id, g_1) = P_2(id, g_2)\}.$$

Thus $Q = Q(\phi_1, \phi_2^c)$ of Lemma 1.7 can be taken as $\ker(\pi)$, and the projections $h_i(g_i) = P_i(id, g_i)$. Notice that $\ker(\pi)$ is Abelian, and as $\overline{H}/h_i(\Gamma_i)$ is the common point spectrum of $(R_{h_0, \phi_i}, H \times G_i/\Gamma_i)$, which is simply (R_{h_0}, H), we must have $h_i(\Gamma_i) = \ker(\pi)$, precisely the criteria for Proposition 1.8.

Theorem 1.9. *Suppose that $(X, \mathcal{F}, \mu, T)$ and $(Y, \mathcal{G}, \nu, S)$ are ergodic, and (R_{h_0}, H) is a version of their common point spectrum. Let $\psi_1 : X \to H$ and $\psi_2 : Y \to H$ be the corresponding factor maps, and $\mathcal{H}_1 = \psi_1^{-1}(\mathcal{B})$ and $\mathcal{H}_2 = \psi_2^{-1}(\mathcal{B})$ the corresponding factor algebras. Now $\psi_1 \times \psi_2 : X \times Y \to H \times H$, and we know a.e. ergodic component of $T \times S$ is a relatively independent joining over a graph*

$$Y_c = \{(h_1, h_2) \in H \times H \mid h_2 = h_1 c\}.$$

Call this joining $\hat{\mu}_c$ and the version of the common point spectrum in it $\hat{\mathcal{H}}_c$. For a.e. c,

$$\mathcal{A}\hat{\mathcal{H}}_c = \mathcal{A}\mathcal{H}_1 \times \mathcal{A}\mathcal{H}_2, \ \hat{\mu}_c\text{-a.s.} .$$

Proof. First, as $\hat{\mu}_c$ is an ergodic measure for a.e. c, Lemma 1.6 tells us that

$$\mathcal{K}\hat{\mathcal{H}}_c = \mathcal{K}\mathcal{H}_1 \times \mathcal{K}\mathcal{H}_2, \ \hat{\mu}_c\text{-a.s.} .$$

Hence $\mathcal{A}\hat{\mathcal{H}}_c \subseteq \mathcal{K}\mathcal{H}_1 \times \mathcal{K}\mathcal{H}_2, \ \hat{\mu}_c - \text{a.s.} .$

This allows us to restrict our attention to the actions of T and S on $\mathcal{K}\mathcal{H}_1$ and $\mathcal{K}\mathcal{H}_2$, respectively. These we write as $(R_{h_0, \phi_1}, H \times G_1/\Gamma_1)$ and $(R_{h_0, \phi_2}, H \times G_2/\Gamma_2)$, where we assume both representations are fully ergodic.

The discussion preceding the Theorem and Proposition 1.8 now finish the result.

Corollary 1.10. *All eigenfunctions of the direct product of two ergodic systems, $T \times S$, are measurable with respect to the product*

$$\mathcal{A}\mathcal{H}_1 \times \mathcal{A}\mathcal{H}_2,$$

where $\mathcal{H}_1 \subseteq \mathcal{K}(T)$ and $\mathcal{H}_2 \subseteq \mathcal{K}(S)$ are the common point spectrum of the two systems.

Proof. From Lemma 1.5 we know that an eigenfunction of $T \times S$ must, for a.e. ergodic component, be measurable with respect to $\mathcal{A}\hat{\mathcal{H}}_c$, hence measurable with respect to $\mathcal{A}\mathcal{H}_1 \times \mathcal{A}\mathcal{H}_2$, by Theorem 1.9. As this is true for a.e. ergodic component, it is true for $\mu \times \nu$ itself, as the algebra of invariant sets is $\mathcal{H}_1 \times \mathcal{H}_2$ measurable.

Definition 1.3. If G and G_1 are both compact Abelian groups and $\phi : G \to G_1$, we call $(R_{g_0, \phi}, G \times G_1)$ a *2-step Abelian action*.

Corollary 1.10 tells us we need only consider 2-step Abelian actions in investigating eigenfunctions of direct products.

Theorem 1.11. *Suppose* $(R_{g_0,\phi_1}, G \times G_1)$ *and* $(R_{g_0,\phi_2}, G \times G_2)$ *are two 2-step Abelian actions, with* (R_{g_0}, G), *their common point spectrum. Any eigenfunction of* $R_{g_0,\phi_1} \times R_{g_0,\phi_2}$ *has the form*

$$e(g, cg, g_1, g_2) = C_1(g_1)C_2(g_2)f(g, cg),$$

where $c \in A \in \mathcal{I}$, *the* $R_{g_0} \times R_{g_0}$-*invariant algebra,* $C_1 \in \hat{G}_1$, $C_2 \in \hat{G}_2$, *and* f *is* $G \times G$-*measurable.*

Proof. Suppose e is an eigenfunction of $R_{g_0,\phi_1} \times R_{g_0,\phi_2}$ with eigenvalue $\lambda(c)$ for ergodic component $\hat{\mu}_c$. For each $g_1' \in G_1$, set

$$e^{g_1'}(g, cg, g_1, g_2) = e(g, cg, g_1 g_1', g_2).$$

One computes that $e^{g_1'}$ is also an eigenfunction with eigenvalue $\lambda(c)$. By ergodicity of the action on the ergodic component, we conclude that

$$e^{g_1'}(g, gc, g_1, g_2) = C_1(g_1')e(g, gc, g_1, g_2).$$

One also computes that C_1 is a measurable, hence continuous, character of G_1. Arguing symmetrically we conclude that

$$e(g, gc, g_1, g_2) = C_1(g_1)C_2(g_2)e(g, gc, 1, 1),$$

completing the result.

As both $\hat{G}_1$ and $\hat{G}_2$ are countable, an eigenfunction as above can always be split up as defined on disjoint invariant sets on each of which both C_1 and C_2 are constant.

2. THE CONZE-LESIGNE ALGEBRA

The previous section has shown that we will find all eigenfunctions of a direct product in the product of their maximal Abelian extensions of their common point spectrum. This has reduced our work to considering Abelian group extensions of group rotations. What we will do now is to see that the characters that arise as C_1 and C_2 in Theorem 1.11 must satisfy a functional equation we call the *Conze-Lesigne equation*.

As a preliminary we will investigate when a compact group extension of a group rotation still has pure point spectrum. Here we will find the characterization is via a reduced form of the Conze- Lesigne equation we call the *Kronecker equation*. In this case our work is more general in two senses. First, we need not assume the second coordinate group is Abelian, the existence of solutions to the Kronecker equation will force it. Second, we can work directly with the group and not its irreducible representations, as the Kronecker algebra itself always represents a compact Abelian group rotation. The Conze-Lesigne algebra, on the other hand, will just be an inverse limit of order-two nil-rotations, and hence the group structure will exist only on finite-dimensional representations.

The if direction of the following result parallels precisely our argument for the Conze-Lesigne equation and can be taken as a warm-up for it.

Theorem 2.1. *Suppose $(R_{g_0, \phi}, G \times G_1 / \Gamma)$ is a fully ergodic isometric extension of a group rotation. Then $R_{g_0, \phi}$ has pure point spectrum iff there is a measurable $f : G \times G \to G_1$ satisfying, for a measurable set of u of positive measure in G,*

$$f(u, gg_0)\phi(g)\Gamma = \phi(ug)f(u, g)\Gamma. \qquad \text{(Kronecker equation at } u\text{)}$$

We will write $f(u, g) = f_u(g)$. In particular, the existence of such a solution forces Γ to be normal and G_1/Γ to be Abelian.

Proof. For the only if direction, if $R_{g_0, \phi}$ has pure point spectrum then there is an ergodic group rotation $(R_{\bar{g}_0}, \overline{G})$ and a conjugation

$$\psi : \overline{G} \to G \times G_1/\Gamma, \quad \text{with } \psi \circ R_{\bar{g}_0} = R_{g_0, \phi} \circ \psi.$$

Assume, without loss of generality, that Γ contains no normal subgroups of G_1. Then one can calculate that an action of the form $R_{\text{id}, f}$ commutes with $R_{g_0, \phi}$ if and only if $f = z \in Z(G_1) = \text{center}(G_1)$.

As the centralizer of the action of $\overline{G}$ is itself, i.e. the closure of the powers of $R_{g_0, \phi}$, and acts transitively on $G \times G_1/\Gamma$, G_1 must be Abelian. Removing the assumption that Γ contains no normal subgroups of G_1, we have Γ is normal and G_1/Γ is Abelian. Hence the action of G_1/Γ pulls back to a subgroup of $\overline{G}$.

The product rule on $\overline{G}$ projects to a product rule on $G \times G_1/\Gamma$ which must take the form

$$(u_1, v_1)(u_2, v_2) = (u_1 u_2, c(u_1 u_2)c(u_1)^{-1}c(u_2)^{-1}v_1 v_2).$$

Thus for $f(u_1, u_2) = f_{u_1}(u_2) = c(u_1 u_2)c(u_1)^{-1}c(u_2)^{-1}$, $R_{u, f_u} \circ \psi = \psi \circ R_{\psi^{-1}(u_1, id)}$. This forces R_{u, f_u} to commute with $R_{g_0, \phi}$, and this commutativity is precisely the Kronecker equation.

For the if direction, to show $R_{g_0, \phi}$ has pure point spectrum we will essentially construct $\overline{G}$. Again assume Γ contains no normal subgroups of G_1. As noted above, for fixed u the Kronecker equation means precisely that R_{u, f_u} and $R_{g_0, \phi}$ commute. We want to show that

$$\overline{G} = \{R_{u, f} \mid R_{u, f} \text{ and } R_{g_0, \phi} \text{ commute}\}$$

is a weakly-closed compact and Abelian group. Define

$$F_u = \{f : G \to G_1 \mid R_{u, f} \text{ and } R_{g_0, \phi} \text{ commute}\}.$$

For $f : G \to G_1$ set $f^u(g) = f(ug)$. One computes that if $f_1 \in F_u$ and $f_2 \in F_v$, then

$$f_2^u f_1 \in F_{uv}.$$

Thus if both F_u and F_v are not empty, neither is F_{uv} and

$$F_{uv} = \{f_2^u f_1 \mid f_2 \in F_v, f_1 \in F_u\}.$$

As F_u is nonempty for an $R_{g_0,\phi}$-invariant set of u of positive measure and is now known to be invariant under differences ($F_{u^{-1}} = \{(f^{u^{-1}})^{-1} \mid f \in F_u\}$), this implies all F_u are nonempty. Moreover, one can easily extend f to be defined on all of $G \times G$.

As $R_{g_0,\phi}$ is fully ergodic and Γ contains no normal subgroups of G_1, the group
$$F_{id} = \{f(g) = z \mid z \in Z(G_1), \text{ the center of } G_1\}.$$

Let $\|\ \|$ represent a symmetric norm on either G or G_1. Define a distance function

$$D(u,v) = \inf\{\int_G \|f_1(g)f_2(g)^{-1}\| \, dm(g) \mid f_1 \in F_u, f_2 \in F_v\}.$$

As $Z(G_1)$ is compact, this inf is in fact a minimum. One checks that $D(u,v)$ depends only on uv^{-1}.

As $f : G \times G \to G_1$ is measurable, for any $\varepsilon > 0$ there is a $\delta > 0$ so that if $\|u_1\| < \delta$, then for some $u, v \in G$ with $uv^{-1} = u_1$,

$$\int_G \|f_u(g)f_v(g)^{-1}\| \, dm(g) < \varepsilon, \text{ as}$$

$$\lim_{\|u_1\|\to 0} \iint_{G\times G} \|f(vu_1,g)f(v,g)^{-1}\| \, dm(v)dm(g) = 0.$$

Thus once $\|uv^{-1}\| < \delta$, $D(u,v) < \varepsilon$, i.e D is continuous.
We conclude that if $g_0^{n_i} \to u$, then there must be $z_i \in Z(G_1)$ with

$$\int_G \|\phi^{(n_i)}(g)(z_i f_u(g))^{-1}\| \, dm(g) \to 0.$$

As $Z(G_1)$ is compact, we conclude that $\{R_{g_0,\phi}\}$ generates a compact group of transformations under pointwise a.s. convergence, and we are finished.

With a little thought it is possible to weaken the fully ergodic hypothesis here to just ergodic. As we will not need this strengthening, we leave it to the interested reader.

Definition 2.1. Suppose $(R_{g_0,\phi}, G \times G_1)$ is an ergodic group extension of R_{g_0} and $\chi : G \to \mathrm{U}(n,\mathbb{C})$ is an irreducible representation of G_1. We say χ is *Conze-Lesigne* for $R_{g_0,\phi}$ on a subset $A \subseteq G$ if there are measurable functions

$$f : A \times G \to \mathrm{U}(n,\mathbb{C}) \text{ and } \lambda : A \to S^1$$

satisfying the *non-Abelian Conze-Lesigne equation*

$$\chi(\phi(gc))f(c,g) = \lambda(c)f(c,gg_0)\chi(\phi(g)).$$

If all irreducible representations of G_1 are Conze-Lesigne for $R_{g_0,\phi}$ we call the extension a *Conze-Lesigne extension*.

Corollary 2.2. *If $R_{g_0,\phi}$ is an ergodic Conze-Lesigne extension, then G_1 is Abelian. Hence the Conze-Lesigne equation can be rewritten for just characters as*

$$f : A \times G \to S^1, \text{ and } \lambda : A \to S^1$$

satisfying for each $c \in A$

$$\frac{\chi(\phi(gc))}{\chi(\phi(g))} = \lambda(c)\frac{f(c, gg_0)}{f(c, g)}. \qquad \text{(Conze-Lesigne or C-L equation)}$$

Proof. The diagonal matrices $\Delta(n, \mathbb{C})$ are the center of $U(n, \mathbb{C})$, so modding out by this center, set

$$\overline{G}_1 = U(n, \mathbb{C})/\Delta(n, \mathbb{C})$$

and

$$\overline{\phi} = \chi \circ \phi\Delta(n, \mathbb{C}).$$

The Conze-Lesigne equation for χ reduces to the Kronecker equation for $\overline{\phi}$. As $R_{g_0,\overline{\phi}}$ is a factor of $R_{g_0,\phi}$, it is ergodic, and Theorem 2.1 tells us $\overline{G}_1$ is Abelian, and hence $n = 1$ and we are done.

Thus the only irreducible representations that can ever satisfy the C-L equation are one-dimensional.

Definition 2.2. Suppose $(X, \mathcal{F}, \mu, T)$ is ergodic and $(R_{g_0,\phi}, G \times G_1/\Gamma)$ is a fully ergodic representation of $\mathcal{K}^2(T)$. Any irreducible representation of G_1 satisfying the non-Abelian Conze-Lesigne equation we call a *C-L representation* of T.

Fact. All C-L representations of T are in fact C-L characters, in particular are characters of $H = G_1/G_1^*\Gamma$, where $(R_{g_0,\phi}, G \times H)$ represents the algebra $\mathcal{A}\mathcal{K}(T)$.

We have included the above material because the original work of Conze and Lesigne concerned the non-Abelian C-L equation, and our work here only concerns the Abelian case. The above shows these to be equivalent. This fact has been pointed out some time ago by Furstenberg and Weiss, although not at the level of the equation itself, but rather at the level of the limiting behavior of certain multi-recurrence averages.

Notice that the characters of G_1 satisfying the Kronecker equation are closed under products, hence represent some factor G_1/H which actually belongs, as we know, in the first coordinate as part of the Kronecker algebra. It is the characters of H which satisfy the C-L equation that really interest us here. Thus we will in general pull off the Kronecker characters, rewriting our representation so that the first coordinate is actually the full Kronecker algebra.

Refining our vocabulary a little, if we want to consider the C-L equation at one point $c \in A$, we call it the *C-L equation at c*. The pair (f, λ) we refer to as the *solution of the C-L equation on A*.

For $c \in G$ we say a function $f : G \to S^1$ is a *c-coboundary* if

$$f(g) = \frac{h(cg)}{h(g)}$$

for some measurable $h : G \to S^1$.

Notice that for each $c \in A$ the left side of the C-L equation is a c-coboundary and the right side is a constant $(\lambda(c))$ times a g_0-coboundary.

Suppose χ is a C-L character relative to ϕ at c, and we modify ϕ by a g_0-coboundary to $\psi(g) = \phi(g) \dfrac{h(g_0 g)}{h(g)}$. We compute that

$$\frac{\chi(\psi(gc))}{\chi(\psi(g))} = \lambda(c) \frac{f(c, gg_0)}{f(c, g)} \frac{\chi(h(g_0 gc))\chi(h(g_0 g))^{-1}}{\chi(h(gc))\chi(hg)^{-1}},$$

and we see that χ is still a C-L character relative to ψ. Thus, replacing $R_{g_0, \phi}$ by the cohomologous action $R_{g_0, \psi}$ does not change the list of C-L characters, or moreover the sets A on which they satisfy the C-L equation, nor the values of $\lambda(c)$.

Lemma 2.3. *Suppose (f_1, λ_1) and (f_2, λ_2) are both solutions of the C-L equation for χ on the same set A. Then $\lambda_1^{-1}(c)\lambda_2(c) = \gamma(c)$ is an eigenvalue of (R_{g_0}), and $f_1 \circ f_2^{-1}(c, g) = h(c)C(c, g)$, where h is arbitrary and $C(c, g)$ is, for each c, a character of G with $C(c, g_0) = \gamma(c)$. On the other hand, if (f, λ) is one solution of the C-L equation on A, and γ and C are as above, then $(f \cdot h \cdot C, \lambda\gamma^{-1})$ is another.*

Proof. We compute

$$\frac{f_1(c, gg_0)}{f_2(c, gg_0)} = \frac{\lambda_2(c)}{\lambda_1(c)} \frac{f_1(c, g)}{f_2(c, g)},$$

and so $f_1 \cdot f_2^{-1}$ is an eigenfunction for each c of (R_{g_0}, G) with eigenvalue $\gamma(c)$. As this group rotation is assumed ergodic, the rest follows. The second half is the same computation in reverse.

Lemma 2.4. *If χ is a C-L character on $A \subseteq G$ of $R_{g_0, \phi}$, where $m(A) > 0$, then it is also on AA^{-1}.*

Proof. Note that for $c_1, c_2 \in A$,

$$\frac{\chi(\phi(gc_1 c_2^{-1}))}{\chi(\phi(g))}$$

$$= \frac{\chi(\phi(gc_1 c_2^{-1}))}{\chi(\phi(gc_1))} \frac{\chi(\phi(gc_1))}{\chi(\phi(g))}$$

$$= \frac{\lambda(c_1)}{\lambda(c_2)} \frac{f(c_1, g_0 g)f(c_2, gc_1 g_0)^{-1}}{f(c_1, g)f(c_2, gc_1)^{-1}}.$$

Fixing c_1 and c_2, this gives a solution of the C-L equation at $c_1 c_2^{-1}$. We must handle measurability in c. Taking a countable dense collection of values $\{c_i\}$ in A, $\bigcup_{i=1}^{\infty}(c_i A^{-1})$ is a subset of full measure in AA^{-1}. We can build up a c-measurable solution of the C-L equation successively on values

$$c \in (c_i A^{-1}) - (\bigcup_{j<i}(c_j A^{-1})).$$

On the remainder of AA^{-1} any solution will do.

Lemma 2.5. *If χ is a C-L character on A for $R_{g_0,\phi}$ with $m(A) > 0$, then it is also on $g_0 A$ and hence on all of G. Hence we simply refer to it as a C-L character.*

Proof. Just notice

$$\frac{\chi(\phi(gg_0c))}{\chi(\phi(g))} = \frac{\chi(\phi(gg_0c))}{\chi(\phi(g_0g))} \frac{\chi(\phi(gg_0))}{\chi(\phi(g))}$$

$$= \lambda(c)\frac{f(c,gg_0)\chi(\phi(gg_0))}{f(c,g)\chi(\phi(g))}$$

for the first part.

As we can assume the solution is on AA^{-1}, which has interior, a finite union of such translates by g_0 will cover G.

Lemma 2.6. *The C-L characters of T are a subgroup of $\widehat{G}$.*

Proof. Just notice that if (f_1, λ_1) is a solution of the C-L equation for χ_1, as is (f_2, λ_2) for χ_2, then $(f_1 f_2^{-1}, \lambda_1 \lambda_2^{-1})$ is for $\chi_1 \chi_2^{-1}$, as

$$\frac{\chi_1(\phi(cg))\chi_2^{-1}(\phi(cg))}{\chi_1(\phi(g))\chi_2^{-1}(\phi(g))} = \lambda_1(c)\lambda_2^{-1}(c)\frac{f_1(c,g_0g)f_2^{-1}(c,g_0g)}{f_1(c,g)f_2^{-1}(c,g)}.$$

Definition 2.4. Suppose $(X, \mathcal{F}, \mu, T)$ is ergodic, and $(R_{g_0,\phi}, G \times G_1)$ is a representation of its maximal 2-step Abelian factor. Let L be the factor group of G_1 dual to the Conze-Lesigne characters of T.

Let $\mathcal{L}(T)$ be the factor algebra of T given by $(R_{g_0,\phi}, G \times L)$ i.e. generated by the C-L characters over the Kronecker algebra.

Lemma 2.7. *If χ is a C-L character of $T = R_{g_0,\phi}$, where (R_{g_0}, G) is the Kronecker factor of $R_{g_0,\phi}$, then the range of χ is all of S^1, i.e. $\mathcal{L}(T)$ is an Abelian toral extension of (R_{g_0}, G).*

Proof. Suppose χ is finite valued, in particular

$$\chi : G_1 \to \{e^{2\pi i j/n}\}$$

for some n.

In the C-L equation we see $\lambda(c)f(c, gg_0)/f(c, g)$ takes values in the range of χ, and so $\lambda^{-n}(c)$ will be an eigenvalue of (R_{g_0}, G) with eigenfunction $f^n(c, g)$. As (R_{g_0}, G) has only countably many eigenvalues, $\lambda(c)$ takes on at most countably many values. Restrict to a subset A, where $\lambda(c) = \lambda$ is a constant. Now for $c \in A$,

$$\frac{\chi(\phi(gc))}{\chi(\phi(g))} = \lambda \frac{f(c, gg_0)}{f(c, g)},$$

and so for $c \in A$,

$$e(g, gc, h_1, h_2) \overset{\text{def}}{=} \chi(h_1)\chi^{-1}(h_2)f(c, g)$$

is an eigenfunction of $T \times T$ relative to $(\mu \times \mu)_c$ with fixed eigenvalue λ. Hence there is an atom in the spectral measure of e relative to $\mu \times \mu$ at λ, and λ is an eigenvalue of (R_{g_0}, G), as this is the Kronecker algebra of the action. But then e is in fact (R_{g_0}, G)-measurable and χ is trivial.

Returning to our study of eigenfunctions of $T \times S$, remember from Theorem 1.11 that if T and S are 2-step Abelian then any such had the form

$$C_1(g_1)C_2(g_2)f(\bar{g}, c\bar{g}),$$

where $c \in A \in \mathcal{I}$, and for each c, C_1 is a character of G_1 and C_2 is a character of G_2.

Theorem 2.9. *Let $(X, \mathcal{F}, \mu, T)$ and $(Y, \mathcal{G}, \nu, S)$ be ergodic, and*

$$(R_{g_0, \phi_2}, G \times G_1) \text{ and } (R_{g_0, \phi_2}, G \times G_2)$$

represent the maximal Abelian extensions of their common point spectrum in each. If a character χ of G_1 appears as a C_1 in some eigenfunction of $T \times S$ then it is a C-L character of T. On the other hand, if χ is a C-L character of T, then it appears as C_1 in an eigenfunction of $T \times T$.

Proof. Write the eigenfunction as

$$e(g, cg, g_1, g_2) = C_1(g_1)C_2(g_2)f(g, cg).$$

Computing:

$$f(gg_0, cgg_0)C_1(\phi_1(g)g_1)C_2(\phi_2(cg)g_2) = \lambda(c)C_1(g_1)C_2(g_2)f(g, cg),$$

or

$$\frac{f(gg_0, cgg_0)}{f(g, cg)} = \lambda(c)(C_1(\phi_1(g))C_2(\phi_2(g)))^{-1}.$$

Thus for all $c_1, c_2 \in A$,

$$\frac{C_1(\phi_1(c_2 c_1^{-1} g))}{C_1(\phi_1(g))} = \frac{\lambda(c_1)}{\lambda(c_2)} \left(\frac{f(c_2 c_1^{-1} g g_0, c_2 g g_0)}{f(g g_0, c_2 g g_0)} \right) \left(\frac{f(c_2 c_1^{-1} g, c_2 g)}{f(g, c_2 g)} \right)^{-1}.$$

Thus for any $c = c_2 c_1^{-1} \in AA^{-1}$ we get a solution to the C-L equation. Measurability of the solution follows exactly as in Lemma 2.4.

For the other direction, just note that if χ is a C-L character for T, then for all $c \in G$

$$e(g, cg, g_1, g_1') = \frac{\chi(g_1)}{\chi(g_2')} f(g, cg)$$

will be an eigenfunction of $T \times T$.

It is interesting to note in this last theorem that if χ arises in an eigenfunction for some product it arises in an eigenfunction of the Cartesian square. This is a natural extension of the fact that if the direct product of T with some other ergodic system is nonergodic, then its Cartesian square is nonergodic. In Section 4 we will see that in a very precise sense it is really only in a Cartesian square that eigenfunctions of $T \times S$ can arise, in that, restricting the two extensions of $\overline{G}$ to just the circle extensions generated by $C_1 \circ \phi_1$ and $C_2 \circ \phi_2$, these two circle extensions are, up to a constant, cohomologous.

Now a 2-step Abelian action can be completely Conze-Lesigne, i.e. all characters of the second component can satisfy the C-L equation, without the second-coordinate group being toral. For this to happen, of course, the first coordinate cannot be the full Kronecker algebra, i.e. some subgroup of characters of the second coordinate actually satisfy the Kronecker equation, and thus a factor group could be split off and added on to the first coordinate. As we continue it will be important for us to know that the second-coordinate group is toral, so we embody this in a definition.

Definition 2.3. If (R_{g_0}, G) is an ergodic group rotation, G_1 is toral, $\phi : G \to G_1$, and all characters of G_1 are C-L characters of $R_{g_0, \phi}$, then we call $R_{g_0, \phi}$ a *C-L action* on $G \times G_1$. We do not require here that (R_{g_0}, G) is necessarily its Kronecker algebra.

Put in this form we gain two things; first, the second coordinate group is an inverse limit of finite-dimensional tori; and second, we will not be constantly fussing around to move all of the Kronecker algebra to the first coordinate. Of course we could, so any ergodic group extension of a group rotation, where every irreducible representation of the second coordinate group satisfies the C-L equation, can be rewritten as a C-L action.

We will need the following final lemma in this section. It indicates further the value of having G_1 a torus. It is false for any other choice, including G_1 just toral. This is an indication of the parallel problem in inverse limits for the C-L algebra.

Lemma 2.10. *Suppose G_1 is a torus, either finite- or infinite- dimensional. If $(R_{g_0,\phi}, G \times G_1)$ is ergodic and of pure point spectrum, then ϕ is cohomologous to a constant.*

Proof. It is enough here to consider $G_1 = S^1$, as the result extends to countable products of G_1's. The result is equivalent to saying that in any compact Abelian group, a subgroup conjugate to S^1 must have a complementary subgroup, i.e. the group is a direct product with the S^1 subgroup as one of the terms. Dualizing, this is equivalent to saying that in any countable discrete Abelian group, a factor group conjugate to $\mathbb{Z}$ has a choice of coset representatives forming a group. But as $\mathbb{Z}$ is singly generated and torsion free, this is trivial.

3. ORDER-TWO NIL-FACTORS

We now give a third characterization of the C-L algebra as the span of all order-two nil-factors of $(X, \mathcal{F}, \mu, T)$. In this section we will show that order-two nil-rotations are always Conze-Lesigne. In the following sections we will see that The C-L algebra is not necessarily an order-two nil-rotation but is the span of an increasing sequence of such. All the results presented in this section are simply the author's reporting of ideas he learned from H. Furstenberg during his recent visit to Maryland. Much of this is also to be found in the treatment of these issues in [Lesigne, 1993].

Definition 3.1. Suppose $\mathcal{N}$ is a locally compact separable metric group, its center, $C(\mathcal{N})$, is compact and toral, and $\mathcal{N}/C(\mathcal{N})$ is Abelian. We call such an $\mathcal{N}$ an *order-two nilpotent group* (with compact toral center.) If Γ is a cocompact subgroup of $\mathcal{N}$ and $n_0 \in \mathcal{N}$, then we call the action $(R_{n_0}, \mathcal{N}/\Gamma)$ an *order-two nil-rotation* and if it arises as a factor of $(X, \mathcal{F}, \mu, T)$, we call it an *order-two nil-factor* of T.

All that interests us is the action $(R_{n_0}, \mathcal{N}/\Gamma)$. Hence we can always assume $\Gamma \cap C(\mathcal{N}) = \{1\}$. The condition that $C(\mathcal{N})$ is compact can be replaced with $C(\mathcal{N}/\Gamma \cap C(\mathcal{N}))$ is closed, hence compact, as we can always mod out by the normal subgroup $\Gamma \cap C(\mathcal{N})$.

Theorem 3.1. *An ergodic rotation on an order-two nilpotent group is a Conze-Lesigne action, i.e. $\mathcal{L}$ is the full algebra.*

Proof. First notice $(R_{n_0}, \mathcal{N}/C(\mathcal{N})\Gamma)$ is an Abelian group rotation. The cosets $C(\mathcal{N})\gamma$, $\gamma \in \Gamma$, are all compact and disjoint. Define a Borel section

$$s : \mathcal{N}/C(\mathcal{N})\Gamma \to \mathcal{N}.$$

Thus

$$s(nC(\mathcal{N})\Gamma)^{-1}n \in C(\mathcal{N})\Gamma.$$

Any such element has a unique representation as a product

$$s(nC(\mathcal{N})\Gamma)^{-1}n = z(n)\gamma(n),$$

and
$$s(nC(\mathcal{N})\Gamma) = nz(n)^{-1}\gamma(n)^{-1}$$
depends only on the coset $nC(\mathcal{N})\Gamma$. Notice also that $z(n\Gamma) = z(n)$ and $z(zn) = z(n)z$ (we are using z's here to mean points of $C(\mathcal{N})$.)

Thus
$$n\Gamma = s(nC(\mathcal{N})\Gamma)C(\mathcal{N})\Gamma.$$

Define a conjugation $f : \mathcal{N}/\Gamma \to \mathcal{N}/C(\mathcal{N})\Gamma \times C(\mathcal{N})$ by

$$f(n\Gamma) = (nC(\mathcal{N})\Gamma, z(n\Gamma)),$$

and so

$$f^{-1}(nC(\mathcal{N})\Gamma, z) = s(nC(\mathcal{N})\Gamma)z\Gamma.$$

We compute for any $d \in \mathcal{N}$,

$$fR_d f^{-1}(nC(\mathcal{N})\Gamma, z)$$
$$= (dnC(\mathcal{N})\Gamma, z(ds(nC(\mathcal{N})\Gamma)z\Gamma))$$
$$= (dnC(\mathcal{N})\Gamma, z(ds(nC(\mathcal{N})\Gamma)\Gamma)z)$$

is an Abelian group extension of $(R_d, \mathcal{N}/C(\mathcal{N})\Gamma)$ with generating function

$$\phi_d(nC(\mathcal{N})\Gamma) = z(ds(nC(\mathcal{N})\Gamma)\Gamma)$$
$$= z(dnz(n)^{-1}\gamma(n)^{-1}\Gamma)$$
$$= z(dn)z(n)^{-1}.$$

Notice this is not a d-coboundary as z is not well-defined on $\mathcal{N}/C(\mathcal{N})\Gamma$ but only on $\mathcal{N}/\Gamma$. The value $z(dn)z(n)^{-1}$ depends only on the coset $nC(\mathcal{N})\Gamma$. This is true for all values $d \in \mathcal{N}$. Hence for any $c \in \mathcal{N}$,

$$\phi_{n_0}(cnC(\mathcal{N})\Gamma)\phi_{n_0}(nC(\mathcal{N})\Gamma)^{-1} = z(n_0cn)z(cn)^{-1}z(n_0n)^{-1}z(n).$$

Now $n_0c = cn_0k(c)$ for some $k(c) \in C(\mathcal{N})$, and we further compute

$$= k(c)(z(cn_0n)z(n_0n)^{-1})(z(cn)z(n)^{-1})^{-1}.$$

For any character χ of $C(\mathcal{N})$, setting

$$\lambda(c) = \chi(k(c))$$

and

$$f(c, n) = \chi(z(cn)z(n)^{-1}),$$

we get

$$\frac{\chi(\phi(cnC(\mathcal{N})\Gamma))}{\chi(\phi(nC(\mathcal{N})\Gamma))} = \lambda(c)\frac{f(c,n_0n)}{f(c,n)}.$$

The computation we went through in proving this last theorem points out the connection between the C-L equation and nilpotent behavior and indicates how we should proceed to show that in fact the C-L algebra must come from order-two nil-factors.

Our basic problem is to find an order-two nilpotent group of which $R_{g_0,\phi}$ is an element, acting transitively on $G \times G_1$. We will accomplish this for G_1 a finite-dimensional torus. We know that in general G_1 is an inverse limit of such tori, and so T acting on $\mathcal{L}(T)$ is the increasing span of factor algebras on which it acts as an order two nil-rotation. It is not until the next section, where we see more precisely what C-L actions can look like that we will see the nature of the inverse limit of groups G_1. We will see there that even though the groups G_1 project backward as an inverse limit, the corresponding order-two nilpotent groups will not. The projections will not necessarily be onto, and without more precise information we will not know that any particular element lifts backward to some element of the inverse limit.

To see where the nilpotent group comes from, notice that $(R_{g_0,\phi}, G \times G_1)$ is just one element of a large group of maps. Let

$$\mathcal{G} = \{R_{c,f} \mid c \in G, \ f : G \to G_1\},$$

a separable metric group topologized by

$$d(R_{c,f}, R_{c',f'}) = \|c,c'\| + \|ff'^{-1}\|_2,$$

where $\|c,c'\|$ is some invariant metric on G.

If $G_1 = \mathbb{T}^n$, then regard $\mathbb{T}^n = (S^1)^n \subseteq \mathbb{C}^n$, with metric normalized so all points of $\mathbb{T}^n$ have length 1.

Definition 3.2. If G and G_1 are compact Abelian metric groups, denote by $\mathrm{Hom}(G, G_1)$ the set of all group homomorphisms of G into G_1 metrized by $d(H_1, H_2) = \|H_1 H_2^{-1}\|_2$.

Lemma 3.3. $\mathrm{Hom}(G, \mathbb{T}^n)$ *is at most countable and is discrete.*

Proof. An element in $\mathrm{Hom}(G, \mathbb{T}^n)$ is determined by choosing n characters of G to become the n coordinates of $\mathbb{T}^n$. There are at most countably many such choices (all are in fact in $\mathrm{Hom}(G, \mathbb{T}^n)$.) If $H_1 \neq H_2$ then they differ on some coordinate, and

$$d(H_1, H_2) \geq \|\chi_i - \chi_j\|_2 = 2/\sqrt{n}.$$

Definition 3.3. For $(R_{g_0,\phi}, G \times G_1)$ a C-L action, let

$$F(c) = \{f : G \to G_1 \mid \phi(cg)\phi(g)^{-1} = \lambda f(g_0 g) f(g)^{-1}\}$$

and

$$F^{c_1}(c) = \{f^{c_1} \mid f^{c_1}(g) = f(c_1 g), \text{ where } f \in F(c)\}.$$

Lemma 3.4. *If $F(c) \neq \emptyset$, then for any fixed $f_0 \in F(c)$,*

$$F(c) = f_0 G_1 \mathrm{Hom}(G, G_1);$$

i.e., any $f \in F(c)$ is of the form $\gamma f_0 H$, where $\gamma \in G_1$ and $H \in \mathrm{Hom}(G, G_1)$.
This is topologized as $G_1 \times \mathrm{Hom}(G, G_1)$. Further, if $F(c_1)$ and $F(c_2) \neq \emptyset$ then

$$F(c_1 c_2) = F(c_1) F^{c_1}(c_2).$$

Proof. Supposing $F(c) \neq \emptyset$, let f_1, $f_2 \in F(c)$.
Then

$$\lambda_1 \lambda_2^{-1} (f_1(g_0 g) f_2^{-1}(g_0 g)) (f_1(g) f_2^{-1}(g))^{-1} = 1,$$

or

$$(f_1(g_0 g) f_2^{-1}(g_0 g)) = \lambda_1^{-1} \lambda_2 (f_1(g) f_2^{-1}(g)).$$

As g_0 acts ergodically, for any $g' \in G$, choose $g_0^{n_i} \to g'$ and

$$f_1(g_0^{n_i} g) f_2^{-1}(g_0^{n_i} g) f_1^{-1}(g) f_2(g) \longrightarrow f_1(g' g) f_2^{-1}(g' g) f_1^{-1}(g) f_2(g) \text{ a.s. in } g.$$

So

$$\lambda_1^{-n_i} \lambda_2^{n_i} \longrightarrow f_1(g' g) f_2^{-1}(g' g) f_1^{-1}(g) f_2(g)$$

for a.e. g and is independent of g. Set

$$H(g') = f_1(g' g) f_2^{-1}(g' g) f_1^{-1}(g) f_2(g),$$

and it is a computation that $H \in \mathrm{Hom}(G, G_1)$.
Now

$$f_1(g) f_2^{-1}(g) = H(g) f_1(1) f_2^{-1}(1) = \gamma H(g).$$

If $f_0 \in F(c)$, it is an easy check that $\gamma f_0 H$ is also.
Now if $f_1 \in F(c_1)$ and $f_2 \in F(c_2)$, then

$$\phi(c_1 c_2 g) \phi^{-1}(g) = (\phi(c_1 c_2 g) \phi^{-1}(c_1 g))(\phi(c_1 g) \phi^{-1}(g))$$
$$= \lambda_1 \lambda_2 (f_1(g_0 g) f_2(c_1 g_0 g)) (f_1(g) f_2(c_1 g))^{-1}.$$

If $G_1 = \mathbb{T}^n$ and $f_0 \in F(c)$, then for any f_1, $f_2 \in F(c)$ we can compute

$$d(f_1, f_2) = \|\gamma_1 H_1 - \gamma_2 H_2\|_2 = \begin{cases} |\gamma_1 - \gamma_2|, & \text{if } H_1 = H_2 \\ \geq \dfrac{2}{\sqrt{n}}, & \text{if } H_1 \neq H_2. \end{cases}$$

This estimate will play a critical role in Section 4.
We will quite often be considering a function $f : G \times G \to G_1$ as a function of its second variable, when its first is fixed at a value c. We will write this as $f(c)$, by which we mean $f(c)(g) = f(c, g)$.

Definition 3.4. We say $(R_{g_0,\phi}, G \times G_1)$ has a *C-L section* if all $F(c) \neq \emptyset$ and there is a measurable section $f : G \times G \to G_1$ so that for all $c \in G$, $f(c) \in F(c)$.

If $\mathcal{N}$ is an order-two nilpotent group, Γ a cocompact subgroup, as in Theorem 3.1, notice

$$f(d_1 C(\mathcal{N})\Gamma, d_2 C(\mathcal{N})\Gamma) = \phi_{s(d_1 C(\mathcal{N})\Gamma)}(d_2 C(\mathcal{N})\Gamma)$$

is a C-L section for the action. We want to prove the converse of this fact, that the existence of a C-L section forces a C-L action to be just one element of an order-two nilpotent group of actions.

Lemma 3.5. *If $(R_{g_0,\phi}, G \times G_1)$ is an ergodic C-L action on $G \times G_1$, then factoring onto any finite-dimensional toral factor of G_1, we get an extension with a C-L section.*

Proof. Remember we know G_1 must be toral. Writing the finite-dimensional toral factor as

$$\vec{\chi} = (\chi_1, \chi_2, \ldots, \chi_n),$$

we know

$$\vec{\chi}(\phi(cg))\vec{\chi}(\phi^{-1}(g)) = \vec{\lambda}\vec{f}(c, g_0 g)\vec{f}^{-1}(c, g),$$

and $\vec{f}$ is a C-L section.

It is easy to see that if G_1 is a direct product of perhaps countably many groups each of which has a C-L section, then so does G_1. What is not true is that if G_1 is an inverse limit of groups with C-L sections, then G_1 also has one. The reason is that as we work back through the inverse limit, elements of $\mathrm{Hom}(G, G_{1,i})$ need not lift back to $\mathrm{Hom}(G, G_{1,i+1})$, and so the C-L section may not lift. In Section 5 we will see that in fact this can happen and a C-L action need not be a nil-rotation. For now we will work with $G_1 = \mathbb{T}^n$ and identify the order-two nilpotent group.

Lemma 3.6. *Suppose $(R_{g_0,\phi}, G \times \mathbb{T}^n)$ has a C-L section f_0. Given any $\varepsilon > 0$ there is an $r > 0$ and a subset $E \subseteq N_r(1)$ with $m(E) \geq (1 - \varepsilon)m(N_r(1))$ so that for all $c_0 \in G$ there is a C-L section f_{c_0} so that for any $c_1 \in E$*

$$\|f_{c_0}(c_1 c_0) - f_{c_0}(c_0)\|_2 < \varepsilon.$$

(Remember: f_0, f_{c_0} are functions of two variables and $f(c)(g) = f(c, g)$.)

Proof. That f_0 is measurable implies for any $\varepsilon > 0$ there is a point $\bar{c} \in G$ and $r > 0$ and a subset $E \subseteq N_r(1)$ with $m(E) \geq (1 - \varepsilon)m(N_r(1))$ so that if $c_1 \in E$ and $c_2 \in N_r(1)$, then

$$\|f_0^{c_2}(c_1 \bar{c}) - f_0(\bar{c})\|_2 < \varepsilon/2$$

and

$$\|f_0^{c_2}(c_1 \bar{c}^{-1}) - f_0(\bar{c}^{-1})\|_2 < \varepsilon/2.$$

Now

$$F(c) = F(c_0) F^{c_0}(c_0^{-1} c\bar{c}) F^{c\bar{c}}(\bar{c}^{-1}),$$

and in particular

$$f_{c_0}(c, g) = f_0(c_0, g) f_0(c_0^{-1} c\bar{c}, c_0 g) f_0(\bar{c}^{-1}, c\bar{c}g)$$

is a C-L section.

For $c_1 \in E$, $c_2 \in N_r(1)$,

$$\begin{aligned}
\|f_{c_0}(c_1 c_0) - f_{c_0}(c_0)\|_2 &= \|f_0(c_0, g) f_0^{c_0}(c_1 \bar{c}) f_0^{c_1 c_0 \bar{c}}(\bar{c}^{-1}) \\
&\quad - f_0(c_0) f_0^{c_0}(\bar{c}) f_0^{c_0 \bar{c}}(\bar{c}^{-1})\|_2 \\
&= \|f_0^{c_0}(c_1 \bar{c}) f_0^{c_1 c_0 \bar{c}}(\bar{c}^{-1}) - f_0^{c_0}(\bar{c}) f_0^{c_0 \bar{c}}(\bar{c}^{-1})\|_2 \\
&\le \|f_0^{c_0}(c_1 \bar{c}) f_0^{c_1 c_0 \bar{c}}(\bar{c}^{-1}) - f_0^{c_0}(\bar{c}) f_0^{c_1 c_0 \bar{c}}(\bar{c}^{-1})\|_2 \\
&\quad + \|f_0^{c_0}(\bar{c}) f_0^{c_1 c_0 \bar{c}}(\bar{c}^{-1}) - f_0^{c_0}(\bar{c}) f_0^{c_0 \bar{c}}(\bar{c}^{-1})\|_2 \\
&= \|f_0(c_1 \bar{c}) - f_0(\bar{c})\|_2 + \|f_0^{c_1}(\bar{c}^{-1}) - f_0(\bar{c}^{-1})\|_2 \\
&\le \varepsilon.
\end{aligned}$$

Lemma 3.7. *Suppose $(R_{g_0,\phi}, G \times \mathbb{T}^n)$ has a C-L section. If $c_i \to c$ in G and $f_i \in F(c_i)$, then there are $\gamma_i \in \mathbb{T}^n$ and $H_i \in \mathrm{Hom}(G, \mathbb{T}^n)$ so that $\gamma_i H_i f_i \to f \in F(c)$.*

Proof. Given $\varepsilon > 0$, by Lemma 3.6 there is an $r > 0$ and set $E \subseteq N_r(1), m(E) > (1-\varepsilon)m(N_r(1))$. Once i, j are sufficiently large $m(c_i c_j^{-1} E \cap E) > 0$, i.e. there are $c, c' \in E$ with $c_i c = c_j c'$. Now

$$\|f_{c_i}(c_i c) - f_{c_i}(c_i)\|_2 < \varepsilon$$

and

$$\|f_{c_j}(c_j c') - f_{c_j}(c_j)\|_2 < \varepsilon.$$

Further, for i, j fixed,

$$f_{c_i}(c_i c) f_{c_j}^{-1}(c_j c') = \gamma_0 H_0,$$

$$f_i^{-1} f_{c_i}(c_i) = \gamma_1 H_1,$$

and

$$f_j^{-1} f_{c_j}(c_j) = \gamma_2 H_2.$$

Thus

$$\|f_i - \gamma_1^{-1} \gamma_2 \gamma_0 H_1^{-1} H_2 H_0 f_j\|_2 < 2\varepsilon.$$

A simple inductive construction now gives a Cauchy sequence in $L^2(m)$ of the form $\hat{f}_i = \gamma_i H_i f_i$ which converges to $f : G \to \mathbb{T}^n$.

As

$$\phi(c_i g)\phi(g)^{-1} = \lambda_i f_i(g_0 g) f_i^{-1}(g) = \lambda_i H_i(g_0) \hat{f}_i(g_0 g) \hat{f}_i^{-1}(g)$$

and $\mathbb{T}^n$ is compact, we can assume $\lambda_i H_i(g_0) \to \lambda$ and

$$\phi(cg)\phi^{-1}(g) = \lambda f(g_0 g) f^{-1}(g)$$

and $f \in F(c)$.

Assuming $(R_{g_0,\phi}, G \times \mathbb{T}^n)$ has a C-L section, define a subgroup

$$\mathcal{N} = \{R_{c,f} \in \mathcal{G} \mid f \in F(c)\}.$$

That it is a subgroup follows from

$$R_{c_1,f_1} R_{c_2,f_2} = R_{c_1 c_2, f_1^{c_2} f_2}.$$

The argument at the very end of Lemma 3.7 shows it is closed.

Theorem 3.8. *Suppose $(R_{g_0,\phi}, G \times \mathbb{T}^n)$ is an ergodic C-L action on $G \times \mathbb{T}^n$. The group $\mathcal{N}$ is a locally compact, separable order-two nilpotent group. Its center is*

$$C(\mathcal{N}) = \{R_{1,\gamma} \mid \gamma \in \mathbb{T}^n\}.$$

It acts transitively on $G \times \mathbb{T}^n$ and the isotropy subgroup of (1,1) is

$$\Gamma = \{R_{1,H} \mid H \in \mathrm{Hom}(G, \mathbb{T}^n)\}.$$

Proof. We first identify the center. Suppose $R_{c,f} \in C(\mathcal{N})$. Then for all $f_1 \in F(c_1)$,

$$f(c_1 g) f^{-1}(g) = f_1(cg) f_1^{-1}(g),$$

and in particular for $c_1 = 1, f_1 = H \in \mathrm{Hom}(G, \mathbb{T}^n)$; we conclude $H(c) = 1$ for all such H, and so $c = 1$. But now $f(c_1 g) f^{-1}(g) = 1$ for all c_1, and $f = \gamma$ is a constant. Thus $C(\mathcal{N})$ is $\{R_{1,\gamma} \mid \gamma \in \mathbb{T}^n\}$.

For local compactness, suppose

$$d(R_{c_i,f_i}, R_{c_j,f_j}) < \frac{1}{2n}.$$

First we can assume $c_i \to c$. By Lemma 3.7 we have γ_i, H_i with $f_i \gamma_i H_i \to f$, $f \in F(c)$. Again we can assume $\gamma_i \to \gamma$ and $f_i H_i \to \gamma^{-1} f$.

But now as $\|f_i - f_j\|_2 < 1/2n$, and if $H_i \neq H_j$,

$$\|f_i H_i - f_j H_j\|_2 \geq \|H_i - H_j\|_2 - \|f_i - f_j\|_2 > \frac{1}{2n},$$

which is not true, implying H_i must be asymptotically constant. Hence on a subsequence R_{c_i, f_i} converges, and $\mathcal{N}$ is locally compact.

Consider the subgroup generated by the powers of $R_{g_0, \phi}$, $\mathrm{Hom}(G, \mathbb{T}^n)$, and $\mathbb{T}^n$,

$$\mathcal{N}_0 = \{R_{ng_0, H\gamma\phi^{(n)}} \mid n \in \mathbb{Z}, H \in \mathrm{Hom}(G, \mathbb{T}^n), \gamma \in \mathbb{T}^n\}.$$

Notice for any $R_{c,f} \in \mathcal{N}$,

$$R_{ng_0, H\gamma\phi^{(n)}} R_{c,f} = R_{1, H(c)\lambda^n} R_{c,f} R_{ng_0, H\gamma\phi^{(n)}},$$

i.e.

$$\mathcal{N}_0 / C(\mathcal{N}) \subseteq C(\mathcal{N}/C(\mathcal{N})).$$

But $\mathcal{N}_0$ is dense in $\mathcal{N}$, so $\mathcal{N}/C(\mathcal{N})$ is Abelian and $\mathcal{N}$ is a locally compact, separable order-two nilpotent group.

That it contains all $R_{1,\gamma}$ and all $R_{c, f(c, \cdot)}$ tells us it acts transitively on $G \times \mathbb{T}^n$. The isotropy subgroup is Γ.

Corollary 3.9. *Suppose $(R_{g_0, \phi}, G \times G_1)$ is a C-L action on $G \times G_1$ and has a C-L section $f : G \times G \to G_1$. Setting*

$$F(c) = \{g : G \to G_1 \mid g = \gamma H f(c), \gamma \in G_1, H \in \mathrm{Hom}(G, G_1)\}$$

and now

$$\mathcal{N} = \{R_{c,g} \mid g \in F(c)\},$$

$\mathcal{N}$ is an order-two nilpotent group, acting transitively on $G \times G_1$, and as the isotropy subgroup of $(1, 1)$ is $\mathrm{Hom}(G, G_1)$,

$$(R_{g_0, \phi}, G \times G_1) \simeq (R_{N_0}, \mathcal{N}/\mathrm{Hom}(G, G_1)),$$

where $N_0 = R_{g_0, \phi}$ is an element of $\mathcal{N}$.

Proof. We know that G_1 is an inverse limit of finite-dimensional tori, on each of which the conclusion of this theorem holds. In particular there are compact subgroups $K_i \subseteq G_1$ so that $K_{i+1} \subseteq K_i$, $\cap_i K_i = \{1\}$ and G_1/K_i is a finite-dimensional torus. Now the C-L section f factors to a C-L section for each extension by G_1/K_i of G. Thus

$$\mathcal{N}_i = \{R_{c,g} \mid g = \gamma H(f(c)K_i), \gamma \in G_1, H \in \mathrm{Hom}(G, G_1/K_i)\}$$

is a locally compact metric order-two nilpotent group by Theorem 3.8.

Now any element $H \in \mathrm{Hom}(G, G_1)$ projects to an element $HK_i \in \mathrm{Hom}(G, G_1/K_i)$, so consider the closed subgroups

$$\mathcal{N}_i' = \{R_{c,g} \mid g = \gamma H f(c)K_i, \gamma \in G_1, H \in \mathrm{Hom}(G, G_1)\} \subseteq \mathcal{N}_i.$$

Each is an order-two nilpotent group, and each is the factor action of $\mathcal{N}$ acting on the quotient space $G \times G_1 / K_i$. Thus for any $A, B \in \mathcal{N}$,

$$ABA^{-1}B^{-1} = R_{1,\lambda_i(A,B)}{}^{k_i}$$

for some $k_i \in K_i$. Along a subsequence the values $\lambda_i(A, B)$ can be chosen to converge, hence the values k_i also converge. But they can only converge to 1, and hence $\mathcal{N}$ is order-two nilpotent. As no $F(c)$ is empty, $\mathcal{N}$ acts transitively on $G \times G_1$, and so $G \times G_1$ must be a homogeneous space of the locally compact group.

This Corollary has made precise our statements earlier that the remaining obstacle to the existence of a maximal order-two nil-rotation factor was the question of whether a C-L section could be lifted through an inverse limit of tori.

What we have now seen is that if $R_{g_0,\phi}$ is a Conze-Lesigne action on $G \times G_1$ and has a C-L section, then it is actually just one particular element of an order-two nilpotent group $\mathcal{N}$ of transformations. In trying to understand exactly what such extensions look like there is no reason to distinguish this particular element $R_{g_0,\phi}$. It is more natural instead to speak of a C-L extension of G_0 given by a function

$$f_0 : G \times G \to G_1$$

satisfying the C-L equation

$$\left(\frac{f_0(c_1, c_2 x)}{f_0(c_1, x)} \right) \left(\frac{f_0(c_2, c_1 x)}{f_0(c_2, x)} \right)^{-1} = \lambda(c_1, c_2).$$

The elements of $\mathcal{N}$ are then of course the group of all transformations of the form

$$\{ R_{c,f} \mid c \in G, \; f(x) = f_0(c, x) H(x) \gamma, \; H \in \mathrm{Hom}(G_0, G_1) \, \gamma \in G_1 \}.$$

Corollary 3.9 tells us that $\mathcal{N}$ must be an order-two nilpotent group, and that f_0 is a C-L section for it.

To return to an actual transformation we simply choose some ergodic element of $\mathcal{N}$. Our particular interest of course will be when (R_{g_0}, G) is the Kronecker algebra, but this is not essential and not necessarily universal in the nilpotent group.

Definition 3.3. We will call such a group of transformations $\mathcal{N}$ a *C-L extension* of G_0. If we want to stress the structure of the second coordinate, we will call it a *C-L extension of G_0 by G_1*.

In the next two sections we will investigate more precisely the structure of C-L actions and C-L extensions. We will find all C-L extensions by finite-dimensional tori. This will tell us precisely when the extensions lift to the inverse limit. Moreover we will associate invariants with each C-L character χ which will determine $\chi \circ \phi$ up to cohomology to a constant.

4. δ-HOMOTOPY AND QUASI-ABELIAN EXTENSIONS

In this section, building on our work of the previous section, we will identify invariants of C-L extensions. Let G be a compact, metric, Abelian and monothetic group. We will show precisely what all C-L extensions of G by finite-dimensional tori are by investigating a δ-fundamental group of $\mathcal{N}/C(\mathcal{N})$. This will lead to our invariants and to an understanding of when such finite-dimensional extensions can project to an inverse limit. In particular we will associate to each C-L character a skew-symmetric bilinear form (matrix) on the universal cover of some finite-dimensional toral factor of G. As we work back through the inverse limit to G_1, we will build up a larger and larger vector of matrices over the terms of an inverse limit of tori. tori in G satisfying certain natural equivariance rules of evolution. We will obtain from this a number of facts. First, if all the bilinear forms are trivial, then the action on the full C-L algebra is an order-two nil-rotation. These are what we will call the *quasi-Abelian Conze-Lesigne actions*, as they correspond to Abramov's notion of quasi-discrete spectrum. Second, we will see that whether the Conze-Lesigne algebra is an order-two nil-rotation is a property solely of the array of bilinear forms; i.e., if we have two C-L actions with the same array and for one the C-L algebra is an order-two nil-rotation, then so is the other. Third, we will see that for any array of bilinear forms satisfying the equivariance rules of evolution, there is a C-L action with this as its array.

This will then allow us to exhibit in the next section a C-L action which is not an order-two nil-rotation.

To begin, consider the following two examples of C-L extensions.

Let $\overline{G}$ be a compact, Abelian, metric and monothetic group, and $G_1 \subseteq \overline{G}$ a toral subgroup. Let $c : \overline{G}/G_1 \to \overline{G}$ be a choice of coset representative. Now for g_0 a generator of $G = \overline{G}/G_1$, and $\phi(g) = c(g_0)c(g)c(g_0 g)^{-1}$, the transformation $R_{g_0,\phi}$ has pure point spectrum, being isomorphic to the rotation $R_{c(g_0)}$. In fact, the map $\psi(\overline{g}) = (\overline{g}G_1, \overline{g}c(\overline{g}G_1)^{-1})$ conjugates $\overline{G}$ to a compact Abelian group of extensions of G by G_1.

Now consider the order-two nilpotent group of all maps of the form

$$\mathcal{N} = \{ R_{1,H}(\psi R_{\overline{g}} \psi^{-1}) : \overline{g} \in \overline{G},\ H \in \operatorname{Hom}(G, G_1) \}.$$

If there exists an $H \in \operatorname{Hom}(G, G_1)$ that is onto, then there will be an element $R_{g_0,\phi} \in \mathcal{N}$ with (R_{g_0}, G) as its Kronecker algebra. We can see this by simply showing that any character of G_1 has continuous spectrum for this action. This reduces the problem to assuming $G_1 = S^1$. In this case, Lemma 1.9 tells us in fact ϕH is cohomologous to $\gamma \chi$, where $\chi \in \hat{G}$ is onto S^1. But now, projecting to a factor action by χ on the first coordinate, we are looking at the classical $(z, z_1) \to (\alpha z, \gamma z z_1)$ action, where it is well known that the second-coordinate function has continuous spectrum (this is the classical example of quasi-discrete spectrum.)

In order for there to be an $H \in \operatorname{Hom}(G, G_1)$ that is onto, G must possess a toral factor that has a version of G_1 as a factor. If G_1 is too big for this, then every element of $\mathcal{N}$ will have a Kronecker algebra larger than its first coordinate.

Notice that for this example $f(g, g') = c(g)c(g')c(gg')^{-1}$ is a C-L section, so that the lift of G to this section generates a compact Abelian subgroup of $\mathcal{N}$.

We will call an extension of this form a *quasi-Abelian (q-A) extension*. It simply extends in a trivial way a compact, Abelian, metric and monothetic group. Notice from Lemma 2.10 that if we have a q-A extension of G by $\mathbb{T}^n$, then $f \equiv 1$ is a C-L section for the extension. This of course is not true of all q-A extensions, as arbitrary toral subgroups do not necessarily have complementary subgroups.

For a second example, we know from the last section that C-L extensions come from order-two nilpotent groups, and from such a group one can get another example. Let $D : \mathbb{R}^n \times \mathbb{R}^n \to \mathbb{R}$ be bilinear and define a group action on $\mathbb{R}^n \times S^1$ by

$$(\vec{u}_1, z_1)(\vec{u}_2, z_2) = (\vec{u}_1 + \vec{u}_2, \exp(2\pi i D(\vec{u}_1, \vec{u}_2)) z_1 z_2).$$

It is a computation that this is a group with

$$(\vec{u}, z)^{-1} = (-\vec{u}, \exp(2\pi i D(\vec{u}, \vec{u})) z^{-1}).$$

We call this group $\mathcal{H}_D$.

One computes that

$$(\vec{u}_1, z_1)(\vec{u}_2, z_2)(\vec{u}_1, z_1)^{-1}(\vec{u}_2, z_2)^{-1}$$
$$= (\vec{0}, \exp(2\pi i (D(\vec{u}_1, \vec{u}_2) - D(\vec{u}_2, \vec{u}_1)))),$$

which is in $C(\mathcal{H}_D)$.

Notice $D(\vec{u}_1, \vec{u}_2) - D(\vec{u}_2, \vec{u}_1) = B(\vec{u}_1, \vec{u}_2)$ is a skew-symmetric bilinear form.

Thus this group is (at worst) order-two nilpotent.

Suppose Γ is a lattice subgroup of $\mathbb{R}^n$ with $D(\gamma_1, \gamma_2) \in \mathbb{Z}$ for all $\gamma_i \in \Gamma$. Then

$$\Gamma_0 = \{(\gamma, 1) \mid \gamma \in \Gamma\}$$

is a closed cocompact subgroup of $\mathcal{H}_D$.

Notice $(\vec{u}, z)(\gamma, 1) = (\vec{u} + \gamma, \exp(2\pi i D(\vec{u}, \gamma)) z)$. Suppose $F \subseteq \mathbb{R}^n$ is a fundamental domain for $\Gamma \in \mathbb{R}^n$, i.e. there is a 1-1 Borel map $c_0 : \mathbb{R}^n / \Gamma \to F$ with

$$c_0(\vec{u} + \Gamma) = \vec{u} - \gamma_F(\vec{u}) \in F,$$

where $\gamma_F(\vec{u})$ translates $\vec{u}$ into F.

Then $c : \mathcal{H}_D / \Gamma_0 \to F \times S^1$ given by

$$c((\vec{u}, z)\Gamma_0) = (c_0(\vec{u} + \Gamma), \exp(-2\pi i D(\vec{u}, \gamma_F(\vec{u})z)$$

is a bijection.

The map c gives us a choice of coset representative, the element in $F \times S^1 \subseteq \mathbb{R}^n \times S^1$. Hence, as in Theorem 3.1, it gives us a C-L section for the action of $\mathcal{H}_D$ on $\mathcal{H}_D / \Gamma$ viewed as $\mathbb{R}^n / \Gamma \times S^1$. In particular,

$$f_{F,\Gamma,D}(\vec{u} + \Gamma, \vec{v} + \Gamma) = \exp\big(2\pi i \big(D(\vec{u}, \vec{v}) + D(\vec{v}, \gamma_F(\vec{v}))$$
$$+ D(\vec{u}, \gamma_F(\vec{u})) + D(\vec{v}, \gamma_F(\vec{u})) - D(\gamma_F(\vec{u}), \vec{v}) - D(\vec{u} + \vec{v}, \gamma_F(\vec{u} + \vec{v})))\big) =$$
$$\exp 2\pi i \big(D(\vec{u}, \vec{v}) + B(\vec{v}, \gamma_F(\vec{u})) + D(\vec{v}, \gamma_F(\vec{v})) - D(\vec{u} + \vec{v}, \gamma_F(\vec{u} + \vec{v}))\big)$$

is a C-L section. Now $D(\vec{v}, \gamma_F(\vec{v}))$ is not well-defined mod Γ, so $f_{F,\Gamma,D}(\vec{u}, \cdot)$ is not $\vec{u}$-cohomologous to $D(\vec{u}, \cdot) + B(\cdot, \gamma_F(\vec{u}))$, but it almost is. In particular, if $\gamma_F(\vec{u}) = \vec{0}$ then it is almost $\vec{u}$-cohomologous to linear.

To be even more specific, as Γ is a lattice subgroup, there is a basis $[\vec{e}_1, \vec{e}_2, \ldots, \vec{e}_n]$ for $\mathbb{R}^n$ of elements which form a generating set for Γ_0.

Write $\vec{u} = \sum_j u_j \vec{e}_j$, $\vec{v} = \sum_j v_j \vec{e}_j$, and $D_{[e]} = [d_{s,t}]$ for the representations of these in this basis. That $D(\gamma_1, \gamma_2) \in \mathbb{Z}$ implies $D_{[e]}$ is an integer matrix. Choose for fundamental domain

$$F_{[e]} = \{\sum_j u_j \vec{e}_j \mid 0 \le u_j < 1\},$$

and now

$$f_{F_{[e]}, \Gamma, D} = \exp\left(2\pi i \sum_{s,t} d_{s,t}(\langle u_s \rangle \langle v_t \rangle - [\langle u_t \rangle + \langle v_t \rangle]\langle u_s + v_s \rangle)\right),$$

where $\langle u \rangle$ is the fractional part of u and $[u]$ is its integer part.

We call examples of the form of $\mathcal{H}_D / \Gamma$ *Heisenberg* examples generalizing from the 3-dimensional Heisenberg group of matrices

$$\begin{bmatrix} 1 & a & c \bmod 1 \\ 0 & 1 & b \\ 0 & 0 & 1 \end{bmatrix}.$$

Lemma 4.1. *For Γ a lattice subgroup of $\mathbb{R}^n$, F a fundamental domain for Γ and D a bilinear form on $\mathbb{R}^n$ with $D(\gamma_1, \gamma_2) \in \mathbb{Z}$ for all γ_1 and $\gamma_2 \in \Gamma$, $f_{F,\Gamma,D}$ is a C-L section from $\mathbb{R}^n / \Gamma_0 \times \mathbb{R}^n / \Gamma_0$ to S^1. In particular*

$$f_{F,\Gamma,D}(\vec{u}, \vec{v} + \vec{w})\big(f_{F,\Gamma,D}(\vec{u}, \vec{w})\big)^{-1} =$$
$$\exp\big(2\pi i(D(\vec{u}, \vec{v}) - D(\vec{v}, \vec{u}))\big) f_{F,\Gamma,D}(\vec{v}, \vec{u} + \vec{w})\big(f_{F,\Gamma,D}(\vec{v}, \vec{w})\big)^{-1}.$$

Proof. A computation.

When $D = D^T$, the group $\mathcal{H}_D$ is Abelian, and so all its actions on $\mathcal{H}_D/\Gamma$ have pure point spectrum. By Lemma 2.10, $f_{F,\Gamma,D}(\vec{u})$ must be $\vec{u}$-cohomologous to a constant, and the extension is q-A. If $D \neq D^T$, the connected component of the identity in $\mathcal{H}_D$ will not be compact. It is via this connected component that we will analyze C-L extensions in general, leading us to the conclusion that they all arise, basically, from these two constructions.

In terms of this last example, our basic problem in this section is, for each C-L character of a C-L action, to find a finite-dimensional toral factor of G and a skew-symmetric bilinear form over this factor so that the extension by this one character is conjugate to the corresponding Heisenberg extension.

We begin with two lemmas. Notice that if $\mathcal{N}$ is a C-L extension of G by G_1, then for any $A, B \in \mathcal{N}$, $ABA^{-1}B^{-1} = R_{1,\lambda(A,B)} \in C(\mathcal{N})$. The map

$$\lambda : \mathcal{N} \times \mathcal{N} \to G_1$$

is a constant on cosets of $C(\mathcal{N})$, hence actually maps

$$\mathcal{N}/C(\mathcal{N}) \times \mathcal{N}/C(\mathcal{N}) \to G_1.$$

Lemma 4.2. *The map*

$$\lambda : \mathcal{N}/C(\mathcal{N}) \times \mathcal{N}/C(\mathcal{N}) \to G_1$$

is a skew-symmetric bi-homomorphism in that fixing either coordinate make it a group homomorphism in the other and, $\lambda(A,B) = \lambda(B,A)^{-1}$.

Proof. As $\mathcal{N}$ is an order-two nilpotent group, we know, for $A, B, C \in \mathcal{N}$,

$$ABA^{-1}B^{-1}ACA^{-1}C^{-1} = A(BC)A^{-1}(BC)^{-1},$$

so $\lambda(A,B)\lambda(A,C) = \lambda(A,BC)$.

Furthermore $(ABA^{-1}B^{-1})^{-1} = BAB^{-1}A^{-1}$, so $\lambda(A,B)^{-1} = \lambda(B,A)$.

If $K \subseteq G$ is a closed subgroup and $\mathcal{N}_0$ is a C-L extension of G/K by G_1, then $\mathcal{N}_0$ lifts to a C-L extension of G by G_1 simply by taking

$$\mathcal{N} = \{R_{c,f} \mid c \in G, \ f(g) = f_0(gK)H(g)\gamma, \text{ where}$$
$$R_{cK,f_0} \in \mathcal{N}_0, H \in \mathrm{Hom}(G, G_1), \gamma \in G_1\}.$$

If $\mathcal{N}$ is cohomologous to such a lift, we simply say that it *lifts from a C-L extension of* G/K.

Lemma 4.3. *Suppose* $\mathcal{N}$ *is a C-L extension of* G *by* G_1, $K \subseteq G$ *is a closed subgroup, and there is a homomorphic lift* $L : K \to \mathcal{N}$, $L(g) = R_{g,f(g)}$, $L(g_1 g_2) = L(g_1)L(g_2)$.
Then $\mathcal{N}$ *is the lift to* G *of a C-L extension of* G/K *by* G_1.

Proof. Choose coset representatives $c(gK) \in gK$ measurably with $c(K) = 1$. Let f_1 be a C-L section with $f_1(1, \cdot) = 1$. Define a new C-L section

$$f_2(g_1, g_2) = f_1(g_1 c(g_1 K)^{-1}, c(g_1 K)g_2)f_1(c(g_1 K), g_2),$$

which is still f_1 for $g_1 \in K$.

Define $h(g) = f_2^{-1}(g, 1)$ and modify $\mathcal{N}$ to the cohomologous

$$\mathcal{N}' = \{R_{g,f'} \mid f'(g_1) = f(g_1)h(gg_1)h(g_1)^{-1}, \ R_{g,f} \in \mathcal{N}\}.$$

This has

$$f'(g_1, g_2) = f_2(g_1, g_2)h(g_1 g_2)h(g_2)^{-1}$$

as a C-L section.

Notice for $g_1 \in K$,

$$\begin{aligned}
f'(g_1, g_2) &= f_2(g_1, g_2)\frac{h(g_1 g_2)}{h(g_2)} \\
&= f_1(g_1, g_2)\frac{f_2(g_2, 1)}{f_2(g_1 g_2, 1)} \\
&= f_1(g_1, g_2)\frac{f_1(g_2 c(g_2 K)^{-1}, c(g_2 K))}{f_1(g_1 g_2 c(g_2 K)^{-1}, c(g_2 K))} \\
&= f_1(g_1, g_2)\frac{1}{f_1(g_1, g_2)} = 1.
\end{aligned}$$

But now for arbitrary g_1, and $c \in K$,

$$\left(\frac{f'(g_1, cg)}{f'(g_1, g)}\right)\left(\frac{f'(c, gg_1)}{f'(c, g)}\right)^{-1} = \frac{f'(g_1, cg)}{f'(g_1, g)} = H_g(c),$$

where Lemma 4.2 tells us $H_g \in \mathrm{Hom}(K, G_1)$.

As G_1 is toral, H_g can be extended to an element of $\mathrm{Hom}(G, G_1)$. Define a new C-L section

$$f_1'(g_1, g_2) = H_{g_1}^{-1}(g_2)f'(g_1, g_2),$$

and for $g_2 \in K$ we can still have $f'_1(g_1, g_2) = 1$, so

$$\left(\frac{f'_1(g_1, cg)}{f'_1(g_1, g)}\right)\left(\frac{f'_1(c, g_1 g)}{f'_1(c, g)}\right)^{-1} = \frac{f'_1(g_1, cg)}{f'_1(g_1, g)} = 1,$$

and $f'_1(g_1, \cdot)$ is a constant on cosets of K. Returning to our coset representative $c(gK)$, set

$$f'_2(g_1, g_2) = f'_1(g_1 c(g_1 K)^{-1}, c(g_1 K)g_2)f'_1(c(g_1 K), g_2) = f'_1(c(g_1 K), g_2),$$

and f'_2 is a constant on cosets of K in both variables. Hence f'_2 is actually a C-L section for a C-L extension of G/K by G_1, and $\mathcal{N}'$ is the lift of this extension to G.

We now remind the reader of the structure of compact, metric, Abelian and monothetic groups. Our main purpose here is to familiarize the reader with our choice of notation. To begin, we will see that such a group splits as a direct product of two groups

$$G \simeq G_{\mathrm{con}} \times G_{\mathrm{dis}}$$

Where G_{con} is toral (topologically connected) and G_{dis} is totally disconnected. We begin by describing each of these kinds of groups.

G_{dis} is an inverse limit of finite cyclic groups. Consider the following construction. For each prime p let C_{p^j} be the cyclic group $\{1, \omega, \dots, \omega^{p^j-1}\}$ of p^j'th roots of unity. There is a natural projection

$$\pi_p : C_{p^{j+1}} \to C_{p^j}, \pi_p(z) = z^p.$$

The inverse limit of the C_{p^j}'s under this projection is the p-adic adding machine $\mathcal{A}_p$.

Any product of the form

$$\underset{i \in I_1}{\otimes} C_{p_i^{j_i}} \underset{i \in I_2}{\otimes} \mathcal{A}_{p_i},$$

where I_1 and I_2 are disjoint subsets of $\mathbb{N}$ and $p_1 < p_2 < \dots$ is a listing of the primes, is compact, metric, Abelian, and monothetic.

Lemma 4.4. *If G is compact, Abelian, monothetic, and metric and all its characters are to finite subgroups of S^1, then*

$$G \simeq \underset{i \in I_1}{\otimes} C_{p_i^{j_i}} \underset{i \in I_2}{\otimes} \mathcal{A}_{p_i}.$$

Proof. We know G is an inverse limit of cyclic groups G/K_j, each of order n_j. A cyclic Abelian group is the direct product of subgroups of prime power order

$$G/K_j \simeq \underset{i \in I_j}{\otimes} C_{p_i^{j_i}}.$$

The primes in I_j must be distinct, as G is monothetic. It is a simple argument that the distinct prime factors of each G/K_j must lift to each other through the inverse limit to either a cyclic group or an adding machine component in a product representation of G.

For G_{con} the situation is more interesting. G_{con} is an inverse limit of finite-dimensional tori,

$$G/K_j \simeq \mathbb{T}^{n_j},$$

where the K_j are nested and intersect to $\{1\}$.

Consider the following description. Let $1 = n_1 \le n_2 \le \dots$ be an increasing sequence of integers with $n_{i+1} - n_i \le 1$. Let $\Gamma_i \subset \mathbb{Z}^{n_i}$ be lattice subgroups with the property that

1) $\Gamma_{i+1} \subseteq \Gamma_i$, Γ_i/Γ_{i+1} cyclic of prime order q_i if $n_i + 1 = n_i$, and

2) $\Gamma_{i+1} = \Gamma_i \times \mathbb{Z} \subseteq \mathbb{Z}^{n+1}$ if $n_{i+1} = n_i + 1$. In this context $q_i = 1$ is a prime.

Set $G_i = \mathbb{R}^{n_i}/\Gamma_i \simeq \mathbb{T}^{n_i}$.
Set $\pi_{i+1} : G_{i+1} \to G_i$ to be

1) $\pi_{i+1}(\vec{v}\Gamma_{i+1}) = \vec{v}\Gamma_i$ if $n_{i+1} = n_i$, and

2) $\pi_{i+1}((v_1, \ldots, v_{n_i+1})\Gamma_{i+1}) = (v_1, \ldots, v_{n_i})\Gamma_i$ if $n_{i+1} = n_i + 1$.

The inverse limit of the groups G_i under the projections π_i is of course a compact metric toral group. We can make the structure of this limit more algorithmic as follows. Each Γ_i is generated by n_i linearly independent vectors $B_i = \{\vec{e}_{i,1}, \ldots, \vec{e}_{i,n_i}\}$. If Γ_i/Γ_{i+1} is cyclic of prime order q_i, i.e. $n_{i+1} = n_i$, we can choose the basis $\{\vec{e}_{i+1,1}, \ldots, \vec{e}_{i+1,n_i}\}$ so that $\{\frac{1}{q_i}\vec{e}_{i+1,1}, \vec{e}_{i+1,2}, \ldots, \vec{e}_{i+1,n_i}\}$ is a basis for Γ_i. That is to say, we can inductively choose the generators for Γ_i so that

0)$\{\vec{e}_{1,1}\} = \{1\}$

1) if $n_{i+1} = n_i$, then there is an $A_i \in \mathrm{SL}(n_i, \mathbb{Z})$ with

$$[q_i^{-1}\vec{e}_{i+1,1}, \vec{e}_{i+1,2}, \ldots, \vec{e}_{i+1,n_i}] = A_i[\vec{e}_{i,1}, \ldots, \vec{e}_{i,n_i}] \text{ and}$$

2) if $n_{i+1} = n_i + 1$, then

$$\vec{e}_{i+1,j} = \begin{cases} \vec{e}_{i,j} & \text{, if } j \le n_i \\ \vec{e}_{i,n_i+1} & = (0, \ldots, 1). \end{cases}$$

Thus this inverse limit group can be parametrized by choosing first a sequence $\vec{s} \in \{0,1\}^{\mathbb{N}}$ with $s_1 = 1$ to define the dimensions

$$n_i = \sum_{j=1}^{i} s_j$$

and now a sequence of pairs indexed over those values of i where $s_i = 0$,

$$\{\{q_i, A_i\} \mid s_i = 0, q_i \text{ is prime}, A_i \in \mathrm{SL}(n_i, \mathbb{Z})\}.$$

Statements 0), 1), and 2) above now construct inductively a set of generators for each Γ_i. We will write the inverse limit obtained this way as $G(\vec{s}, \{q_i, A_i\})$, as this gives sufficient information to construct the inverse limit. Of course this representation is not at all unique.

Lemma 4.5. *Any group of the form* $G(\vec{s}, \{q_i, A_i\})$ *is compact, Abelian, metric and monothetic.*

Proof. The only issue is to check it is monothetic. To find a generator, each time $n_{i+1} = n_i + 1$ lift the generator constructed at step i to an element whose last coordinate is irrationally related to all the other coordinates. Kronecker's theorem tells us such an element in $G(\vec{s}, \{a_i, A_i\})$ will have dense orbit in each $\mathbb{R}^{n_i}/\Gamma_i$, hence in the inverse limit.

Lemma 4.6. If (R_{g_0}, G) is ergodic, i.e. g_0 generates G, and for all nontrivial $\chi \in \hat{G}$, $\chi(g_0)$ is not a root of unity, then for some $\vec{s}$, $\{q_i, A_i\}$,

$$G \simeq G(\vec{s}, \{q_i, A_i\}).$$

Proof. We can exhaust $\hat{G}$ by an increasing sequence of finitely generated subgroups

$$\hat{G}_i = \langle \chi_1, \ldots, \chi_i \rangle$$

so that either

1) $\hat{G}_{i+1}/\hat{G}_i$ is cyclic of prime order q_i, or

2) $\chi_{i+1}(g_0)$ is not a rational power of any element of $\langle \chi_1(g_0), \ldots, \chi_i(g_0) \rangle$.

Each $G/\Gamma_i = G_i = \langle \widehat{\chi_1, \ldots, \chi_i} \rangle$ is a torus, and either

1) $G_{i+1}/G_i \simeq \Gamma_i/\Gamma_{i+1}$ is cyclic of prime order q_i, or

2) $G_{i+1} \simeq G_i \times S^1 = \langle \widehat{\chi_1, \ldots, \chi_i} \rangle \times \langle \widehat{\chi_{i+1}} \rangle$.

Interpreted additively on $\mathbb{R}^{n_i}$, this tells us that $G \simeq G(\vec{s}, \{q_i, A_i\})$ for the parameters generated by the above description.

Theorem 4.7. If G is compact, Abelian, metric, and monothetic then

$$G \simeq G_{\mathrm{con}} \times G_{\mathrm{dis}}, \text{ where}$$
$$G_{\mathrm{dis}} \simeq \underset{i \in I_1}{\otimes} C_{p_i^{j_i}} \underset{i \in I_2}{\otimes} A_{p_i}, \text{ and}$$
$$G_{\mathrm{con}} \simeq G(\vec{s}, \{q_i, A_i\}).$$

Proof. G is certainly the inverse limit of a sequence G_i, where each $\hat{G}_i$ is finitely generated, $\pi_{i+1} : G_{i+1} \to G_i$.

Now $\hat{G}_i$ has a finite subgroup Q_i consisting of all characters taking values in the roots of unity. Hence G_i has a connected subgroup $K_i = \widehat{\hat{G}_i/\hat{H}_i}$ which must be a finite-dimensional torus. Of course G_i/K_i is finite, and in fact cyclic as letting g_i be the image in G_i of some fixed generator of G, the powers of g_i are coset representatives. Thus

$$G_i = \bigcup_{j=0}^{k_i-1} g_i^j K_i, \text{ a disjoint union.}$$

As K_i is a torus, there is a $\kappa_i \in K_i$ with $\kappa_i^{k_i} = g_i^{k_i}$, and now $c_i = g_i \kappa_i^{-1}$ generates a cyclic group C_i of order k_i, and

$$G_i \simeq C_i \times K_i.$$

As K_i is the connected component of the identity in G_i, π_{i+1} must map K_{i+1} onto K_i and C_{i+1} onto some finite subgroup of G_i. We want to see that C_{i+1} can be chosen so that $\pi_{i+1}(C_{i+1}) = C_i$.

Consider the subgroup $\pi_{i+1}^{-1}(1)$ and its list of cosets $\pi_{i+1}^{-1}(c_i^j)$. Because $g_{i+1} \pi_{i+1}^{-1}(\kappa_i^{-1}) = \pi_{i+1}^{-1}(c_i)$ is in this list of cosets, their union hits all of the cosets of K_{i+1}, in particular the coset $D = g_{i+1} K_{i+1} \cap \pi_{i+1}^{-1}(c_i)$ is not empty. As $D^{k_{i+1}} \subseteq K_{i+1}$ is closed under products, it is a subgroup and contains 1. Thus for some $c_{i+1} \in D$, $c_{i+1}^{k_{i+1}} = 1$. Thus we can choose κ_{i+1} with $c_{i+1} = g_{i+1} \kappa_{i+1} \in D$, and we get

$$\pi_{i+1} : C_{i+1} \to C_i$$
$$\pi_{i+1} : K_{i+1} \to K_i,$$

and now Lemmas 4.4 and 4.6 complete the result.

We want to metrize G in a precise way to help with our later work. Hence we fix a particular representation as

$$G = G(\vec{s}, \{q_i, A_i\}) \underset{i \in I_1}{\otimes} C_{p_i^{j_i}} \underset{i \in I_2}{\otimes} A_{p_i}.$$

To metrize $G_{\mathrm{con}} = G(\vec{s}, \{q_i, A_i\})$, remember that it is constructed explicitly as an inverse limit of tori $\mathbb{R}^{n_i}/\Gamma_i$.

Put on this torus the standard Euclidean norm $|\cdot|$ normalized so that $\mathbb{R}^{n_i}/\Gamma_i$ has diameter 1.

Any point $g \in G$ can be written as a sequence in the inverse limit, $g = (g_1, g_2, \dots)$, $g_i \in \mathbb{R}^{n_i}/\Gamma_i$. Set

$$\|g\| = \sum \frac{1}{2^i} |g_i|.$$

On each cyclic group $C_{p_i^{j_i}}$ put the discrete $\{0,1\}$-valued norm, and on the product

$$\|(c_i)_{i \in I_1}\| = \sum_{i \in I_1} \frac{\|c_i\|}{2^i}.$$

A point in $\mathcal{A}_p$ is of the form $(k_0, k_1, \dots)$, $0 \le k_j < p$ with the group action given as termwise addition $\mathrm{mod}(p)$ with carry. Set $\|(k_i)\| = 3^{-j}$, where j is the index of the first nonzero term in (k_i). Thus $\mathcal{A}_p$ has diameter 1. As usual, for $(a_i)_{i \in I_2} \in \otimes_{i \in I_2} \mathcal{A}_{p_i}$, set

$$\|(a_i)_{i \in I_2}\| = \sum_{i \in I_2} \frac{\|a_i\|}{2^i}.$$

The norm $\| \cdot \|$ on G is set to be the sum of the norms on the three components.

From now on G will always be a compact, metric, Abelian and mono-thetic group represented and metrized in this way. In our C-L extensions notice that, as the second coordinate is always toral, it too comes under this description.

At this point we can strengthen Lemma 2.10 in a critical way.

Definition 4.1. We say two elements $u, v \in G$ are δ-*connected*, $\delta > 0$, if there is a sequence $u = g_0, g_1, \ldots, g_k = v$ with $\|g_i g_{i+1}^{-1}\| < \delta$. This is an equivalence relation. We say a metric group G is δ-*connected* if any two $u, v \in G$ are δ-connected. We say G is 0-*connected* if it is δ-connected for all $\delta > 0$. From this one can define the δ-*connected component of 1 in* G, which will be a subgroup, and the 0-*connected component of* 1, which is their intersection as $\delta \to 0$. Clearly in a compact, Abelian, metric and monothetic group, the 0-connected component of 1 is G_{con}. Hence to be toral, to be 0-connected, and to be topologically connected are the same.

Lemma 4.8. *Suppose G is monothetic and totally disconnected, G_1 is toral, which is to say 0-connected, (R_{g_0}, G) is ergodic and $(R_{g_0,\phi}, G \times G_1)$ has pure point spectrum. Then ϕ is cohomologous to a constant.*

Proof. We know G_1 is an inverse limit of finite-dimensional toral factors G_1/K_i, and Lemma 2.10 tells us $\phi K_i : G \to G_1/K_i$ is cohomologous to a constant. As G_1/K_i is a torus, for a.e. $\gamma \in G_1$, $(R_{g_0,\gamma\phi K_i}, G \times G_1/K_i)$ is ergodic and has pure point spectrum. Hence, going to the inverse limit, for a.e. $\gamma \in G_1$, $(R_{g_0,\gamma\phi}, G \times G_1)$ is ergodic and has pure point spectrum. Fix such a choice of γ. Now $R_{g_0,\gamma\phi}$ generates a compact, Abelian, monothetic and metric group of transformations $\overline{G}$ acting transitively on $G \times G_1$. The group of maps $\overline{G}_1 = \{R_{1,\gamma} \mid \gamma \in G_1\}$ must be a connected subgroup of $\overline{G}$, as $\overline{G}$ acts transitively and is the connected component of the identity in $\overline{G}$, as $\overline{G}/\overline{G}_1 \simeq G$ is totally disconnected. Theorem 4.7 tells us $\overline{G}$ contains a totally disconnected subgroup K complementary to $\overline{G}_1$. As K projects 1-1 to K/G_1,

$$K = \{R_{g,h(g)} \mid g \in G\}, \quad \text{where } h(g_1, g)h(g_2, g_1 g) = h(g_1 g_2, g).$$

Set $g = 1$ and fix g_2 to get

$$h(g_2, g_1) = h(g_2 g_1, 1)h(g_1, 1)^{-1}$$

is a g_2-coboundary. Hence, as $\gamma\phi = \gamma_0 h(g_0)$, ϕ is g_0-cohomologous to the constant $\gamma_0 \gamma^{-1}$.

Lemma 4.9. *If G is totally disconnected and monothetic and G_1 is toral, then any $H \in \mathrm{Hom}(G, G_1)$ is of the form*

$$H(g) = c(g_0)c(g)c(g_0 g)^{-1}$$

for any generator g_0 for G. Moreover, for all $g' \in G$, $c(g')c(g)c(g' g)^{-1} = H'(g) \in \mathrm{Hom}(G, G_1)$.

Proof. From Lemma 4.8 it is sufficient to show that $R_{g_0,H}$ has pure point spectrum. For this it is sufficient to assume G_1 is S^1. In this case $H \in \hat{G}$ takes values in a finite subgroup of S^1. In particular its range is of the form

$$\{H(1), H(g_0), \ldots, H(g_0)^{k-1}\} \text{ for some } k \in \mathbb{N}.$$

Let $c(1) = 1$ and leave $c(g_0)$ undefined for the moment. The equation

$$c(g_0^{j+1}) = c(g_0)c(g_0^j)H(g_0^j)^{-1}$$

now inductively defines

$$c(g_0^j) = c(g_0)^j H(g_0^{\frac{j^2-j}{2}}).$$

The only question is if this remains correct at $j = k$, i.e. if

$$c(g_0)^k H(g_0^{\frac{k^2-k}{2}}).$$

If k is odd, then $(k^2 - k)/2$ is divisible by k, and we can set $c(g_0) = 1$. If k is even, then $(k^2 - k)/2 = k/2 \pmod{k}$, and, as $H(g_0^{\frac{k}{2}}) = -1$, choosing $c(g_0)$ as a k'th root of -1 will give the result. We conclude that $R_{g_0,H}$ has pure point spectrum, and so, by Lemma 4.8, $R_{g_0,\gamma H}$ generates a compact Abelian group $\overline{G}$ of transformations which we know to be isomorphic to $G \times G_1$; which is to say, there is a function $c : G \to G_1$ so that

$$\overline{G} = \{R_{u,f} \mid f(g) = \gamma c(ug)c(g)^{-1}\},$$

and $\{R_{u,f} \mid f(g) = c(ug)c(g)^{-1}\}$ is a subgroup isomorphic to G, and so

$$c(g_1g_2g_3)c(g_1g_2)^{-1}c(g_2g_3)^{-1}c(g_1g_3)^{-1}c(g_1)c(g_2)c(g_3) = 1,$$

which implies $c(g')c(g)c(g'g)^{-1}$ is a group homomorphism.

Lemma 4.10. *Suppose G is 0-connected and monothetic, G_1 is toral, and $H \in \text{Hom}(G, G_1)$. If $R_{g_0,H}$ has pure point spectrum for some generator $g_0 \in G$ then, $H \equiv 1$.*

Proof. It is enough to evaluate this on each character of G_1, hence to assume $G_1 = S^1$. In this case $H = \chi \in \hat{G}$. Either χ is trivial or it is circle-valued, in which case we would conclude that the action $(z, z_1) \to (\chi(g_0)z, zz_1)$ has pure point spectrum, which it does not.

The previous two lemmas are the essential ingredients for pushing our work on finite-dimensional toral extensions to the inverse limit for q-A extensions. This will become evident later. For now we return to finite-dimensional toral extensions.

Definition 4.2. A sequence of values in G of the form $u = g_0, g_1, \ldots, g_k = v$ with $\|g_i g_{i+1}^{-1}\| < \delta$ we call a δ-*chain from u to v*. If $u = 1$, we call it a δ-*chain to v* and if we also have $v = 1$, then we call it a δ-*loop*. Set

$$G_\delta = \{g \in G \mid g \text{ is } \delta\text{-connected to } 1\}.$$

This is a closed subgroup.

Lemma 4.11. *For $\delta > 0$, let*

$$I_1(\delta) = \{i \in I_1 \mid 2^{-i} < \delta\},$$

and for each $i \in I_2$, let

$$t_i = \max(0, [-\log_3(2^i\delta)] + 1)$$

and define the subgroup

$$\mathcal{A}_{p_i}(\delta) = \{(k_i) \mid k_j = 0 \text{ for } j \le t_i\}.$$

Then

$$G_\delta = G_{\text{con}} \underset{i \in I_1(\delta)}{\otimes} C_{p_i^{j_i}} \underset{i \in I_2}{\otimes} \mathcal{A}_{p_i}(\delta) \text{ is}$$

a closed subgroup of G of finite index

$$\prod_{i \in I_1 \setminus I_1(\delta)} p_i^{j_i} \prod_{i \in I_2} p_i^{t_i}$$

(remember $t_i = 0$ for all but finitely many $i \in I_2$). Further, the first coordinate G_{con} is connected in the usual sense and the latter two coordinates each have diameter less than δ.

Proof. The form of $\| \cdot \|$ tells us G_δ is the direct product of the δ-connected components of 1 in each of the three components of G. The first is 0-connected. In $\otimes_{i \in I_1} C_{p_i^{j_i}}$, two elements that differ at an index not in $I_1(\delta)$ are at least δ apart, hence are not δ-connected. Two points that differ only on coordinates in $I_1(\delta)$ are δ-connected.

The diameter of $\mathcal{A}_{p_i}$ as normalized in G is 3^{-i}, and two points in $\mathcal{A}_{p_i}$ which differ in a coordinate $j \le t_i$ are at least δ apart and cannot be δ-connected. Those which differ only after the t_i'th coordinate are within δ of each other.

The importance of δ-connectedness to us is the following result about C-L extensions of G.

Lemma 4.12. *Suppose $\mathcal{N}$ is a C-L extension of G by $\mathbb{T}^n$. For any $\varepsilon \leq (4\sqrt{n})^{-1}$ there is a $\delta = \delta(\mathcal{N}, \varepsilon)$ so that if $\|c_1 c_2^{-1}\| < \delta$ and $R_{c_1, f_1} C(\mathcal{N})$ is some lift of c_1 to $\mathcal{N}/C(\mathcal{N})$, then there is a unique lift $R_{c_2, f_2} C(\mathcal{N})$ of c_2 with*

$$\|R_{c_1, f_1} C(\mathcal{N}), R_{c_2, f_2} C(\mathcal{N})\| < \varepsilon.$$

That is to say, the factor map from $\mathcal{N}/C(\mathcal{N})$ to G has local homeomorphic inverses in neighborhoods of G of diameter $\delta(\mathcal{N}, (4\sqrt{n})^{-1}) = \delta_0(\mathcal{N})$.

Proof. By Lemma 3.7, there is a δ so that if $\|c_1 c_2^{-1}\| < \delta$, then there is an $f_2 \in F(c_2)$ with

$$\|R_{c_1, f_1} C(\mathcal{N}), R_{c_2, f_2} C(\mathcal{N})\| < \varepsilon.$$

For any other $f'_2 \in F(c_2)$,

$$f'_2 = \gamma H f_2,$$

and if $H \neq 1$ then

$$\|R_{c_1, f_1} C(\mathcal{N}), R_{c_2, f'_2} C(\mathcal{N})\| \geq \|R_{c_2, f_2} C(\mathcal{N}), R_{c_2 \cdot f'_2} C(\mathcal{N})\| - (4\sqrt{n})^{-1}$$
$$= \min_{\gamma \in \mathbb{T}^n} d(f_2, \gamma H f_2) - (4\sqrt{n})^{-1} = \min_{\gamma \in \mathbb{T}^n} \|1 - \gamma H\|_2 - (4\sqrt{n})^{-1} \geq (4\sqrt{n})^{-1}.$$

Thus if $\delta < \delta_0(\mathcal{N})$ and u is δ-connected to v, then along any δ-chain of elements connecting u to v there is a unique way to continuously extend a lift of u to $\mathcal{N}/C(\mathcal{N})$ to a lift of v along the chain.

Suppose $\mathcal{N}$ is a C-L extension of G by $\mathbb{T}^n$ and $\delta \leq \delta_0(\mathcal{N})$. For any δ-loop in G $\ell = \{1, g_1, \ldots, g_{k-1}, 1\}$ we can start with the identity lift $R_{1,1}$ and extend around the loop to reach a second lift of 1 to $\mathcal{N}/C(\mathcal{N})$, which of necessity is of the form $HC(\mathcal{N})$, $H \in \operatorname{Hom}(G, \mathbb{T}^n)$. We call this element $H_\mathcal{N}(\ell) \in \operatorname{Hom}(G, \mathbb{T}^n)$.

We will first show that the values $H_\mathcal{N}(\ell)$ determine $\mathcal{N}$ up to cohomology. We will then show that all nontrivial $H_\mathcal{N}(\ell)$ are generated by Heisenberg extensions.

We begin with two lemmas.

Lemma 4.13. *Suppose $\mathcal{N}$ is a C-L extension of G by $\mathbb{T}^n$ and $\delta < \delta_0(\mathcal{N})$. Suppose for all δ-loops ℓ in G, $H_\mathcal{N}(\ell) = 1$. Then $\mathcal{N}$ is cohomologous to an $\mathcal{N}'$ with a C-L section f' with the property that for all $g \in G_\delta$, $f(g, \cdot)$ is a constant on cosets of G_δ.*

Proof. For any $g \in G_\delta$, let $\{1, g_1, \ldots, g_k = g\}$ be a δ-chain to g. Starting from the lift $L(1) = R_{1,1} C(\mathcal{N})$, we can extend continuously along the chain to a lift of g. As all $H_\mathcal{N}(\ell) = 1$, this lift $L(g) \in \mathcal{N}/C(\mathcal{N})$ is independent of the choice of δ-chain.

One easily sees that $L(g_1 g_2) = L(g_1) L(g_2)$, and L maps G homeomorphically to a compact Abelian subgroup of $\mathcal{N}/C(\mathcal{N})$.

Let

$$\overline{G} = \{R_{c,f}R_{1,\gamma} \mid \gamma \in \mathbb{T}^n, \ R_{c,f}C(\mathcal{N}) \in L(G)\}$$

be the lift of $L(G)$ to $\mathcal{N}$.

This is a compact order-two nilpotent subgroup of $\mathcal{N}$. Now G_δ is still monothetic. Let g_0 be a generator and $R_{g_0,\sigma} \in \overline{G}$. Consider the compact Abelian subgroup

$$K = \overline{\{(R_{g_0,\sigma})^n R_{1,\gamma} \mid \gamma \in \mathbb{T}^n\}} \subseteq \overline{G}.$$

As g_0 generates G_δ, K projects to all of G_δ. Any measurable action of a compact Abelian group as measure-preserving transformations of a standard probability space always fibers into ergodic components which are isomorphic to translations on homogeneous spaces of the group. In particular, $R_{g_0,\sigma}$ must have pure point spectrum. Now G_δ has finite index in G, so G breaks into cosets $\{G_\delta, c_1 G_\delta, \ldots, c_k G_\delta\}$ on each of which g_0 acts ergodically and over which the extension to $R_{g_0,\sigma}$ has pure point spectrum. Lemma 2.10 tells us σ must be cohomologous to a constant. Hence σ is cohomologous to a σ', i.e.

$$\sigma' = h(g_0 g)h(g)^{-1}\sigma(g),$$

which is constant on cosets of G_δ. Let

$$\mathcal{N} = \{R_{c,f'} \mid f'(g) = f(g)h(cg)h^{-1}(g), \ f \in F(c)\},$$

and $\mathcal{N}'$ is cohomologous to $\mathcal{N}$.

Further, setting

$$H = \overline{\{(R_{g_0,\sigma'}^n\}} \subseteq \mathcal{N}',$$

we conclude $R_{g,f'} \in H$ has f' a constant on cosets of G_δ. A measurable selection

$$f' : G_\delta \times G \to \mathbb{T}^n$$

with $R_{g,f'(g)} \in H$ starts a C-L section for $\mathcal{N}'$ which extends by translation to the finite list of cosets of G_δ in G, finishing the result.

Our next theorem is once more based on the observation that G_δ is a subgroup of finite index, and G/G_δ is cyclic. Letting q be this index, for any generator c of G, $1, c^2, \ldots, c^{q-1}$ are a set of coset representatives for G_δ.

Theorem 4.14. *Suppose $\mathcal{N}$ is a C-L extension of G by $\mathbb{T}^n$, $\delta < \delta_0(\mathcal{N})$, and f is a C-L section with the property that for any $g \in G_\delta$, $f(g, \cdot)$ is a constant on cosets of G_δ. Then $\mathcal{N}$ is cohomologous to a quasi-Abelian extension.*

Proof. Remember we can always multiply a function $f(g, \cdot)$ by a constant or an element of $\text{Hom}(G, \mathbb{T}^n)$ and get a new choice for a C-L section. To begin, then, we can assume $f(g, g_1) = 1$ for both $g, g_1 \in G_\delta$.

Notice

$$\left(\frac{f(g, c^{k+1}g_1)}{f(g, c^k g_1)} \right) \left(\frac{f(c, c^k g g_1)}{f(c, c^k g_1)} \right)^{-1} = \lambda(g, c), \text{ so}$$

$$\frac{f(c, c^k g g_1)}{f(c, c^k g_1)} \lambda(g, c) = \gamma_k(g, c)$$

is independent of g_1. Thus

$$\frac{f(c, c^k g g_1)}{f(c, c^k g) f(c, c^k g_1)} = \gamma_k$$

is independent of both g and $g_1 \in G_\delta$.

As $\mathbb{T}^n$ is a divisible group, we can replace $f(c, \cdot)$ by an $f_1(c, \cdot) = \gamma f(c, \cdot)$, where γ has the property that

$$\gamma_0 \gamma_1 \dots \gamma_{q-1} \gamma^q = 1.$$

In this case,

$$f_1(c, c^k g g_1) = \gamma_{1,k} f_1(c, c^k g) f_1(c, c^k g_1) \text{ where } \prod_{k=0}^{q-1} \gamma_{1,k} = 1.$$

For $c^k g \in G$ define $h(c^k g) = \prod_{i=0}^{k-1} \gamma_{1,i}^{-1}$ and the group cohomologous to $\mathcal{N}$,

$$\mathcal{N}_1 = \{ R_{g,f_1} \mid f_1(g_1) = h(gg_1)h(g_1)^{-1} f(g_1), R_{g,f} \in \mathcal{N} \},$$

with C-L section satisfying

$$f_1(g, \cdot) = f(g, \cdot) \text{ if } g \in G_\delta \text{ and } f_1(c, c^k g) = f(c, c^k g) \gamma_k^{-1}.$$

Thus we now conclude

$$\frac{f_1(c, c^k g g_1)}{f_1(c, c^k g) f_1(c, c^k g_1)} = 1,$$

and so

$$f_1(c, c^k \cdot) = H_k \in \text{Hom}(G_\delta, \mathbb{T}^n).$$

Noticing that $G_\delta = \{ g^q \mid g \in G \}$,

$$\text{Hom}(G_\delta, \mathbb{T}^n) = \{ H^q \mid H \in \text{Hom}(G, \mathbb{T}^n) \}.$$

Choose $H \in \text{Hom}(G, \mathbb{T}^n)$ with

$$H^q = H_0 H_1 \dots H_{q-1}.$$

Replacing $f_1(c, \cdot)$ by $H^{-1}f_1(c, \cdot)$, we can assume

$$H_0 H_1 \ldots H_{q-1} \equiv 1.$$

Now define $h_1(c^k g) = (H_0 H_1 \ldots H_{k-1})^{-1}(g)$ to get the cohomologous group

$$\mathcal{N}' = \{R_{g,f'} \mid f'(g_1) = h(gg_1)h(g)^{-1}f(g_1),\ R_{g,f} \in \mathcal{N}_1\}.$$

This has a C-L section f' with $f'(g, g_1) = 1$ if both g and g_1 are in G_δ, but moreover $f'(c, \cdot) = 1$.

This tells us

$$\frac{f'(g, c^{k+1}g_1)}{f'(g, c^k g_1)} = \lambda(g, c),$$

so $f'(g, c^k g_1) = (\lambda(g, c))^k$, a constant on the coset $c^k G_\delta$.

But this says $(\lambda(g, c))^q = 1$ and $f'(g, \cdot) \in \mathrm{Hom}(G, \mathbb{T}^n)$ for each $g \in G_\delta$, and so all $R_{g,1} \in \mathcal{N}'$ and 1 is a C-L section for $\mathcal{N}'$.

Corollary 4.15. *If $\mathcal{N}$ is a C-L extension of G by $\mathbb{T}^n$ and $\delta < \delta_0(\mathcal{N})$ has $H_\mathcal{N}(\ell) = 1$ for all δ-loops ℓ, then $\mathcal{N}$ is a q-A extension.*

Theorem 4.16. *Suppose $(R_{g_0, \phi}, G \times G_1)$ is a Conze-Lesigne action, i.e. all characters of G_1 satisfy the C-L equation. Further, suppose for any finite-dimensional toral factor G_1/K, $(R_{g_0, \phi}, G \times G_1/K)$ is q-A. Then $(R_{g_0, \phi}, G \times G_1)$ itself is an element of an q-A extension of G by G_1.*

Proof. Write G_1 as the inverse limit of $G/K_i \simeq \mathbb{T}^{n_i}$ where, $K_{i+1} \subseteq K_i$ and $\cap_i K_i = \{1\}$. Now restricting to G/K_i, Lemma 4.14 tells us ϕK_i is cohomologous to $\gamma_i H_i K_i$, where $\gamma_i \in G_1$ and $H_i \in \mathrm{Hom}(G, G_1/K_i)$.

Now $H_i = H_i^{\mathrm{con}} \otimes H_i^{\mathrm{dis}}$, where

$$H_i^{\mathrm{con}} \in \mathrm{Hom}(G_{\mathrm{con}}, G/K_i) \text{ and}$$

$$H_i^{\mathrm{dis}} \in \mathrm{Hom}(G_{\mathrm{dis}}, G/K_i).$$

From Lemma 4.9 we can assume $H_i^{\mathrm{dis}} \equiv 1$.

Now $\gamma_{i+1} H_{i+1}^{\mathrm{con}} K_i$ is g_0-cohomologous to $\gamma_i H_i^{\mathrm{con}} K_i$, since both are g_0-cohomologous to ϕK_i. But this says $H_{i+1}^{\mathrm{con}}(H_i^{\mathrm{con}})^{-1} K_i$ is g_0-cohomologous to $\gamma_i \gamma_{i+1}^{-1} K_i$. But Lemma 4.10 now implies that $H_{i+1}^{\mathrm{con}} K_i = H_i^{\mathrm{con}}$, and so $\gamma_i \gamma_{i+1}^{-1} K_i$ is a g_0-coboundary.

As $H_{i+1}^{\mathrm{con}} K_i = H_i^{\mathrm{con}}$, the sequence of group homomorphisms H_i lifts through the inverse limit to an $H^{\mathrm{con}} \in \mathrm{Hom}(G_{\mathrm{con}}, G_1)$. Now $R_{g_0, \phi H^{-1}}$ on each factor $(G \times G_1/K_i)$ of the inverse limit has pure point spectrum, hence on the inverse limit has pure point spectrum. By Lemma 1.11 it is a group rotation on some twisted extension $\overline{G}$ of the factor group $\overline{G}/K \simeq G$ by the subgroup $G_1 \subseteq \overline{G}$. Thus $R_{g_0, \phi}$ is an element of a q-A extension of the type described at the beginning of this section.

Theorem 4.17. *Suppose R_{g_0,ϕ_1} and R_{g_0,ϕ_2} are two C-L actions extending (R_{g_0}, G) by G_1. Then $(R_{g_0,\phi_1^{-1}\phi_2})$ is also a C-L action. Furthermore, for each factor $G_1/K \simeq \mathbb{T}^n$, in the three C-L extensions $\mathcal{N}_1$, $\mathcal{N}_2$ and $\mathcal{N}_3 = \{R_{c,h_1h_2} \mid R_{c,h_1^{-1}} \in \mathcal{N}_1, R_{c,h_2} \in \mathcal{N}_2\}$ obtained by projecting the three actions onto this toral factor, for any $\delta < \min(\delta_0(\mathcal{N}_1), \delta_0(\mathcal{N}_2), \delta_0(\mathcal{N}_3))$, for any δ-loop ℓ,*

$$H_{\mathcal{N}_3^K}(\ell) = H_{\mathcal{N}_1^K}(\ell)^{-1} H_{\mathcal{N}_2^K}(\ell).$$

Thus if in particular we have

$$H_{\mathcal{N}_1^K}(\ell) = H_{\mathcal{N}_2^K}(\ell),$$

then $R_{g_0,\phi_1^{-1}\phi_2}$ is an element of a q-A extension, and if one of the two, say R_{g_0,ϕ_1} was an element of a C-L extension, then the other R_{g_0,ϕ_2} must also be an element of a C-L extension.

Proof. If ϕ_1 and ϕ_2 both satisfy the C-L equation, so does $\phi_1^{-1}\phi_2$. Also, if $\mathcal{N}_1$ and $\mathcal{N}_2$ are two C-L extensions of G by G_1, with C-L sections f_1 and f_2, then

$$\mathcal{N} = \{R_{c,h_1h_2} \mid R_{c,h_1^{-1}} \in \mathcal{N}_1, R_{c,h_2} \in \mathcal{N}_2\}$$

is also a C-L extension with C-L section $f_1^{-1}f_2$.

For any $\delta < \min(\delta_0(\mathcal{N}_1^K, \delta_0(\mathcal{N}_2^K), \delta_0(\mathcal{N}^K))$, if ℓ is a δ-loop in G, as we follow the three lifts of the loop around in the three nilpotent groups, as the lifts are local homeomorphisms the only possible choice for the skewing function of the lift to $\mathcal{N}_3/C(\mathcal{N}_3)$ is the product of the skewing functions of the lifts of the other two. Hence

$$H_{\mathcal{N}_3^K}(\ell) = H_{\mathcal{N}_1^K}(\ell)^{-1} H_{\mathcal{N}_2^K}(\ell).$$

The rest follows directly, as we know all q-A actions are elements of q-A extensions.

This now tells us that whether or not a C-L action is actually an element of a C-L extension is determined completely by the nontrivial values of $H_{\mathcal{N}^K}(\ell)$ for finite-dimensional toral factors G_1/K. We will not come to a complete understanding of this question. What we will see, though, is what the nontrivial structure of $H_{\mathcal{N}^K}(\ell)$ can be, and that it can indeed stop a C-L action from being an element of a C-L extension.

5. δ-HOMOTOPY AND NON-TRIVIAL CONZE-LESIGNE ACTIONS

We develop now some very simple homology for δ-loops in order to show that the nontrivial values $H_{\mathcal{N}}(\ell)$ must come from a finite-dimensional torus in G as a Heisenberg-like extension. We give three basic kinds of changes in a δ-chain $\{g_0, g_1, \dots, g_k\}$ which will generate 'δ-homotopies'.

Definition 5.1. We say two δ-chains $\{g_0, g_1, \ldots, g_k\}$ and $\{h_0, h_1, \ldots, h_t\}$ are *one-step δ-homotopic* if either

1) $t = k + 1$ and for some $j \le k$

$$\{h_0, h_1, \ldots, h_{k+1}\} = \{g_0, g_1, \ldots, g_j, g_j, g_{j+1}, \ldots, g_k\}, \text{ or}$$

2) $k = t + 1$ and for some $j \le t$

$$\{g_0, g_1, \ldots, g_{t+1}\} = \{h_0, h_1, \ldots, h_j, h_j, h_{j+1}, \ldots, h_k\}, \text{ or}$$

3) $k = t$, $g_0 = h_0$, $g_t = h_t$ and $\|g_i h_i^{-1}\| \le \delta$.

We say $\{g_0, g_1 \ldots, g_l\}$ and $\{h_0, h_1, \ldots, h_t\}$ are *δ-homotopic* if you can reach $\{h_0, \ldots, h_t\}$ from $\{g_0, \ldots, g_k\}$ by a finite sequence of one-step δ-homotopic δ-chains.

Lemma 5.1. *Suppose $\mathcal{N}$ is a C-L extension of G by $\mathbb{T}^n$, $\delta < \delta_0(\mathcal{N})$, and ℓ_1 and ℓ_2 are δ-homotopic δ-loops in G. Then $H_\mathcal{N}(\ell_1) = H_\mathcal{N}(\ell_2)$.*

Proof. This is easily seen for one-step δ-homotopies.

Definition 5.2. For G a compact, Abelian, metric and monothetic group and $\delta > 0$ let

$$K_\delta = \{g \in G_{\text{con}} = G(\vec{s}, \{q_i, A_i\}) \,|\, g = (\vec{v}_1, \vec{v}_2, \ldots), \vec{v}_i \in \mathbb{R}^{n_i} / \Gamma_i$$
$$\text{and } \vec{v}_i = \vec{0} \text{ for } i - 3 \le \log_2(\delta)\} \otimes G_{\delta, \text{dis}},$$

a closed subgroup of G_δ.

Lemma 5.2. *Suppose $\ell_1 = (g_0, g_1, \ldots, g_k)$ and $\ell_2 = (h_0, h_1, \ldots, h_k)$ are two δ-loops in G and*

$$g_i K_\delta = h_i K_\delta, \, 0 \le i \le k.$$

Then ℓ_1 and ℓ_2 are δ-homotopic.

Proof. First, all δ-loops lie in G_δ. The diameter of each $\mathcal{A}_{p_j}(\delta)$ is less than δ, so replacing in each g_i and h_i the $\mathcal{A}_{p_j}(\delta)$ term with a 1 is a δ-homotopy. The same is true for each $C_{p_k^{j_k}}$, $k \in I_1(\delta)$. Hence, one by one we can δ-homotope these coordinates to 1. After finitely many steps, the diameter of the remaining terms is less than δ, and we can δ-homotope them all simultaneously to 1. Hence we can assume both ℓ_1 and ℓ_2 are in $G_{\delta, \text{con}}$. As this group is 0-connected, ℓ_1 and ℓ_2 are δ-homotopic to $\delta/2$-loops ℓ_1' and ℓ_2' with $\ell_1' K_\delta = \ell_2' K_\delta$. As K_δ has diameter less than $\sum_{j=\log_2(\delta)+2}^{\infty} 2^{-j} < \frac{\delta}{4}$, ℓ_1' and ℓ_2' are in fact $\delta/2$-homotopic.

Now G_δ / K_δ is a finite-dimensional torus. The elements of its fundamental group

$$\Pi_1(G_\delta / K_\delta) \overset{\text{def}}{=} \Pi_\delta(G)$$

are closed curves, up to homotopy.

Setting $\dim(G_\delta/K_\delta) = n_\delta$,

$$\Pi_\delta(G) \simeq \mathbb{Z}^{n_\delta}$$

via a choice of n_δ generators for Γ_δ, where

$$G_\delta/K_\delta \simeq \mathbb{R}^{n_\delta}/\Gamma_\delta.$$

Any choice $\vec{m} = (m_1, m_2, \ldots, m_{n_\delta}) \in \mathbb{Z}^{n_\delta}$ corresponds in $\Pi_\delta(G)$ to a curve which winds around the j'th basis dimension m_j times. There is a unique oriented geodesic representative $S_{\vec{m}}$ of $\vec{m}$ passing through $\vec{0}$. If the basis for Γ_δ is $(\vec{e}_1, \vec{e}_2, \ldots, \vec{e}_{n_\delta}) \subseteq \mathbb{Z}^{n_\delta}$, the geodesic representative of $\vec{m}$ is the projection to G_δ/K_δ of the line

$$\left[\vec{e}_1, \vec{e}_2, \ldots, \vec{e}_{n_\delta} \right] \vec{m} t.$$

If we choose a new basis $[\vec{f}_1, \ldots, \vec{f}_{n_\delta}] = [\vec{e}_1, \ldots, \vec{e}_{n_\delta}] A$, $A \in \mathrm{SL}(n_\delta, \mathbb{Z})$, then a particular geodesic S representing $\vec{m}$ relative to $[\vec{e}_1, \ldots, \vec{e}_{n_\delta}]$ represents the vector $A^{-1}\vec{m}$ relative to the basis $[\vec{f}_1, \ldots, \vec{f}_{n_\delta}]$.

Lemma 5.3. *For any $\delta > 0$, for a geodesic circle $S \in G_\delta/K_\delta$ suppose $\ell = (y_0, y_1, \ldots, y_k)$ is a $\delta/2$-loop of points in G_δ/K_δ, running in order around S. Then*

1) ℓ can be lifted to a δ-loop of points in G_δ,

2) any two such loops lifted from the same S are δ-homotopic and

3) any δ-loop in G_δ is δ-homotopic to some such lift.

Proof. To lift a point gK_δ back to G_δ, choose 1 as coset representative in $G_{\delta,\mathrm{dis}}$. Lifting through the inverse limit starting at $G_{\delta,\mathrm{con}}$ to G_δ, always choose the representative in the next term nearest to $\vec{0}$. That is to say, when the dimension $n_{i+1} = n_i + 1$, choose the next coordinate to be 0. When $n_{i+1} = n_i$, and there are q_i possible values for the next term in the inverse limit, choose the one nearest to 0. We call this point in G_δ the *standard lift* of the original gK_δ. If two points in $G_{\delta,\mathrm{con}}/K_\delta$ differ by less than $\delta/2$ then their standard lifts differ by less than δ, as $K_{\delta,\mathrm{con}}$ has diameter less than $\delta/4$. Thus the standard lift of a $\delta/2$-loop in G_δ/K_δ form a δ-loop in G_δ, and the standard lifts of a $\delta/2$-homotopy in G_δ/K_δ is a δ-homotopy in G.

To argue 2), it is not difficult to see that any two $\delta/2$-loops lying ordered around the same geodesic circle in a finite-dimensional torus are $\delta/2$-homotopic, as each is $\delta/2$-homotopic to the union. The above remarks now tell us the standard lifts are δ-homotopic.

To argue 3), notice first that for any $g \in G_\delta$, if g_1 is the standard lift of gK_δ, then $\|gg_1^{-1}\| < \delta$. Thus it suffices to show 3) for a δ-loop ℓ

consisting of standard lifts. Consider points in the projection ℓ_1 of ℓ to G_δ/K_δ. Connect these points in the torus in order by line segments to form a closed piecewise geodesic curve, which is homotopic to some geodesic circle S. Along this curve, add in sufficiently many points to get a $\delta/2$-loop ℓ_2. Along the homotopy to S, by splitting and shifting points appropriately, we can construct a $\delta/2$-homotopy of ℓ_2 to some $\delta/2$-loop ℓ' along S. The standard lift of ℓ_1 is δ-homotopic to ℓ, and the $\delta/2$-homotopy now lifts to a δ-homotopy to the standard lift of ℓ'.

Thus, if we fix a basis $[\vec{e}_1,\ldots,\vec{e}_{n_\delta}]$ for Γ_δ and regard $\vec{m}$ as representing a geodesic circle, we get a unique element

$$H_{\mathcal{N}}(\vec{m}) \in \text{Hom}(G,\mathbb{T}^n).$$

That is to say we can regard

$$H_{\mathcal{N}} \mid \widehat{G_\delta/K_\delta} \to \text{Hom}(G,\mathbb{T}^n).$$

One easily observes this is a group homomorphism. In $\widehat{G_\delta/K_\delta}$ the group action is concatenation, mod homotopy. If one follows the concatenated loops around in order in the lift to $\mathcal{N}$, one must arrive at the product of the two elements in $\text{Hom}(G,\mathbb{T}^n)$. This group structure already exists on the level of δ-loops mod δ-homotopy, where the group operation is concatenation. If δ is small enough, one concludes that

$$H_{\mathcal{N}}(\ell_1 \circ \ell_2) = H_{\mathcal{N}}(\ell_1)H_{\mathcal{N}}(\ell_2).$$

Rather than work from the outset at this level, we have instead shown first that the only nontrivial values for $H_{\mathcal{N}}$ lie over a finite-dimensional torus, and here we are working with the classical fundamental group.

Suppose $\mathcal{N}'$ is a C-L extension of G/K by G_1, where G is a compact, Abelian, metric, and monothetic group and K is a closed subgroup. Remember, $\mathcal{N}'$ lifts to a C-L extension of G of the form

$$\mathcal{N} = \{R_{c,f} \mid f(g) = Hf'(gK), H \in \text{Hom}(G,G_1), R_{cK,f} \in \mathcal{N}'\}.$$

Theorem 5.4. *Suppose $\mathcal{N}$ is a C-L extension of G by $\mathbb{T}^n$ and $\delta < \delta_0(\mathcal{N})$. For any δ-loop ℓ in G,*

$$H_{\mathcal{N}}(\ell) \in \text{Hom}(G_\delta/K_\delta,\mathbb{T}^n).$$

Moreover, $\mathcal{N}$ is a lift of a C-L extension of $G/(K_\delta G_{\text{dis}})$ by $\mathbb{T}^n$.

Proof. Restrict $\mathcal{N}$ to

$$\mathcal{N}_c = \{R_{g,f} \in \mathcal{N} \mid g \in K_\delta\}$$

acting on the coset $cK_\delta \times \mathbb{T}^n$. For a.e. c this is a C-L extension of K_δ by $\mathbb{T}^n$.

For $\delta < \delta_0(\mathcal{N})$, any δ-loop ℓ in K_δ is a δ-loop in $\mathcal{N}$, hence for a.e. c has a unique continuous lift to $\mathcal{N}_c$ given by restricting the lift to $\mathcal{N}$ to its action on $K_\delta \times G_1$. We know $H_{\mathcal{N}}(\ell) = 1$, as $\ell \subseteq K_\delta$. Further, we know up to δ-homotopy there are at most countably many δ-loops in K_δ. Hence, a.s. in c, all δ-loops ℓ in K_δ have $H_{\mathcal{N}_c}(\ell) = 1$, i.e. for a.e. c, $\mathcal{N}_c$ is cohomologous to the trivial lift of K_δ to $\mathbb{T}^n$. It is not difficult to show that these cohomologies can be measurably chosen in c, so that $\mathcal{N}$ is cohomologous to an $\mathcal{N}'$ with a C-L section f', where $f'(g, \cdot) \equiv 1$ for $g \in K_\delta$. By Lemma 4.3, we conclude $\mathcal{N}$ is a lift to G of a C-L extension of $G/(K_\delta G_{\mathrm{dis}})$ by $\mathbb{T}^n$. For such a lift,

$$H_{\mathcal{N}}(\ell) \in \mathrm{Hom}(G/K_\delta G_{\mathrm{dis}}).$$

Corollary 5.5. *All C-L extensions of G by $\mathbb{T}^n$ are obtained by taking a C-L extension of some $G/K \simeq \mathbb{T}^m$ and lifting to G.*

Thus we have reduced our problem to understanding C-L extensions of finite-dimensional tori by finite-dimensional tori. This reduces to just considering extensions by S^1.

As we are now working completely in finite-dimensional tori. We will regard the first coordinate group as $\mathbb{R}^n/\Gamma$ for some lattice subgroup $\Gamma \subseteq \mathbb{Z}^n$, our group operation will be $+$, and the identity will be indicated by $\vec{0}$. The second-coordinate group will be a product of circles $S^1 \subseteq \mathbb{C}$ with group operation $\cdot$. When it is more convenient for us to represent the second coordinate additively we will say so, but we will almost always exponentiate our additive expressions back to S^1.

Lemma 5.6. *Suppose $\mathcal{N}$ is a C-L extension of $\mathbb{R}^n/\Gamma$ by S^1. Then there is a basis $[\vec{e}_1, \ldots, \vec{e}_n]$ for Γ so that relative to this basis the connected component of $\vec{0}$ in $\mathcal{N}/C(\mathcal{N})$ is isomorphic to $\mathbb{Z}^a \times \mathbb{R}^{n-a}$.*

Proof. Let $\Gamma_1 \subseteq \Gamma$ be the subgroup of elements of $\mathbb{R}^n$ for which the corresponding geodesic circles S have

$$H_{\mathcal{N}}(S) = \{\vec{0}\}.$$

We first show Γ/Γ_1 is torsion free. If not, then for some $\vec{m} \in \Gamma \backslash \Gamma_1$ there is a q with $q\vec{m} \in \Gamma_1$. But this would imply $H_{\mathcal{N}}(\vec{m}) \in \widehat{\mathbb{T}^n}$ had as range a finite subgroup of S^1. But as $\mathbb{T}^n$ is connected, $\vec{m} \in \Gamma_1$.

This now tells us that choosing a minimal set of generators $\{\vec{e}_1, \ldots, \vec{e}_a\}$ for Γ_1, we can extend it to a minimal set of generators $\{\vec{e}_1, \ldots, \vec{e}_n\}$ for Γ, and the result follows.

Lemma 5.7. *Suppose $\mathcal{N}$ is a C-L extension of $G = \mathbb{R}^n / \Gamma$ by S^1, where Γ is a lattice subgroup of $\mathbb{Z}^n$ and the connected component of $\vec{0}$ in $\mathcal{N}/C(\mathcal{N})$ is*

$$Q \simeq \mathbb{R}^a \times \mathbb{T}^{n-a}.$$

The bihomomorphism $\lambda : Q \times Q \to S^1$ lifts to a skew-symmetric bilinear form

$$B : \mathbb{R}^n \times \mathbb{R}^n \to \mathbb{R}.$$

For a choice of basis $[\vec{e}_1, \ldots, \vec{e}_n] = [e]$ for $\mathbb{R}^n$ which generates Γ, if we represent $\vec{u} \in \Gamma$ as $\vec{u} = \sum u_i \vec{e}_i$, then

$$B(\vec{u}, \vec{v}) = (u_i)^T B_{[e]}(v_j) = \sum_{i,j=1}^{m} b_{i,j} u_i v_j, \quad \text{where } b_{i,j} = -b_{j,i} \in \mathbb{Z}^n.$$

For $\vec{u} \in \mathrm{Hom}(\mathbb{T}^n, S^1) = \mathbb{Z}^n$,

$$H_{\mathcal{N}}(\vec{u})(\vec{v}\,\Gamma) = \exp(2\pi i B(\vec{u}, \vec{v})).$$

Lastly,

$$B_{[f]} = [e]^{-1}[f] B_{[e]} [f]^{-1}[e],$$

where $[e]^{-1}[f] \in SL(n, \mathbb{Z})$, as both $[e]$ and $[f]$ generate Γ.

Proof. It is sufficient to demonstrate this when $\Gamma = \mathbb{Z}^n$, as the change of variable formulae are standard. First, λ lifts to a skew-symmetric bihomomorphism $\mathbb{R}^n \times \mathbb{R}^n \to S^1$. This lifts uniquely to the covering space $\mathbb{R}$ of S^1 as a skew-symmetric bilinear form

$$B : \mathbb{R}^n \times \mathbb{R}^n \to \mathbb{R}.$$

For any $\vec{u} \in \Gamma$, the line from $\vec{0}$ to $\vec{u}$ projects to a loop in $\mathbb{T}^n$. Let $\exp(r) = e^{2\pi i r}$. We can compute, for $f \in F(\vec{v}\,\Gamma)$, that

$$R_{\vec{v}\,\Gamma, f} R_{1, H_{\mathcal{N}}(\vec{u})} R_{\vec{v}\,\Gamma, f}^{-1} R_{1, H_{\mathcal{N}}(\vec{u})}^{-1} = R_{1, \exp(B(\vec{v}, \vec{u}))} = R_{1, H_{\mathcal{N}}(\vec{u})}(-\vec{v}\,\Gamma).$$

Hence $\exp(B(\vec{u}, \vec{v})) = H_{\mathcal{N}}(\vec{u})(\vec{v}\,\Gamma)$.

This forces B to be an integer matrix, as $B(\vec{u}, \cdot)$ must be well-defined mod $\mathbb{Z}$ for all $\vec{u} \in \mathbb{Z}^n$.

We now know all nontrivial values for $H_{\mathcal{N}}(\ell)$ are projections to an $H_{\mathcal{N}}(S)$ where S is a geodesic circle in a finite-dimensional toral factor of G which, relative to a basis, we can write as $H_{\mathcal{N}}(\vec{m})$. All toral factors of G lie in

$$G_{\mathrm{con}} = G(\vec{s}, \{q_j, A_j\}),$$

and of course

$$G_1 = G(\vec{s}^1, \{q_i^1, A_i^1\}).$$

Fix these representations of these groups.

We now examine how the bilinear forms defining $H_\mathcal{N}(\vec{m})$ via Lemma 5.7 evolve as we move through the inverse limits to both G_{con} and G_1. These will all be expressed as vectors of matrices relative to the bases $[e_j]$ and $[e_i^1]$ of these representations.

Consider a C-L extension of $\mathbb{R}^{n_j}/\Gamma_j$ by S^1. We have associated with Γ_j a generating basis $[\vec{e}_{j,1}, \ldots, \vec{e}_{j,n_j}] = [e_j]$. Associated with this extension is a bilinear form, represented by a matrix $B_{[e_j]}$ relative to this basis. Suppose now we move one step further in the inverse limit to G_{con}. There are two possibilities for how the basis changes depending, on the value of s_j, given by the following rules of evolution.

Rule 1:. *If $s_j = 1$, the dimension of the torus increases by one, and we simply add on a new generator. In this case*

$$B_{[e_{j+1}]} = \begin{bmatrix} B_{[e_i]} & & 0 \\ & & \vdots \\ 0 & \cdots & 0 \end{bmatrix}.$$

If $s_j = 0$ then

$$[e_{j+1}] = [e_j]A_i \begin{bmatrix} q_j & 0 & \cdots & 0 \\ 0 & 1 & & \\ \vdots & & \ddots & \vdots \\ 0 & & \cdots & 1 \end{bmatrix},$$

and so

$$B_{[e_{j+1}]} = \begin{bmatrix} q_j & 0 & \cdots & 0 \\ 0 & 1 & & \\ \vdots & & \ddots & \vdots \\ 0 & & \cdots & 1 \end{bmatrix} A_j^T B_{[e_j]} A_j \begin{bmatrix} q_j & 0 & \cdots & 0 \\ 0 & 1 & & \\ \vdots & & \ddots & \vdots \\ 0 & & \cdots & 1 \end{bmatrix}.$$

If we consider an extension of $G/\Gamma \simeq \mathbb{T}^m$ by a finite-dimensional torus $G_1/\Gamma^1 \simeq \mathbb{T}^n$, relative to a

basis $[e^1]$ each coordinate of the torus $\mathbb{T}^n$ will be assigned a bilinear form, hence the extension is determined, up to cohomology, by a vector of bilinear forms. This is again dependent on the choice of basis made for the torus $\mathbb{T}^n$, so we write it as

$$[B_1^{[e^1]}, B_2^{[e^1]}, \ldots, B_n^{[e^1]}].$$

If we make a change of basis to $[f^1]$, then

$$[B_1^{[f^1]}, B_2^{[f^1]}, \ldots, B_n^{[f^1]}]^T = [f^1]^{-1}[e^1][B_1^{[e^1]}, B_2^{[e^1]}, \ldots, B_n^{[e^1]}]^T.$$

Suppose $R_{g_0, \phi}$ is a C-L action on $G \times G_1$. Having chosen a term $\mathbb{R}^{n_i^1}/\Gamma_i^1$ in the inverse limit to G_1, we obtain a C-L extension $\mathcal{N}_1$ of G by $\mathbb{T}^{n_i^1}$. By Corollary 5.5 and Lemma 5.7, there is a first term $\mathbb{R}^{n_{j(i)}}/\Gamma_{j(i)}$ in the inverse limit to G_{con} over which the bilinear forms associated with $H_{\mathcal{N}_i}(\ell)$ are of the form $H_{\mathcal{N}_i}(\vec{m})$, $\vec{m} \in \mathbb{Z}^{n_{j(i)}}$.

Relative to the bases $[e_i^1]$ and $[e_j]$, $j \geq j(i)$, we write these as a vector of matrices

$$\vec{B}_{[e_j]}^{[e_i^1]} = [B_{[e_j]}^{[e_i^1],1}, \ldots, B_{[e_j]}^{[e_i^1],n_i^1}].$$

As j increases beyond $j(i)$ and i remains fixed, these matrices evolve according to Rule 1. We also need to examine their evolution in i.

Suppose now $j \geq j(i + 1)$; we ask how $\vec{B}_{[e_j]}^{[e_{i+1}^1]}$ and $\vec{B}_{[e_j]}^{[e_i^1]}$ are related.

Rule 2:. *If $s_i^1 = 1$, i.e. the dimension n_{i+1}^1 is one larger than n_i^1, for $k \leq n_i$,*

$$B_{[e_j]}^{[e_{i+1}^1],k} = B_{[e_j]}^{[e_i^1],k} \text{ and } B_{[e_j]}^{[e_{i+1}^1],n_{i+1}^1}$$

is the new skew-symmetric matrix defined by the new bilinear form associated with the new dimension of the torus.

If $s_i^1 = 0$, i.e. $n_{i+1} = n_i$, then as

$$[e_{i+1}^1] = [e_i^1]A_i^1 \begin{bmatrix} q_i^1 & 0 & \cdots & 0 \\ 0 & 1 & & \\ \vdots & & \ddots & \\ 0 & & \cdots & 1 \end{bmatrix},$$

$$\vec{B}_{[e_j]}^{[e_{i+1}^1]} = \begin{bmatrix} (q_i^1)^{-1} & 0 & \cdots & 0 \\ 0 & 1 & & \\ \vdots & & \ddots & \\ 0 & & \cdots & 1 \end{bmatrix} (A_i^1)^{-1} \vec{B}_{[e_j]}^{[e_i^1]}.$$

Perhaps the most important observation to make here is that it implies the first coordinate of $(A_i^1)^{-1} \vec{B}_{[e_j]}^{[e_i^1]}$ has all its integer entries divisible by q_i^1.

Definition 5.3. Suppose G is compact, metric, Abelian and monothetic, and G_1 is compact, metric and 0-connected with

$$G_{con} = G(\vec{s}, \{q_j, A_j\}) \text{ and } G_1 = G(\vec{s}^1, \{q_i^1, A_i^1\}),$$

and we have $j : \mathbb{N} \to \mathbb{N}$ nondecreasing and vectors of skew-symmetric matrices

$$\mathcal{B} = \{\vec{B}_{[e_j]}^{[e_i^1]} \mid B_{[e_j]}^{[e_i^1],k} \text{ is a skew-symmetric } n_j \times n_j \text{ matrix, } j \geq j(i)\}$$

evolving in j according to Rule 1, and in i according to Rule 2. We call such a $\mathcal{B}$ a *bilinear array compatible with* (G, G_1). If $R_{g_0, \phi}$ is a C-L action on $G \times G_1$, then we let $\mathcal{B}(R_{g_0, \phi})$ be the associated bilinear array.

We will spend some time with such arrays with a 'staircase' lower edge $E = \{(j, i) \mid j(i) \leq j \leq j(i+1)\}$. Notice that Rules 1 and 2 completely force the terms of the array once you know just the new last terms added when moving down from row i to row $i+1$ at column $j(i+1)$ where $n_{i+1}^1 = n_i^1 + 1$.

If we have two such arrays which agree at all indices (i, j) where they both exist, we will consider them to be equal.

We now restate Theorem 4.16 in terms of bilinear arrays.

Lemma 5.8. *If R_{g_0, ϕ_1} and R_{g_0, ϕ_2} are two C-L actions on $G \times G_1$, then we know $R_{g_0, \phi_1^{-1} \phi_2}$ is also, and its bilinear array is*

$$\mathcal{B}(R_{g_0, \phi_1^{-1} \phi_2}) = \{\vec{B}_{[e_j, 2]}^{[e_i^1]} - \vec{B}_{[e_j, 1]}^{[e_i^1]}\},$$

where

$$\mathcal{B}(R_{g_0, \phi_1}) = \{\vec{B}_{[e_j, 1]}^{[e_i^1]}\} \text{ and } \mathcal{B}(R_{g_0, \phi_2}) = \{\vec{B}_{[e_j, 2]}^{[e_i^1]}\}.$$

If R_{g_0, ϕ_1} is an element of a C-L extension of G by G_1, then any other C-L action R_{g_0, ϕ_2} with $\mathcal{B}(R_{g_0, \phi_1}) = \mathcal{B}(R_{g_0, \phi_2})$ is also an element of a C-L extension.

Proof. The associated bilinear arrays $\mathcal{B}(R_{g_0, \phi_t})$ determine, by Lemma 5.7, all values $H_{\mathcal{N}_i}(\ell)$, for all δ-loops ℓ, δ small enough, where $\mathcal{N}_i$ is the C-L extension of G by $\mathbb{R}^{n_i^1} / \Gamma_i^1$ that is the corresponding factor of R_{g_0, ϕ_t}. Theorem 4.16 now completes the result.

Thus the bilinear system $\mathcal{B}$ alone determines whether or not a given C-L action is part of a C-L extension. We will now show that for any bilinear array $\mathcal{B}$ compatible with G, G_1 there is a C-L action $R_{g_0, \phi}$ with $\mathcal{B} = \mathcal{B}(R_{g_0, \phi})$. The generator g_0 will not be arbitrary in this, but will have to belong to the arcwise-connected component of the identity when projected to G_{con}.

Theorem 5.9. *If $G = G_{\mathrm{con}} = G(\vec{s}, \{q_j, A_j\})$, $G_1 = G(\vec{s}^{\,1}, \{q_i^1, A_i^1\})$, $\mathcal{B} = \{\vec{B}_{[e_j]}^{[e_i^1]} \mid k \geq j(i)\}$ is a bilinear array compatible with (G, G_1), and g_0 is a generator for G in the arcwise-connected component of the identity, then there is a $\phi: G \to G_1$ so that $R_{g_0, \phi}$ is a C-L action with $\mathcal{B}(R_{g_0, \phi}) = \mathcal{B}$.*

There are two issues to handle in proving Theorem 5.9. We have to create C-L sections over tori with a given bilinear form, and we have to maintain the evolution of the C-L action over g_0 through to the inverse limit of these sections. As we will see later, the entire section cannot always be lifted. Both these issues are handled via the Heisenberg example described earlier.

To be completely explicit we will fix our choice of fundamental domain for each Γ_j. We will select them inductively, $F_j \subseteq \mathbb{R}^{n_j}$.

As $\Gamma_1 = \mathbb{Z}$ and $n_1 = 1$, let $F_1 = [0, 1)$. Supposing F_j is constructed, build F_{j+1} as follows.

1) If $n_{j+1} = n_j + 1$, then let $F_{j+1} = F_j \times [0, 1)$.

2) If $n_{j+1} = n_j$, then Γ_j / Γ_{j+1} is cyclic of order q_j, generated in particular by $q_j^{-1} \vec{e}_{j+1,1}$. Set

$$F_{j+1} = \cup_{k=0}^{q_j - 1} (F_j + k q_j^{-1} \vec{e}_{j+1,1}).$$

This is a disjoint union, of course.

For each matrix component $B_{[e_j]}^{[e_i^1], k}$ of a term in $\mathcal{B}$, we define $D_{i,j,k}$ to be the bilinear form on $\mathbb{R}^{n_j}$ whose matrix representation in the basis $[e_j]$ is the upper-triangular matrix $D_{[e_j]}^{[e_i^1], k}$ satisfying

$$D_{[e_j]}^{[e_i^1], k} - \left(D_{[e_j]}^{[e_i^1], k} \right)^T = B_{[e_j]}^{[e_i^1], k}.$$

Let $\pi_j : G = G_{\text{con}} \to \mathbb{R}^{n_j} / \Gamma_j$ be the projection to the j'th component of the inverse limit, and $\vec{u}_j(g) \in F_j$ be the choice of coset representative, $\pi_j(g) = \vec{u}_j(g) \Gamma_j$.

This now gives us C-L sections from $G \times G$ to S^1 of the form

$$f_{i,j,k}(g_1, g_2) = f_{F_j, \Gamma_j, D_{i,j,k}}(\pi_j(g_1), \pi_j(g_2)).$$

Lemma 4.1 tells us the matrix representation in the basis $[e_j]$ of the bilinear form associated with this C-L extension of G by S^1 is $B_{[e_j]}^{[e_i^1], k}$.

Lemma 5.10. *If g_0 is arcwise connected to the identity in $G = G_{\text{con}}$, then for all j sufficiently large there is an $s_{i,j,k} : G \to S^1$ and a constant $\lambda_{i,j,k}$ so that*

$$\left(f_{i,j+1,k}(g_0, g) \right) \left(f_{i,j,k}(g_0, g) \right)^{-1} = \lambda_{i,j,k} s_{i,j,k}(g_0 g) s_{i,j,k}(g)^{-1},$$

i.e. $f_{i,j+1,k}$ and $f_{i,j,k}$ are within a constant of being g_0-cohomologous.

Proof. To say g_0 is arcwise connected to the identity is to say, once j is large enough, if $n_{j+1} = n_j$ then $\vec{u}_{j+1}(g_0) = \vec{u}_j(g_0)$ (the coset representatives of g_0 do not tend to infinity in $\mathbb{R}^{n_j}$.)

Assume j is this large. From Rule 1, if $n_{j+1} = n_j + 1$ then $f_{i,j+1,k} = f_{i,j,k}$, so we need only consider those j with $n_{j+1} = n_j$.

Fix the values of i, j, k, and relative to the basis $[e_{j+1}]$, $D_{[e_j]}^{[e_i^1], k}$ transforms to

$$\begin{bmatrix} q_j & 0 & \cdots & 0 \\ 0 & 1 & & \\ \vdots & & \ddots & \\ 0 & & \cdots & 1 \end{bmatrix} A_i^T D_{[e_j]}^{[e_i^1], k} A_i \begin{bmatrix} q_j & 0 & \cdots & 0 \\ 0 & 1 & & \\ \vdots & & \ddots & \\ 0 & & \cdots & 1 \end{bmatrix} \stackrel{\text{def}}{=} [D'].$$

Let D_j and D_{j+1} be the bilinear forms whose representations in $[e_{j+1}]$ are $D_{[e_j]}^{[e_i^1],k}$ and $[D']$ respectively. Notice both differ from their transpose by $B_{[e_{j+1}]}^{[e_i^1],k}$, and so $D = D_{j+1} - D_j$ is a symmetric bilinear form. We compute

$$
\left(f_{i,j+1,k}(g_0,g)\right)\left(f_{i,j,k}(g_0,g)\right)^{-1} =
$$
$$
\exp\big(2\pi i\big(D(\vec{u}_{j+1}(g_0),\vec{u}_{j+1}(g)) - D_j(\vec{u}_{j+1}(g),\vec{u}_j(g) - \vec{u}_{j+1}(g))
$$
$$
+ D_j(\vec{u}_{j+1}(g_0) + \vec{u}_{j+1}(g), \gamma_{F_{j+1}}(\vec{u}_{j+1}(g_0) + \vec{u}_{j+1}(g)) - \gamma_{F_j}(\vec{u}_{j+1}(g_0)
$$
$$
+ \vec{u}_{j+1}(g))) - D_j(\vec{u}_{j+1}(g_0) + \vec{u}_{j+1}(g), \gamma_{F_{j+1}}(\vec{u}_{j+1}(g_0) + \vec{u}_{j+1}(g)))\,\big)\big)
$$
$$
= f_{F_{j+1},\Gamma_{j+1},D}(\vec{u}_{j+1}(g_0),\vec{u}_{j+1}(g))s(g_0 g)s(g)^{-1},
$$

where

$$
s(g) = \exp(2\pi i(\ D_j(\vec{u}_{j+1}(g), \gamma_{F_{j+1}}(\vec{u}_{j+1}(g)) - \gamma_{F_j}(\vec{u}_{j+1}(g)))\ )).
$$

Notice that $\gamma_{F_{j+1}}(\vec{u}) - \gamma_{F_j}(\vec{u})$ is well-defined on $\mathbb{R}^{n_j}/\Gamma_{j+1}$.
But now, as D is symmetric, Lemma 4.1 completes the result.

This tells us, for any $\lambda \in S^1$, $R_{g_0,\lambda f_{i,j,k}}(g_0)$ and $R_{g_0,\lambda \lambda_{i,j,k} f_{i,j+1,k}}(g_0)$ are conjugate by the map $(g,z) \to (g, s_{i,j,k}(g)z)$.
This tells us we can define an array of constants

$$
\{c_{i,j,k} \mid j \geq j(i),\ 1 \leq k \leq n_i^1\}
$$

with $R_{g_0,c_{i,j,k} f_{i,j,k}}(g_0)$ conjugate to $R_{g_0,c_{i,j+1,k} f_{i,j+1,k}}(g_0)$. To be precise,

1) When $n_{i+1}^1 = n_i^1 + 1$ set $c_{i,j(i+1),n_{i+1}^1} = 1$, i.e. start each new row of constants at value 1.

2) If $c_{i,j,k}$ is already defined, extend along the row by setting

$$
c_{i,j+1,k} = \begin{cases} c_{i,j,k}\lambda_{i,j,k} & \text{when } n_j = n_{j+1} \\ c_{i,j,k} & \text{when } n_{j+1} = n_j + 1. \end{cases}
$$

3) If $c_{i,j(i+1),k}$ is already defined (i.e $k < n_{i+1}^1$), when $n_{i+1}^1 = n_i^1 + 1$ set $c_{i+1,j,k} = c_{i,j,k}$ and when $n_{i+1}^1 = n_i^1$ then

$$
(c_{i+1,j,1},\ldots,c_{i+1,j,n_i^1})^T = \begin{bmatrix} (q_i^1)^{-1} & & & 0 \\ & 1 & & \\ & & \ddots & \\ 0 & & & 1 \end{bmatrix} [A_i^1]^{-1}(c_{i,j,1},\ldots,c_{i,j,n_i^1})^T.
$$

Define cocycles

$$
\phi_{i,j} : G \to (S^1)^{n_i^1} \text{ by } \phi_{i,j}(g) = \{c_{i,j,k}\big(\exp(2\pi i(\vec{u}_j(g_0)D_{[e_j]}^{[e_i^1],k}\vec{u}_j(g))\big)\}_{k=1}^{n_i^1}.
$$

Corollary 5.11. *As g_0 is in the arcwise-connected component of the identity in G, once j is large enough,*

1) for i fixed and $j \geq j(i)$, $\phi_{i,j}$ and $\phi_{i,j+1}$ are g_0-cohomologous and hence $R_{g_0,\phi_{i,j}} \simeq R_{g_0,\phi_{i,j}}$,

2) and for $j = j(i+1)$, if $n^1_{i+1} = n^1_i + 1$, then $R_{g_0,\phi_{i+1,j}}$ factors onto $R_{g_0,\phi_{i,j}}$ by deleting the last component of $(S^1)^{n^1_{i+1}}$; and if $n^1_{i+1} = n^1_i$, then $R_{g_0,\phi_{i+1,j}}$ factors onto $R_{g_0,\phi_{i,j}}$ by the projection $\mathbb{R}^{n^1_i}/\Gamma_{i+1} \to \mathbb{R}^{n^1_i}/\Gamma_i$ on the second coordinate viewed additively in the exponent.

Now consider the sequence of cocycle extensions along the upper edge of the array, $E = \{(j,i) \mid j(i) \leq j \leq j(i+1)\}$. Corollary 5.11 tells us that $\phi_{i,j}$, $j(i) \leq j \leq j(i+1)$ are all g_0-cohomologous and at the 'steps' the actions $R_{g_0,\phi_{i+1,j(i+1)}}$ factor onto $R_{g_0,\phi_{i,j(i+1)}}$ by modding out by a subgroup on the second coordinate.

Thus, lifting upward along this staircase, we can construct an inverse limit cocycle $\phi : G_0 \to G$ which, when projected onto the i'th torus term in the inverse limit to G, becomes cohomologous to $\phi_{i,j(i)}$.

Corollary 5.12. $R_{g_0,\phi}$ *is Conze-Lesigne and the corresponding array* $\mathcal{B}(R_{g_0,\phi})$ *is* $\mathcal{B}$.

Proof. As the projections $\pi_i \circ \phi$ are cohomologous to $\phi_{i,j(i)}$, the associated bilinear form is $\vec{B}^{[e^1_j]}_{[e_j(i)]}$ as required, and Rules 1 and 2 force the rest of the entries.

This has now completed the proof of Theorem 5.9. Notice that what we have shown is that if g_0 is in the arcwise-connected component of the identity it has the universal property of being able to realize any bilinear array compatible with (G, G_1).

Corollary 5.13. *If G is a torus, either finite- or infinite- dimensional, then any C-L action $R_{g_0,\phi}$ on $G \times G_1$ is an element of a C-L extension.*

Proof. The array $\mathcal{B}(R_{g_0,\phi})$ is realizable by the construction of Theorem 5.9. Now, as tori are arcwise connected, all the R_g actually lift through this construction to the inverse limit, and so this construction is actually of a C-L extension, and Theorem 4.16 completes the result.

We now give an example showing that when G is not a torus a C-L action need not belong to a C-L extension. Let

$$S = \{(z_1, z_2, \ldots) \mid z_i \in S^1, \ z^2_{i+1} = z_i\}$$

be the 2-solenoid, and set

$$G = S \times S^1 \text{ and } G_1 = S.$$

For choice of bases, let $[e_j] = \begin{bmatrix} 2^j & 0 \\ 0 & 1 \end{bmatrix}$ and $[e_i^1] = [2^i]$. Of course $n_j = 2$ and $n_i^1 = 1$.

Set

$$B_{[e_j]}^{[e_i^1]} = \begin{bmatrix} 0 & 2^{j-i} \\ -2^{j-i} & 0 \end{bmatrix}$$

for $j \geq j(i) \overset{\text{def}}{=} i$.

This is a bilinear array compatible with (G, G_1). Now for g_0 any generator of G in the arcwise-connected component of the identity, this bilinear array is realizable for a C-L action of the form $R_{g_0,\phi}$. On the other hand, the next theorem shows such an action cannot be an element of a C-L extension.

Theorem 5.14. *Any C-L extension of $G = S \times S^1$ by $G_1 = S$ is quasi-Abelian.*

Proof. Suppose $\mathcal{N}$ is a C-L extension of G by G_1. Then $R_{1,\phi} \in \mathcal{N}$ iff $\phi \in \text{Hom}(G, G_1)$. For $H \in \text{Hom}(G, G_1)$

$$H(((z_i), z)) = H_1((z_i)) H_2(z),$$

where $H_2 \in \text{Hom}(S^1, S)$.

But as S has no circle subgroups, $H_2 \equiv 1$.

In the i'th term of the inverse limit $(\mathcal{N}_i)$, from Lemma 5.7,

$$H_{\mathcal{N}_i}(\vec{u})(\vec{v} + \Gamma_i) = \exp(2\pi i B(\vec{u}, \vec{v})).$$

As $H_2 \equiv 1$ we know $H_{\mathcal{N}_i}(\vec{u})((0, t) + \Gamma_i) = 1$, and so $B_i(\vec{u}, (0, t)) = 0$, or

$$B_i = \begin{bmatrix} * & * \\ 0 & 0 \end{bmatrix}.$$

But as B_i is skew symmetric, $B_i = [0]$, and we are done.

Given two groups G and G_1, both 0-connected, and a generator g_0 we can ask what are the bilinear arrays that are realizable by C-L actions $R_{g_0,\phi}$. We know this collection of arrays is closed under componentwise addition and always contains the $[0]$ array, and so is a group.

We will not push any further with this area but will end this section with some questions.

1) Are the $g_0 \in G$ arcwise connected to the identity the only elements over which all compatible bilinear arrays can be realized?

2) If a bilinear array is realizable from a C-L extension of $g_0 \in G$, can it be realized by a Heisenberg example as in the proof of Theorem 5.9?

3) How can one characterize the bilinear arrays which are realizable by a C-L action $R_{g_0,\phi}$ in terms of g_0?

4) Are there $g_0 \in G$ with the property that any C-L action $R_{g_0,\phi}$ must be an element of a C-L extension?

5) If a bilinear array can be realized over all $g_0 \in G$, then must any C-L action with this as its bilinear array be an element of a C-L extension?

6. BACK TO EIGENFUNCTIONS OF $T \times S$

We now return to our work of Section 1 to ask when C_1, a C-L character of T, and C_2, a C-L character of S, arise together in an eigenfunction of $T \times S$. We can assume $T = R_{g_0,\phi}$ and $S = R_{g_0,\phi'}$ are both C-L actions on $(G \times G_1)$ and $(G \times G_1')$ respectively, with (R_{g_0}, G) their common Kronecker factor. Thus for C a character of G_1, $(R_{g_0,C\circ\phi}, G \times S^1)$ is an element of a C-L extension of G by S^1.

We will associate with $R_{g_0,C\circ\phi}$ a pair of invariants relative to a representation of G as $G(\vec{s}, \{q_j, A_j\}) \times G_{\text{dis}}$.

The first is the array (here just a sequence) of matrices

$$\mathcal{B}(R_{g_0,C\circ\phi}) = \{ B_{[e_j]}^{[1]} \mid j \geq j(1) \}$$

representing the bilinear form associated with this extension.

To get the second invariant, construct a Heisenberg extension $R_{g_0,\overline{\phi}}$ of G by S^1 by lifting an extension of $\mathbb{R}^{n_{j(1)}}/\Gamma_{j(1)}$ with bilinear form D, the upper triangle of $B_{[e_{j(1)}]}^{[1]}$, explicitly as in the construction proving Theorem 5.9.

Because this extension has the same bilinear array as $R_{g_0,C\circ\phi}$, we see that $(R_{g_0,\overline{\phi}^{-1}C\circ\phi}, G \times S^1)$ is a quasi-Abelian extension, i.e. $\overline{\phi}^{-1}C \circ \phi$ is g_0-cohomologous to some $\lambda\chi$, $\chi \in \widehat{G_{\text{con}}}$. In this form λ is not unique, as there are constants (eigenvalues) that are coboundaries, but χ is unique by Lemma 4.9. The second invariant is

$$\chi = \chi(R_{g_0,C\circ\phi}).$$

Lemma 6.1. *Suppose $(R_{g_0,\phi}, G \times G_1)$ is a C-L action C-L characters, then*

$$\mathcal{B}(R_{g_0,C_1^{-1}C_2\circ\phi}) = \mathcal{B}(R_{g_0,C_2\circ\phi}) - \mathcal{B}(R_{g_0,C_1\circ\phi}) \text{ and}$$

$$\chi(R_{g_0,C_1^{-1}C_2\circ\phi}) = \chi(R_{g_0,C_1\circ\phi})^{-1}\chi(R_{g_0,C_2\circ\phi}).$$

Furthermore, if (R_{g_0}, G) is the Kronecker algebra of this action, then $C_1 = C_2$ iff

1) $\mathcal{B}(R_{g_0,C_2\circ\phi}) = \mathcal{B}(R_{g_0,C_1\circ\phi})$ and

2) $\chi(R_{g_0,C_1\circ\phi}) = \chi(R_{g_0,C_2\circ\phi}).$

Proof. The first equality follows from Lemma 5.8. The second follows from noting that the Heisenberg constructions for $\mathcal{B}$ and $-\mathcal{B}$ are inverses of one another. One direction of the latter statement is trivial. For the other, just note that 1) and 2) imply $R_{g_0,C_1^{-1}C_2 \circ \phi}$ has pure point spectrum as $C_1^{-1}C_2 \circ \phi$ is cohomologous to a constant. But as the Kronecker algebra of $R_{g_0,\phi}$ is (R_{g_0}, G), this implies $C_1^{-1}C_2$ must be the trivial character.

Theorem 6.2. *Suppose* $(R_{g_0,\phi}, G \times G_1)$ *and* $(R_{g_0,\phi'}, G \times G_1')$ *are two C-L actions. Characters* $C_1 \in \widehat{G_1}$ *and* $C_2 \in \widehat{G_1'}$ *arise in an eigenfunction*

$$e(g, cg, g_1, g_1') = C_1^{-1}(g_1)C_2(g_1')f(g, cg)$$

of $R_{g_0,\phi} \times R_{g_0,\phi'}$ *iff*

1) $\mathcal{B}(R_{g_0,C_2 \circ \phi'}) = \mathcal{B}(R_{g_0,C_1 \circ \phi})$ *and*

2) $\chi(R_{g_0,C_1 \circ \phi}) = \chi(R_{g_0,C_2 \circ \phi'})$.

Proof. Consider the family of S^1 extensions of R_{g_0} given by the functions

$$\phi_c(g) = (C_1^{-1} \circ \phi(g))(C_2 \circ \phi'(cg)).$$

Each is a C-L extension, as both the first and second terms are C-L characters.

For the forward direction, that $C_1^{-1}(g_1)C_2(g_1')f(g, cg)$ is an eigenfunction of $R_{g_0,\phi} \times R_{g_0,\phi'}$ tells us $g_c(g, z) = zf(g, cg)$ is an eigenfunction of R_{g_0,ϕ_c}. This eigenfunction separates points on the S^1 fiber over each $g \in G$. Hence R_{g_0,ϕ_c} has pure point spectrum and ϕ_c is cohomologous to a constant. Hence

1) $\mathcal{B}(R_{g_0,\phi_c}) = [0]$ *and*

2) $\chi(R_{g_0,\phi_c}) = 1$

completing one direction.

For the backward direction, 1) and 2) imply that for each c, ϕ_c is cohomologous to a constant $\lambda(c)$, which is to say the circle extension

$$(g, cg, z) \to (g_0g, g_0cg, \phi_c(g)z)$$

of $R_{g_0} \times R_{g_0}$ has pure point spectrum on all its ergodic components, and its eigenfunctions must span L^2 of the S^1 fiber over almost every $(g, cg) \in G \times G$.

Restricted to a fiber (g, cg), an eigenfunction looks like $z^n f(g, cg)$. In order to separate points, for some eigenfunction we must have $n = 1$, which is to say there is an eigenfunction of the form

$$zf(g, cg) = C_1^{-1}(g_1)C_2(g_1')f(g, cg).$$

This completes our work, associating to each C-L character two invariants, a sequence of bilinear forms encoding its Heisenberg structure, and then a character of the connected component of G encoding its quasi-Abelian structure.

At the end of Section 5 we posed several questions concerning the precise structure of possible C-L actions. We now can add to those questions. There are two directions in which to try to extend this work. The C-L algebra has been seen here to arise from eigenfunctions of a Cartesian square and to be the inverse limit of order-two nil-rotations. What happens when we consider Cartesian products of more terms? Do the eigenfunctions of these products come from higher-order Conze-Lesigne algebras which are inverse limits of higher-order nil-rotations? Such an investigation seems directly approachable with the methods and results constructed here.

REFERENCES

1. J. Bourgain, *Pointwise ergodic theorems for arithmetic sets, with an Appendix by J. Bourgain, H. Furstenberg, Y. Katznelson, and D. Ornstein*, Pub. Math. I.H.E.S. **69** (1989), 5–45.

2. J.-P. Conze, E. Lesigne, *Théorèmes ergodiques pour des mesures diagonales*, Bull. Soc. Math. France, **112** (1984), 143-145.

3. J.-P. Conze, E. Lesigne, *Sur un théorème ergodique pour des mesures diagonales*, C.R. Acad. Sci. Paris **t.306, serie I** (1988), 491–493.

4. A. del Junco, M. Lemańczyk, and M. Mentzen, *Semisimplicity, joinings and group extensions*, (preprint).

5. H. Furstenberg, *The structure of distal flows*, Amer. J. Math. **85** (1963), 477-515.

6. M. Lemańczyk, M. Mentzen, *Ergodic Abelian group extensions of rotations*, Torun (1990).

7. E. Lesigne, *Équations fonctionnelles, couplages de produits gauches et théorèmes ergodiques pour mesures diagonales*, Bull. Soc. Math. France **121** (1993), 315–351.

8. D. Newton, *On canonical factors of ergodic dynamical systems*, J. London Math. Soc. **19** (1978), 129–136.

9. D. J. Rudolph, *Fundamentals of Measurable Dynamics*, Oxford University Press, New York, 1990.

10. W. A. Veech, *A criterion for a process to be prime*, Monatshefte Math. **94** (1982), 335–341.

11. R. Zimmer, *Extensions of ergodic group actions*, Illinois J. of Math. **20** (1976 (a)), 373–409.

12. R. Zimmer, *Ergodic actions with generalized discrete spectrum*, Illinois J. of Math. **20** (1976 (b)), 555–588.

DEPARTMENT OF MATHEMATICS, UNIVERSITY OF MARYLAND, COLLEGE PARK MD 20742

E-mail address: djr@math.umd.edu

CONFERENCE PROGRAM

Survey Lecture. Jean-Franqis Méla, *Singular measures on the circle and spectral theory*

Mahendra Nadkarni, L^∞ *and* L^2 *spectra of rank one transformations*

Emmanuel Lesigne, *Ergodic properties of q-multiplicative sequences*

Jacob Feldman, *Decreasing sequences of σ-fields and a question about Brownian motion*

Poornima Raina, *Maximal order of the Fourier transforms of a function in* $L^p(\mathbb{R})$

Martin Bluemlinger, *Rajchman measures on locally compact groups*

Survey Lecture. Jean-Paul Thouvenot, *Some properties and applications of joinings in ergodic theory*

Geoffrey Goodson, *The weak closure property and ergodic transformations isomorphic to their inverses*

Tomasz Downarowicz, *Strictly nonpointwise Markov operators*

Jonathan King, *On the Affine Problem of Erdös*

Mariusz Lemanczyk, *A natural family of factors of a measure-preserving transformation*

Sebastien Ferenczi, *Spectral multiplicity and rank*

Jan Kwiatkowski, *Rank and spectral multiplicity in ergodic theory*

Survey Lecture. Mate Wierdl, *Fourier methods and almost everywhere convergence*

Chris Bose, *Ergodic baker's transformations*

Jim Olsen, *Besicovitch sequences as weights in the weighted pointwise ergodic theorem*

Meir Smorodinsky, *On the ergodicity of the Lévy transformation*

Ryszard Jajte, *Almost sure convergence of projections to selfadjoint operators*

Christian Skau, *Topological orbit equivalence and dimension groups*

Jean-Marc Derrien, *Théorème ergodique polynomial ponctuel pour les K-systèmes*

Doğan Çömez, *Weighted ergodic theorems for mean ergodic* L^1 *contractions*

Survey Lecture. Ralf Spatzier, *Non-commutative harmonic analysis and rigidity phenomena in dynamics*

Livio Flaminio, *Entropy rigidity for deformation of hyperbolic geodesic flows*

Eli Glasner, *Some new results concerning topological entropy*

Fernando Oliveira, *Sums of certain Cantor sets*

Yves Derriennic, *Random walks in random environment: Some ergodic problems*

Survey Lecture. Vitaly Bergelson, *Relations with number theory and combinatorics*

Mike Keane, *The size of some simple Cantor-like sets*

Jon Aaronson, *Cocycles*

Alan Forrest, *Deducing Szemerédi's Theorem without using van der Waerden's Theorem*

Hitoshi Nakada, *Ergodic theory of Rosen's continued fraction*

CONFERENCE PARTICIPANTS

Jon Aaronson
Department of Mathematics
Tel Aviv University
Tel Aviv 69978, Israel
aaro@math.tau.ac.il

Martin Bluemlinger
Inst. 114-1
Techn. Universität Wien
Wiedner Hauptstrasse 8-10/114
A-1040 Wien, Austria
mbluemli@email.tuwien.ac.at

Doğan Çömez,
Department of Mathematics
North Dakota State University
Fargo, ND 58105
USA
comez@plains.nodak.edu

Jean-Marc Derrien
Département de Mathématiques
Université F. Rabelais
Parc de Grandmont
37200 Tours, France
derrien@univ-tpours.fr

Tomasz Downarowicz
Department of Mathematics
Tech. University of Wroclaw
50-370 Wroclaw
Poland
downar@math.im.pwr.wroc.pl

Jacob Feldman
Department of Mathematics
University of California
Berkeley, CA 94720, USA
feldman@math.berkeley.edu

Vitaly Bergelson
Department of Mathematics
Ohio State University
Columbus, OH 43210, USA
vitaly@math.ohio-state.edu

Chris Bose
Department of Mathematics
University of Victoria
POB 3045
Victoria, BC, Canada V8W 3P4
cbose@entropy.uvic.ca

Sameh Daoud
Department of Mathematics
Faculty of Science
Ain Shams University
Abbasiya, Cairo, Egypt
daoud@egfrcuvx.bitnet

Yves Derriennic
Department of Mathematics
Université de Bretagne Occidentale
6 Av. V. Le Gorgeu
29287 Brest, France
derrienn@kelenn-gw.univ-brest.fr

Nashat Faried
Department of Mathematics
Faculty of Science
Ain Shams University
Abbasiya, Cairo, Egypt
faried@egfrcuvx.bitnet

Sebastien Ferenczi
Laboratoire de Mathématiques Discrètes
CNRS–UPR 9016
Case 930–163 av. de Luminy
F13228 Marseille Cedex 9, France
ferenczi@lmd.univ-mrs.fr

Livio Flaminio
Department of Mathematics
University of Florida
Gainsville, FL 32611
USA
flaminio@math.ufl.edu

Alan Forrest
Department of Math. and Stat.
University of Edinburgh
King's Buildings, Mayfield Road
Edinburgh EH9 3JZ, United Kingdom
forrest@mathematics.edinburgh.ac.uk

Eli Glasner
School of Mathematics
Tel Aviv University
Tel Aviv 69978, Israel
glasner@math.tau.ac.il

Geoffrey Goodson
Department of Mathematics
Towson State University
Towson, MD 21204, USA
e7m2grg@toe.towson.edu

Kesh Govinder
Department of Mathematics
University of Natal
Durban 4001
South Africa
govinder@ph.und.ac.za

Ryszard Jajte
Institute of Mathematics
Łódź University
ul. Stefana Banacha 22
90-238 Łódź, Poland
rjajte@plunlo51.bitnet

Michael Keane
Technical University of Delft
Mekelweg 4
2628 CD Delft
The Netherlands
m.s.keane@twi.tudelft.nl

Jonathan King
Department of Mathematics
University of Florida
Gainesville, FL 32611
USA
squash@math.ufl.edu

Jan Kwiatkowski
Instytut Matematyki
Uniwersytet M. Kopernika
Ui. Chopina 12/18
87-100 Torun, Poland
jkwiat@pltumk11.bitnet

Peter Leach
Department of Mathematics
University of Natal
Durban 4001
South Africa
leach@ph.und.ac.za

Mariusz Lemanczyk
Instytut Matematyki
Uniwersytet M. Kopernika
U1. Chopina 12/18
87-100 Torun, Poland
mlem@pltumk11.bitnet

Emmanuel Lesigne
Département de Mathématiques
Université F. Rabelais
Parc de Grandmont
37200 Tours, France
lesigne@univ-tours.fr

Jean Francois Méla
Départment de Mathématiques
Université de Paris-Nord
93430 Villetaneuse
Paris, France
mela@math.univ-paris13.fr

Hitoshi Nakada
Department of Mathematics
Keio University
Hiyoshi 3-14-1
Kohoku-ku, Yokohama 223
Japan
nakada@math.keio.ac.jp

James Olsen
Department of Mathematics
North Dakota State University
Fargo, ND 58105, USA
jolsen@plains.nodak.edu

Poornima Raina
Department of Mathematics
University of Bombay
Bombay 400 098
India
poornima@tifrvax.tifr.res.in

Christian Skau
Department of Mathematics
University of Trondheim
N-7055 Dragvoll, Norway
christian.skau@avh.unit.no

Ralf Spatzier
Department of Mathematics
University of Michigan
Ann Arbor, MI 48109, USA
spatzier@math.lsa.umich.edu

Mate Wierdl
Department of Mathematics
Northwestern University
Evanston, IL 60208, USA
mate@math.nwu.edu

Mahendra Nadkarni
Center of Adv. Study in Math.
University of Bombay
Bombay 400098
India
dani@tifrvax.tifr.res.in

Fernando Oliveira
ICEX, Univ. de Minas Gerais
Av. Antonio Carlos 6627
30000 Belo Horizonte
Brazil
dynamics@brufmg.bitnet

Karl Petersen
Department of Mathematics, CB 3250
University of North Carolina
Chapel Hill, NC 27599, USA
karl_petersen@unc.edu

Ibrahim Salama
School of Business
North Carolina Central University
Durham, NC 27707, USA
Department of Biostatistics, UNC
Chapel Hill, NC 27599, USA
salama@cs.unc.edu

Meir Smorodinsky
Department of Mathematics
Tel Aviv University
Tel Aviv 69978, Israel

Jean-Paul Thouvenot
Laboratoire de Probabilités
Université de Paris 6
4 Place Jussieu
75252 Paris, France

Printed in the United States
By Bookmasters